AF832316

Cattle, Their Predators and Geomatics Research

Michael O'Neal Campbell

Cattle, Their Predators and Geomatics Research

Springer

Michael O'Neal Campbell
Simon Fraser University
Burnaby, BC, Canada

ISBN 978-3-031-97362-8 ISBN 978-3-031-97363-5 (eBook)
https://doi.org/10.1007/978-3-031-97363-5

© The Editor(s) (if applicable) and The Author(s), under exclusive license to Springer Nature Switzerland AG 2025

This work is subject to copyright. All rights are solely and exclusively licensed by the Publisher, whether the whole or part of the material is concerned, specifically the rights of translation, reprinting, reuse of illustrations, recitation, broadcasting, reproduction on microfilms or in any other physical way, and transmission or information storage and retrieval, electronic adaptation, computer software, or by similar or dissimilar methodology now known or hereafter developed.
The use of general descriptive names, registered names, trademarks, service marks, etc. in this publication does not imply, even in the absence of a specific statement, that such names are exempt from the relevant protective laws and regulations and therefore free for general use.
The publisher, the authors and the editors are safe to assume that the advice and information in this book are believed to be true and accurate at the date of publication. Neither the publisher nor the authors or the editors give a warranty, expressed or implied, with respect to the material contained herein or for any errors or omissions that may have been made. The publisher remains neutral with regard to jurisdictional claims in published maps and institutional affiliations.

This Springer imprint is published by the registered company Springer Nature Switzerland AG
The registered company address is: Gewerbestrasse 11, 6330 Cham, Switzerland

If disposing of this product, please recycle the paper.

Preface

This book examines the relationship between wild cattle, domesticated cattle, sympatric large mammal carnivores, and complementary and competitive human actors from the perspective of the biological and geographical sciences. These sciences include the biogeography and ecology of wild cattle and their adversarial carnivores, the agricultural geography and political ecology of cattle domestication, and the potential role of geomatics technologies in assessing, monitoring, and mapping these relationships. This novel, multidisciplinary approach is justified due to the complexity of the topic, which cannot be meaningfully covered by one academic discipline. The selected topic is also crucial to human existence, since cattle are one of the most important aspects of agricultural production and their industrial development has significant environmental, ecological, socioeconomic, and conservation impacts.

The core theme of this book is that the domestication of cattle is one of the most important aspects of human development and civilization and that the sustainability of this development requires a better knowledge of background factors such as the ecology of wild and domestic cattle and competing food sources and environmental impacts as well as the ecology of cattle predators. Pre-domesticated wild cattle had and still have a complex ecological relationship with natural and human-modified ecosystems, which include large mammalian carnivores that also form important relationships with humans through predation and protection. Some conservation efforts have challenged the killing of large carnivores to protect livestock, exacerbating conflicts between agricultural production and environmental protection.

Cattle farming systems are also being challenged by new human cultures, such as vegetarianism and environmental movements concerned about the environmental impacts of intensive cattle farming. Research methods on these topics were initially based on approaches from the biological and social sciences, but recently computer-based systems have developed into important research tools. This includes systems based on geomatics as well as information derived from Big Data and Big Tech systems. The topics covered in this book reflect the author's background knowledge based on multidisciplinary and interdisciplinary research.

This book is divided into two parts and nine chapters. The part "The Development of Cattle Domestication and Ecology" addresses the basic background for the study of cattle and associated large carnivore ecology, including modern wild cattle, wild cattle predators, the domestication of cattle from ancient breeds, and the prehistoric predators of these cattle. The part "Modern Cattle and Their Predators" examines the predator ecology of domesticated cattle in Europe, Asia, Africa, and North and South America.

The chapter "Wild Cattle and Their Ecology" examines the past and current ecology of wild cattle species in the Americas, Eurasia, and Africa (subfamily Bovinae J. E. Gray, 1821; and the genera Bubalus C. H. Smith, 1827; Bos Linnaeus, 1758; Pseudoryx Dung, Giao, Chinh, Tuoc, Arctander & MacKinnon, 1993; Syncerus Hodgson, 1847; and bison Hamilton Smith, 1827), some of which are partial ancestors of domestic cattle or related to them (*Bos taurus*, Linnaeus, 1758). Research methods, based on the tools of geomatics, are also introduced, these having a wide range of applications in the sciences related to the ecology of wild cattle and carnivores, particularly landcover assessments for habitats, deforestation, savanna expansion, animal tracking for foraging and migration, and population density mapping. Geomatics-based methods facilitate the quantification of animal interactions and animal-land relationships and even the predictive modeling of human-animal relationships. It is argued that such computational studies represent an important contribution to the future of ecological studies of cattle.

The chapter "Wild Cattle and Their Carnivores" examines the wild cattle species in the context or their relationship with large carnivores, including bears (Ursus, Linnaeus, 1758), big cats (Panthera Oken, 1816, Puma Jardine, 1834, and Acinonyx Brookes, 1828), and larger dogs (Canis Linnaeus, 1758 and Lycaon Brookes, 1827) and hyenas (Crocuta Kaup, 1828 and Hyaena Brisson, 1762). Each of the large carnivores are examined in terms of their ecology, conservation status, and current adaptation to landcover change, and their abilities in the predation of cattle. The ability of wild cattle to defend themselves from carnivores, based on their organization, adaptation, size, aggressive behavior, and defensive tools such as horns, thick skins, and speed is also examined. The relationship between wild cattle and their carnivores is also examined in the context of human and livestock presence and habitat conversion. At the intra- and intercontinental levels, the conflicts between conservation and human land use are also discussed. Geomatics-based research methods and applications are also examined.

The chapter "Cattle Ancestry, Domestication and Carnivores" examines the domestication of cattle as a crucial development in human history, starting with the ancient species of wild aurochs in Southwest Asia, which later diverged with *Bos taurus* (taurine cattle) in West Asia and Europe and *Bos indicus* (Zebu indicine cattle) in South, Southeast and East Asia, and later in Africa. Later inbreeding has been documented between aurochs/domesticated cattle breeds and other wild species such as gaur, banteng, water buffalo, and yak in Asia, European aurochs and bison in Europe, and to a lesser extent American bison in North America. This examination of the ancestry of the two main genetic strands of cattle, the taurine and indicine breeds, is a background for knowledge of the problems of livestock domestication and is crucial to the current scenarios of political, economic, socio-cultural, and environmental cooperation and conflict that enable and constrain more effective sustainable development.

The chapter "Cattle Ancestry and Ancient Carnivores" examines the largely extinct predators of ancient and wild cattle, including the ancestral subspecies of tigers of Eurasia, Asiatic and African lions, Asiatic leopards, Eurasian brown bears, hyenas, and prehistoric and contemporary wolves. The relationship between these species and their ancestral cattle is being studied, with some researchers assessing the adult aurochs to be too large to be hunted by some predators and that the predators focus primarily on juveniles. Further problems include the reduction in cattle size through the development of taurine and indicine cattle from the larger aurochs. Also examined is the role of humans in cattle farming in terms of killing large carnivores, changing habitats and further breeding cattle, and the resulting effects on cattle ecology.

The chapter "Predators of Cattle in Asia" examines the Asian continent as the continent with the largest number predatory species capable of killing domestic cattle, comprising tigers, leopards, snow leopards, brown and black bears, wolves and dholes, and possibly the continent of cattle origins. Northwestern Asia is similar to Europe in carnivore species, with the main larger carnivores being brown bears and wolves, but eastern and southern Asia includes tigers, leopards, dholes and black bears, with snow leopards in cold mountainous areas. There are also semi-domesticated cattle, such as the gayal, and domesticated water buffalo and yak, indicine and taurine breeds. There are serious conflicts between large predators and cattle managers, and between political elites and the local cattle herders, over the conservation of large predators and cattle herding. Such conflicts vary regionally, especially between northern Asia, south and southeast Asia, and southwestern Asia. The status of the tiger is possibly the most important issue, especially in South Asia. The predator-cattle issues are increasingly, effectively managed, and geomatics-based research methods are increasingly applied, especially in China and India.

The chapter "Predators of Cattle in Europe" examines the relationship between domestic cattle and their predators, within the larger scenario of conservation and agricultural development, and the development trend of this relationship, taking the European continent as an example. Europe has a very large domestic cattle population, and rivals North America as having the most effective predator conservation laws, counterpointed the well-managed agricultural systems. Current evidence indicates that large predator populations are increasing across Europe, due to effective and successful conservation policies and management, with important consequences for agricultural development. For the main predators, wolves and brown bears, there have been significant population increases, with the largest predatory impacts on sheep, and to a lesser extent on goats. For cattle, there is a lesser effect despite the large population, and largely only from brown bears and wolves, and minimally from lynxes, and few if any records of attacks from wolverines, the last only on calves. Simultaneously, intensive agricultural development increased numbers and protections for cattle have conflicted with the increasingly effective conservation policies and the consequent population increases of predators. Current evidence suggests that domestic cattle, similar to wild Eurasian bison, are often immune to wolf and brown bear predation, but calves may be targeted. In addition, conflicts may emerge between cattle owners and conservation managers which may have significant impacts on future of agricultural and conservation policy. Geomatics-based research methods and applications are also examined.

The chapter "Predators of Cattle in Africa" examines the taurine and indicine breeds and their predators, especially the African lion and African leopard, and to a lesser extent hyenas, wild dogs, and cheetahs. Lions prey on all sizes of cattle, while leopards, hyenas, and other smaller predators take mostly calves, sheep, and goats. Cattle herding is mostly practiced in savanna regions outside the forest, including the Guinea, Sudan, and Sahel savanna belts, and cattle may be susceptible to some diseases of the moist forests. Serious conflicts exist between conservation of large carnivores and expanding cattle herding, and environmental issues include deforestation and human population increase and generally higher levels of desertification and human poverty than in other continents. Agricultural and conservation and the geomatics-based research methods have supported more detailed research on these issues in sub-Saharan Africa.

The chapter "Predators of Cattle in North America" examines the status of cattle and predators in North America, including Mexico and Central America, which similar to South America, has no indigenous domestic cattle, and domestic breeds based on taurine, indicine, and mixed breeds, and introduced water buffalo. Two predatory species, wolves and brown bears, play the same ecological roles as in Europe, but in North America, there are two big cats, the common cougars and rare jaguar in the southern part of the continent, and a second bear species, the American black bear, all of which can cattle. Similar to Europe, the human-carnivore conflicts are advanced, with payments to cattle for carnivore-based cattle losses, and conservation movements are based on powerful political support. Effective management of both cattle herding and carnivore conservation exist. The current evidence suggests that the main cattle predator is the gray wolf, followed by brown bears and cougars, and a lesser extent black bears and jaguars. Geomatics-based methods are also used for supportive analyses, generally at a more advanced level than in other continents. The advanced nature of cattle, carnivore, and protected area management and geomatics-based management suggests that North America may be a flexibly assessed role model for the other continents.

The chapter "Predators of Cattle in South America" examines domestic cattle in South America, a continent that lacks a local wild cattle species, and cattle predation is dominated by two felids, the jaguar and the cougar, and to a much lesser extent the other large carnivore, the spectacled bear. South America lacks an indigenous canine social predator, as the maned wolf is not powerful and social enough to prey on cattle. Taurine and indicine cattle and domestic water buffaloes have been introduced to South America. The predominant cattle predator is reputed to be the jaguar, although in some cases cougar predation may be blamed on the jaguar, and cougar and spectacled bear predation is a key factor of social conflict. The current assessment is of increased attention to large carnivore conservation and development of protected areas, which conflicts with increasing cattle populations and retaliatory actions against cattle predation. Geomatics-based research methods and applications are also increasingly applied to these topics, especially in the context of deforestation and farmland expansion.

The "Conclusions" chapter examines the possible futures of cattle and carnivore relations, based on current trajectories of social, economic, political, technological, and biological research and events. Currently, cattle represent the most important mammals domesticated for human consumption, but cattle maintenance has been challenged by the environmental outcomes of dense cattle populations, assessments of human quality of life, the increased medical and social influences on vegetarianism, and the demand for meat products conditioned by human health outcomes. These may affect the size of cattle populations and consequently their relationship with wildland ecologies. Research methods, including geomatics and other computer-based methods, have important implications for future agricultural management. Successful carnivore and forest protection has also competed with cattle management. Ecotourism increasingly competes with field agriculture and deforestation.

These topics illustrate the complexity of the relationship between humans, domestic and wild cattle, and large mammalian carnivores in highly dynamic, local, regional, and changing environments, related to ecological frameworks, political power, access to economic resources, and cultural differences. It is concluded that integrated research methods are needed to study these topics. The author hopes that this book will raise awareness of the diversity of relationships between cattle, large carnivores, and humans.

Burnaby, BC, Canada Michael O'Neal Campbell

Contents

The Development of Cattle Domestication and Ecology

This part consists of four chapters examining the ecology of wild cattle and their associated large carnivores, the domestication of wild cattle, and extinct bovine predators. The premise is that understanding the ecology of wild and ancestral cattle, particularly with regard to their relationship to their predators, is fundamental to assessing the current ecology and management of domesticated cattle, even as the domesticated cattle move into new regions and continents are introduced. This approach considers the integration of multiple disciplinary approaches, including wildlife ecology (the participation of wild cattle and carnivores in localized and regionalized food webs), biogeography (the spatial configuration of ecosystems based on systemic, areal factors), cultural ecology (the historical, human engagements with the nature, which have changed ecosystems), agricultural sciences including domestication (the human interventions that changed the genetics of wild species), and conservation management (political economy/ecology, environmental sociology, and environmental biology of ecosystem management). These four chapters identify the areas of wildlife ecology and contemporary domestication that have produced the current context of domesticated cattle and large carnivores.

A major contextual issue is the position of cattle as natural, dominant herbivores in most regional ecosystems. In Eurasia, Africa, and North America, the large cattle species rank only behind the pachyderms (modern and extinct elephants, mammoths, mastodons, rhinos, and hippos) and before the smaller ungulates (antelopes, gazelles, deer, sheep and goats) in terms of biomass). Current research suggests that domestication has not and is not significantly changing the relationship between cattle and their natural enemies. However, a key issue concerns the vulnerability of domestic cattle compared to their wild counterparts and ancestors, as the former may be smaller, less powerful and have lost some of their predator control abilities. Therefore, large carnivores may pose a greater threat to domestic cattle because they can affect the proximity between grazer and carnivore habitats and lead to conflicts between agricultural development and carnivore conservation.

This phenomenon may alter human perceptions of large, charismatic carnivores such as big cats and bears, thereby exacerbating conflicts between farmers and conservationists. This may shift the core themes of this relationship from the wildlife biology of large mammal relationships to more complex issues related to human groups, power, and negotiation. These four chapters explore these themes by examining the ecology of wild cattle and then the issues of domestication. A starting point is the physical characteristics of cattle, as domesticated breeding can result in a reduction in physical size, musculature, and/or horns in some breeds. The aggressive temperament of some wild cattle, such as the water buffalo, can also be reduced. These changes, primarily aimed at making it easier for human pastoralists to cope, may also increase vulnerability. These are important themes from which to develop further studies on cattle domestication and predation and to identify regional differences, which will be examined in part "Modern Cattle and Their Predators".

Wild Cattle and Their Ecology

Introduction

Wild cattle, including free ranging domesticated and semi domesticated breeds, are among the most important mammals in the modern world, being part of some of the ancestry to domestic cattle (Table 1), among the most important meat source for human societies, and still important sources of food in their own right (Verkaar et al., 2004; MacEachern, 2007; Gurunathan et al., 2009; MacEachern et al., 2009; Castelló, 2016a, b; Sahu et al., 2020). Wild cattle are large to huge, very strong and muscular, heavily built, generally social herbivores, second in size only to the "pachyderms" (elephants, rhinoceroses and hippopotamuses) (Table 2), and they dominate those ecosystems where they are present (Mloszewski, 1983; Lenstra & Bradley, 1999; Hassanin & Ropiquet, 2004; Grubb, 2005a, b, c, d; MacEachern et al., 2009; Castelló, 2016a, b). Ahrestani (2018: 3) describes subfamily Bovinae as having "large and have stout bodies, hollow horns, relatively short legs, long tails with a terminal tuft of hair, broad muzzles, and no facial, pedal, or inguinal glands."

The largest wild cattle, the gaur, wild yaks, and American and European bison, can reach weights of over 1000 kg, and shoulder heights of over 2 meters. Similarly sized domestic cattle include the Italian Chianina, which is reputed to be the world's largest domestic breed and similar in size to the gaur (Nel, 2012). There are also small species of wild cattle, such as the Tamaraw and the Anoa, which may weigh less than 300 kg, and the antelope-like Saola (Tables 1 and 2). The San Diego Wildlife Alliance (2024) describes interspecies differences in body weight (the heaviest being the Asian wild buffalo 1200 kilograms, and the lightest as the 299 kilograms, although elsewhere the gaur is listed as the heaviest at about 1500 kilograms; see Ahrestani, 2018 and Table 2), body length (the longest as the American bison and yak at 3–3.8 meters and the shortest the anoa at 1.2–1.8 meters), and shoulder height (the tallest as the gaur at 1.6–2.2 meters and the shortest as the anoa at 160–100 centimeters) (Ahrestani, 2018).

Wild cattle usually exhibit sexual dimorphism, with males having larger body and horn sizes as well as larger necks, humps, and sometimes a dewlap, which is a

© The Author(s), under exclusive license to Springer Nature Switzerland AG 2025

M. O. Campbell, *Cattle, Their Predators and Geomatics Research*,
https://doi.org/10.1007/978-3-031-97363-5_1

Table 1 Tribe: Bovini (Gray, 2021) or wild cattle, including domesticated and semi-domesticated breeds

Subtribe	Genus	Species and description
Bovina (Gray, 2021)	*Bos* (Linnaeus, 1758)	**Wild** Gaur (*Bos gaurus*, Smith, 1827) Banteng (*Bos javanicus*, d'Alton, 1823) Wild yak (*Bos mutus*, Przewalski, 1883), Kouprey (*Bos sauveli*, Urbain, 1937) **Domesticated** Domestic cattle (*Bos taurus*, Linnaeus, 1758) Gayal (*Bos frontalis*, Lambert, 1804), Domestic yak (*Bos grunniens*, Linnaeus, 1766)
	Bison (Hamilton-Smith, 1827)	**Wild** American bison (*Bison bison*, Linnaeus, 1758, or *Bos bison*, Linnaeus, 1758), European bison (*Bison bonasus*, Linnaeus, 1758, or *Bos bonasus* Linnaeus, 1758)
Bubalina (Rütimeyer, 1865)	*Syncerus* (Hodgson, 1847)	**Wild** African buffalo (*Syncerus caffer* (Sparrman, 1779))
	Bubalus (Hamilton-Smith, 1827)	**Wild** Wild water buffalo (*Bubalus arnee*, Kerr, 1792), Lowland anoa (*Bubalus depressicornis*, Smith, 1827) Mountain anoa (*Bubalus quarlesi*, Ouwens, 1910) Tamaraw (*Bubalus mindorensis*, Heude, 1888) **Domestic** Domestic water buffalo (*Bubalus bubalis*, Linnaeus, 1758),
Pseudorygina (Hassanin & Douzery, 1999)	*Pseudoryx* (Dung et al., 1993)	**Wild** Saola (*Pseudoryx nghetinhensis*, Dung)

Sources: Boyles et al. (2016), Burton et al. (2016a, b), Buzzard and Berger (2016), Duckworth et al. (2016), Gardner et al. (2016), Aune et al. (2017), Kaul et al. (2019), IUCN SSC Antelope Specialist Group (2019), Plumb et al. (2020), Timmins et al. (2020)

longitudinal flap of skin that hangs below the lower jaw or neck. For example, in the case of the gaur, Ahrestani (2018: 4) notes that recorded weights of adult males were 590, 782, 864, 864, 931, and 941 kg. For adult females, the weights were 440 kg and 703 kg. Polák and Frynta (2010: 1255) refer to Rensch's rule, which holds that for species where males are larger than females, the sexual size dimorphism (SSD) increases with increasing species size. Polák and Frynta (2010: 1255) found that among domestic cows, yaks, buffaloes, and other bovines and 18 wild species/subspecies of the subfamily Bovinae, the male to female body mass ratio in domestic cows (1.48) was similar to that of other species of domestic and wild bovines except domestic buffaloes (1.19). However, although there was a clear

Table 2 Wild cattle and other larger herbivores

Species name	Scientific name	Weight (kg)	Length (m)	Shoulder height (m)
African bush elephant	*Loxodonta africana* (Blumenbach, 1797)	2000–6100	2.2–4	2.2–4
Asiatic elephant	*Elephas maximus indicus* (Cuvier, 1798)	2000–6000	5.5–6.4	1.95–3.00
Forest elephant	*Loxodonta cyclotis* (Matschie, 1900)	2700–6000	1.6–2.86	1.44–1.55
White rhinoceros	*Ceratotherium simum* (Burchell, 1817)	1000–3600	3.92–4.54	1.71–2.85
Black rhinoceros	*Diceros bicornis* (Linnaeus, 1758)	700–1400	3.5–4.45	1.5
Indian rhinoceros	*Rhinoceros unicornis* (Linnaeus, 1758)	1500–2400	3.8–4.6	1.48–1.86
Javan rhinoceros	*Rhinoceros sondaicus* (Desmarest, 1822)	1500–2000	2.5–4.7	1.7
Sumatran rhinoceros	*Dicerorhinus sumatrensis* (Fischer, 1814)	600–1000	2.5–4.7	1–1.5
Gaur	*Bos gaurus* (Smith, 1827)	700–1500	3.2–4.35	1.65–2.2
Banteng	*Bos javanicus* (d'Alton, 1823)	600–800	2.60–295	1.60
Wild yak	*Bos mutus* (Przewalski, 1883)	500–1200	3.60–5.80	1.6–2.05
Kouprey	*Bos sauveli* (Urbain, 1937)	680–910	3.10–3.23	170–190
American bison	*Bison bison* (Linnaeus, 1758)	318–900	2.43–4.75	1.52–1.86
European bison	*Bison bonasus* (Linnaeus, 1758)	800–1000	3.10–4.22	1.8–1.95
African buffalo	*Syncerus caffer* (Sparrman, 1779)	425–870	2.40–4.5	1.05–1.60
Wild water buffalo	*Bubalus arnee* (Kerr, 1792)	800–1200	3.0–4.0	1.5–1.9
Lowland anoa	*Bubalus depressicornis* (Smith, 1827)	90–225	2.20	0.80–0.90
Mountain anoa	*Bubalus quarlesi* (Ouwens, 1910)	150–300	149–180	0.70–0.75
Tamaraw	*Bubalus mindorensis* (Heude, 1888)	180–300	280	1.00
Saola	*Pseudoryx nghetinhensis* (Dung)	80–100	173	0.80–0.90

Sources: Van Zyll de Jong (1986), Robichaud (1998), Lundrigan and Zachariah (2000), Saari (2002), Newell and Sorin (2003), Oliphant (2003), Gesch (2004), Roth (2004), Connor (2009), Kurnit (2009), Hendricks (2013), Jianlin et al. (2014), Semenov (2014), Ng (2015), Karkala (2016), Dulal (2017), Howard (2017), Crider (2019), Arias (2023), Huffman (2023a, b, 2024), Succop (2023), Dinerstein (2024)

All the elephants and rhinoceroses are classified into the Orders Proboscidea (Illiger, 1811) and Perissodactyla (Owen, 1848), respectively. The Bovines are classified under the Order Artiodactyla (Owen, 1848). For all species, the length may include the tail, which is usually about 40–100 centimeters relation of male to female body mass ratio (slope 1.21) as predicted by Rensch's rule, there was no similar relation when body mass was replaced by shoulder height (Polák & Frynta, 2010: 1255).

There may also be differences in coloration, with the males' colors being darker, but the general coloration consists of varying shades of black to reddish-brown and sometimes lighter markings (Boitani, 1984). D'Ammando et al. (2022: 541) argue that despite exceptions, "among mammals, bovids provide some of the most striking examples of sexual dimorphism in colouration and pelage appendages, such as beards and manes." For example, for the gaur, Ahrestani (2018: 4) points out that males attain a black pelage with age and adult males greater than 5 years are termed "black bulls," while females have a "brown pelage." Huffman (2023a, b) notes that male African buffalo are significantly larger than females and are usually darker than females. For the banteng, Castelló (2016a, b) describes adult males as blackish to dark brown, with young males browner and females paler brown or even reddish chestnut brown.

Wild cattle are generally social animals. Their tendency to live and move in large, socially dynamic herds underscores their dominance (Conybeare, 1980; Prins, 1989, 1996; Dora, 2004; Melletti et al., 2007; Korte, 2008a; Winnie et al., 2008; Cornélis et al., 2011; Naidoo et al., 2014; Caron et al., 2016; Chigwenhese et al., 2016; Bennitt et al., 2018). Wild cattle are classified within the Family Bovidae (Gray, 2021) within the Order Artiodactyla (Owen, 1848). Caron et al. (2023 153) note that for the African buffalo, the main social units are mixed or breeding herds, which are made up of adult females, weanlings, and subadults of both sexes (subadults are between 3 and 5 years old), with a small percentage of adult males. Bachelor groups are made up of males (two or more) who are 4 years of age and older. These groups may join mixed herds for mating and leave the groups to access resources until the new mating opportunities.

Another example is the American bison, which gathers in seasonal herds. These changes are called seasonal aggregation or seasonal segregation. The former occurs when male bison join groups of mixed ages and sexes to form larger groups, particularly during the rutting or breeding season. Seasonal segregation occurs during the non-rutting season when bulls leave the group to form smaller, separate groups. Herd size is influenced by feeding conditions, seasonal behavior, population size, landscape composition, and environmental conditions. Hierarchical social dominance structures occur, with larger groups during the summer breeding season and a decline in numbers in the fall after the rutting season. The number of chickens decreases rapidly in the fall after the rut. There may be a segregation between adult females, calves, and immature males on the one hand and large bull males on the other, possibly because the latter group requires more food for their larger size (National Park Service, 2017).

All the wild cattle species have been assessed for conservation possibilities (Table 3), and their populations are much lower than those of the domesticated or semi-domesticated species (Table 4). According to the San Diego Wildlife Alliance (2024), the factors for the decline in wild cattle populations include overhunting ("Just 100 years ago, there were 40–60 million American bison in North America.

Table 3 Conservation status of tribe: Bovini (Gray, 2021) or wild cattle (IUCN Sources)

Species	Conservation status	Population trend	Population
Gaur	Vulnerable	Decreasing	6000–21,000
Banteng	Endangered	Decreasing	4000–8000
Wild yak	Vulnerable	Decreasing	7500–9999
American bison	Near threatened	Stable	11,248–13,123
European bison	Near threatened	Increasing	2518
African buffalo	Near threatened	Decreasing	398,000–401,000
Wild water buffalo	Endangered	Decreasing	2500
Lowland anoa	Endangered	Decreasing	2499
Mountain anoa	Endangered	Decreasing	0–2500
Tamaraw	Critically endangered	Decreasing	220–300
Saola	Critically endangered	Decreasing	25–750, 50–300
Kouprey	Critically endangered	Decreasing	0–50

Sources: Boyles et al. (2016), Burton et al. (2016a, b), Buzzard and Berger (2016), Duckworth et al. (2016), Gardner et al. (2016), Timmins et al. (2016), Aune et al. (2017), Kaul et al. (2019), IUCN SSC Antelope Specialist Group (2019), Plumb et al. (2020), Timmins et al. (2020)

Table 4 Conservation status of tribe: Bovini (Gray, 2021) or domesticated cattle

Species	Conservation status	Population trend	Population
Domestic cattle	Not threatened	Stable	942, 630,000
Gayal	Endangered	Declining	Unknown
Domestic yak	Not threatened	Stable	14,000,000
Domestic water buffalo	Not threatened	Stable	202,000,000

Sources: Qi et al. (2010), Wang et al. (2017), Zhang et al. (2020), San Diego Wildlife Alliance (2024), Statistica (2024)

But over-hunting wiped them out. By 1903, fewer than 2000 survived in zoos and private management, and in isolated wild populations," with similar problems for the wild yak in the Himalayan Mountain region), grazing competition from increased domestic cattle herds (they may be interbreeding, reducing the number of pure-bred wild cattle), and diseases of domestic cattle (San Diego Wildlife Alliance, 2024). An example was the rinderpest epidemic, which killed millions of wild and domestic cattle and was eradicated in 2010. Normile (2010: 435) notes that "Although nearly forgotten in much of the West, as recently as the early 1900s, outbreaks of rinderpest—from the German for 'cattle plague'—regularly ravaged cattle herds across Eurasia, often claiming one-third of the calves in any herd. The virus, a relative of those that cause canine distemper and human measles, spreads through exhaled droplets and feces of sick animals, causing fever, diarrhea, dehydration, and death in a matter of days. It primarily affects young animals; those that survive an infection are immune for life." The disease began in the late nineteenth century: "When the virus hit previously unexposed herds, the impact was horrific. In less than a decade after the virus was inadvertently introduced to the horn of Africa in 1889, it spread throughout sub-Saharan Africa, killing 90% of the cattle and a large proportion of domestic oxen used for plowing and decimating wild buffalo, giraffe, and wildebeest populations," and by 1994, "rinderpest was entrenched in central Africa, the Arabian Peninsula, and a swath stretching from Turkey through India and to Sri Lanka." (Normile, 2010: 435).

These tables show the difference in population status of wild cattle species and domesticated breeds. Given the wide range of wild cattle species, the low population numbers indicate dispersed, low population densities, which increase population vulnerability. Despite the very large populations of domesticated cattle, the impacts of predation by large carnivores would still be severe for human stakeholders, including pastoralists and consumers of cattle products. Factors for additional problems include the increasing intrusion of animal activities into carnivore habitats and local increases in carnivore populations as a result of successful conservation management. Wild cattle face similar scenarios. Wild cattle are hunted primarily by large carnivores and smaller carnivores, which either hunt in social, cooperative groups to kill large prey or kill smaller young cattle and calves (Campbell, 2017). These predators are listed by continent in Tables 5 and 6.

The presence of large carnivores varies on a continental scale. Although there are numerous human-caused factors in the extinction or reduction of wild cattle populations, the role of large carnivores on a continental scale is generally uncertain. In Eurasia, Africa, and North and South America, there are large carnivores, namely bears, big cats, wolves, and hyenas, which are capable of killing young cattle and, in some cases, even adult wild and domesticated cattle. The role of such carnivores in predation assumes important dimensions as the carnivores are also threatened and successful conservation efforts have in some cases contributed to an increase in the carnivore population, leading to further threats to cattle. The sizes of these carnivores are listed in Table 7.

The Importance of the Cattle-Carnivore Relationship

The relationship between domesticated cattle their carnivores is an important economic issue, considering the economic value of cattle in modern economies (Campbell, 2017; Campbell & Jha, 2021). As noted by Braczkowski et al. (2023), carnivore predation on domestic cattle can have devastating impacts on poorer people in poor countries, "more than half of the world's largest carnivore species have over one third of their habitat overlapping the most economically vulnerable human communities," and these predatory occurrences are a crucial aspect of the global attempts at sustainable development. This provides the primary justification for the study of the relationship between wild cattle and their carnivores, which share similar carnivores with the generally smaller and more vulnerable domestic cattle (Campbell, 2017; Campbell & Jha, 2021) (Tables 8 and 9 show the characteristics of large carnivores).

Treves et al. (2003: 1491) note that "as a group, carnivores exert a profound influence on biological communities via predation and interspecific competition. Carnivores often regulate or limit the numbers of their prey, thereby altering the structure and function of entire ecosystems." Conflicts between people and carnivores emerge mainly through the threat to human life and activities, and economic security, with the latter including threats to economically valuable animals, such as

Table 5 Predators of cattle in Asia

Continent	Species	Scientific species name, including subspecies	Hunting method
Asia	Tiger	***Panthera tigris* (Linnaeus, 1758** (*Panthera tigris tigris*, Linnaeus, 1758; *Panthera tigris sondaica*, Temminck, 1844)	Ambush
	Lion	***Panthera leo* (Linnaeus, 1758)** *Panthera leo leo* (Linnaeus, 1758)	Ambush and pursuit
	Leopard	***Panthera pardus* (Linnaeus, 1758)** (*P. p. fusca*, Meyer, 1794; *P. p. kotiya*, Deraniyagala, 1956; *P. p. delacouri*, Pocock, 1930; *P. p. orientalis*, Schlegel, 1857; *P. p. tulliana*, Valenciennes, 1856; *P. p. nimr*, Hemprich and Ehrenberg, 1830; *P. p. melas*, Cuvier, 1809)	Ambush and pursuit
	Clouded leopard	***Neofelis nebulosa* (Griffith, 1821)**	Ambush, stalking
	Sunda clouded leopard	***Neofelis diardi* (G. Cuvier, 1823)** (*Neofelis diardi borneensis*, Wilting, Christiansen, Kitchener, Kemp, Ambu and Fickel, 2007; *Neofelis diardi diardi* (Cuvier, 1823)	Ambush, stalking
	Brown bear	***Ursus arctos arctos* (Linnaeus, 1758)** (*Ursus arctos beringianus*, Middendorff, 1851; *Ursus arctos collaris*, F. G. Cuvier, 1824; *Ursus arctos isabellinus*, Horsfield, 1826; *Ursus arctos lasiotus*, Gray, 1867; *Ursus arctos pruinosus*, Blyth, 1854)	Frontal, ambush, pursuit
	Black bear	***Ursus thibetanus* (G. Cuvier, 1823)** (*Ursus thibetanus ussuricus*, Heude, 1901; *Ursus thibetanus thibetanus*, Cuvier, 1823; *Ursus thibetanus mupinensis*, Heude, 1901; *Ursus thibetanus laniger*, Pocock, 1932; *Ursus thibetanus japonicus*, Schlegel, 1857; *Ursus thibetanus gedrosianus*, Blanford, 1877; *Ursus thibetanus formosanus*, R. Swinhoe, 1864)	Frontal, ambush, pursuit
	Sloth bear	***Melursus ursinus* (Shaw, 1791)** (*M. u. ursinus*) (Shaw, 1791) (*M. u. inornatus*) Pucheran, 1855	Frontal, ambush, pursuit
	Wolf	***Canis lupus* (Linnaeus, 1758)** (*Canis lupus lupus*, Linnaeus, 1758; *Canis lupus arabs*, Pocock, 1934; *Canis lupus campestris*, Dwigubski, 1804; *Canis lupus chanco*, Gray, 1863; *Canis lupus chanco*, Gray, 1863; *Canis lupus dingo*, Meyer, 1793; *Canis lupus pallipes*, Sykes, 1831)	Pursuit, group attack
	Dhole	***Cuon alpinus* (Pallas, 1811)** (*Cuon alpinus alpinus*, Pallas, 1811; *C. a. adjustus*, Pocock, 1941; *C. a. fumosus*, Pocock, 1936; *C. a. hesperius*, Afanasjev and Zolotarev, 1935; *C. a. laniger*, Pocock, 1936; *C. a. lepturus*, Heude, 1892; *C. a. sumatrensis*, Hardwicke, 1821)	Pursuit, group attack

The smaller carnivores, wolves and dholes, are social hunters, hence can take down much larger cattle

Table 6 Predators of cattle in Africa, North and South America, and Europe

Africa	Lion	***Panthera leo* (Linnaeus, 1758)** (*Panthera leo leo*, Linnaeus, 1758; P. l. melanochaita (Smith, 1842)	
	Leopard	***Panthera pardus* (Linnaeus, 1758)** (*P. p. pardus*, Linnaeus, 1758)	
	Hyena	***Crocuta crocuta* (Erxleben, 1777)**	
	Cheetah	***Acinonyx jubatus* (Schreber, 1775)** (*A. j. jubatus*, Schreber, 1775; *A. j. soemmeringii*, Fitzinger, 1855; *A. j. hecki*, Hilzheimer, 1913)	Pursuit
	Ethiopian wolf	***Canis simensis* (Ruppell, 1840)**	
	Wild dog	***Lycaon pictus* (Temminck, 1820)** (*L. p. pictus*, Temminck, 1820; *L. p. lupinus*, Thomas, 1902; *L. p. somalicus*, Thomas, 1904; *L. p. sharicus*, Thomas and Wroughton, 1907; *L. p. manguensis*, Matschie, 1915)	Pursuit
North America	Brown bear	***Ursus arctos* (Linnaeus, 1758)** (*U. arctos horribilis*, Linnaeus, 1758; *U. arctos middendorffi*, Merriam, 1896)	
	Black bear	***Ursus americanus* (Pallas, 1780)** (*U. americanus americanus*, Pallas, 1780; *U. americanus californiensis*, Miller, 1900; *U. americanus carlottae*, Osgood, 1901; *U. americanus vancouveri*, Hall, 1928, etc.)	
	Cougar	***Puma concolor* (Linnaeus, 1771)** (*P. c. cougar*, Kerr, 1792; *P. c. costaricensis*, Merriam, 1901)	
	Wolf	***Canis lupus* (Linnaeus, 1758)** (*C. l. arctos*, Pocock, 1935; *C. l. columbianus*, Goldman, 1941; *C. l. crassodon*, Hall, 1932; *C. l. baileyi*, Nelson and Goldman, 1929)	
	Coyote	***Canis latrans* (Say, 1823)** (*C. l. latrans*, Say 1823; *C. l. texensis*, Bailey 1905; *C. l. incolatus*, Hall 1934; *C. l. var.*, Lawrence & Bossert 1969; *C. l. mearnsi*, Merriam 1897)	
South America	Jaguar	***Panthera onca* (Linnaeus, 1758)**	
	Cougar	***Puma concolor* (Linnaeus, 1771)** (*Puma concolor concolor*, Linnaeus, 1771; *Puma concolor discolor*, Schreber, 1777)	
	Coyote	***Canis latrans* (Say, 1823)** (*C. l. mearnsi*, Merriam 1897; *C. l. hondurensis*, Goldman 1936; *C. l. dickey*, Nelson 1932; *C. l. goldmani*, Merriam 1904; *C. l. vigilis*, Merriam 1897; *C. l. impavidus*, Alen 1903)	
Europe	Brown Bear	***Ursus arctos* (Linnaeus, 1758)** (*Ursus arctos arctos*, Linnaeus, 1758; *Ursus arctos marsicanus/arctos*, Altobello, 1921; *Ursus arctos pyrenaicus*)	
	Wolf	***Canis lupus* (Linnaeus, 1758)**	

The smaller carnivores, wolves, wild dogs, and coyotes, are social hunters, hence can take down much larger cattle. Sources: Campbell (2017, 2018)

livestock and companion animals (Campbell, 2017; Campbell & Jha, 2021). Treves et al. (2003: 1491) further argue that "many larger carnivore species are specialized for ungulate predation; therefore, some individuals readily kill domesticated ungulates when opportunities arise."

Table 7 Carnivore size

Species	Weight (kg)	Body length (m)	Shoulder height (m)	Tail (m)
Tiger	91–423	1.98–3.7	90–110	85–110
Lion	126–272	2.4–3.3	1.20	100
Jaguar	68–136	1.5–1.85	75	70–90
Leopard	17–65 kg	1.6–2.3	60–70	66–102
Cougar	29–120	0.86–1.54	60–90	63–95
Cheetah	21–72	1.12–1.50	67–94	65–85
Show leopard	25–75	1–1.3	56	80–105
Clouded leopard	11–23	1.23–2.00	50–55	61–91
Sunda clouded leopard	15–30	1.15–2.01	50	55–91
Brown bear	80–600	1–2.8	90–150	6–22
American black bear	39–409	1.2–2	70–110	8–14
Asiatic black bear	65–200	1.2–1.9	70–100	6.5–10.6
Spectacled bear	60–200	1.3–2.0	64	7
Sloth bear	55–140	1.5–1.9	60–90	7–12
Gray wolf	23.0–80	0.87–1.30	60–90	0.35–0.52
Ethiopian wolf	11.2–19.3	0.84–1.0	53–62	0.27–0.40
Coyote	7–23	0.75–1.3	60	0.30–0.50
African wild dog	18–36	0.75–1.1	60–75	30–40 cm
Dhole	17–21	1	50	40–45

Sources: Hemmer (1972), Rosevear (1974), Grzimek (1990), Nowak (1991), Estes (1992), Brown (1993), Sillero-Zubiri and Gottelli (1994), Chacon (2000), Clouded leopard (Neofelis nebulosa) Husbandry Guidelines (2000), Tokar (2001), Macdonald (2001), Ballenger (2002), Bies (2002), Mulheisen et al. (2002), Smith (2002), Durbin et al. (2004), Goodness (2004), Harrington (2024), Nowak (2005), Skinner and Chimimba (2005), Wozencraft (2005a, b, c), Bunker (2007), Dacres (2024), Kronk (2007), Holmes (2009), Nogueira (2009), Hunt (2011), Fenner (2012), Hancock (2012), Kamler et al. (2012), Lehnert (2013), Montsion (2014), Kingdon (2015), Zoo Idaho (2016), Castelló (2018), International Society for Endangered Cats (ISEC) Canada (2021), Britannica (2023), Britannica (2024a, b), Johnson (2024), Kays (2024), Lariviere (2024), San Diego Zoo Wildlife Alliance (2024), Zoological Wildlife Foundation (2024)

Large herbivores (e.g., buffalo) are usually associated with large carnivores (e.g., big cats) in ecological studies because the latter hunt and can be killed by the former and both require large areas of land that may be needed by humans and can be perceived as a threat to human life, habitats, and agricultural needs (Campbell, 2017; Campbell & Jha, 2021; Jha & Campbell, 2021). Durant (2022) asserts that the coexistence of humans with large carnivores poses challenges to human welfare, livelihoods, development, resource management, and politics. Even where humans and carnivores have coexisted in the past, traditional patterns of behavior toward large carnivores can be disrupted by broader processes of economic, social, political, and climatic change, and in such areas of cooperation and conflict, conservation actions have typically focused on human actions near large carnivores to promote sustainable practices.

Ghosal et al. (2015) also examine the social construction of landscapes as a key concept for studying people's responses to large carnivores in the context of ecological, economic, social, and cultural changes in the case study areas, and add that human responses to large carnivores may vary from support and coexistence within or oustide protected areas, to resistance, conflict, and/or extermination in the face of

Table 8 Big cat conservation status

Species	Population size	Population trend, status	Habitat variables	Conservation actions
Tiger	2608–3905	Decreasing, endangered	Forest, savanna/grass	Habitats, local politics
Lion	23,000	Decreasing, vulnerable	Savanna, shrubland, grassland, forest	Habitats, local politics
Jaguar	64,000	Decreasing, near threatened	Forests, also more open vegetation	Habitats, local, international politics
Leopard	700,000+	Decreasing, vulnerable	Forest, savanna, shrub/grassland	Habitats, parks, local, national politics
Cougar	50,000+	Decreasing, least concern	Forest, savanna, shrub/grassland	Habitats, national/local politics
Cheetah	6517	Decreasing, vulnerable	Savanna, shrubland, grassland	International policies, parks, local politics
Show leopard	2710–3386	Decreasing, vulnerable	Forest, shrub/grass/rocky land	International policies, parks, local politics
Clouded leopard	3700–5580	Decreasing, vulnerable	Forest, shrubland	Local studies, parks, monitoring
Sunda clouded leopard	4500	Decreasing, vulnerable	Forest	Local studies, parks, monitoring

Note that the lion, leopard, and cougar have very wide ranges; hence, there are substantial variations in the figures above. Cougar populations are increasing in Canada and parts of the United States (Thompson & Jenks, 2010; LaRue et al., 2012; Wilson, 2022). Other sources: Hearn et al. (2015), Nielsen et al. (2015), De la Torre, et al. (2017), McCarthy et al. (2017), Quigley et al. (2017), Farhadinia et al. (2019), Gray et al. (2021), Goodrich et al. (2022), Durant et al. (2023), Nicholson et al. (2023), Stein et al. (2023), Britannica (2024c)

real or imagined attacks of humans and/or livestock (Grumbine, 1990; Purvis et al., 2000; Woodroffe, 2000; Carbone & Gittleman, 2002; Cardillo et al., 2004; Berger et al., 2013; Ghosal et al., 2015; Kuijper et al., 2016; Wolf & Ripple, 2016). Some extermination efforts of carnivores are successful, either deliberately prosecuted or indirectly due to habitat degradation and failures of conservation policy (Karanth & Stith, 1999; Carbone & Gittleman, 2002; Karanth et al., 2004, 2010a, b; Karanth & Chellam, 2009; Campbell & Torres Alvarado, 2011; Kansky et al., 2014; Campbell, 2017). Examples of extinct or very rare large carnivores include the Javan tiger (*Panthera tigris spondaica*, Temminck, 1844), the Bali tiger (*Panthera tigris balica*, Schwarz, 1912) and the Caspian tiger (*Panthera tigris virgate*, Illiger, 1815) (Seidensticker et al., 1999; Mazák & Groves, 2006; Jackson & Nowell, 2008), the Eastern cougar (Puma concolor, Linnaeus, 1771), declared extinct in 2011 (United States Fish and Wildlife Service, 2011), and possibly the jaguar in selected areas (Hatten et al., 2005).

Recent economic problems have exacerbated the externalities of human carnivore conflict over cattle, as conservation of large carnivores, expansion of livestock area, and exploitation of wild cattle increase. Treves et al. (2003: 1492) point out that there are currently three main policy measures for managing wild carnivores, which include the removal, conservation, or selective killing of carnivore

Table 9 Bear and canine conservation status

Species	Population size	Population trend, status	Habitat variables	Conservation actions
Brown bear	110,000	Stable, least concern	Forest, shrubland, grassland, wetlands	Habitats, national politics, parks
American black bear	850,000–950,000.	Increasing, least concern	Forest, shrubland, grassland, wetlands	Habitats, national politics, parks
Asiatic black bear	60,000	Decreasing, vulnerable	Forest, shrubland, grassland, wetlands	Habitats, parks, national/local politics
Andean bear	2500–10,000	Decreasing, vulnerable	Forest, shrubland, grassland	Local studies, parks, monitoring
Sloth bear	<10,000 to >20,000	Decreasing, vulnerable	Forest, savanna, shrubland, grassland	Habitats, parks, local politics
Gray wolf	112,100	Stable, least concern	Forest, shrub and grassland, wetlands	Habitats, legislation, national, local politics
Ethiopian wolf	197	Decreasing, endangered	Shrubland, grassland, wetlands, rocky areas	Habitats, protection, parks, vaccination
Coyote	<10,000, 000	Increasing, least concern	Forest, shrubland, grassland, desert	Ubiquitous, in urban area, monitoring
African wild dog	1409	Decreasing, endangered	Forest, savanna, shrubland, grassland, desert	Habitats, migration, diseases, conflicts
Dhole	949–2215	Decreasing, endangered	Forest, shrubland, grassland	Habitats, few parks, local conflicts/politics

Note in some cases the number represents the mature individuals, rather than the total population. Sources: Marino and Sillero-Zubiri (2011), Kamler et al. (2015), Garshelis, et al. (2016), McLellan et al. (2017), Velez-Liendo and García-Rangel (2017), Kays (2018), Dharaiya et al. (2020), Garshelis and Steinmetz (2020), Woodroffe and Sillero-Zubiri (2020), Boitani et al. (2023), World Population Review (2024a, b)

populations. These policies may overlap in application and content and may be supported or opposed by various interest groups, including conservationists and conservationists, livestock and companion animal owners, and the general public. For the removal or eradication of carnivores, Treves et al. (2003: 1492) point out that campaigns to eradicate carnivore populations have now largely ceased, with the exception of a few campaigns against exotic carnivores that threaten native fauna. Extermination of carnivores is usually justified by preventing agricultural losses or protecting other species. However, these measures face political opposition in the United States based on animal welfare, conservation, tourism, and scientific concerns. In conservation or preservation, the main goal—and sometimes the result—has been to restore carnivore populations, sometimes increasing negative problems such as "problematic" carnivores that may attack humans, human habitats, or associated animals, with legal protections (successful or political dynamics). The eradication of such carnivores and their conservation require large investments of personnel, time, and resources in interaction with the public, as managers demarcate and patrol borders, ban and prosecute poachers, encourage community participation, or claims for damages have to be checked (Treves et al. (2003: 1492).

Human-wildlife conflict is one of the key outcomes of carnivore conservation, with the main dynamics centered on large carnivores, human livelihoods, and associated resources such as livestock (Campbell, 2017; Campbell & Jha, 2021; Braczkowski et al., 2023). Braczkowski et al. (2023) cited the relationship between large carnivores and domesticated cattle and found that in the distribution of 18 large carnivores, the economic vulnerability to losses from predation (measured by the impact on annual per capita income) for households was twofold and was eight times higher in transition and developing countries compared to developed economies. Issues include the fact that ranchers in less developed countries generally produce less meat per animal, and most carnivore ranges lie outside protected areas, and five threatened carnivores have more than a third of their range in the most economically important areas, impacting rural areas sustainable development, including the elimination of human underdevelopment (Braczkowski et al., 2023).

Braczkowski et al. (2023) identify one of the biggest obstacles to sustainable development as the relationship between domestic cattle rearing and the needs of large carnivore conservation and preservation. They note that while large carnivores—species that range from wolves and hyenas to lions and bears—are essential for controlling ecosystem processes that enhance human well-being and significantly boost tourism economies, protecting them is difficult because they frequently kill valuable livestock in both community lands and areas next to protected areas. Crucially, the economic shocks resulting from livestock loss to large carnivores can be extremely high. A single livestock predation event can cost a household up to two-thirds of its annual income, and the world's poorest people may be most affected by the cost of coexisting with large carnivores. This is because the economic vulnerability of less developed countries to carnivore predation on cattle can be up to eight times greater than that of developed countries (Braczkowski et al., 2023). There are examples given of lion and leopard predation on domestic cattle near Tanzania's Serengeti National Park and tiger and leopard predation on domestic cattle in Bhutan, with losses in each case amounting to almost 20% of the owners' yearly income (see Wang & Macdonald, 2006; Holmern et al., 2007), as well as of lion-human disputes in Queen Elizabeth National Park and WAZA National Park in Cameroon.

Politicians from different interest groups have clashed over the problems presented by the struggle to preserve big carnivores while also providing economic opportunities for the underprivileged (Campbell, 2017; Campbell & Jha, 2021). Braczkowski et al. (2023) argue that charismatic carnivore species, like African lions and tigers, have a significant value to societies, especially at the national or international level, but that their value is frequently negligible or even negative at the local level (see also Good et al., 2017; Courchamp et al., 2018; Heydinger et al., 2019; Jacobsen et al., 2022). One of the biggest challenges facing biodiversity conservation today, according to Dickman et al. (2011: 13937), is making it easier to protect species that are valuable on a global scale but have little or no value locally. Endangered species, like large carnivores, can impose significant financial costs locally, especially in rural areas that are already impoverished and where households are least able to bear the costs. They can also impede the efforts of local populations, especially traditional pastoralists, to escape poverty.

It is arguable that Western stakeholders are disengaged from the impacts of large carnivores on local, rural stakeholders, who lose income from predation events. Therefore, local communities bear a greater share of the responsibility for protecting wildlife, and in some cases, conflicts may emerge when western conservation agencies and associated stakeholders intrude into local livelihood activities (Bond & Mkutu, 2018; Campbell, 2017, 2018; Campbell & Jha, 2021; Jha & Campbell, 2021). Bans on trophy hunting are an example, which may remove an important local source of income, and increase populations of large carnivores, which may lead to psychological repercussions, such as fear of carnivore attacks, hesitancy to move in the dark, grief over livestock loss, and post-traumatic stress disorder, and carnivore predation on local people (Braczkowski et al., 2023; see also Lindsey et al., 2012; Dickman et al., 2019; Di Minin et al., 2016; Gulati et al., 2021; Bruskotter et al., 2022).

The conservation of large carnivores, wild cattle, and the management of domesticated cattle are therefore crucial competing strands within the larger political ecology and environmental management of agriculture and conservation, and important issues within the scope of global sustainable development. The rest of this chapter will examine the conservation of wild cattle on a continent-by-continent basis. At the continental level, Asia has the largest number of wild cattle species, with Africa, North America, and Europe having one species each.

Asian Wild Cattle Conservation Status

Gaur

The natural, previous habitat of the gaur was mostly forested land across from throughout mainland south and southeast Asia and Sri Lanka, but it is now extinct in Sri Lanka and arguably in Bangladesh, ranging in scattered areas across Bhutan, Cambodia, China, India, Lao PDR, Malaysia (Peninsular Malaysia only), Myanmar, Nepal, Thailand, and Vietnam (Duckworth et al., 2016). The gaur is declining in population regionally, especially in Indochina and Malaysia, and possibly to a lesser extent in Myanmar and China, and in India and Nepal (Duckworth et al., 2016). The factors for these variations are the existence and distribution of protected areas, cultural structures, and various protection issues in unprotected areas (Karanth et al., 2010a, b; Duckworth et al., 2016). Overall, the key issues for direct population reductions include wild meat and horn hunting in Southeast and South Asia. Duckworth et al. (2016) note that "although the proportion of the global population which is in relatively well protected habitats in India is increasing, and thus the averaged global decline rate ought to decrease, an ongoing decline of 30% (in the next three generations) or more is adjudged likely. The recent history in Malaysia proves that directed Gaur poaching can induce rapid declines in regions previously judged relatively safe." The example is cited of Malaysia, where the poaching is carried out by international gangs, likely to relocate to other countries if their earnings in Malaysia decrease, a possibility without current evidence, but which has occurred for other species such as otters from Southeast Asia (Duckworth et al., 2016) (Table 10).

Threats against the gaur populations vary regionally, as according to Duckworth et al. (2016), there are broad regional generalizations between south and central India, on the one hand, and Southeast Asia, China, and northeast India on the other.

Table 10 Gaur conservation issues

Variable	Classification
Population size	15,000–35,000
Mature individuals	6000–21,000
Population trend	Decreasing
Habitat	Forest, savanna, shrubland, grassland
Range	Bhutan; Cambodia; China; India; Lao People's Democratic Republic; Malaysia (Peninsular Malaysia); Myanmar; Nepal; Thailand; Vietnam
Known diseases	Rinderpest and foot-and-mouth disease
Conservation status	Vulnerable
Conservation actions	Wildlife refuges and protection activities, limited, broad, unfocused

Sources: Krishnan (1972), Salter (1983), Pal and Guin (1986), Choudhury (2002), Pasha et al. (2004), Ahrestani and Karanth (2014), Duckworth et al. (2016)

Note: "Most population figures given for numbers of Gaur are little more than guesses" (Duckworth et al., 2016)

For the latter region, hunting is important as a factor for gaur extinction, alongside habitat destruction over the past decades (Duckworth et al., 1999; Duckworth & Hedges, 1998; Timmins & Rattanak, 2001; Choudhury, 2002; Steinmetz, 2004, Steinmetz et al., 2007). Other factors include the international trade in gaur horns and body parts (Srikosamatara et al., 1992; Baird, 1993; Srikosamatara & Suteethorn, 1994, 1995; Martin & Phipps, 1996; Srifa La-Ong et al., 1997; Duckworth et al., 1999). In these areas, there is hunting, coupled with human population growth, infrastructural and commercial agricultural and mining expansion, and locally ineffective conservation management (Duckworth et al., 2016). Duckworth et al. (2016) note that despite the effective conservation of tiger and Asian elephant in Malaysia, this country has a comparatively declining gaur population, possibly due to actions such as direct hunting and habitat conversion.

For south and central India, some factors for gaur population decline may be unclear, but hunting and livestock are possible factors during the early twentieth century. Duckworth et al. (2016) notes that in this region, "hunting is much less prevalent than in Southeast Asia, reflecting the general enforcement of wildlife protection laws." Legal protections of gaur are however ineffective, as declining gaur populations are recorded in several protected areas, such as the Bhadra Tiger Reserve, Nagarahole National Park, Thattekad Wildlife Sanctuary, and Kanger Valley National Park (Jathanna et al., 2003; Madhusudan & Karanth, 2000, 2002; Pasha et al., 2004; Madhusudan, 2004, 2005). Duckworth et al. (2016) further note that a factor may be competition between gaur and domestic livestock, arguing that the primary factor presently restricting gaur populations in southern and central India is likely livestock grazing, which is extensively practiced even in otherwise well-managed protected areas. However, the threat posed by interaction with native stocks is underestimated. In the Bandipur livestock grazing areas, the gaur density decreased sharply with increasing livestock density; in the pasture area under study, halving the livestock density allowed the number of gaurs to increase by a factor of 57. The average gaur density was 132 times higher in the livestock-free areas than in the nearby livestock grazing areas.

Diseases are also important for gaur survival (Pasha et al., 2004), these including rinderpest, anthrax, and foot-and-mouth disease. Mohanarangan et al. (2010) note that another common cause of death in Gaur is epidemic diseases like rinderpest. There have been reports of the illness in this species from central and southern India. Gaur's deaths from anthrax and foot-and-mouth disease were also mentioned. These diseases transmitted by domestic cattle are another threat, as foot-and-mouth is the most common, but rinderpest has affected India the most dramatically; in 1968, the disease almost completely destroyed several subpopulations of gaur, and in the late 1980s, with anthrax an important threat in north and central Myanmar (Duckworth et al., 2016; see also Krishnan, 1972; Salter, 1983; Pal & Guin, 1986; Choudhury, 2002). Young (2002) describes a case of foot-and-mouth disease occurred in Nawegaon National Park in Maharashtra, where "Cattle-transmitted diseases are the prime cause of death in free-ranging gaur" and the gaur where the only wild ungulates affected.

Duckworth et al. (2016) state that "predation is by far the major cause of Gaur mortality, perhaps followed by disease" and describe the example of Bandipur, India, where "Gaur is the most important prey in terms of the biomass taken by Tiger (42%) and Leopard (15%)" with gaur remains detected in tiger (24%) and leopard (9%) feces (Andheria et al., 2007). In another case, adjoining Nagarahole, India, gaur comprised 45% of tiger kills, with remains in 17% of tiger feces (Karanth & Sunquist, 1995). In Tadoba-Andhari, India, gaur and sambar (*Rusa unicolor* Kerr, 1792) comprised almost 70% of the tiger prey biomass, and 80% in Pench-Maharashtra, India (Karanth & Kumar, 2005). These findings, among others, suggest that gaur may be selectively killed by tigers, even with other prey available (Karanth & Sunquist, 1995).

Banteng

The natural original habitat of the banteng is forests across southern China (Yunnan) to northeast India and mainland southeast Asia, including Peninsular Malaysia to Borneo, Java, and possibly Bali (Gardner et al., 2016). The banteng favors mixed semi-open dry deciduous forest and secondary forest, and grassy openings within evergreen closed forest, usually near water but may avoid evergreen forest and closed canopy semi-evergreen forest (Wharton, 1968; Hoogerwerf, 1970), and is affected by human presence (Gardner et al., 2016). The banteng was a widely distributed species, but the population has declined by possibly more than 90% after the 1960s, in its current main population area in Cambodia. However, according to Gardner et al. (2016), there is uncertainty concerning the population and status of banteng in several locations and regions, including Cambodia, Bali, Kalimantan, Vietnam, Sabah, Myanmar, and Lao PDR, as well as extinction and presence possibilities in Sarawak, Brunei, and parts of west Malaysia, Bangladesh, India, and Yunnan, China.

The majority of the population (60% of a total of about 8000) currently occurs in the eastern Cambodian forest. Current populations of Banteng are recorded in Huay Kha Khaeng, Khao Ang Rue Nai and Dong Yai Wildlife Sanctuaries, and Kui Buri, Pangsida, Ta Phraya National Parks, in Thailand (Gardner et al., 2016). Other populations are located in protected areas in Mondulkiri Province in eastern Cambodia, and parts of Sabah, including Tabin Wildlife Reserve, Malua Forest Reserve, and Maliau Basin Conservation Area. It may also occur in Kalimantan areas such as Kutai National Park and Bukit Baka Bukit Raya National Parks and Hutan Kapur Sangkulirang Nature Reserve (Gardner et al., 2016) (Table 11).

The population of banteng is decreasing, and the species is considered endangered, and may soon be considered critically endangered, due to further declines (Gardner et al., 2016). These declines are largely due to meat hunting and horn trading, and habitat degradation across the species' range in southeast Asia. Threats include habitat losses and hunting, despite protection schemes. Gardner et al. (2016) cite the case of Cambodia, which is the location of the largest population. Here, a significant portion of the species' former Cambodian range has probably been lost

Table 11 Banteng conservation issues

Variable	Classification
Population size	8000
Mature individuals	4000–8000
Population trend	Decreasing
Habitat	Forest, Grassland
Range	Cambodia; Indonesia (Kalimantan, Jawa, Bali); Malaysia (Sabah); Myanmar; Thailand; Vietnam
Known diseases	Cattle diseases,
Conservation status	Endangered
Conservation actions	Protected areas, limited control of illegal hunting, habitat destruction

Sources: Gardner et al. (2016)

to hunting as a result of the acceleration of deforestation and fragmentation of the once vast lowland forests over the past five years, primarily as a result of the development of agribusiness concessions and the proliferation of small-scale smallholdings and other land claims. In other population centers, including Java, Kalimantan, Sawarak, Sabah, Lao PDR, and Thailand, factors for the marked population declines include habitat losses through agricultural intensification, human disturbance, road development for logging, increased hunting, especially the use of firearms, possible interbreeding with domestic cattle, disease transmission from domestic cattle, and ineffective protection (Duckworth & Hedges, 1998).

Gardner et al. (2016) note that, in comparing the impacts of habitat destruction and hunting (for meat, trophies and traditional medicines), the primary concern with habitat loss in the Democratic People's Republic of Vietnam and Laos, where deforestation is also on the rise, is the future viability of national populations, as these species were driven out of forests long before deforestation. Additionally, hunting has a greater impact in Indochina than habitat loss and fragmentation, which primarily have an indirect effect by making hunting easier for the banteng populations that still exist. The species has been eliminated from much of its former range in mainland Southeast Asia, even in areas where suitable habitat is still present. Hunting is the main direct threat to the species (Gardner et al., 2016; see also Duckworth & Hedges, 1998, Duckworth et al., 1999; Pudyatmoko, 2004; Steinmetz, 2004; Tordoff et al., 2005). Other factors may include use of banteng parts for traditional medicine, as for the gaur, but this is difficult to assess, and cattle diseases are also an important threat (Wharton, 1968; Gardner et al., 2016). Habitat and population fragmentation, and logistical problems have increased difficulties with the implementation of conservation policies (Gardner et al., 2016).

Wild Yak

The historic range of the wild yak covered the Tibetan Plateau, which includes parts of China, northern India, and Nepal (Schaller & Liu, 1996; Buzzard & Berger, 2016). Other regions possibly populated by yaks included Kazakhstan, Mongolia, and southern Russia, with extinctions occurring possibly from 800 to 300 BP (Smith & Xie, 2008; Buzzard & Berger, 2016). The favored habitats include alpine tundra, grasslands, and desert mostly in the northern Tibetan plateau, mostly between 4000 and 6100 meters altitude (Wiener et al., 2006). These environments are normally ecologically harsh, with precipitation of hail and snow of about 100–300 mm, with little surface water and saline lakes, and in some cases temperatures below −40 °C, as in the case of the Chang Tang Reserve in northwestern Tibet. Grasses, sedges, forbs, and small shrubs are the dominant vegetation of this alpine, or high cold steppe ecosystem (Schaller & Gu, 1994). Yaks are seasonally migratory, favoring lower elevations in winter (Miller et al., 1994, Smith & Xie, 2008; Berger et al., 2014).

Currently, the wild yak is considered extirpated from Nepal and Bhutan, but there are uncertain reports of its presence in Nepal (Miller et al., 1994; Buzzard & Berger, 2016). Populations are reported from India, in the Ladakh region of Jammu and Kashmir (Rawat & Sankar, 2011) and China, where it ranges in scattered populations in the Tibetan Plateau, especially in the Chang Tang Reserve, in northern Tibet (Schaller & Liu, 1996, Fox et al., 2004), in southeastern Xinjiang, and the Kekexili Nature Reserve in Qinghai and near the Kunlun Mountains (Harris et al., 1999, Harris & Loggers, 2004, Schaller et al., 2007), with some scattered populations "east and south of the main population, in the west central Tibet, south-central Qinghai, and western Gansu" (Buzzard & Berger, 2016).

Wild yaks have been protected in China since 1962, and are currently protected by the central government, especially within nature reserves, but there may be issues with habitat loss or poaching. Wild yaks are also protected in India (Buzzard & Berger, 2016). Current assessments estimate about 8000–8500 wild yaks in Tibet, with a breakdown of 7000–7500 in the Chang Tang Reserve, China, which is in the northern Tibetan Plateau, and is the third-largest land protected nature reserve in the world, 3200–3700 in Qinghai Province, China, and 2000–2500 in Xinjiang, China (Miller et al., 1994; Schaller & Liu, 1996; Schaller, 1998). Buzzard and Berger (2016) note that these figures suggest a population of about 15,000 in 1995, but the accuracy of some figures may be uncertain. Due to evidence of population declines, more recent figures may hover around 10,000 mature individuals, the assessment given by Harris and Leslie (2008; see also Buzzard & Berger, 2016). Wild yak in India are much rarer, as Buzzard and Berger (2016) note that there are only about 110 wild yak left, and some of them migrate seasonally from Chinese-controlled areas into Jammu and Kashmir's Ladakh region, and these populations possibly increasing but interbreeding and competing with domestic cattle and yak (see also Schaller & Liu, 1996, Rawat & Sankar, 2011) (Table 12).

Threats to the wild yak populations include diseases transmitted from domestic cattle, some type being undocumented (Buzzard & Berger, 2016), low birthrates possibly due to diseases such as brucellosis, currently undocumented (see Schaller & Gu, 1994; Buzzard & Berger, 2016), poaching and habitat destruction, and interbreeding with domestic cattle and domestic yak (Schaller & Gu, 1994; Qi et al., 2009). Buzzard and Berger (2016) report that the main threat to wild yak

Table 12 Wild yak conservation issues

Variable	Classification
Population size	15,000
Mature individuals	<10,000 (7500–9999)
Population trend	Decreasing
Habitat	Alpine tundra, grasslands, cold desert mountainous regions
Range	China; India
Known diseases	Cattle diseases
Conservation status	Vulnerable
Conservation actions	Habitat conservation, restriction of poaching

Sources: Miller et al. (1994), Schaller and Liu (1996), Schaller (1998), Harris and Leslie (2008), Buzzard and Berger (2016)

populations is thought to be poaching, especially commercial poaching for meat. Because they tend to avoid the high ridges and hill bases that seem to be favored by females, males are generally more vulnerable to hunting, especially from motorized hunters. But the threat of poaching to the yak population has decreased since most of the Qinghai-Tibet Plateau saw the confiscation of weapons (see also Miller & Schaller, 1997; Harris et al., 1999; Schaller & Gu, 1994; Harris, 2007; Berger et al., 2014).

Conservation actions have limited poaching, with some positive impacts, despite limitations due to factors such as competition from domestic cattle (Harris, 2007; Buzzard & Berger, 2016). Examples are cited of the Arjin Shan Reserve in Xinjiang, China, with possibly 2000 yaks (Buzzard et al., 2010), in Qinghai, China, where nearly 2000 wild yaks were reported, with possible population increases (Harris et al., 1999, Harris & Loggers, 2004, Harris, 2007), and in Chang Tang, Kekexili and the Sanjiangyuan Nature Reserves, where increased populations were reported or suggested (Schaller et al., 2007; Schaller, 2012).

Another threat is increased human presence, based on the wild yak's low tolerance of human and livestock presence, and such avoidance reduces habitat availability for wild yak (Buzzard & Berger, 2016). Buzzard and Berger (2016) note that interactions and conflicts between domestic herders and wild yaks have recently increased in areas where wild yaks have persisted or multiplied. These conflicts have included the abduction of domestic yaks into wild herds and, in certain cases, damage to people or their property (Tsering et al., 2006; Buzzard, 2011). Although it's rare right now, this could lead to more retaliatory killings. However, because of the advantages of hybridization from breeding wild yak bulls with domestic cows, wild yaks are common in some villages within the Sanjiangyuan Reserve in Qinghai. This interbreeding between wild yaks and domestic cattle is a serious threat to the domestic yaks, which may also include transmission of diseases (Schaller & Gu, 1994; Schaller & Liu, 1996, Harris et al., 1999, Smith & Xie, 2008, Rawat & Sankar, 2011).

Wild Water Buffalo

The wild water buffalo historically ranged across South and South-east Asia, from occurring from the west of Pakistan to Indochina. This species occurs principally in the presence of fresh water, including low-lying alluvial grasslands, with ox-bow lakes and river courses, sandy islands, riparian forests, and woodlands (Kaul et al., 2019). Kaul et al. (2019) argue that "the present distribution of Wild Water Buffalo should not be necessarily regarded as indicating its preference for particular eco-types because human activities have largely excluded it from many areas." Wild water buffalo remnant populations may occur in a few locations in southern Nepal, southern Bhutan, western Thailand, eastern Cambodia, and northern Myanmar. They may occur in multiple locations in India, mainly in central and northeastern Assam. Possibly, wild water buffalo have become extinct in Bangladesh, the Malaysian Peninsula, and the islands of Sumatra, Java, and Borneo (Grubb, 2005a, b, c, d). Though wild water buffalo may be extinct in Vietnam, Myanmar, and Cambodia, recent studies report buffaloes in Cambodia and Myanmar (Corbet & Hill, 1992; Choudhury, 2010). Wild water buffalo presence is uncertain in Sri Lanka, especially after the rinderpest outbreak and later genetic mixing with feral and domestic buffaloes (Kaul et al., 2019).

Currently, the wild water buffalo population is endangered and declining, and Kaul et al. (2019) note that "an inferred population reduction of at least 50% over the last three generations (generation length estimated at 8–10 years) seems likely given the severity of the threats, especially hybridization; it is projected to continue into the future." The population of wild water buffalo is difficult to assess. Kaul et al. (2019) cite the difficulty of differentiating between wild domestic buffalo and wild buffalo, and hybrids, so population assessments are mostly educated guesses. Some natural populations of buffalo are similar or identical to wild populations. For example, in Cambodia, traditional buffalo husbandry practices permit herds to roam freely in forested areas, and it can be difficult to differentiate between wild and domestic buffalo (Kaul et al., 2019). Current habitats include seasonally flooded grasslands and mixed forests in Nepal, alluvial grasslands in Bhutan and Assam, dry deciduous forests with grassy understory in Madhya Pradesh, India, and grasslands, mixed deciduous, and dry evergreen forests in Thailand (Kaul et al., 2019). In general, wild water buffalo are intolerant of human disturbance and will seek cover in isolated areas. One exception is the lone bulls who come to mate with domestic buffaloes. Although buffalo are typically diurnal, in certain places they are nocturnal to avoid people (Kaul et al., 2019) (Table 13).

The main threats to wild water buffalo are hunting, diseases from domestic livestock, habitat loss and degradation, legal protection issues, and loss of genetic diversity due to introgression with domestic and feral buffaloes (Kaul et al., 2019). Hunting is an important history activity, usually for meat, and also as retaliation for buffalo encroachment and for sport and trophy hunting. Wild water buffalo are legally protected in Thailand, Nepal, Bhutan, and India, which may be a factor for the recent decline in hunting activity. Trophy hunting and poaching for wild meat are threats in Southeast Asia, especially in Cambodia and Myanmar (Hedges et al.,

Table 13 Wild water buffalo conservation issues

Variable	Classification
Population size	3690
Mature individuals	2500
Population trend	Decreasing
Habitat	Forest, Savanna, Grassland, Wetlands
Range	Bhutan; Cambodia; India; Myanmar; Nepal; Thailand
Known diseases	Cattle diseases
Conservation status	Endangered
Conservation actions	Legal protection, protected areas, Indian Conservation Action Plan

Sources: Kaul et al. (2019)

2008; Kaul et al., 2019). There is also traditional community hunting, which is practiced in central India (Mishra, 2002). Wild water buffalo have also been known to invade cultivated land and raid crops, as a result of prolonged droughts, habitat fragmentation, and degradation. These actions may invite retaliation from cultivators, and result in killing of buffaloes (Kaul et al., 2019).

Other threats are diseases, habitat loss and degradation, and local community and insurgency (Kaul et al., 2019). Wild water buffaloes are vulnerable to several cattle diseases, which can result in local extinction where small, regional groups exist. There is also habitat loss and degradation, mostly losses of suitable grazing ground and water resources, national and local legal and social problems, lack of effective protective policies (Choudhury, 2010), and trade in food, sport, and handicraft products. There are also genetic factors, especially the variable genetic introgression and hybridization from domestic buffaloes (Muley, 2001; Flamand et al., 2003). Kaul et al. (2019) note that "introgression is likely to take place throughout the range wherever wild and domestic buffalo come into contact" and over 90% of the population is thought to reside close to sizable domestic water buffalo populations. Additionally, the risk of hybridization rises exponentially over a few generations as wild populations decline and fragment. The truly pure wild water buffalo population may be half of the current estimated population or less due to population fragmentation and the close proximity of domestic water buffalo (Kaul et al., 2019; see also Lau et al., 1998; Flamand et al., 2003; Kumar et al., 2007; Yindee et al., 2010; Babar et al., 2012).

Kaul et al. (2019) contend that this may be especially true in Indochina, where the animals were allowed to roam freely for most of the year due to traditional methods of water buffalo husbandry, frequently in natural habitats over five kilometers from villages. In fact, wild and domesticated buffalo in forest areas may be difficult to distinguish. Because of this, and because of the male-dominated mating system of the water buffalo, there may have been frequent contact and potential for introgression between the wild and domestic breeds before hunting decimated the wild population (Timmins & Rattanak, 2001; Kaul et al., 2019). In certain regions, locals and skilled hunters were unable to differentiate between domestic and wild water buffalo based solely on appearance (Kaul et al., 2019). Inbreeding within the wild buffalo's declining populations also poses a serious threat to its survival, although the introduction of domestic buffalo genes form has compromised the genetic integrity of the wild water buffalo (Kaul et al., 2019).

Conservation measures for wild water buffalo include legal protection in Bhutan, India, Nepal, and Thailand, and protected areas in Cambodia such as the Mondulkiri Protection Forest and the Srepok Wilderness Area Project (Goodman et al., 2003; Kaul et al., 2019). Most wild buffalo inhabit protected areas, with favorable habitats including lowland floodplains (e.g., in the Hukaung Valley, Myanmar). Priorities for conservation, according to Kaul et al. (2019), include protection from hunting (especially in Thailand, Cambodia, and Myanmar), but a more important threat in South Asia is the mixing with domestic cattle, especially domestic water buffalo. A conservation action plan created to restore the wild water buffalo in central India's states of Chhattisgarh and Maharashtra, where most of the remaining Indian water buffalo are found (Kaul et al., 2019).

Lowland Anoa

Historically, the lowland anoa has ranged across forests on Indonesian islands (Weber, 1890, Sarasin & Sarasin, 1901; Mohr, 1921; Harper, 1945; Groves, 1969; Burton et al., 2005, 2016a, b). Burton et al. (2016a, b) note that there is little knowledge on the ecology and life history of the two species of anoas, with primary factors being human land use and proximity rather than the primary or secondary nature of the forest (see also Mustari, 1995; O'Brien & Kinnaird, 1996; Mustari, 2003; Burton et al., 2005; Dwiyahreni, 2006; Burton & MacDonald, 2009; Wheeler et al., 2009; Okarda, 2010). Other factors may be riverine and lowland forests with food plants and fruit-bearing trees, and in some cases, coastal proximity, as anoas may drink seawater (Mustari, 2003). The anoa is solitary, but adults may move in pairs. Unlike some other cattle species, it is a browser rather than a grazer (Foead, 1992; Whitten et al., 2002).

Currently, the lowland anoa is endemic to Indonesia, and found where it is found only on Sulawesi and Buton Island, in subpopulations under 250 mature animals (Burton et al., 2005). The distribution is unclear, but the population is declining, possibly after the 1980–2005 period, but no surveys have certified these changes. Population fragmentation is more certain, and threats may be more certain than those faced by the mountain anoa (Burton et al., 2005, 2016a, b). There are even possibilities of local extinction due to hunting. One report by O'Brien and Kinnaird (1996) reported a 50–95% population decline in Tangkoko Nature Reserve in Northern Sulawesi over 10 years, leading to extinction reported in later surveys (Burton et al., 2005) (Table 14).

According to Burton et al. (2016a, b), the two biggest threats to this species are hunting, mostly for food, and land conversion for agriculture, with additional threats from gold mining and other operations associated with the acquisition of non-timber forest products as per the latest reports, hunting poses the greatest threat, and the international trade in live animals or body parts exists but does not present a

Table 14 Lowland anoa conservation issues

Variable	Classification
Population size	5000
Mature individuals	2499
Population trend	Decreasing
Habitat	Forest
Range	Indonesia (Sulawesi and Buton Island)
Known diseases	Unknown
Conservation status	Endangered
Conservation actions	The Indonesian Ministry of Forestry published a Strategy and Conservation Action Plan for Anoa (2013–2022), involving breeding programs, high-priority sites for in situ conservation. Protection from hunting and habitat losses. High-priority sites for in situ conservation of anoa have been identified

Sources: Miller (2002), Mustari (2003), Indonesian Ministry of Forestry (2013), Burton et al. (2016a, b)

significant risk (Burton et al., 2016a, b). Local people may hunt in unprotected forests, but they also hunt illegally in nature reserves. To kill anoas, local hunters typically use dogs, spears, or snares (Miller, 2002). Miller (2002) notes that the lowland anoa has no predatory threats, except from people, but infants may be preyed upon by pythons (*Malayopython reticulatus*, Schneider, 1801 or *Python molurus*, Linnaeus, 1758) and the endemic Sulawesi palm civet (*Macrogalidia musschenbroekii*, Schlegel, 1877).

Mountain Anoa

The historic range of the mountain anoa, like that of the lowland anoa, includes most of the forests of Sulawesi and Buton Island (Weber, 1890; Sarasin & Sarasin, 1901; Mohr, 1921; Harper, 1945; Groves, 1969; Burton et al., 2005). Their current population distribution and ecology of the mountain anoa are largely unknown (Burton et al., 2005). Burton et al. (2016b) note that there is too little information available to determine the precise number of Mountain Anoa that exist today. In Sulawesi, anoa appear to be fairly common within their current known range. However, their numbers have declined since the late nineteenth century, with serious local decreases between 1980 and 2000 (roughly three generations) (Burton et al., 2016b). This species has declined all over Sulawesi, but especially in the southern and northeastern peninsulas. Even in protected areas, habitat loss and meat hunting have been blamed for these declines. The majority of populations are fragmenting quickly, with most subpopulations numbering less than 250 mature animals, indicating that metapopulation management may soon be necessary to maintain viable populations (Burton et al., 2016b).

The mountain anoa is assessed as a dense, mountain forest species, especially in forests with dense understory and near waters and low or no human presence (Foead,

1992; Sugiharta, 1994; Schilz, 2004; Burton et al., 2016b). Water proximity is important, as anoas may wallow in water and muddy environments, and may drink seawater. The mountain anoa is a solitary browser and grazer (Foead, 1992). Currently, the mountain anoa is endemic to Indonesia, on Sulawesi and Buton Island (Burton et al., 2005). Although it is principally recorded at elevations from 1000 to 2300 m, it has been reported near sea level (Burton et al., 2016b) (Table 15).

The mountain anoa has declined across Sulawesi, especially in the southern and northeastern peninsulas (Thornback, 1983). According to Burton et al. (2005), the two main threats facing this species are habitat degradation (mostly due to agriculture and gold mining) and hunting for food. The illegal international trade in live animals or body parts may pose a minor risk, but this is not seen as a threat (Melisch, 1995; Indonesian Ministry of Forestry, 2013; Burton et al., 2005). The majority of populations are fragmenting quickly, indicating that metapopulation management may soon be necessary to maintain viable populations (Burton et al., 2005). Populations within protected areas such as Lore Lindu National Park are declining, due to extensive hunting pressure. In general, it is unknown how quickly the decline is occurring.

Indonesian law provides complete protection for mountain anoa (Jahja, 1987; Burton, 2001; Burton & Macdonald, 2009; Burton et al., 2005), largely through protected areas such as Tanjung Peropa Nature Reserve, Bogani Nani-Wartabone National Park, and Lore Lindu National Park (Indonesian Ministry of Forestry, 2013, Burton et al., 2005). The Indonesian Ministry of Forestry (2013) released Anoa 2013–2022: A Strategy and Conservation Action Plan in 2013, the stated vision of which was "Stable population of anoa in its wild habitat through reduced poaching (hunting) and illegal trade, maintenance of the existing habitat, through active involvement of the stakeholders." Recently, 11 high-priority sites in mainland Sulawesi and 2 in Buton were set up for the conservation of anoa (Indonesian Ministry of Forestry, 2013). The sites were based on the ecology of subpopulations, forest cover and forest habitat connectivity, and local management capabilities

Table 15 Mountain anoa conservation issues

Variable	Classification
Population size	3000–5000
Mature individuals	<2500
Population trend	Decreasing
Habitat	Forest
Range	Sulawesi and Buton Island
Known diseases	Unknown
Conservation status	Endangered
Conservation actions	The Indonesian Ministry of Forestry published a Strategy and Conservation Action Plan for Anoa 2013–2022 involving breeding programs, high-priority sites for in situ conservation. Protection from hunting and habitat losses. High-priority sites for in situ conservation of anoa have been identified

Sources: Schilz (2004), Burton et al. (2016a, b)

based on the organization of national parks, protected forests, restoration concessions, or other variably managed sites. There were also protections from hunting and habitat loss, research on population status, local education, training, and partnerships, as well as genetic and morphological research.

Tamaraw or Mindoro Dwarf Buffalo

Taraw were historically found in low- to medium-elevation forests, dense bamboo jungles, swampy river valleys, and open grasslands or forest clearings throughout the Philippine Island of Mindoro (9735 km^2 in area), from sea level to high peaks (up to over 1800 m) (Rabor, 1977). Pleistocene records from Luzon indicate that the species was more common prehistorically in the Philippines (Beyer, 1957 in Kuehn, 1986). The species is currently limited to isolated locations above 200 meters and is primarily found in mixed forests and secondary forests (Kuehn, 1986; Custodio et al., 1996, Heaney et al., 2002). In Mindoro, the current range includes the three known surviving subpopulations located in the western part of the island and occupy less than 10,000 hectares of land, with the largest protected areas being Mount Iglit-Baco National Park in the center of the island, the Aruyan-Malati Tamaraw Reserve (Sabalayan Municipality), and the Mount Calavite Wildlife Sanctuary in the north of the island (Custodio et al. 1996; De Leon et al., 1996). There are very few other isolated groups outside of these subpopulations (Schütz, 2015). Pleistocene records from Luzon indicate that the species was more common prehistorically in the Philippines (Beyer, 1957 in Kuehn, 1986).

The minimum population is estimated at 430 animals, of which just over half are matured animals (Boyles et al., 2016). The subpopulation of Mount Iglit-Baco National Park, as estimated in 2016, was estimated to be about 413 individuals. Population increases have been hinted, but it is possible that such increases are due to more accurate estimates and immigration from outside the counting areas (Schütz, 2015). Despite protection from poaching and disturbance, the range has been decreasing over two decades from 5000 to 2000 hectares (Schütz 2015; Boyles et al., 2016).

The subpopulation at Aruyan-Malati was estimated at 15–20 animals in the early twenty-first century. More recent surveys suggest about 12 animals within less than 1000 hectares (Schütz, 2015). The subpopulation on Mount Calavite was estimated at 15 animals, possibly fewer more recently. Boyles et al. (2016) notes that the population losses could be as high as 25% over the next three generations (approximately 30 years, 1990–2016), indicating a critical scenario.

The tamaraws are solitary, with associations including females and (Talbot & Talbot, 1966) and temporary associations between males and females (Custodio et al., 1996; Kuehn, 1986). The tamaraw is a grazer feeding in open grasslands, and sometimes resting within denser vegetation (Talbot & Talbot, 1966). Due to interference and disturbance from humans, tamaraws, which were formerly diurnal, are now mostly nocturnal (Talbot & Talbot, 1966). Because of the relative protection, it enjoys within its "core zone" in Mount Iglit-Baco National Park, tamaraw behavior is more diurnal and it can be observed in larger groups (Talbot & Talbot, 1966; Boyles et al., 2016). Despite this protection, the population still appears to be declining, and over 95% of individuals are estimated within one subpopulation in the 5000 hectare Mount Iglit-Baco National Park (Table 16).

The primary threat to the tamaraw is habitat loss due to agriculture based on increased human populations (Boyles et al., 2016). There are also agricultural fires that destroy habitat. Cattle ranching and agricultural activities introduce diseases and reduction of suitable grazing habitat. Tamaraws are also hunted for subsistence and sport (Rabor, 1977). Although hunting was regulated before the Second World War, more recently controls have been relaxed due to human population increase, logging, livestock farming, and availability of firearms (Talbot & Talbot, 1966). Sport hunting has decreased since the 1980s as a result of increased patrol and reconnaissance since the refuge's establishment, a decrease in the tamaraw population, and the closure of neighboring ranches. There has been no information reported on the international trade in this species or its byproducts (Boyles et al., 2016).

The associated threats are based on human activity, and the lack of natural corridors prevents the tamaraw populations from dispersal within their range (Schütz, 2015; Boyles et al., 2016). Wildlife in the lowland parts of Mindoro is seriously threatened by poaching and illegal activities. Nevertheless, even in cases where the species is not expressly controlled (trapping, snare trapping), the customary hunting practices of the nearby indigenous communities in tamaraw areas constitute a significant source of disturbance and potential harm to tamaraw. Additionally, as a result of population growth and their incapacity to expand into lowland areas now inhabited by new settlers, indigenous communities in Mindoro are placing more and more pressure on land use decisions (Boyles et al., 2016). The tamaraw receives protection under Philippine law, and there has been an unsuccessful captive breeding program. The Aruyan-Malati area has been designated as "Critical Habitat" and a Protected Area Management Plan for Mount Iglit-Baco National Park is planned, with a strictly protected zone (Boyles et al., 2016).

Table 16 Tamaraw conservation issues

Variable	Classification
Population size	494
Mature individuals	220–300
Population trend	Decreasing
Habitat	Forest, shrubland, grassland
Range	Philippines
Known diseases	Diseases spreading from livestock
Conservation status	Critically endangered
Conservation actions	Total protection under Philippine law. "Critical Habitat" areas were developed, supported by sland-wide population surveys, ethnoecological studies

Sources: Gesch (2004), Boyles et al. (2016), Peña (2022)

Saola

The historic range of the Saola has declined significantly (Timmins et al., 2020). Timmins et al. (2020) note that the Saola has a very specific habitat of moist evergreen forests, based on knowledge of its historical and present range. This type of forest, which primarily occurs on the eastern (Vietnamese) slopes of the Annamite Mountains, has little to no dry season and rarely has monthly rainfall below an average of 40 mm (see also Duckworth et al., 1999). Currently, the range include the Annamite Mountains region of Laos and Vietnam (Vu Van Dung et al., 1994, Kemp et al., 1997a, b), but Timmins et al. (2020) note that the "precise northern and southern limits to the species range in Laos, as in Vietnam, are still uncertain" and "because of regional climatic factors, the majority of suitable habitat is, or was, probably more abundant in Vietnam than in Laos, but Vietnam's much higher human population density has severely reduced both habitat and Saola numbers in the habitat that remains." The soala appears to favor larger forest fragments, above 100 km^2, with lower hunting pressure (Timmins et al., 2020). Although some of the original forested saola habitat remains, it is argued that some areas of suitable habitat may not have any resident saola, due to various pressures (Table 17).

Hence, the "total Saola population is undoubtedly less than 750, and likely much less" and "population declines of more than 80% over three generations are projected for the past, present, and future, and rates of decline are more likely to rise than fall" (Timmins et al., 2020). The total population numbers are unknown (Tham Ngoc Diep et al., 2004). Sightings are rare: the first field photograph of saola was taken in 1998, in Pu Mat National Park in Vietnam, and later in Bolikhamxay Province in Laos, and Hue-Quang Nam in Vietnam (Whitfield, 1998; Robichaud, 1999; Hance, 2013). Throughout its range, the saola is the rarest and least sighted ungulate in the environment, and is currently declining (Weir & Dinh Van Cuong, 2004; Timmins et al., 2020).

The main threats are intensive non-specific bushmeat and traditional medicine hunting, road and track construction, the latter increasing the dispersal and

Table 17 Saola conservation issues

Variable	Classification
Population size	<750
Mature individuals	25–750, 50–300
Population trend	Decreasing
Habitat	Forest
Range	Lao People's Democratic Republic; Vietnam
Known diseases	Unknown
Conservation status	Critically endangered
Conservation actions	Protected by national law in Vietnam, National protected areas, Provincial protected areas, National parks, Nature reserves, and weak laws to stop hunting

Sources: Holcomb (2005), Timmins et al. (2020)

fragmentation of saola populations, and habitat losses (Timmins et al., 2020). There is very little to no protection against hunting in some parts of the species' range, with few sightings from local hunters. Habitat loss is rated as a lesser threat to the saola than hunting, with such losses largely due to construction, such as the Ho Chi Minh Road through Vietnam's Annamite Mountains, and linked branches, which may increase human access to the habitat areas (Timmins et al., 2020). Regarding hunting, Timmins et al. (2020) note that the range is threatened by commercial hunting, but the specific demand for Saola is rare as it is unknown in the traditional Chinese pharmacopoeia, and its meat is not preferred to other herbivores. In rare occasions, soala horns are traded. One source indicates that in Pu Huong Nature Reserve in Vietnam, 30 saolas were killed between 1995 and 2003 (Quang & Trung, 2004; Timmins et al., 2020). For habitat destruction, human activities include small-scale subsistence and commercial agriculture, timber extraction, roads, mining, and hydro-power development. The low populations also may preclude effective breeding, and saolas are usually unsuccessful in captivity (Stone, 2006; Robichaud & Long, 2014).

Conservation actions include the 2010 IUCN SSC Asian Wild (AWCSG) Specialist Group Regional conservation strategy for wild cattle and buffaloes in South-east Asia, including Saola (IUCN-SSC Asian Wild Cattle Specialist Group, 2010), with updates (Saola Working Group, 2013). Protected areas include national and provincial protected areas, but some are not fully protected, and national parks with more protection (Timmins et al., 2020). Priorities for further conservation include effective patrolling against snaring and other types of hunting, assessment of Saola status and conservation, more nature reserves for Saola, and increased international support (Timmins et al., 2020). Timmins et al. (2020) note that despite the increased efforts for saola conservation, intensive hunting and other pressures are still causing the species to decline, and it is unclear if the greater focus on saola protection will prevent the extinction of the species.

Kouprey

Historically, the kouprey was found in open, deciduous dipterocarp forests mixed with grasslands in western Vietnam, southeastern Thailand, southern Lao People's Democratic Republic, and Cambodia (Grubb, 2005a, b, c, d). However, with no recent records from these countries, the species is possibly extinct (Timmins et al., 2016). Old records suggest that this species is both a grazer and browser and moved in small herds (Wharton, 1957). Ideal habitat for the kouprey is now limited, comprising patches of mixed deciduous and semi-evergreen forest, within the larger deciduous dipterocarp forest, with pools and mineral licks (Timmins & Rattanak, 2001; Tordoff et al., 2005; AWCSG, 2011). It is also suggested that Kouprey may have been domesticated in Cambodia, possibly in hybrid form with domestic breeds, but this position is contested (Hassanin et al., 2006; Galbreath et al., 2006; Grigson, 2007; Hassanin & Ropiquet, 2007; Hedges et al., 2007; Timmins et al., 2016).

The kouprey is listed as critically endangered (possibly extinct), with more than 40 years without confirmed sightings, and "the species is most likely to be extinct" (Timmins et al., 2016). The possible current extinction of the kouprey is however contested (Timmins et al., 2016). For example, in a 1988 Vietnam workshop, reports suggested 100–300 Koupreys in Vietnam, Lao PDR, Cambodia, and Thailand, but these figures "were little more than guesses" and "unfortunately, it now seems that these figures were too optimistic" (Timmins et al., 2016), with the last published records of Koupreys in 1963/64 (Wharton, 1966, 1968). Timmins et al. (2016) note that should the kouprey continue to exist, it will most probably be as solitary individuals or very small populations, rather than as functional groups and priority areas for other wildlife species should be protected, and additional camera trap photo analysis concentrated on areas of possible kouprey habitat, rather than additional surveys (Table 18).

Hunting, for local consumption and for trade (meat and body parts, especially horns and skulls), is the main for kouprey local or total extinction (Duckworth & Hedges, 1998; Timmins & Rattanak, 2001; Winker, 2004; Tordoff et al., 2005; Timmins et al., 2016). Kouprey horns were reported in Cambodia and Thailand (Salter et al., 1990). Diseases can also be transmitted by domestic cattle, and habitat loss results from increased land clearance for cultivation, mining, and logging, which are important but less important than hunting (Tordoff et al., 2005; Timmins, 2006; Bezuijen et al., 2008). Conservation measures for large mammal communities in eastern Cambodia are currently the best approach, and it is unlikely that specific survey work for kouprey would yield more conclusive evidence than what has already been documented. Suggested approaches include strengthening and expanding the current initiatives in the Siema Biodiversity Conservation Area and the Mondulkiri Protection Forest's Srepok Wilderness Area (AWCSG, 2011; Timmins et al., 2016).

Table 18 Kouprey conservation issues

Variable	Classification
Population size	0–50
Mature individuals	0–50
Population trend	Decreasing
Habitat	Forest, savanna, grassland
Range	Cambodia; Lao People's Democratic Republic (Possibly Extinct)
Known diseases	Cattle transmission
Conservation status	Critically endangered (possibly extinct)
Conservation actions	Conservation measures for large mammals, surveys

Sources: Timmins et al. (2016)

American Wild Cattle Conservation Status

American Bison

The North American bison possessed the broadest natural distribution of any herbivore in North America during its historical range, from the arid grasslands of northern Mexico's Chihuahua state to the grasslands of the United States and Canada's Great Plains to the riparian meadows of inland Alaska (Reynolds et al., 2003; Wilson & Reeder, 2005; Redford et al., 2009; Aune et al., 2017). Before European settlers arrived on the continent, the plains bison engaged in seasonal migrations. Due to changes in land use, bison no longer migrate, which restricts their range and causes depopulation. During the breeding and calving season (May to August), the bison shows seasonal aggregation (Aune et al., 2017). Bison ecology is based on grassland and meadow vegetative communities, including sedges, flowering plants, woody plant leaves, and lichens (Boyd, 2003; Boyd & Gates, 2006; Brodie, 2008; Fuhlendorf et al., 2010; Plumb et al., 2013; Aune et al., 2017).

Excessive hunting in the nineteenth century led to a drastic decline in the North American bison population, with later partial recovery under effective conservation management (Stephenson et al., 2001; List et al., 2007). The number of populations managed for species conservation and ecological restoration has increased slightly. However, all adults are part of ongoing management programs; if these ended, the species would be classified as endangered (Aune et al., 2017). Approximately 97% of the bison population is kept in private captivity for commercial reproduction; very few of these herds serve primarily to conserve the species or for public interest. Small herds (less than 400) are typically managed for conservation purposes in the public interest (Aune et al., 2017). Populations are widely distributed, and few geographic locations allow natural migrations between subpopulations. It is estimated that there are between 11,248 and 13,123 adults in total in free-ranging and semi-free-ranging populations (Aune et al., 2017). The species is not currently in decline, but arguably if the current management systems are changed or abolished, the number of wild adult individuals could decline dramatically, with natural supports for the population and human exploitation or interference. Protection is therefore essential for this species (Freese et al., 2007; Aune et al., 2017). Due to these issues of conservation protection, the American bison is listed as near threatened, or critically depleted, but with a stable population (Rogers et al., 2022) (Table 19).

Conservation actions supported by government agencies and non-government organizations, and management by federal and state/provincial agencies, and public tolerance and support of wild, free-ranging bison managed as wildlife are limited in large-scale landscapes. Currently, 6 of the 20 wild herds representing 11,956 animals (63.7%) are anchored by national parks, refuges, or sanctuaries. The species' current range is restricted by land use and wildlife management policies in the southern area and by wildlife and reportable disease management policies in the northern portion of the North American range. Bison functioning as wild currently

Table 19 Wild yak conservaticn issues

Variable	Classification
Population size	31,000
Mature individuals	11,248–13,123
Population trend	Stable
Habitat	Forest, savanna, shrubland, grassland, wetlands
Range	Canada, United States, Mexico
Known diseases	Bovine tuberculosis and brucellosis, genetic diversity issues
Conservation status	Near threatened, critically depleted
Conservation actions	Recovery programs, captive herds, Plains Bison reintroductions

Sources: Gates et al. (1995, 2001, 2010), Stephenson et al. (2001); Reynolds et al. (2003), Bison Gross and Wang (2005), Halbert and Derr (2007, 2008), List et al. (2007), Sanderson et al. (2008), Hedrick (2009), Plumb et al. (2013), Aune et al. (2017), Rogers et al. (2022)

occupy less than 1.2% of their original range (Sanderson et al., 2008). Populations are considered viable in the long term if they exceed 1000 individuals (Gates et al., 2010). Managing herds for both socioeconomic and conservation purposes on Indigenous people's land may also present opportunities. Yet, incorporating Western, scientifically grounded conservation strategies into tribal communities presents considerable cultural, social, and financial obstacles (Aune et al., 2017; Rogers et al., 2022).

Due to commercial, subsistence, and recreational hunting, the bison was almost completely eradicated throughout its range in North America in the nineteenth century. Limited recovery has been observed in captive herds and the wild as a result of recent conservation efforts (Fuller et al., 2007). While significant population recovery has occurred as a result of private commercial production, the Red List does not consider private commercial production to be conservation of bison as a wild animal. Currently, there are a number of threats to the bison population, including habitat loss, genetic manipulation of commercial bison for market traits, gene introgression in cattle, loss of genetic non-interchangeability through hybridization between bison subspecies, low population impact in most maintenance herds, and the threat of depopulation as a management response to the infection of some wild populations harboring reportable cattle diseases (Gates et al., 1995, 2001, 2005, 2010; Wilson et al., 2002; Gross & Wang, 2005; Geremia et al., 2014). The bison is categorized as a wild and livestock animal in Canada, the United States, and Mexico, with local classifications varying. Culling is another threat to the populations of this species because it stops the spread of brucellosis and tuberculosis in cows. There may be chances in some areas for the ecological restoration of herds that are mainly kept for conservation on federal, state, and local lands (Sanderson et al., 2008; Freese et al., 2007).

European Wild Cattle Conservation Status

European Bison

The natural habitat of the European buffalo is open and mixed forest and grassland (Plumb et al., 2020). Plumb et al. (2020) note that "contemporary habitat selection indicates the species' preference for open and wet/open habitats and abandoned fields over the forests, especially coniferous ones" (see also Kuemmerle et al., 2010, 2018, Kowalczyk et al., 2019, Zielke et al., 2019). Research indicates that the European bison had a wide range across western and northern Europe, from France and Sweden to the Ural and Caucasian mountains (Plumb et al., 2020; see also Pucek et al., 2004; Soubrier et al., 2016; Węcek et al., 2017; Hofman-Kamińska et al., 2019). Open habitats were reduced by forest expansion around 12,000 to 8000 BP, and historic human land use contributed to declines in bison populations, with near extinction between 9500 and 7000 years BP, with some recovery in Eastern Europe, especially in Poland and western Belarus, and Caucasus mountains (Pucek et al., 2004; Hofman-Kamińska et al., 2019). The lowland bison and Caucasian bison became extinct in the early twentieth century (Pucek et al., 2004).

Most European bison populations are composed of reintroduced individuals, which may more open terrain or forests (Gębczyńska et al., 1991; Kowalczyk et al., 2011; Kerley et al., 2012; Bocherens et al., 2015, Merceron et al., 2015; Hofman-Kamińska et al., 2018; Kowalczyk et al., 2019, Zielke et al., 2019; Kerley et al., 2020). Home ranges of bison cover from several dozen to two hundred km2, in relation to habitat structure and seasonal migrations (Kowalczyk & Plumb, 2020). Seasonal migrations occur, based on food conditions, and may include movements of several kilometers (Perzanowski et al., 2012; Kowalczyk et al., 2013; Krasińska & Krasiński, 2013). Dispersal by mixed groups is beginning to be observed in expanding populations, as the result of density-dependent intra-specific competition (Kowalczyk et al., 2013; Krasińska et al., 2014) (Table 20).

Threats to the European bison include low genetic variability, small and fragmented population groups, suboptimal habitats, diseases, poaching, and conservation inconsistency (Plumb et al., 2020). Plumb et al. (2020) note that the "contemporary genetic variability of European Bison is relatively low, being an effect of population bottleneck due to species extirpation in the wild in the early 20th century and subsequent restoration from limited number of captive survivors" and that the "entire contemporary species population derives from only 12 individual genotypes" (see also Pucek et al., 2004; Wójcik et al., 2009; Tokarska et al., 2011). Tokarska et al. (2009) also add that, for example, within the subpopulation of 462 in the Polish part of the Białowieża Forest, the actual population size genetically may be less than 30. The degree of fragmentation is also important, as the majority of recorded subpopulations are isolated; hence, "we can infer further genetic drift of inherently low genetic diversity to be amplified through long-term isolation" (Plumb et al., 2020).

There is also the issue of "refugee species" in which the European bison is assessed as a species generally outside its natural habitat and adapting to changing

Table 20 European bison conservation issues

Variable	Classification
Population size	6244
Mature individuals	2518
Population trend	Increasing
Habitat	Forest, shrubland, grassland
Range	Belarus, Bulgaria, Germany, Latvia, Lithuania, Poland, Romania, Russian Federation, Slovakia, Ukraine
Known diseases	Blue tongue virus, foot-and-mouth disease, respiratory viruses, and bovine tuberculosis, 90 species of parasites
Conservation status	Near threatened
Conservation actions	Bison population viability analyses, suggesting increasing subpopulation size, effective gene flow among subpopulations, restoration of natural movements between populations. Collaborative conservation planning

Sources: Daleszczyk and Bunevich (2009); Radwan et al. (2010), Kerley et al. (2012), Karbowiak et al. (2014a, b), Kołodziej-Sobocińska et al. (2016a, b, c), Hartway et al. (2020), Larska and Krzysiak (2019), Plumb et al. (2020)

environments (Plumb et al., 2020; see also Mendoza & Palmqvist, 2008; Kerley et al., 2012; Bocherens et al., 2015; Hofman-Kamińska et al., 2019). This definition acknowledges the negative impacts of this adaptation, including health and population issues (Kerley et al., 2012; Hofman-Kamińska, et al. 2019). Plumb et al. (2020) argue that restricting a species adapted to open habitats to forest habitats can counteract natural selection pressures, impose additional feeding, increase management complexity and costs, increase susceptibility to disease and parasites, and increase the risk of human-animal conflict when bison are trying to get out of the forests (see also Hofman-Kamińska & Kowalczyk, 2012; Kerley et al., 2012; Kowalczyk et al., 2013; Kołodziej-Sobocińska et al., 2016a, b, c; Samojlik et al., 2019). There is a general scarcity of accessible suitable mixed habitats in densely populated Europe (Alcantra et al., 2013; Plumb et al., 2020).

Diseases are also prevalent, such as blue tongue virus, foot-and-mouth disease, respiratory viruses, and bovine tuberculosis, and almost 90 species of infectious parasites (Drożdż et al., 1998; Karbowiak et al., 2014a, b; Larska & Krzysiak, 2019). A factor for parasitic infection is the supplementary feeding and unnatural grouping of bison in winter (Radwan et al., 2010, Kerley et al., 2012, Karbowiak et al., 2014b, Kołodziej-Sobocińska et al., 2016b). Poaching and culling are additional problems, for example, in Ukraine in the 1990s (WWF, 2019). Inconsistent conservation strategies exist across its range, including management of habitats, culling, and distributions (Pucek et al., 2004; Alcantara et al., 2013). Most bison currently live under protection schemes, and therefore are vulnerable to protection policies. The bison is not currently threatened by trade and use, and their population viability for small, fragmented populations is studied and monitored (Daleszczyk & Bunevich, 2009; Hartway et al., 2020).

African Wild Cattle Conservation Status

African Buffalo

The historic range of the African buffalo was across most of vegetated sub-Saharan Africa, including savanna, forest, and mixed vegetation (IUCN SSC Antelope Specialist Group, 2019). Cornelis et al. (2023) point out that regarding biomass, abundance, and geographic distribution, buffalo rank among the most successful mammals. Nearly every natural habitat south of the Sahara is included in its distribution area. Currently, African buffalo inhabit semi-arid shrublands, acacia woodland, miombo-brachystegia woodland, mountain grasslands and forests (up to elevations well over 4000 m above sea level), coastal savannas, and moist lowland rainforests. They are not found in deserts and sub-deserts such as the Namib/Kalahari, the dry Horn of Africa, and the Sahara/Sahel transition zone (Prins & Sinclair, 2013; IUCN SSC Antelope Specialist Group, 2019; Cornélis et al., 2014, 2023). Although it can live at elevations above 2500 m, as it does in Kenya's Aberdare National Park, it primarily inhabits savannas with high herbaceous biomass. It also occurs in dry shrubland and grassy clearings in dense tropical rainforests, and dry biomes with constant surface water availability. It is uncertain if the range included North Africa or the Nile Valley north of Khartoum (Prins & Sinclair, 2013). Current research findings suggest that the buffalo requires annual rainfall levels above 250 mm (the Sahel savanna and more moist climatic zones) (possibly aurochs, which are the ancestors of domestic cattle may have occupied a similar, competitive scenarios (Cornélis et al., 2014; IUCN SSC Antelope Specialist Group, 2019).

The ideal habitats vary for the four buffalo subspecies. The forest buffalo (*Syncerus caffer nanus* Boddaert, 1785) occurs in two rainforest and mixed vegetation areas with annual rainfall mostly above 1500 mm, in West Africa and the Congo Basin and Lower Guinea forests (Cornélis et al., 2014). Current research indicates significant population declines in both regions, largely due to habitat losses and poaching (Cornélis et al., 2014). The West African Savanna Buffalo (*Syncerus caffer brachyceros*) occurs in savannas and gallery forests, mostly in isolated patches. The Central African Savanna Buffalo (*Syncerus caffer aequinoctialis*) inhabits Sahelo-Sudanian savannas and gallery forests. The Cape Buffalo (*Syncerus caffer caffer*) is found throughout the savannas, grasslands, and mixed, denser vegetation of East and Southern Africa (IUCN SSC Antelope Specialist Group, 2019; Cornélis et al., 2023).

Since the nineteenth century, the African buffalo has been declining in population and habitat areas, due to anthropogenic factors (land conversion, poaching, disease outbreaks), as well as natural occurrences like droughts. There was also culling to limit the transmission of pathogens to cattle, including foot-and-mouth disease (FMD) and trypanosomiasis. Hence, most populations occur in or near protected areas, such as the Kruger National Park in South Africa (Cornélis et al., 2014; IUCN SSC Antelope Specialist Group, 2019). Variable estimates exist for the population of African buffaloes vary, from 830,000 and range of 500,00–1000,000, or >627,000 for the savanna buffalo (East, 1999) or lower figure of about >513,000 for the savanna buffalo subspecies (Cornélis et al., 2014). All the subspecies are believed to be declining across their ranges. The IUCN SSC Antelope Specialist Group (2019) estimated the population of buffalo based on the estimates of East (1999) and Cornélis et al. (2014). They added the estimate of 60,000 forest buffalo (which could have declined at the same rate as the savanna species, at 15% down to 56%) to the 513,000 savanna estimate of Cornélis et al. (2014), who give total a of 573,000; "Thus a very approximate global estimate of the buffalo population is 569,000–573,000 (398,000–401,000 mature individuals)" (IUCN SSC Antelope Specialist Group, 2019) (Table 21).

Estimates support a continual decline in African buffalo populations, with factors for the decline in forest buffalo populations including marked deforestation and increased bushmeat extraction in West Africa (Christie et al., 2007, Norris et al., 2010, Hansen et al., 2013). Factors across Africa for the other savanna subspecies include poaching, habitat conversion, and increased human and domestic livestock populations: based on the decline of buffalo populations between East (1999) and Cornélis et al. (2014), a population decline of approximately 30% is estimated by 2027 (1999–2027 represents three generations). Although the African buffalo's current status is Near Threatened, this may degenerate to Vulnerable with further threat impacts (IUCN SSC Antelope Specialist Group, 2019). Habitat losses are mostly due to urbanization, agriculture, protected area encroachment, and hunting may occur even in national parks. The buffalo is also a prized target of bushmeat and trophy hunters and may be bred on game farms. Conservation is determined largely by protected areas and managed hunting zones (Cornélis et al., 2014). Social and military conflicts have had serious impacts on buffalo habitats and populations.

Table 21 African buffalo conservation issues

Variable	Classification
Population size	>513,000
Mature individuals	398,000–401,000
Population trend	Decreasing
Habitat	Forest, savanna, shrubland, grassland
Range	Angola; Benin; Botswana; Burkina Faso; Burundi; Cameroon; Central African Republic; Chad; Congo; Congo, The Democratic Republic of the; Côte d'Ivoire; Equatorial Guinea; Ethiopia; Gabon; Ghana; Guinea; Guinea-Bissau; Kenya; Liberia; Malawi; Mali; Mozambique; Namibia; Niger; Nigeria; Rwanda; Senegal; Sierra Leone; Somalia; South Africa; South Sudan; Sudan; Tanzania; Togo; Uganda; Zambia; Zimbabwe
Known diseases	Rinderpest, pleuro-pneumonia
Conservation status	Near Threatened
Conservation actions	Protected areas, managed hunting zones, game farms

Sources: Cornélis et al. (2014), IUCN SSC Antelope Specialist Group (2019)

Other threats include drought and several infectious diseases that are both imported and indigenous. The severe rinderpest (now eradicated) epidemic of the 1890s and pleuropneumonia killed up to 95% of livestock and wild ungulates, and nearly eradicated local buffalo populations (Winterbach, 1998; Morens et al., 2011).

Geomatics Applications to Cattle-Carnivore Research

Geomatics-based applications to the study and assessment of cattle-carnivore relations include more accurate information and more effective analyses of land cover change, representation of linear and multidirectional habitat change, population densities, tracking for species presence assessments, and predictive modeling. Computer-aided systems are a significant technological advancement with broad applications in conservation biogeography, economics, anthropology, sociology, and human geography. In North America, they are commonly referred to as geomatics, and in Europe and Asia, as geoinformatics. The field of geomatics encompasses a wide range of disciplines, such as remote sensing, geodesy, cartography, global positioning systems (GPS), geographic information science (GIS), and geographic information systems (GIS) (Campbell, 2017, 2018; Campbell & Jha, 2021; Jha & Campbell, 2022, 2023).

According to current research findings, geomatics-based systems allow for more accurate and dependable computations for environmental research with outcomes that are more applicable to practical scenarios. Methods based on geomatics make it simple to quantify changes in land cover and animal populations and habitats that have an impact on conservation management and policy. Predictive modeling of human behavior in conservation applications, in the context of human decision-making and the assessment of the socioeconomic, political, cultural, and even psychological effects of biogeographical shifts, however, may be more subjective and problematic, with an important role for field and social surveys. Measuring issues related to applied fields like cultural studies, urban planning, and environmental conservation presents additional challenges, especially in relation to conservation of rare and ecologically demanding species. Evaluating these issues may help with the application of geomatic techniques in conservation (Campbell, 2017, 2018; Campbell & Jha, 2021; Jha & Campbell, 2022, 2023).

Remote sensing is an important aspect of geomatics (Campbell, 2018, 2022). Jensen (2007: 4) defines remote sensing as "the acquiring of data about an object without touching it" (Jensen, 2007: 4). Gupta (2003: 1) offers a more detailed definition: "Remote sensing, in the simplest words, means obtaining information about an object without touching the object itself. It has two facets: the technology of acquiring data through a device which is located at a distance from the object, and analysis of the data for interpreting the physical attributes of the object, both these aspects being intimately linked with each other." The term remote sensing therefore refers to visual data that has been captured, ranging from early photography to more sophisticated space-based technology that is presently computerized. Applications of remote sensing that are pertinent to mammal ecology and conservation include detailed identification of living forms (variably sized species, over variable distances and scales), behavior recording in animals (e.g., migrations, mating, breeding, and feeding habits), changes in the land cover (including vegetation, temperature, moisture, and species shifts), and the examination of these factors for the evaluation of ecosystems (Campbell, 2018).

McDermid et al. (2010: 193) contend that remote sensing methods enable modeling of several environmental attributes including physiognomy, structure, nutrient concentrations and biomass, and associations with wildlife ecology and habitat. For wildlife studies, Neuman, et al. (2015: 8) point out that "animal movement patterns in space and time are a central aspect of animal ecology... Remotely-sensed environmental indices can play a key role in understanding movement patterns by providing contiguous, relatively fine-scale data that link animal movements to their environment," and "animal movement studies incorporate remotely-sensed products that provide time series of environmental response variables. This would facilitate wildlife management and conservation efforts, as well as the predictive ability of movement analyses." Such methods would be particularly useful for the study of large mobile mammals (Campbell, 2017, 2018).

Two particularly important advancements in remote sensing technology are LiDAR and drones, the former mostly for vegetation and habitat assessments, the latter for tracking animals (Campbell, 2022). LiDAR, or light detection and ranging, is a new, highly effective technology. This method makes use of light pulses to provide precise, three-dimensional data regarding surface features. Global positioning system (GPS) receivers, lasers, and scanners are all common components of LiDAR devices, which are typically built on aircraft platforms. They employ visible, ultraviolet, or near-infrared light to target objects like chemical compounds, vegetation, precipitation, clouds, and aerosols. Comparing LIDAR to older sensing techniques, higher resolution models and maps can be produced with greater accuracy (Campbell, 2018). Ecological assessments are among a few of the topics for which LIDAR products have been developed and applied (Campbell, 2018).

Another relatively recent development is drone-based research, which is based on drones (also called unmanned aerial vehicles, or UAVs, or remotely piloted aircraft, or RPA), especially the more sophisticated models that are equipped with global positioning systems (GPS) (Campbell, 2022). These drones are operated by either onboard computers or a human operator stationed on the ground. Although drones were first developed for military use, they are now being used for commercial, agricultural, and environmental purposes in the civilian world (Campbell, 2018, 2022). Drones are variably sized, from insect-sized to airplane-sized devices. Most of sophisticated drones are equipped to have an autopilot system including a computer, a GPS, a compass, a barometric altimeter, and in some cases a video camera or other photography technology, and physical sensors (for monitoring temperature, humidity, etc.), optical sensors (capable of infrared and/or ultraviolet light detection), or auditory sensors (for sound detection) (Ditmer et al., 2015; Ivošević et al., 2015). "Ecodrones" or "conservation drones" are drones used for ecological and conservation studies (Ivošević et al., 2015: 113). Hodgson et al. (2016) observe that drones are being used by ecologists more frequently to monitor wildlife, and

Manfreda et al. (2018) discuss how drones are used in environmental research. Ivošević et al. (2015) assert that environmental managers operating in diverse ecosystems can benefit from using drones as a low-impact and cost-effective solution.

Another important aspect of geomatics is geographical information systems/science or GIS. Geographical information science is the discipline that uses geographical information systems (GIS) to create computer-generated maps, images, analyses, and scientific supports (Campbell, 2018). Bartelme (2012, 145) defines GIS is a "collection of spatial data plus the corresponding functions for storage and retrieval, a collection of algorithmic and functional tools, a set of hardware and software components needed for the handling of geospatial data, a special type of information technology, a gold mine for answers to geospatial questions, and a model of spatial relationships and spatial reconnaissance." GIS is used for "capturing, storing, analysis, and visualization of data that describe a part of the Earth's surface, the technical and administrative entities, as well as findings of geoscience, economics, and ecological applications" (Bartelme, 2012: 148).

GIS-based analyses enable integrative studies that combine social and environmental research, which can include urban and transport planning, forestry development, agricultural land use, and ecological impacts such as habitat change (Jha & Campbell, 2021, 2022). GIS may be used to assess the integration of the inputs and spatial imprints scaled actors, including individual people, communities, NGOs, business groups, and governments, and their relevance to ecological issues (Akcakaya, 2000; Subak, 2000; Trakolis, 2000; Zumeta, 2000; Campbell, 2022). Important GIS applications to conservation include the compilation and analysis of areas with and near protected areas (Burroughs & Clark, 1995; Salwasser et al., 1995; Martin et al., 2001; Jha & Campbell, 2021, 2022, 2023). GIS-related conservation parameters that are pertinent to the conservation of large mammals include the spatial distribution and patterns of land use conflicts, the physical backdrop of competition for preferred spaces, the social and psychological effects on humans (such as fear, interests, and conservation preferences), the political and economic conflicts (such as alternative land uses, restrictions on industrial and urban land uses to preserve habitats), and the historical relationships between humans and animals within anthropological systems (Campbell, 2022). These varied and intricate relationships require a variety of geomatic techniques for analytic studies, depending on the needs of time, space, interrelationship complexity, and human variables such as economic and socio-cultural structural frameworks and actor-oriented microstudies (Campbell & Jha, 2021; Jha & Campbell, 2021, 2022, 2023).

GIS methods are also used for predictive modeling in conservation planning, habitat analysis, and environmental threats (Altaweel, 2017; Campbell, 2018). The application of GIS predictive modeling to environmental threats, habitat analysis, and wildlife conservation is discussed by Altaweel (2017). Predictive modeling with GIS is one development that evaluates land cover regions appropriate for long-term wildlife communities and potential migration paths (Perkl, 2016). GIS techniques can be used to investigate issues such as the likelihood of wildlife-vehicle collisions caused by transportation networks in natural habitats, and the extent of habitat areas designated for wildlife conservation (Shanee et al., 2013), as well as natural water sources for habitat preservation (Yang et al., 2014; Altaweel, 2017). Social issues are also included; this may be termed public participation in conservation planning with GIS (Altaweel, 2017). According to Altaweel (2017), a novel approach has been to increase public involvement given that protection and mitigation are needed to combat threats to wildlife. Public involvement is needed to determine public support for the protection of particular animal and plant species, and required ecosystems, with expert supports (see also Cox et al., 2014).

Conclusions

Wild cattle are dominant land mammals, second only to large pachyderms in size and similar to the largest land carnivores as a charismatic species relative to humans. Similar to these other large mammals, wild cattle require extensive habitats that may conflict with prime human lands used for crop farming, livestock raising and grazing, hunting, and rural and suburban human habitation. The populations of most wild cattle species are declining, and they require legal and effective protection, which is limited in many parts of the Global South. Large carnivores exist within similar limits, and their populations are generally declining due to habitat destruction, human encroachment into habitats, and limited conservation efforts. Investigating the status of wild cattle and large carnivores requires comprehensive social and ecological integrated analyses, which may require the use of advanced geomatic technologies. This chapter has presented the basic information for an in-depth study of wild cattle and associated mammal-carnivore ecology as a basis for the larger study of the relationship between domestic cattle and larger carnivores, which is critically important due to the importance of domestic cattle as global food sources and large carnivores as charismatic animals in the focus of global nature conservation. The next chapter examines the relationship between species of wild cattle and large carnivores in Asia, Africa, North America, and Europe.

References

Ahrestani, F. S. (2018). Bos frontalis and Bos gaurus (Artiodactyla: Bovidae). *Mammalian Species, 50*(959), 34–50. https://doi.org/10.1093/mspecies/sey004

Ahrestani, F., & Karanth, K. U. (2014). Gaur Bos gaurus. In M. Melletti & J. Burton (Eds.), *Ecology, evolution and behavior of wild cattle: Implications for conservation*. Cambridge University Press.

Akcakaya, A. R. (2000). Conservation and Management for Multiple Species: Integrating Field Research and Modelling into Management Decisions. *Environmental Management, 26*, S75–S83.

Alcantara, C., Kuemmerle, T., Baumann, M., Bragina, E. V., Griffiths, P., Hostert, P., Knorn, J., Müller, D., Prishchepov, A. V., Schierhorn, F., Sieber, A., & Radeloff, V. C. (2013). Mapping the extent of abandoned farmland in Central and Eastern Europe using MODIS time series satellite data. *Environmental Research Letters, 8*, 035035.

Altaweel, M. (2017). The role of GIS in wildlife research and conservation. *Goegraphy Realm*. Retreived from https://www.geographyrealm.com/role-gis-wildlife-research-conservation/#_ftn7.

Andheria, A., Karanth, K. U., & Kumar, N. S. (2007). Diet and prey profiles of three sympatric large carnivores in Bandipur Tiger Reserve in India. *Journal of Zoology (London), 273*, 169–175.

Arias, D. (2023). Rhinoceros unicornis. Indian Rhinoceros (On-line), Animal Diversity Web. https://animaldiversity.org/accounts/Rhinoceros_unicornis/

Aune, K., Jørgensen, D., & Gates, C. (2017). Bison bison (errata version published in 2018). The IUCN Red List of Threatened Species 2017: e.T2815A123789863. https://doi.org/10.2305/IUCN.UK.2017-3.RLTS.T2815A45156541.en.

AWCSG. (2011). Searching for the last kouprey: Final report to Critical Ecosystem Partnership Fund. IUCN SSC AWCSG.

Babar, M. E., Hussain, T., Imran, M., Nagarajan, M., & Kumar, S. (2012). Mitochondrial DNA diversity patterns in Pakistani buffalo. *Animal Genetics, 43*(3), 315–317.

Baird, I. (1993). Wildlife trade between the southern Lao PDR provinces of Champassak, Sekong, and Attapeu and Thailand, Cambodia, and Vietnam. TRAFFIC Southeast Asia Field Report No. 3. Kuala Lumpur, Malaysia.

Ballenger, L. (2002). "Ursus arctos". Brown Bear (On-line), Animal Diversity Web. https://animaldiversity.org/accounts/Ursus_arctos/

Bartelme, N. (2012). Geographic Information Systems. In Kresse, W. & Danko, D. M. (Eds.) *Springer Handbook of Geographic Information*, 145–174. New York: Springer.

Bennitt, E., Bonyongo, M. C., Harris, S., & Barrett, L. (2018). Cape buffalo (Syncerus caffer caffer) social dynamics in a flood-pulsed environment. *Behavioral Ecology, 29*(1), 93–105. https://doi.org/10.1093/beheco/arx138

Berger, J., Buuveibaatar, B., & Mishra, C. (2013). Globalization of the cashmere market and the decline of large mammals in Central Asia. *Conservation Biology, 27*, 679–689. https://doi.org/10.1111/cobi.12100

Berger, J., Cheng, E., Kang, A., Krebs, M., Li, L., Lu, Z. X., Buqiong, B., & Shaller, G. B. (2014). Sex differences in ecology of wild yaks at high elevation in the Kekexili Reserve, Tibetan Qinghai Plateau, China. *Journal of Mammalogy, 95*(3), 638–645.

Beyer, H. O. (1957). *New finds of fossil mammals from the Pleistocene strata of the Philippines*. National Research Council of the Philippines. University of the Philippines, Bulletin 41:1–23.

Bezuijen, M. R., Timmins, R., & Teak, S. (Eds.). (2008). *Biological surveys of the Mekong River between Kratie and Stung Treng Towns, northeast Cambodia, 2006–2007*. WWF Greater Mekong – Cambodia Country Programme, Cambodia Fisheries Administration and Cambodia Forestry Administration, Phnom Penh.

Bies, L. (2002). "Melursus ursinus" Sloth bear (On-line), Animal Diversity Web. https://animaldiversity.org/accounts/Melursus_ursinus/

Bocherens, H., Hofman-Kamińska, E., Drucker, D. G., Schmölcke, U., & Kowalczyk, R. (2015). European bison as a refugee species? Evidence from isotopic data on early holocene bison and other large herbivores in Northern Europe. *PLoS One, 10*(2), e0115090.

Boitani, L. (1984). *Simon & Schuster's guide to mammals*. Simon & Schuster/Touchstone Books.

Boitani, L., Phillips, M., & Jhala, Y. (2023). Canis lupus (amended version of 2018 assessment). The IUCN Red List of Threatened Species 2023: https://doi.org/10.2305/IUCN.UK.2023-1.RLTS.T3746A247624660.en.

Bond, J., & Mkutu, K. (2018). Exploring the hidden costs of human–wildlife conflict in northern Kenya. *African Studies Review, 61*(1), 33–54. https://doi.org/10.1017/asr.2017.134

Boyd, D. P. (2003). *Conservation of North American bison: Status and recommendations*. Faculty of Environmental Design, University of Calgary.

Boyd, D., & Gates, C. C. (2006). A brief review of the status of Plains Bison in North America. *Journal of the West, 45*(2), 15–21.

Boyles, R., Schutz, E., & de Leon, J. (2016). Bubalus mindorensis. The IUCN Red List of Threatened Species 2016: https://doi.org/10.2305/IUCN.UK.2016-2.RLTS.T3127A50737640.en.

Braczkowski, A. R., O'Bryan, C. J., Lessmann, C., Rondinini, C., Crysell, A. P., Gilbert, S., Stringer, M., Gibson, L., & Biggs, D. (2023). The unequal burden of human-wildlife conflict. *Communications Biology, 6*(1), 182. https://doi.org/10.1038/s42003-023-04493-y

Britannica. (2023). Spectacled bear. Encyclopedia Britannica. https://www.britannica.com/animal/spectacled-bear

Britannica. (2024a). Asiatic black bear. Encyclopedia Britannica. https://www.britannica.com/animal/Asiatic-black-bear

Britannica (2024b). Sloth bear. Encyclopedia Britannica. https://www.britannica.com/animal/sloth-bear

Britannica (2024c). Leopard. Encyclopedia Britannica. https://www.britannica.com/animal/leopard

Brodie, J. F. (2008). *A review of American bison (Bos bison) demography and population dynamics*. American Bison Society Working Paper 2.

Brown, G. (1993). *The Great Bear Almanac*. Lyons & Burford.

Bruskotter, J. T., Vucetich, J. A., Gilbert, S. L., Carter, N. H., & George, K. A. (2022). Tragic trade-offs accompany carnivore coexistence in the modern world. *Conservation Letters, 15*, e12839. https://doi.org/10.1111/conl.1284

Bunker, A. (2007). "Canis simensis". Ethiopian wolf (On-line), Animal Diversity Web. Accessed March 18, 2024 at https://animaldiversity.org/accounts/Canis_simensis/

Burton, J. A. (2001). *The Anoa 2000 Conservation Project. Field Report: Systematics and Conservation of Large Mammals in South East Sulawesi (Sulawesi Tenggara), with Notes on North and Central Sulawesi*. Unpublished Report.

Burton, J. A., & Macdonald, A. A. (2009). Variation in distribution and abundance of large mammals in Lore Lindu National Park, Central Sulawesi, Indonesia. *Malayan Nature Journal, 61*(4), 273–283.

Burton, J. A., Hedges, S., & Mustari, A. H. (2005). The taxonomic status, distribution and conservation needs of the Lowland anoa Bubalus depressicornis and mountain anoa B. quarlesi. *Mammal Review, 35*(1), 25–50.

Burton, J., Wheeler, P., & Mustari, A. (2016a). Bubalus depressicornis. The IUCN Red List of Threatened Species 2016: https://doi.org/10.2305/IUCN.UK.2016-2.RLTS.T3126A46364222.en.

Burton, J., Wheeler, P., & Mustari, A. (2016b). Bubalus quarlesi. The IUCN Red List of Threatened Species 2016: https://doi.org/10.2305/IUCN.UK.2016-2.RLTS.T3128A46364433.en.

Burroughs, R. H., & Clark, T. W. (1995). Ecosystem management: A comparison of greater yellowstone and georges bank. *Environmental Management, 19*(5), 649–663. https://doi.org/10.1007/bf02471947

Buzzard, P. J. (2011). *Conservation of wild yak for the benefits of hybridization with domestic yak on the Tibetan plateau, China*. International Congress for Conservation Biology.

Buzzard, P., & Berger, J. (2016). Bos mutus. The IUCN Red List of Threatened Species 2016: https://doi.org/10.2305/IUCN.UK.2016-2.RLTS.T2892A101293528.en.

Buzzard, P. J., Zhang, H. B., Xü, D. H., & Wong, H. M. (2010). *A globally important wild yak Bos mutus population in the Arjinshan Nature Reserve*, Xinjiang, China. Oryx, 44(4), 577–580. https://doi.org/10.1017/s0030605310000591

Campbell, M. (2017). *Biological conservation in the 21st century: A conservation biology of large wildlife*. Nova Science Publishers.

Campbell, M. (2017). *Conservation Biology in the 21st Century: A Conservation Biology of Large Wildlife*. New York, NY: Nova Science Publishers.

Campbell, M. (2018). *Geomatics and Conservation Biology*. New York, NY: Nova Science Publishers.

Campbell, M. (2022). *The Great Eagles: Their Evolution, Ecology and Conservation*. Taylor and Francis (CRC Press), Boca Raton, Florida.

Campbell, M. & Jha, K. K. (2021). *Critical Research Techniques in Animal and Habitat Ecology: International Examples*. New York, NY: Nova Science Publishers

Campbell, M., & Torres Alvarado, M. E. (2011). Public perceptions of jaguars Panthera onca, pumas Puma concolor and coyotes Canis latrans in El Salvador. *Area, 43*(3), 250–256.

Carbone, C., & Gittleman, J. L. (2002). A common rule for the scaling of carnivore density. *Science, 295*, 2273–2276.

Cardillo, M., Purvis, A., Sechrest, W., Gittleman, J. L., Bielby, J., & Mace, G. M. (2004). Human population density and extinction risk in the world's carnivores. *PLoS Biology, 2*(7), 0909–0914.

Caron, A., Cornélis, D., Foggin, C., et al. (2016). African buffalo movement and zoonotic disease risk across Transfrontier Conservation Areas, southern Africa. *Emerging Infectious Diseases, 22*(2), 277–280. https://doi.org/10.3201/eid2202.140864

Caron, A., Bennitt, E., Wielgus, E., Cornelis, D., Miguel, E., & de Garine-Wichatitsky, M. (2023). African buffalo social dynamics: What Is a buffalo herd? In D. Alexandre Caron, P. C. Cornélis, & H. H. T. Prins (Eds.), *Ecology and management of the African Buffalo* (pp. 153–179). Cambridge University Press.

Castelló, J. R. (2016a). *Bovids of the word*. Princeton University Press.

Castelló, J. R. (2016b). *Genus Bos: Banteng. Bovids of the World: Antelopes, gazelles, cattle, goats, sheep, and relatives*. Princeton University Press.

Castelló, J. R. (2018). *Canids of the world: Wolves, wild dogs, foxes, jackals, coyotes, and their relatives*. Princeton University Press.

Chacon, R. (2000). "Cuon alpinus" Dhole (On-line), Animal Diversity Web. Accessed March 18, 2024 at https://animaldiversity.org/accounts/Cuon_alpinus/

Chigwenhese, L., Murwira, A., Zengeya, F. M., et al. (2016). Monitoring African buffalo (Syncerus caffer) and cattle (Bos taurus) movement across a damaged veterinary control fence at a Southern African wildlife/livestock interface. *African Journal of Ecology, 54*(4), 415–423. https://doi.org/10.1111/aje.12288

Choudhury, A. (2002). Distribution and conservation of the Gaur Bos gaurus in the Indian Subcontinent. *Mammal Review, 32*, 199–226.

Choudhury, A. (2010). *The Vanishing Herds: The wild water buffalo*. Collins Books.

Christie, T., Steininger, D., Juhn, D., & Peal, A. (2007). Fragmentation and clearance of Liberia's forests during 1986-2000. *Oryx, 41*(4), 539–543.

Clouded leopard (Neofelis nebulosa) Husbandry Guidelines. (2000). https://documents.pub/document/husbandry-guidelines-clouded-leopard-clouded-leopard-neofelis-nebulosa-husbandry.html?page=1

Connor, T. (2009). *Loxodonta cyclotis*. African Forest Elephant (On-line), Animal Diversity Web. https://animaldiversity.org/accounts/Loxodonta_cyclotis/

Conybeare, A. B. (1980). Buffalo numbers, home range and daily movement in the Sengwa Wildlife Research Area, Zimbabwe. *South African Journal of Wildlife Research, 10*(3–4), 89–93. https://journals.co.za/doi/abs/10.10520/AJA03794369_2246

Corbet, G. B., & Hill, J. E. (1992). *Mammals of the Indo-Malayan Region: A systematic review*. Oxford University Press.

Cornélis, D., Benhamou, S., Janeau, G., et al. (2011). Spatiotemporal dynamics of forage and water resources shape space use of West African savanna buffalo. *Mammalogy, 92*(6), 1287–1297. https://doi.org/10.1644/10-MAMM-A-397.1

Cornélis, D., Melletti, M., Korte, L., Ryan, S. J., Mirabile, M., Prin, T., & Prins, H. H. T. (2014). African buffalo (Syncerus caffer Sparrman, 1779). In M. Melletti & J. Burton (Eds.), *Ecology, evolution and behaviour of wild cattle: Implications for conservation* (pp. 326–372). Cambridge University Press.

Cornelis, D., Renaud, P. C., Melletti, M., Fonteyn, D., Bonhotal, H., Hauptfleisch, M., et al. (2023). Conservation status of the African buffalo: a continent-wide assessment. In A. Caron, D. Cornélis, P. Chardonnet, & H. H. T. Prins (Eds.), *Ecology and management of the African Buffalo* (pp. 66–128). Cambridge University Press.

Courchamp, F., Jaric, I., Albert, C., Meinard, Y., Ripple, W. J., & Chapron, G. (2018). The paradoxical extinction of the most charismatic animals. *PLoS Biology, 16*(4), e2003997. https://doi.org/10.1371/journal.pbio.2003997

Cox, C., Morse, W., Anderson, C., & Marzen, L. (2014). Applying public participation geographic information systems to wildlife management. *Human Dimensions of Wildlife*. [Online], *19*(2), 200–214.

Crider, M. (2019). Rhinoceros sondaicus. Javan Rhinoceros (On-line), Animal Diversity Web. https://animaldiversity.org/accounts/Rhinoceros_sondaicus/

Custodio, C. C., Lepiten, M. V., & Heaney, L. R. (1996). Bubalus mindorensis. *Mammalian Species, 520*, 1–5.

D'Ammando, G., Franks, D. W., & Bro-Jørgensen, J. (2022). Living in mixed-sex groups limits sexual selection as a driver of pelage dimorphism in bovids. *Behavioral Ecology, 33*(3), 541–548. https://doi.org/10.1093/beheco/arac016

Dacres, K. (2024). Panthera tigris: Tiger. Animal Diversity Web. Retrieved from https://animaldiversity.org/accounts/Panthera_tigris/

Daleszczyk, K., & Bunevich, A. N. (2009). Population viability analysis of European bison populations in Polish and Belarusian parts of Białowieża Forest with and without gene exchange. *Biological Conservation, 142*, 3068–3075.

De la Torre, J. A., Gonzalez-Maya, J. F., Zarza, H., Ceballos, G., & Medellín, R. A. (2017). The Jaguar's spots are darker than they appear: Assessing the global Conservation status of the Jaguar Panthera onca. *Oryx, 51*, 1–16.

de Leon, J., Lawas, N., Escalada, R., Ong, P., Callo, R., Hedges, S., Ballou, J., Armstrong, D., & Seal, U. S. (1996). *Tamaraw (Bubalus mindorensis) Population and Habitat Viability* Assessment Report.

Dharaiya, N., Bargali, H. S., & Sharp, T. (2020). Melursus ursinus (amended version of 2016 assessment). The IUCN Red List of Threatened Species 2020: https://doi.org/10.2305/IUCN.UK.2020-1.RLTS.T13143A166519315.en.

Di Minin, E., Leader-Williams, N., & Bradshaw, C. J. A. (2016). Banning Trophy hunting will exacerbate biodiversity loss. *Trends in Ecology & Evolution, 31*(2), 99–102. https://doi.org/10.1016/j.tree.2015.12.006

Dickman, A. J., Macdonald, E. A., & Macdonald, D. W. (2011). A review of financial instruments to pay for predator conservation and encourage human-carnivore coexistence. *Proceedings of the National Academy of Sciences of the United States of America, 108*(34), 13937–13944. https://doi.org/10.1073/pnas.1012972108

Dickman, A., Cooney, R., Johnson, P. J., Louis, M. P., Roe, D., & 128 signatories. (2019). Trophy hunting bans imperil biodiversity. *Science (New York, N.Y.), 365*(6456), 874. https://doi.org/10.1126/science.aaz0735

Diep, T. N., Long, D. T., & Tuoc, D. (2004). Annex 3 - report of survey on Saola. In J. Hardcastle, S. Cox, N. T. Dao, & A. G. Johns (Eds.), *Rediscovering the Saola - Proceedings of "Rediscovering the Saola - a status review and planning workshop"*. WWF Indochina Programme, Hanoi, Vietnam.

Dinerstein, E. (2024). *Black rhinoceros*. Encyclopedia Britannica. https://www.britannica.com/animal/black-rhinoceros

Ditmer, M. A., Vincent, J. B., Werden, L. K., Tanner, J. C., Laske, T. G., Iaizzo, P. A., Garshelis, D. L., & Fieberg, J. R. (2015). Bears Show a Physiological but Limited Behavioral Response to Unmanned Aerial Vehicles. *Current Biology 25*(17), 2278–2283. https://doi.org/10.1016/j.cub.2015.07.024

Dora, C. A. (2004). *The Influences of Habitat Structure and Landscape Heterogeneity on African Buffalo (Syncerus cgffer) Group Size in Hiuhiuwe-iMfolozi Game Reserve, South Africa*. MSc, Environmental Sciences, Oregon State University.

Dróżdż, J., Demiaszkiewicz, A. W., & Lachowicz, J. (1998). Ashworthius sidemi (Nematoda, Trichostrongylidae) a new parasite of the European bison Bison bonasus (L.) and the question of independence of A. gagarini. *Acta Parasitologica, 43*, 75–80.

Duckworth, J. W., & Hedges, S. (1998). *A review of the status of tiger, Asian elephant, gaur, and Banteng in Viet Nam, Lao, Cambodia, and Yunnan (China), with recommendations for future conservation action*. WWF Indochina Programme, Hanoi, Viet Nam.

Duckworth, J. W., Salter, R. E., & Khounboline, K. (1999). *Wildlife in Lao PDR: 1999 Status Report*. IUCN, Vientiane, Laos.

Duckworth, J. W., Sankar, K., Williams, A. C., Samba Kumar, N., & Timmins, R. J. (2016). Bos gaurus. The IUCN Red List of Threatened Species 2016. e.T2891A46363646. https://doi.org/10.2305/IUCN.UK.2016-2.RLTS.T2891A46363646.en.

Dulal, D. (2017). Ceratotherium simum. White Rhinoceros (On-line), Animal Diversity Web. https://animaldiversity.org/accounts/Ceratotherium_simum/

Dunham, K. M., Ghiurghi, A., Cumbi, R., & Urbano, F. (2010). Human–wildlife conflict in Mozambique: A national perspective, with emphasis on wildlife attacks on humans. *Oryx, 44*(2), 185–193.

Durant, S. M., Marino, A., Linnell, J. D. C., Oriol-Cotterill, A., Dloniak, S., Dolrenry, S., Funston, P., Groom, R. J., Hanssen, L., Horgan, J., Ikanda, D., Ipavec, A., Kissui, B., Lichtenfeld, L., McNutt, J. W., Mitchell, N., Naro, E., Samna, A., & Yirga, G. (2022). Fostering coexistence between people and large carnivores in Africa: Using a theory of change to identify pathways to impact and their underlying assumptions. *Frontiers of Conservation Science, 2*, 698631. https://doi.org/10.3389/fcosc.2021.698631

Durant, S. M., Groom, R., Ipavec, A., Mitchell, N. & Khalatbari, L. (2023). Acinonyx jubatus (amended version of 2022 assessment). The IUCN Red List of Threatened Species 2023: https://doi.org/10.2305/IUCN.UK.2023-1.RLTS.T219A247393967.en.

Durbin, L. S., Venkataraman, A., Hedges, S., & Duckworth, W. (2004). Dhole (Cuon alpinus). In C. Sillero-Zubiri, M. Hoffmann, & D. W. Macdonald (Eds.), *Canids: Foxes, wolves, jackals, and dogs. Status survey and conservation action plan* (pp. 210–219). IUCN/ SSC Canid Specialist Group.

Dwiyahreni, A. A. (2006). *Anoa abundance in Lambusango forests, Buton and the influencing factors*. Report to BKSDA Southeast Sulawesi, Kendari. Department of Biological Science, Scarborough Centre for Coastal Studies. University of Hull.

East, R. (1999). African Antelope Database 1998. IUCN, Gland, Switzerland and Cambridge, UK.

Estes, R. (1992). *The behavior guide to African mammals: Including hoofed mammals, carnivores, primates* (pp. 410–419). University of California Press.

Farhadinia, M., Rostro-García, S., Feng, L., Kamler, J. F., Andrew, S., Elena, S., Khorozyan, I., Al-Duais, M., Ge, J., & Macdonald, D. (2019). Big cats in borderlands: Challenges and implications for transboundary conservation of Asian leopards. *Oryx, 55*, 1–10.

Fenner, K. (2012). "Tremarctos ornatus" Spectacled bear (On-line), Animal Diversity Web. Accessed March 18, 2024 at https://animaldiversity.org/accounts/Tremarctos_ornatus/

Flamand, J. R. B., Vankan, D., Gairhe, K. P., Duong, G., & Barker, J. S. F. (2003). Genetic identification of wild Asian water buffalo in Nepal. *Animal Conservation, 6*, 265–270.

Foead, N. (1992). Studi Habitat dan Pakan Anoa Gunung [Bubalus (Anoa) quarlesi, Ouwen] Di Taman Nasional Lore Lindu, Sulawesi Tengah. Fakultas Kehutanan, Universitas Gadjah Mada.

Fox, J. L., Mathiesen, P., Yangzom, D., Næss, M. W., & Xu, B. R. (2004). Modern wildlife conservation initiatives and the pastoralist/hunter nomads of northwestern Tibet. *Rangifer, 15*, 17–27.

Freese, C. H., Aune, K. E., Boyd, D. P., Derr, J. N., Forrest, J. C., Cormack Gates, C., Gogan, P. J., Grassel, S. M., Halbert, N. D., Kunkel, K., & Redford, K. H. (2007). Second chance for the plains bison. *Biological Conservation, 136*, 175–184.

Fuhlendorf, S. D., Allred, B. W., & Hamilton, R. G. (2010). *Bison as keystone herbivores on the Great Plains: Can cattle serve as proxy for evolutionary grazing patterns?* ABS Working Paper No 4. Wildlife Conservation Society, NY.

Fuller, J. A., Garrott, R. A., White, P. J., Aune, K. E., Roffe, T. J., & ad Rhyan, J. C. (2007). Reproduction and survival of Yellowstone bison. *Journal of Wildlife Management, 71*(7), 2365–2372.

Galbreath, G. J., Mordacq, J. C., & Weiler, F. H. (2006). Genetically solving a zoological mystery: was the kouprey (Bos sauveli) a feral hybrid?. *Journal of Zoology, 270*(4), 561–564. https://doi.org/10.1111/j.1469-7998.2006.00188.x

Gardner, P., Hedges, S., Pudyatmoko, S., Gray, T. N. E., & Timmins, R. J. (2016). Bos javanicus. The IUCN Red List of Threatened Species 2016. https://doi.org/10.2305/IUCN.UK.2016-2.RLTS.T2888A46362970.en.

Garshelis, D., & Steinmetz, R. (2020). Ursus thibetanus (amended version of 2016 assessment). The IUCN Red List of Threatened Species 2020: https://doi.org/10.2305/IUCN.UK.2020-3.RLTS.T22824A166528664.en.

Garshelis, D. L., Scheick, B. K., Doan-Crider, D. L., Beecham, J. J., & Obbard, M. E. (2016). Ursus americanus (errata version published in 2017). The IUCN Red List of Threatened Species 2016: https://doi.org/10.2305/IUCN.UK.2016-3.RLTS.T41687A45034604.en.

Gates, C. C., Elkin, B., & Dragon, D. (1995). Investigation, control and epizootiology of anthrax in an isolated, free-roaming bison population in northern Canada. *Canadian Journal of Veterinary Research, 59*, 256–264.

Gates, C. C., Stephenson, R. O., Reynolds, H. W., Van Zyll De Jong, C. G., Schwantje, H., Hoefs, M., Nishi, J., Cool, N., Chisholm, J., James, A., & Koonz, B. (2001). *National recovery plan for the wood bison (Bison bison athabascae)*. Recovery of Nationally Endangered Wildlife (RENEW), Ottawa, Ontario, Canada.

Gates, C. C., Stelfox, B., Muhly, T., Chowns, T., & Hudson, R. J. (2005). *The ecology of bison movements and distribution in and beyond Yellowstone National Park: A critical review with implications for winter use and transboundary population management*. University of Calgary, Faculty of Environmental Design, Calgary, Alberta, Canada.

Gates, C. C., Freese, C. H., Gogan, P. J. P., & Kotzman, M. (2010). America Bison: Status Survey and Conservation Guidelines 2010. IUCN, Gland, Switzerland.

Gębczyńska, Z., Gębczyński, M., & Martynowicz, E. (1991). Food eaten by free-living European bison. *Acta Theriologica, 36*, 307–313.

Geremia, C., Wallen, R., & White, P. J. (2014). *Population dynamics and adaptive management of Yellowstone bison*. National Park Service, Yellowstone National Park, Mammoth, Wyoming.

Gesch, P. (2004). "Bubalus mindorensis". Tamaraw (On-line), Animal Diversity Web. Accessed March 29, 2024 at https://animaldiversity.org/accounts/Bubalus_mindorensis/

Ghosal, S., Skogen, K., & Krishnan, S. (2015). Locating human-wildlife interactions: Landscape constructions and responses to large carnivore conservation in India and Norway. *Conservation and Society, 13*(3), 265–274. https://doi.org/10.4103/0972-4923.170403

Good, C., Burnham, D., & Macdonald, D. W. (2017). A cultural conscience for conservation. *Animals: An Open Access Journal from MDPI, 7*(7), 52. https://doi.org/10.3390/ani7070052

Goodman, P. S., Conway, A. J., & Timmins, R. J. (2003). *Selection and planning of an intensively managed wildlife protection zone in the dry forest ecoregion of north-eastern Cambodia*. WWF Cambodia, Phnom Penh, Cambodia.

Goodness, T. (2004). "Ursus thibetanus" Asiatic black bear (On-line), Animal Diversity Web. Accessed March 18, 2024 at https://animaldiversity.org/accounts/Ursus_thibetanus/

Goodrich, J., Wibisono, H., Miquelle, D., Lynam, A. J., Sanderson, E., Chapman, S., Gray, T. N. E., Chanchani, P., & Harihar, A. (2022). Panthera tigris. The IUCN Red List of Threatened Species 2022: https://doi.org/10.2305/IUCN.UK.2022-1.RLTS.T15955A214862019.en.

Gray, T., Borah, J., Coudrat, C. N. Z., Ghimirey, Y., Giordano, A., Greenspan, E., Petersen, W., Rostro-García, S., Shariff, M., & Wai-Ming, W. (2021). Neofelis nebulosa. The IUCN Red List of Threatened Species 2021: https://doi.org/10.2305/IUCN.UK.2021-2.RLTS.T14519A198843258.en.

Gross, J. E., & Wang, G. (2005). *Effects of population control strategies on retention of genetic diversity in National Park Service bison (Bison bison) herds*. Final Report, Yellowstone Research Group, USGS-BRD. United State Geological Survey, Bozeman, Montana, USA.

Groves, C. P. (1969). Systematics of the anoa (Mammalia, Bovidae). *Beaufortia, 17*, 1–12.

Grubb, P. (2005a). Family Bovidae. In D. E. Wilson & D. M. Reeder (Eds.), *Mammal species of the world: A taxonomic and geographic reference* (3rd ed., pp. 637–722). Johns Hopkins University Press.

Grubb, P. (2005b). Species Dicerorhinus sumatrensis. In D. E. Wilson & D. M. Reeder (Eds.), *Mammal species of the world: A taxonomic and geographic reference* (3rd ed., p. 635). Johns Hopkins University Press.

Grubb, P. (2005c). Artiodactyla. In D. E. Wilson & D. M. Reeder (Eds.), *Mammal species of the world. A taxonomic and geographic reference* (3rd ed., pp. 637–722). Johns Hopkins University Press.

Grubb, P. (2005d). Order Artiodactyla. In D. E. Wilson & D. M. Reeder (Eds.), *Mammal species of the world: A taxonomic and geographic reference*. Johns Hopkins University Press.

Grumbine, R. E. (1990). Viable populations, reserve size, and federal lands management: A critique. *Conservation Biology, 4*, 127–134.
Grzimek, B. (1990). *Grzimek's Encyclopedia of mammals* (pp. 52–53). McGraw-Hill.
Gulati, S., Karanth, K. K., Le, N. A., & Noack, F. (2021). Human casualties are the dominant cost of human-wildlife conflict in India. *Proceedings of the National Academy of Sciences of the United States of America, 118*(8), e1921338118. https://doi.org/10.1073/pnas.1921338118
Gupta, R. P. (2003). Introduction. In *Remote sensing geology*. Springer. https://doi.org/10.1007/978-3-662-05283-9_1
Gurunathan, K., Biswas, S., & Rajkumar, S. (2009). Buffalo as a potential food animal. *International Journal of Livestock Production, 1*(1), 001–005. https://academicjournals.org/journal/IJLP/article-full-text-pdf/8E6DC251420
Halbert, N. D., & Derr, J. N. (2007). A comprehensive evaluation of cattle introgression into US Federal bison herds. *Journal of Heredity, 98*(1), 1–12.
Halbert, N. D., & Derr, J. N. (2008). Patterns of genetic variation in US federal bison herds. *Molecular Ecology, 17*, 4963–4977.
Hance, J. (2013). Asia's 'unicorn' photographed in Vietnam. Available at: https://news.mongabay.com/2013/11/asias-unicorn-photographed-in-vietnam/
Hancock, C. (2012). "Neofelis diardi". Sunda Clouded Leopard (On-line), Animal Diversity Web. Accessed March 17, 2024 at https://animaldiversity.org/accounts/Neofelis_diardi/
Hansen, M. C., Potapov, P. V., Moore, R., Hancher, M., Turubanova, S. A., Tyukavina, A., Thau, D., Stehman, S. V., Goetz, S. J., Loveland, T. R., Kommareddy, A., Egorov, A., Chini, L., Justice, C. O., & Townshend, J. R. G. (2013). High-resolution global maps of 21st-century forest cover change. *Science, 342*, 850–853.
Harper, F. (1945). *Extinct and vanishing mammals of the old world*. American Committee for International Wildlife Preservation.
Harrington, E. (2024). Panthera leo: Lion. Animal Diversity Web. Retrieved from https://animaldiversity.org/accounts/Panthera_leo/.
Harris, R. B. (2007). *Wildlife conservation in China: Preserving the habitat of China's wild west*. M. E. Sharpe.
Harris, R. B., & Leslie, D. (2008). Bos mutus 2008 IUCN Red List of Threatened Species. http://www.iucnredlist.org.
Harris, R. B., & Loggers, C. O. (2004). Status of Tibetan plateau mammals in Yeniugou, China. *Wildlife Biology, 10*(2), 91–99.
Harris, R. B., Pletscher, D. H., Loggers, C. O., & Miller, D. J. (1999). Status and trends of Tibetan plateau mammalian fauna, Yeniugou, China. *Biological Conservation, 87*, 13–19.
Hartway, C., Hardy, A., Jones, L., Moynahan, B., Traylor-Holzer, K., McCann, B., Aune, K., & Plumb, G. (2020). *Long-term viability of Department or Interior bison under current management and potential metapopulation strategies*. U.S. National Park Service - Natural Resource Report NPS/NRSS/BRD 2020/2097.
Hassanin, A., & Ropiquet, A. (2004). Molecular phylogeny of the tribe Bovini (Bovidae, Bovinae) and the taxonomic status of the Kouprey, Bos sauveli Urbain 1937. *Molecular Phylogenetics and Evolution, 33*(3), 896–907. https://doi.org/10.1016/j.ympev.2004.08.009
Hassanin, A., & Ropiquet, A. (2007). Resolving a zoological mystery: the kouprey is a real species. Proceedings. *Biological sciences, 274*(1627), 2849–2855. https://doi.org/10.1098/rspb.2007.0830
Hassanin, A., Ropiquet, A., Cornette, R., Tranier, M., Pfeffer, P., Candegabe, P., & Lemaire, M. (2006). Has the kouprey (Bos sauveli Urbain, 1937) been domesticated in Cambodia?. *Comptes rendus biologies, 329*(2), 124–135. https://doi.org/10.1016/j.crvi.2005.11.003
Hatten, J. R., Averill-Murray, A., & van Pelt, W. E. (2005). A spatial model of potential jaguar habitat in Arizona. *Journal of Wildlife Management, 69*(3), 1024–1033.
Hayward, M., Hofmeyr, M., O'Brien, J., & Kerley, G. (2006b). Prey preferences of the cheetah (Acinonyx jubatus) (Felidae: Carnivora): Morphological limitations or the need to capture rapidly consumable prey before kleptoparasites arrive? *Journal of Zoology, 270*, 615–627.

Heaney, L., Walker, E.K., Tabaranza, B.R. & Ingle, N.R. (2002). Mammalian diversity in the Philippines: An assessment of the adequacy of current data. *Sylvatrop 10*, 6–27.

Hearn, A., Ross, J., Brodie, J., Cheyne, S., Haidir, I. A., Loken, B., Mathai, J., Wilting, A., & McCarthy, J. (2015). Neofelis diardi (errata version published in 2016). The IUCN Red List of Threatened Species 2015: https://doi.org/10.2305/IUCN.UK.2015-4.RLTS.T136603A50664601.en.

Hedges, S., Baral, H., Timmins, R., & Duckworth, J. (2008). Bubalus arnee. The IUCN Red List of Threatened Species 2008. https://doi.org/10.2305/IUCN.UK.2008.RLTS.T3129A9615891.en.

Hedges, S., Groves, C. P., Duckworth, J. W., Meijaard, E., Timmins, R. J., & Burton, J. A. (2007). Was the kouprey a feral hybrid? A response to Galbreath et al. (2006). *Journal of Zoology, 271*(3), 242–245. https://doi.org/10.1111/j.1469-7998.2006.00293.x

Hedrick, P. W. (2009). Conservation genetics and North American bison (Bison bison). *Journal of Heredity, 100*(4), 411–420.

Hemmer, H. (1972). Uncia uncia. *Mammalian Species, 20*, 1–5. https://doi.org/10.2307/3503882

Hendricks, K. (2013). Bison bonasus. European Buffalo (On-line), Animal Diversity Web. https://animaldiversity.org/accounts/Bison_bonasus/

Heydinger, J. M., Packer, C., & Tsaneb, J. (2019). Desert-adapted lions on communal land: Surveying the costs incurred by, and perspectives of, communal-area livestock owners in northwest Namibia. *Biological Conservation, 236*, 496–504. https://doi.org/10.1016/j.biocon.2019.06.003

Hodgson, J. C., Baylis, S. M., Mott, R., Herrod, A., & Clarke, R. H. (2016). Precision wildlife monitoring using unmanned aerial vehicles. *Scientific reports, 6*, 22574. https://doi.org/10.1038/srep22574

Hofman-Kamińska, E., & Kowalczyk, R. (2012). Farm crops depredation by European bison (Bison bonasus) in the vicinity of forest habitats in northeastern Poland. *Environmental Management, 50*, 530–541.

Hofman-Kamińska, E., Bocherens, H., Borowik, T., Drucker, D. G., & Kowalczyk, R. (2018). Stable isotope signatures of large herbivore foraging habitats across Europe. *PLoS One, 13*, e0190723.

Hofman-Kamińska, E., Bocherens, H., Drucker, D. G., Fyfe, R. M., Gumiński, W., Makowiecki, D., Pacher, M., Piličiauskienė, G., Samojlik, T., Woodbridge, J., & Kowalczyk, R. (2019). Adapt or die - Response of large herbivores to environmental changes in Europe during the Holocene. *Global Change Biology, 29*(9), 2915–2930.

Holcomb, D. (2005). Pseudoryx nghetinhensis (On-line), Animal Diversity Web. https://animaldiversity.org/accounts/Pseudoryx_nghetinhensis/

Holmern, T., Nyahongo, J., & Røskaft, E. (2007). Livestock loss caused by predators outside the Serengeti National Park, Tanzania. *Biological Conservation, 135*(4), 518–526. https://doi.org/10.1016/j.biocon.2006.10.049

Holmes, K. (2009). "Neofelis nebulosa" (On-line), Clouded Leopard. Animal Diversity Web. Accessed March 17, 2024 at https://animaldiversity.org/accounts/Neofelis_nebulosa/

Hoogerwerf, A. (1970). Udjung Kulon. The Land of the Last Javan Rhinoceros. E.J. Brill, Leiden, The Netherlands.

Howard, M. (2017). Loxodonta africana African Bush Elephant (On-line), Animal Diversity Web. https://animaldiversity.org/accounts/Loxodonta_africana

Huffman, B. (2023a). *Syncerus caffer – African buffalo*. Retrieved from https://www.ultimateungulate.com/Artiodactyla/Syncerus_cafferFull.html

Huffman, B. (2023b). Bubalus quarlesi. Mountain anoa. https://www.ultimateungulate.com/Artiodactyla/Bubalus_quarlesi.html.

Huffman, B. (2024). Syncerus caffer African buffalo. http://www.ultimateungulate.com/Artiodactyla/Syncerus_caffer.html.

Hunt, A. (2011). "Panthera pardus". Leopard (On-line), Animal Diversity Web. Accessed March 17, 2024 at https://animaldiversity.org/accounts/Panthera_pardus/

Indonesian Ministry of Forestry. (2013). Strategy and Conservation Action Plan for Anoa (Bubalus depressicornis and Bubalus quarlesi) 2013–2022.

International Society for Endangered Cats (ISEC) Canada. (2021). *Sunda Clouded Leopard*. Retrieved from https://wildcatconservation.org/wild-cats/asia/sunda-clouded-leopard/

IUCN SSC Antelope Specialist Group. (2019). Syncerus caffer. The IUCN Red List of Threatened Species 2019: e.T21251A50195031. https://doi.org/10.2305/IUCN.UK.2019-1.RLTS.T21251A50195031.en.

Ivošević, B., Han, Y. G., Cho, Y., & Kwon, O. (2015). The use of conservation drones in ecology and wildlife research. *Journal of Ecology and Environment, 38*(1), 113–118. https://doi.org/10.5141/ecoenv.2015.012

Jackson, P., & Nowell, K. (2008). *Panthera tigris. IUCN Red list of threatened species. Version 2011.1*. International Union for Conservation of Nature, Gland.

Jacobsen, K. S., Sandorf, E. D., Loveridge, A. J., Dickman, A. J., Johnson, P. J., Mourato, S., Contu, D., & Macdonald, D. W. (2022). What is a lion worth to local people–Quantifying of the costs of living alongside a top predator. *Ecological Economics, 198*, 107431. https://doi.org/10.1016/j.ecolecon.2022.107431

Jahja, M. M. (1987). The possibility of breeding anoa in captivity: An alternative for conservation of the species. *Biotrop Special Publication, 30*, 101–108.

Jathanna, D., Karanth, K. U., & Johnsingh, A. J. T. (2003). Estimation of large herbivore densities in the tropical forests of southern India using distance sampling. *Journal of Zoology (London), 261*, 285.

Jensen, J. R. (2007). *Remote Sensing of the Environment: An Earth Resource Perspective*. 2nd Edition, Pearson Prentice Hall, Upper Saddle River.

Jha, K. K & Campbell, M. (2021). *Critical Research Techniques in Animal and Habitat Ecology: Examples from India*. New York, NY: Nova Science Publisher.

Jha, K. K. & Campbell, M. (2022). *Dynamics and Interrelations between Nature, Science and Society*. New York, NY: Nova Science Publishers.

Jha, K. K. & Campbell, M. (2023). *Vultures of India*. New York, NY: Nova Science Publishers.

Jianlin, H., Melletti, M., & Burton, J. (2014). *Wild yak (Bos mutus Przewalski, 1883), Ecology, Evolution and Behavior of Wild Cattle: Implications for Conservation*. Cambridge University Press.

Johnson, W. (2024, March 17). Cheetah. Encyclopedia Britannica. https://www.britannica.com/animal/cheetah-mammal

Kamler, J. F., Johnson, A., Vongkhamheng, C., & Bousa, A. (2012). The diet, prey selection, and activity of dholes (Cuon alpinus) in northern Laos. *Journal of Mammalogy, 93*(3), 627–633. https://doi.org/10.1644/11-MAMM-A-241.1

Kamler, J. F., Songsasen, N., Jenks, K., Srivathsa, A., Sheng, L., & Kunkel, K. (2015). Cuon alpinus. The IUCN Red List of Threatened Species 2015: https://doi.org/10.2305/IUCN.UK.2015-4.RLTS.T5953A72477893.en.

Kansky, R., Kidd, M., & Knight, A. T. (2014). Meta-analysis of attitudes toward damagecausing mammalian wildlife. *Conservation Biology, 28*, 924–938. https://doi.org/10.1111/cobi.12275. PMID: 24661270.

Karanth, K. U., & Chellam, R. (2009). Carnivore conservation at the crossroads. *Oryx, 43*, 1–2.

Karanth, K. U., & Kumar, N. S. (2005). *Distribution and dynamics of tiger and prey populations in Maharashtra, India*. Final Technical Report. Centre for Wildlife Studies, Bangalore, India.

Karanth, K. U., & Stith, B. M. (1999). Prey depletion as a critical determinant of tiger population viability. In J. Seidensticker, S. Christie, & P. Jackson (Eds.), *Riding the tiger: Tiger conservation in human-dominated landscapes* (pp. 100–113). Cambridge University Press.

Karanth, K. U., & Sunquist, M. E. (1995). Prey selection by tiger, leopard and dhole in tropical forests. *Journal of Animal Ecology, 64*, 439–450.

Karanth, K. U., Chundawat, R. S., Nichols, J. D., & Kumar, N. S. (2004). Estimation of tiger densities in the tropical dry forest of Panna, Central India using photographic capture-recapture sampling. *Animal Conservation, 7*, 285–290.

Karanth, K. K., Nichols, J. D., Karanth, K. U., Hines, J. E., & Christensen, N. L. (2010a). The shrinking ark: Patterns of large mammal extinctions in India. *Proceedings of the Royal Society London-B, 277*, 1971–1979.

Karanth, K. U., Funston, P., & Sanderson, E. (2010b). Many ways of skinning a cat: Tools and techniques for studying wild felids. In D. W. Macdonald & A. J. Loveridge (Eds.), *The biology and conservation of wild felids* (pp. 197–216). Oxford University Press.

Karbowiak, G., Demiaszkiewicz, A. W., Pyziel, A. M., Wita, I., Moskwa, B., Werszko, J., Bień, J., Goździk, K., Lachowicz, J., & Cabaj, W. (2014a). The parasitic fauna of the European bison (Bison bonasus) (Linnaeus, 1758) and their impact on the conservation. Part 1 The summarizing list of parasites noted. *Acta Parasitologica, 59*, 363–371.

Karbowiak, G., Demiaszkiewicz, A. W., Pyziel, A. M., Wita, I., Moskwa, B., Werszko, J., Bień, J., Goździk, K., Lachowicz, J., & Cabaj, W. (2014b). The parasitic fauna of the European bison (Bison bonasus) (Linnaeus, 1758) and their impact on the conservation. Part 2 The structure and changes over time. *Acta Parasitologica, 59*, 372–379.

Karkala, N. (2016). Elephas maximus. Asiatic Elephant (On-line), Animal Diversity Web. https://animaldiversity.org/accounts/Elephas_maximus/

Kaul, R., Williams, A.C., Rithe, K., Steinmetz, R., & Mishra, R. (2019). Bubalus arnee. The IUCN Red List of Threatened Species 2019: https://doi.org/10.2305/IUCN.UK.2019-1.RLTS.T3129A46364616.en.

Kays, R. (2018). Canis latrans (errata version published in 2020). The IUCN Red List of Threatened Species 2018: https://doi.org/10.2305/IUCN.UK.2018-2.RLTS.T3745A163508579.en.

Kays, R. W. (2024). Lion. Encyclopedia Britannica. https://www.britannica.com/animal/lion

Kemp, N., Dilger, M., Burgess, N., & Dung, C. V. (1997a). The Saola Pseudoryx nghetinhensis in Vietnam - new information on distribution and habitat preferences, and conservation needs. *Oryx, 31*, 37–44.

Kemp, N., Dilger, M., Burgess, N., & Van Dung, V. (1997b). Status of the saola Pseudoryx nghetinhensis. *Oryx, 31*, 89–91.

Kerley, G. I. H., Kowalczyk, R., & Cromsigt, J. P. G. M. (2012). Conservation implications of the refugee species concept and the European bison: King of the forest or refugee in a marginal habitat? *Ecography, 35*, 519–529.

Kerley, G. I. H., Cromsigt, J. P. G. M., & Kowalczyk, R. (2020). European bison conservation cannot afford to ignore alternative hypotheses: A commentary on Perzanowski et al., 2019. *Animal Conservation, 23*, 479. https://doi.org/10.1111/acv.12606

Kingdon, J. (2015). Cheetah Acinonyx jubatus. In *The Kingdon Field guide to African mammals* (2nd ed., pp. 403–404). Bloomsbury.

Kołodziej-Sobocińska, M., Demiaszkiewicz, A., Lachowicz, J., Borowik, T., & Kowalczyk, R. (2016a). Influence of management and biological factors on the parasitic invasions in the wild-spread of blood-sucking nematode Ashworthius sidemi in European bison (Bison bonasus). *International Journal for Parasitology: Parasites and Wildlife, 5*, 286–294.

Kołodziej-Sobocińska, M., Demiaszkiewicz, A. W., Pyziel, A. M., Marczuk, B., & Kowalczyk, R. (2016b). Does the blood-sucking nematode Ashworthius sidemi (Trichostrongylidae) cause deterioration of blood parameters in European bison (Bison bonasus)? *European Journal of Wildlife Research, 62*, 781–785.

Kołodziej-Sobocińska, M., Pyziel, A. M., Demiaszkiewicz, A. W., Borowik, T., & Kowalczyk, R. (2016c). Pattern of parasite egg shedding by European bison (Bison bonasus) in the Białowieża Primeval Forest, Poland. *Mammal Research, 61*, 179–186.

Korte, L. (2008a). Variation of group size among African buffalo herds in a forest–savanna mosaic landscape. *Journal of Zoology, 275*(3), 229–236. https://doi.org/10.1111/j.1469-7998.2008.00430.x

Kowalczyk, R., & Plumb, G. (2020). European bison. In K. Hackländer & F. E. Zachos (Eds.), *Handbook of the mammals of Europe*. Springer.

Kowalczyk, R., Taberlet, P., Coissac, E., Valentini, A., Miquel, C., Kamiński, T., & Wójcik, J. M. (2011). Influence of management practices on large herbivore diet-case of European bison in Białowieża Primeval Forest (Poland). *Forest Ecology and Management, 261*, 821–828.

Kowalczyk, R., Krasińska, M., Kamiński, T., Górny, M., Struś, P., Hofman-Kamińska, E., & Krasiński, Z. A. (2013). Movements of European bison (Bison bonasus) beyond the Białowieża Forest (NE Poland): Range expansion or partial migrations? *Acta Theriologica, 58*, 391–401.

Kowalczyk, R., Wójcik, J. M., Taberlet, P., Kamiński, T., Miquel, C., Valenti, A., Crainec, J. M., & Coissac, E. (2019). Foraging plasticity allows a large herbivore to persist in a sheltering forest habitat: DNA metabarcoding diet analysis of the European bison. *Forest Ecology and Management, 449*, 117474. https://doi.org/10.1016/j.foreco.2019.117474

Krasińska, M., & Krasiński, Z. A. (2013). *European Bison: The nature monograph*. Springer.

Krasińska, M., Krasiński, Z., Olech, W., & Perzanowski, K. (2014). European bison. In M. Meletti & J. Burton (Eds.), *Ecology, evolution and behaviour of wild cattle: Implications for conservation* (pp. 115–173). Cambridge University Press.

Krishnan, M. (1972). An ecological survey of the larger mammals of peninsular India. *Journal of the Bombay Natural History Society, 69*, 322–349.

Kronk, C. (2007). "Ursus americanus". American Black Bear (On-line), Animal Diversity Web. Accessed March 18, 2024 at https://animaldiversity.org/accounts/Ursus_americanus/

Kuehn, D. W. (1986). Population and social characteristics of the Tamarao (Bubalus mindorensis). *Biotropica, 18*, 263–266.

Kuemmerle, T., Perzanowski, K., Chaskovskyy, O., Ostapowicz, K., Halada, L., Bashta, A., Hostert, P., Waller, D. M., & Radeloff, V. C. (2010). European bison habitat in the Carpathian Mountains. *Biological Conservation, 143*, 908–916.

Kuemmerle, T., Levers, C., Bleyhl, B., Olech, W., Perzanowski, K., Reusch, C., & Kramer-Schadt, S. (2018). One size does not fit all: European bison habitat selection across herds and spatial scales. *Landscape Ecology, 33*, 1159–1572.

Kuijper, D. P. J., Sahlén, E., Elmhagen, B., Chamaille´-Jammes, S., Sand, H., Lone, K., & Cromsigt, J. P. G. M. (2016). Paws without claws? Ecological effects of large carnivores in anthropogenic landscapes. *Proceedings of the Royal Society of Biology, 283*, 20161625. https://doi.org/10.1098/rspb.2016.1625

Kumar, S., Nagarajan, M., Sandhu, J. S., Kumar, N., Behl, V., & Nishanth, G. (2007). Mitochondrial DNA analyses of Indian water buffalo support a distinct genetic origin of river and swamp buffalo. *Animal Genetics, 38*(3), 227–232.

Kurnit, J. (2009). Diceros bicornis. Black Rhinoceros (On-line), Animal Diversity Web. https://animaldiversity.org/accounts/Diceros_bicornis/

La-Ong, S., Pothieng, D., & Sakon, B. (1997). *Survey on wildlife trading at the frontier of Thailand-Cambodia and People's Republic Democratic of Laos*. WWF-Thailand, Bangkok, Thailand.

Lariviere, S. (2024, February 20). Coyote. Encyclopedia Britannica. https://www.britannica.com/animal/coyote-mammal

Larska, M., & Krzysiak, M. K. (2019). Infectious disease monitoring of European Bison (Bison bonasus). In M. Ferretti (Ed.), *Wildlife population monitoring* (pp. 248–269). IntechOpen.

LaRue, M. L., Nielsen, C. K., Dowling, M., Miller, K., Wilson, B., Shaw, H., & Anderson, C. R., Jr. (2012). Cougars are recolonizing the Midwest: Analysis of cougar confirmations during 1990-2008. *Journal of Wildlife Management, 76*, 1364–1369.

Lau, C. H., Drinkwater, R. D., Yusoff, K., Tan, S. G., Hetzel, D. J., & Barker, J. S. (1998). Genetic diversity of Asian water buffalo (Bubalus bubalis): mitochondrial DNA D-loop and cytochrome b sequence variation. *Animal genetics, 29*(4), 253–264. https://doi.org/10.1046/j.1365-2052.1998.00309.x

Lehnert, E. (2013). "Acinonyx jubatus". Cheetah (On-line), Animal Diversity Web. Accessed March 17, 2024 at https://animaldiversity.org/accounts/Acinonyx_jubatus/

Lenstra, J. A., & Bradley, D. G. (1999). Systematics and phylogeny of cattle. In R. Fries & A. Ruvinsky (Eds.), *The genetics of cattle* (pp. 1–14). CAB International.

Lindsey, P. A., Balme, G. A., Booth, V. R., & Midlane, N. (2012). The significance of African lions for the financial viability of trophy hunting and the maintenance of wild land. *PLoS One, 7*(1), e29332. https://doi.org/10.1371/journal.pone.0029332

List, R., Ceballos, G., Curtin, C., Gogan, J. P., Pacheco, J., & Truett, J. (2007). Historic distribution and challenges to Bison recovery in the Northern Chihuahuan Desert. *Conservation Biology, 21*(6), 1487–1494.

Lundrigan, B., & Zachariah, T. (2000). Bos frontalis. Gaur (On-line), Animal Diversity Web. https://animaldiversity.org/accounts/Bos_frontalis/

Macdonald, D. (2001). *The new Encyclopedia of mammals*. Oxford University Press.

MacEachern, S. (2007). *Molecular evolution of the domesticated cow (Bos taurus)*. PhD thesis. La Trobe University, Genetics Department.

MacEachern, S., McEwan, J., & Goddard, M. (2009). Phylogenetic reconstruction and the identification of ancient polymorphism in the Bovini tribe (Bovidae, Bovinae). *BMC Genomics, 10*, 177. https://doi.org/10.1186/1471-2164-10-177

Madhusudan, M. D. (2004). Recovery of wild large herbivores following livestock decline in a tropical Indian wildlife reserve. *Journal of Applied Ecology, 41*, 858–869.

Madhusudan, M. D. (2005). The global village: Linkages between international coffee markets and grazing by livestock in a south Indian wildlife reserve. *Conservation Biology, 19*, 411–420.

Madhusudan, M. D., & Karanth, K. U. (2000). Hunting for an answer: Is local hunting compatible with large mammal conservation in India? In J. G. Robinson & E. L. Bennett (Eds.), *Hunting for sustainability in the tropics* (pp. 339–355). Columbia University Press.

Madhusudan, M. D., & Karanth, K. U. (2002). Local hunting and the conservation of large mammals in India. *Ambio, 31*(1), 49–54.

Manfreda, S., McCabe, M. F., Miller, P. E., Lucas, R., Madrigal, V. P., Mallinis, G., Dor, E. B., Helman, D., Estes, L., Ciraolo, G., Müllerová, J., Tauro, F., de Lima, M. I., de Lima, J. L. M. P., Maltese, A., Frances, F., Caylor, K., Kohv, M., Perks, M., ... Toth, B. (2018). On the use of unmanned aerial systems for environmental monitoring. *Remote sensing, 10*(4), Article 641. https://doi.org/10.3390/rs10040641

Marino, J., & Sillero-Zubiri, C. (2011). Canis simensis. The IUCN Red List of Threatened Species 2011: https://doi.org/10.2305/IUCN.UK.2011-1.RLTS.T3748A10051312.en.

Martin, E. B., & Phipps, M. (1996). A review of the wild animal trade in Cambodia. *TRAFFIC Bulletin, 16*(2), 45–60.

Martin, L. R. G., Murphy, S. D., & Quon, S. P. (2001). Effective Planning and Implementation of Ecological Rehabilitation Projects: A Case Study of the Regional Municipality of Waterloo (Ontario, Canada). *Environmental Management, 27*(3), 421–433. https://doi.org/10.1007/s002670010159

Mazák, J. H., & Groves, C. P. (2006). A taxonomic revision of the tigers (Panthera tigris). *Mammalian Biology, 71*(5), 268–287.

McCarthy, T., Mallon, D., Jackson, R., Zahler, P., & McCarthy, K. (2017). Panthera uncia. The IUCN Red List of Threatened Species 2017: https://doi.org/10.2305/IUCN.UK.2017-2.RLTS.T22732A50664030.en.

McDermid, G. J., Coops, N. C., Wulder, M. A., Franklin, S. E., & Seitz, N. E. (2010). Critical Remote Sensing Contributions to Spatial Wildlife Ecological Knowledge and Management. *Springer EBooks*, 193–221. https://doi.org/10.1007/978-4-431-87771-4_11

McLellan, B. N., Proctor, M. F., Huber, D., & Michel, S. (2017). Ursus arctos (amended version of 2017 assessment). The IUCN Red List of Threatened Species 2017: https://doi.org/10.2305/IUCN.UK.2017-3.RLTS.T41688A121229971.en.

Melisch, R. (1995). Anoa threatened by souvenir trade in South Sulawesi, Indonesia. *Oryx, 29*(4), 224–225.

Melletti, M., Penteriani, V., Mirabile, M., & Boitani, L. (2007). Some behavioral aspects of forest buffalo (Syncerus caffer nanus): From herd to individual. *Journal of Mammalogy, 88*(5), 1312–1318. https://doi.org/10.1644/06-MAMM-A-240R1.1

Mendoza, M., & Palmqvist, P. (2008). Hypsodonty in ungulates: An adaptation for grass consumption or for foraging in open habitat? *Journal of Zoology, 274*, 134–142.

Merceron, G., Hofman-Kamińska, E., & Kowalczyk, R. (2015). 3D dental microwear texture analysis of feeding habits of sympatric ruminants in the Białowieża Primeval Forest, Poland. *Forest Ecology and Management, 328*, 262–269.

Miller, D. (2002). Bubalus depressicornis (On-line), Animal Diversity Web. https://animaldiversity.org/accounts/Bubalus_depressicornis/

Miller, D. J., & Schaller, G. B. (1997). Conservation threats to the Chang Tang Wildlife Reserve, Tibet. *Ambio, 26*(3), 185–186.

Miller, D. J., Harris, R., & Cai, G. (1994). Wild yaks and their conservation on the Tibetan Plateau. In R. Zhang, J. Han, & J. Wu (Eds.), *Proceedings of the 1st international congress on Yak* (pp. 27–34). Gansu Agricultural University, Lanzhou.

Mishra, R. P. (2002). *Studies on Indian wild buffalo (Bubalus bubalis L.) and its habitat in Udanti wildlife sanctuary*. Pt. R.S. Shukla University.

Mloszewski, M. J. (1983). *The behavior and ecology of the African buffalo*. Cambridge University Press.

Mohanarangan, A., Nagarajan, R., & Desai, A. A. (2010). Gaur- A wild cattle in grave peril. *Science Reporter, 47*, 33–35.

Mohr, E. (1921). Die geographische verbreitung der anoa-arten auf Celebes. *Archiv fuer Naturgeschichte, 87*, 208–214.

Montsion, L. (2014). "Uncia uncia" Snow Leoaprd (On-line), Animal Diversity Web. Accessed March 17, 2024 at https://animaldiversity.org/accounts/Uncia_uncia/

Morens, D. M., Holmes, E. C., Davis, A. S., & Taubenberger, J. K. (2011). Global rinderpest eradication: Lessons learned and why humans should celebrate too. *Journal of Infectious Diseases, 204*, 502–505.

Muley, P. D. (2001). *Genetic and morphometric studies to differentiate between wild and domestic Asian water buffalo (Bubalus bubalis L.) and their hybrids in Kaziranga National Park, Assam, India*. University of Wisconsin.

Mulheisen, M., Allen, C., & Allen, C. (2002). Lycaon pictus. African Wild Dog (On-line), Animal Diversity Web. https://animaldiversity.org/accounts/Lycaon_pictus/

Mustari, A. H. (1995). *Population and Behaviour of Lowland Anoa (Bubalus depressicornis Smith) in Tanjung Amolengu Wildlife Reserve South-East Sulawesi, Indonesia*. M.Sc. Thesis, Faculty of Forestry Science, Georg-August University, Göttingen.

Mustari, A. H. (2003). *Ecology and conservation of lowland anoa Bubalus depressicornis in Sulawesi, Indonesia*. University of New England.

Naidoo, R., Du Preez, P., Stuart-Hill, G., Beytell, P., & Taylor, R. (2014). Long-range migrations and dispersals of African buffalo (Syncerus caffer) in the Kavango–Zambezi Transfrontier Conservation Area. *African Journal of Ecology, 52*(4), 581–584. https://doi.org/10.1111/aje.12163

National Park Service. (2017). Bison Bellows: Group Dynamics. https://www.nps.gov/articles/bison-bellows-7-7-16.htm.

Nel, C. (2012). Chianina cattle: The gentle white Italian giant. Farmer's Weekly. https://www.farmersweekly.co.za/animals/cattle/the-gentle-white-italian-giant/.

Newell, T., & Sorin, A. (2003). Bison bison. American Bison (On-line), Animal Diversity Web. https://animaldiversity.org/accounts/Bison_bison/

Ng, C. (2015). Syncerus caffer. African buffalo (On-line), Animal Diversity Web. https://animaldiversity.org/accounts/Syncerus_caffer/

Nicholson, S., Bauer, H., Strampelli, P., Sogbohossou, E., Ikanda, D., Tumenta, P.F., Venktraman, M., Chapron, G., & Loveridge, A. (2023). Panthera leo. The IUCN Red List of Threatened Species 2023: https://doi.org/10.2305/IUCN.UK.2023-1.RLTS.T15951A231696234.en.

Nielsen, C., Thompson, D., Kelly, M., & Lopez-Gonzalez, C. A. (2015). Puma concolor (errata version published in 2016). The IUCN Red List of Threatened Species 2015: https://doi.org/10.2305/IUCN.UK.2015-4.RLTS.T18868A50663436.en.

Nogueira, J. (2009). Panthera onca: Jaguar. Animal Diversity Web. Retrieved from https://animaldiversity.org/accounts/Panthera_onca/

Norris, K., Asase, A., Collen, B., Gockowksi, J., Mason, J., Phalan, B., & Wade, A. (2010). Biodiversity in a forest-agriculture mosaic – the changing face of West African rainforests. *Biological Conservation, 143*, 2341–2350.

Nowak, R. M. (1991). *Walker's mammals of the world*. Johns Hopkins University Press.

Nowak, R. M. (2005). Cheetah Acinonyx jubatus. In *Walker's carnivores of the world* (pp. 270–272). Johns Hopkins University Press.

O'Brien, T. G., & Kinnaird, M. F. (1996). Changing populations of birds and mammals in North Sulawesi Orby. *Oryx, 30*, 150–156.

O'Brien, T. G., & Kinnaird, M. F. (1996). Birds and mammals of the Bukit Barisan Selatan National Park, Sumatra, Indonesia. Oryx, 30(3), 207–217. https://doi.org/10.1017/s0030605300021657

Okarda, B. (2010). *Potential habitat and spatial distribution of Anoa (Bubalus spp.) in Lore Lindu National Park*. Bogor Agricultural University.

Oliphant, M. (2003). Bos grunniens. Yak (On-line), Animal Diversity Web. https://animaldiversity.org/accounts/Bos_grunniens/

Pal, B. C., & Guin, D. P. (1986). Population structure of gaur herds at Garumara Wildlife Sanctuary, West Bengal, India. *Proceedings of the Zoological Society of Calcutta, 35*(1–2), 89–95.

Pasha, M. K. S., Sankar, K., Qureshi, Q., & Areendran, G. (2004). Indian bison or gaur (Bos gaurus Lambert, 1804). *ENVIS Bulletin (Wildlife Institute of India, Dehra Dun), 7*, 91–102.

Peña, K. D. (2022). 10,000 to less than 500: How people are driving the tamaraw to extinction. https://newsinfo.inquirer.net/1676307/10000-to-less-than-500-how-people-are-driving-the-tamaraw-to-extinction

Perkl, R. M. (2016). Geodesigning landscape linkages: Coupling GIS with wildlife corridor design in conservation planning. *Landscape and Urban Planning*. [Online], *156*, 44–58.

Perzanowski, K., Januszczak, M., & Wołoszyn-Gałęza, A. (2012). Seasonal movements of wisents (Bison bonasus L., 1758) in the Bieszczady Mountains (SE Poland). *Biological Letters, 49*, 11–17.

Plumb, G. E., White, P. J., & Aune, K. (2013). *American Bison. Wild cattle of the world*. Cambridge University Press.

Plumb, G., Kowalczyk, R., & Hernandez-Blanco, J. A. (2020). Bison bonasus. The IUCN Red List of Threatened Species 2020: e.T2814A45156279. https://doi.org/10.2305/IUCN.UK.2020-3.RLTS.T2814A45156279.en.

Polák, J., & Frynta, D. (2010). Patterns of sexual size dimorphism in cattle breeds support Rensch's rule. *Evolutionary Ecology, 24*, 1255–1266. https://doi.org/10.1007/s10682-010-9354-9

Prins, H. H. T. (1989). Buffalo herd structure and its repercussions for condition of individual African buffalo cows. *Ethology, 81*, 47–71. https://doi.org/10.1111/j.1439-0310.1989.tb00757.x

Prins, H. H. T. (1996). *Ecology and behaviour of the African Buffalo. Chapman & Hall Wildlife Ecology and Behaviour Series (WEBS)*. Chapman & Hall.

Prins, H. H. T., & Sinclair, A. R. E. (2013). Syncerus caffer African Buffalo. In J. Kingdon & M. Hoffmann (Eds.), *The mammals of Africa* (Pigs, Hippopotamuses, Chevrotain, Giraffes, Deer and Bovids) (Vol. VI, pp. 125–136). Bloomsbury.

Pucek, Z., Belousova, I. P., Krasińska, M., Krasiński, Z. A., & Olech, W. (2004). *European Bison: Status survey and conservation action plan*. IUCN.

Pudyatmoko, S. (2004). *Does the banteng (Bos javanicus) have a future in Java? Challenges of the conservation of a large herbivore in a densely populated island*. Knowledge Marketplace Reports of the 3rd IUCN World Conservation Congress: 6 pp. Bangkok, Thailand.

Purvis, A., Gittleman, J. L., Cowlishaw, G., & Mace, G. M. (2000). Predicting extinction risk in declining species. *Proceedings of the Royal Society of London. Series B: Biological Sciences, 267*, 1947–1952.

Qi, X. B., Jianlin, H., Wang, G., Rege, J. E. O., & Hanotte, O. (2009). Assessment of cattle genetic introgression into domestic yak populations using mitochondrial and micro-satellite DNA markers. *Animal Genetics, 41*, 242–252.

Qi, X. B., Jianlin, H., Wang, G., Rege, J. E., & Hanotte, O. (2010). Assessment of cattle genetic introgression into domestic yak populations using mitochondrial and microsatellite DNA markers. *Animal Genetics, 41*(3), 242–252. https://doi.org/10.1111/j.1365-2052.2009.01989.x

Quang, H. X., & Trung, C. T. (2004). Status of the Saola in Pu Huong Nature Reserve, Nghe An Province and the potential for conservation. In J. Hardcastle, S. Cox, N. T. Dao, & A. Grieser Johns (Eds.), *Rediscovering the Saola - Proceedings of "Rediscovering the Saola - A Status Review and Planning Workshop"* (pp. 52–55). WWF Indochina Programme, Hanoi, Vietnam.

Quigley, H., Foster, R., Petracca, L., Payan, E., Salom, R., & Harmsen, B. (2017). Panthera onca (errata version published in 2018). The IUCN Red List of Threatened Species 2017: https://doi.org/10.2305/IUCN.UK.2017-3.RLTS.T15953A50658693.en.

Rabor, D. S. (1977). *Philippine Birds & Mammals*. Honolulu: University of Hawaii Press.

Radwan, J., Demiaszkiewicz, A. W., Kowalczyk, R., Lachowicz, J., Kawałko, A., Wójcik, J. M., Pyziel, A. M., & Babik, W. (2010). An evaluation of two potential risk factors, MHC diversity and host density, for infection by an invasive nematode Ashworthius sidemi in endangered European bison (Bison bonasus). *Biological Conservation, 143*, 2049–2053.

Rawat, G. S., & Sankar, K. (2011). *Habitat ecology and conservation status of Wild Ungulates in Northern Parts of Changthang Wildlife Sanctuary, Ladakh*. Wildlife Institute of India.

Redford, K. H., Aune, K., & Fearn, E. (2009). The second recovery of bison: Ecological recovery of North America's largest mammal. *The Wildlife Professional, 3*, 46–50.

Reynolds, H., Gates, C., & Glahoht, R. (2003). Bison. In J. Chapman & G. Feldhamer (Eds.), *Wild mammals of North America, biology, management, and economics* (pp. 1009–1060). Johns Hopkins University Press.

Robichaud, W. G. (1998). Physical and behavioral description of a captive saola, Pseudoryx nghetinhensis. *Journal of Mammalogy, 79*(2), 394–405. https://doi.org/10.2307/1382970

Robichaud, W. (1999). Saola Conservation Action Plan for Lao PDR: 1999 Revision.

Robichaud, W., & Long, B. (2014). aola Pseudoryx nghetinhensis Dung, Giao, Chinh, Touc, Arctander and Mackinnon, 1993. In M. Melletti & J. A. Burton (Eds.), *Ecology, evolution and behaviour of wild cattle: Implications for cosnervation* (pp. 318–325). Cambridge University Press.

Rogers, L. R., Ranglack, D. H., & Plumb, G. (2022). Bison bison (Green Status assessment). The IUCN Red List of Threatened Species 2022: e.T2815A281520221.

Rosevear, D. R. (1974). *The carnivores of West Africa* (pp. 75–91). Trustees of the British Museum (Natural History).

Roth, J. (2004). Bubalus bubalis Wild water buffalo (On-line), Animal Diversity Web. https://animaldiversity.org/accounts/Bubalus_bubalis/

Saari, J. (2002). Bos javanicus. Banteng (On-line), Animal Diversity Web. Accessed March 18, 2024 at https://animaldiversity.org/accounts/Bos_javanicus/

Sahu, B. K., Parganiha, A., & Pati, A. K. (2020). Behavior and foraging ecology of cattle: A review. *Journal of Veterinary Behavior-clinical Applications and Research, 40*, 50–74. https://doi.org/10.1016/j.jveb.2020.08.004

Salter, R. E. (1983). *Summary of currently available information on internationally threatened wildlife species in Burma*. FAO.

Salter, R. E., Phanthavong, B., Sawathvong, S., Souriyakan, S., & Louanglath, K. (1990). *An assessment of the current status of kouprey and other wild cattle in southern Laos*. Forest Resources Conservation Project, Lao/Swedish Forestry Cooperation Programme. Vientiane, Laos.

Salwasser, H., Doyle, H. T. & Kessler, W. B. (1995). Biodiversity generates new viewson old issues. *Forum for Applied Research and Public Policy, 10*(2), 121–123.

Samojlik, T., Fedotova, A., Borowik, T., & Kowalczyk, R. (2019). Historical data on European bison management in Bialowieza Primeval Forest can contribute to a better contemporary conservation of the species. *Mammal Research, 64*, 543–557.

San Diego Wildlife Alliance. (2024). *Wild Cattle*. Retrieved from https://animals.sandiegozoo.org/animals/wild-cattle.

San Diego Zoo Wildlife Alliance. (2024). *Dhole Cuon alpinus*. Retrieved from https://animals.sandiegozoo.org/animals/dhole.

Sanderson, E., Redford, K. H., Weber, B., Aune, K., Baldes, D., Berger, J., Carter, D., Curtin, C., Derr, J., Dobrott, S., Fearn, E., Fleener, C., Forrest, S., Gerlach, C., Gates, C. C., Gross, J., Gogan, P., Grassel, S., Hilty, J. A., Jensen, M., Kunkel, K., Lammers, D., List, R., Minkowski, K., Olson, T., Pague, C., Robertson, P. B., & Stephenson, B. (2008). The ecological future of the North American Bison: Conceiving long-term, large-scale conservation of wildlife. *Conservation Biology, 22*(2), 252–266.

Saola Working Group. (2013). Conservation Through Collaboration: Proceedings of the 3rd Meeting of the Saola Working Group. Saola Working Group, Asian Wild Cattle Specialist Group of the IUCN Species Survival Commission. Vientiane.

Sarasin, P., & Sarasin, F. (1901). *Entwurf einer Geographyraphisch-Geologischen Beschreibung der Insel Celebes*. C.W. Kreidel.

Schaller, G. B. (1998). *Wildlife of the Tibetan Steppe*. University of Chicago Press.

Schaller, G. B. (2012). *Tibet Wild: A naturalist's journey on the roof of the world*. Island Press.

Schaller, G. B., & Gu, B. Y. (1994). Ungulates in Northwest Tibet. *National Geographic Research and Exploration, 10*(3), 266–293.

Schaller, G. B., & Liu, W. L. (1996). Distribution, status, and conservation of wild yak Bos grunniens. *Biological Conservation, 76*, 1–8.

Schaller, G. B., Kang, A. L., Hashi, T. D., & Cai, P. (2007). A winter wildlife survey in the northern Qiangtang of Tibet Autonomous Region and Qinghai Province, China. *Acta Theriologica Sinica, 27*, 309–316.

Schilz, A. (2004). Bubalus quarlesi (On-line), Animal Diversity Web. https://animaldiversity.org/accounts/Bubalus_quarlesi/

Schütz, E. (2015). *Tamaraw population survey and current status*. Unpublished Report.

Seidensticker, J., Christie, S., & Jackson, P. (1999). *Preface. Riding the tiger. Tiger conservation in human-dominated landscapes*. Cambridge University Press.

Semenov, U. A. (2014). *The Wisents of Karachay-Cherkessia. Proceedings of the Sochi National Park (8): 23–24*. KMK Scientific Press.

Shanee, S., Tello-Alvarado, J. C., Vermeer, J., & Bóveda-Penalba, A. J. (2013). GIS risk assessment and GAP analysis for the Andean Titi Monkey (Callicebus oenanthe). *Primate Conservation*. [Online], *26*(1), 17–23.

Sillero-Zubiri, C., & Gottelli, D. (1994). Canis simensis. *Mammalian Species, 485*, 1–6. https://doi.org/10.2307/3504136

Skinner, J. D., & Chimimba, C. T. (2005). Subfamily Acinonychinae Pocock 1917. In *The mammals of the Southern African Subregion* (3rd ed., pp. 379–384). Cambridge University Press.

Smith, J. (2002). "Canis lupus" Gray Wolf (On-line), Animal Diversity Web. Accessed March 18, 2024 at https://animaldiversity.org/accounts/Canis_lupus/

Smith, A. T., & Xie, Y. (2008). *A guide to the mammals of China*. Princeton University Press.

Soubrier, J., Gower, G., Chen, K., Richards, S. M., Llamas, B., Mitchell, K. J., Ho, S. Y. W., Kosintsev, P., Lee, M. S. Y., Baryshnikov, G., Bollongino, R., Bover, P., Burger, J., Chivall, D., Crégut-Bonnoure, E., Decker, J. E., Doronichev, V. B., Douka, K., Fordham, D. A., Fontana, F., Fritz, C., Glimmerveen, J., Golovanova, L. V., Groves, C., Guerreschi, A., Haak, W., Higham, T., Hofman-Kamińska, E., Immel, A., Julien, M., Krause, J., Krotova, O., Langbein, F., Larson, G., Rohrlach, A., Scheu, A., Schnabel, R. D., Taylor, J. F., Tokarska, M., Tosello, G., van der Plicht, J., van Loenen, A., Vigne, J., Wooley, O., Orlando, L., Kowalczyk, R., Shapiro, B., & Cooper, A. (2016). Early cave art and ancient DNA record the origin of European bison. *Nature Communications, 7*, 13158–13158.

Srikosamatara, S., & Suteethorn, V. (1994). Wildlife conservation along the Thai-Lao Border. *Natural History Bulletin of the Siam Society, 42*, 3–21.

Srikosamatara, S., & Suteethorn, V. (1995). Populations of gaur and banteng and their management in Thailand. *Natural History Bulletin of the Siam Society, 43*(1), 55–83.

Srikosamatara, S., Siripholdej, B., & Suteethorn, V. (1992). Wildlife trade in Lao P. D. R. and between Lao P. D. R. and Thailand. *Natural History Bulletin of the Siam Society, 40*, 1–47.

Statistica. (2024). Number of cattle worldwide from 2012 to 2023 (in million head). Statista University.

Stein, A.B., Athreya, V., Gerngross, P., Balme, G., Henschel, P., Karanth, U., Miquelle, D., Rostro-García, S., Kamler, J.F., Laguardia, A., Khorozyan, I., & Ghoddousi, A. (2023). Panthera pardus. The IUCN Red List of Threatened Species 2023: https://doi.org/10.2305/IUCN.UK.2023-1.RLTS.T15954A215195554.en.

Steinmetz, R. (2004). Gaur (Bos gaurus) and banteng (B. javanicus) in the lowland forest mosaic of Xe Pian Protected Area, Lao P.D.R.: Abundance, habitat use, and conservation. *Mammalia, 68*, 141–157.

Steinmetz, R., Seuaturien, N., Chutipong, W., Chamnankit, C., & Phoonjampa, R. (2007). *The ecology and conservation of tigers, other large carnivores, and their prey in Kuiburi National Park, Thailand*. WWF Thailand and Kuiburi National Park, Bangkok, Thailand.

Stephenson, R. O., Gerlach, S. C., Gurthrie, R. D., Harington, C. R., Mills, R. O., & Hare, G. (2001). Wood bison in late holocene Alaska and adjacent Canada; paleontological, archaeological and historical records. In S. C. Gerlach & M. S. Murray (Eds.), *People and wildlife in northern North America: Essays in honor of R. Dale Guthrie* (pp. 125–159). BAR International Series 944.

Stone, R. (2006). The Saola's Last Stand. *Science, 314*, 1380–1383.

Succop, N. (2023). *Dicerorhinus sumatrensis*. Sumatran Rhinoceros (On-line), Animal Diversity Web. https://animaldiversity.org/accounts/Dicerorhinus_sumatrensis/

Subak, S. (2000). Forest Protection and Reforestation in Costa Rica: Evaluation of a Clean Development Mechanism Prototype. *Environmental Management, 26*(3), 283–297. https://doi.org/10.1007/s002670010087

Sugiharta, A. (1994). *Abundance and habitat characterization of mountain anoas in Besoa, Lore Lindu National Park, Indonesia*. M.Sc. Thesis. New Mexico State University, New Mexico, USA.

Talbot, L. M., & Talbot, M. H. (1966). The tamarau (Bubalus mindorensis): Observations and recommendations. *Mammalia, 30*, 1–12.

Thompson, D. J., & Jenks, J. A. (2010). Dispersal movements of subadult cougars from the Black Hills: The notion of range expansion and recolonization. *Ecosphere, 1*, 1–11.

Thornback, J. (1983). *Wild cattle, bison and buffaloes, their status and potential value*. IUCN.

Timmins, R. J. (2006). *An assessment of the biodiversity conservation significance of the Mekong Ramsar site, Stung Treng, Cambodia*. Mekong Wetlands Biodiversity Conservation and Sustainable Use Programme, Vientiane.

Timmins, R. J., & Rattanak, O. (2001). *The importance of Phnom Prich Wildlife Sanctuary and adjacent areas for the conservation of tigers and other key species*. WWF Indochina Programme, Hanoi, Vietnam and Phonm Penh, Cambodia.

Timmins, R. J., Burton, J., & Hedges, S. (2016). Bos sauveli. The IUCN Red List of Threatened Species 2016: https://doi.org/10.2305/IUCN.UK.2016-2.RLTS.T2890A46363360.en.

Timmins, R. J., Hedges, S., & Robichaud, W. (2020). Pseudoryx nghetinhensis (amended version of 2016 assessment). The IUCN Red List of Threatened Species 2020: https://doi.org/10.2305/IUCN.UK.2020-1.RLTS.T18597A166485696.en.

Tokar, E. (2001). "Canis latrans" Coyote (On-line), Animal Diversity Web. Accessed March 18, 2024 at https://animaldiversity.org/accounts/Canis_latrans/

Tokarska, M., Kawałko, A., Wójcik, J. M., & Pertoldi, C. (2009). Genetic variability in the European bison (Bison bonasus) population from Białowieza forest over fifty years. *Biological Journal of the Linnean Society, 97*, 801–809.

Tokarska, M., Pertoldi, C., Kowalczyk, R., & Perzanowski, K. (2011). Genetic status of the European bison Bison bonasus after extinction in the wild and subsequent recovery. *Mammal Review, 41*, 151–162.

Tordoff, A. W., Timmins, R. J., Maxwell, A., Keavuth, H., Vuthy, L., & Hourt, K. E. (Eds.). (2005). *Biological assessment of the Lower Mekong Dry Forests Ecoregion, 192*. WWF Greater Mekong Programme, Phnom Penh, Cambodia.

Trakolis, D. (2000). Biodiversity and Conservation Actions on Mount Voras, Greece. *Environmental Management, 26*(2), 145–151. https://doi.org/10.1007/s002670010077

Treves, A., & Karanth, K. U. (2003). Human-Carnivore Conflict and Perspectives on Carnivore Management Worldwide. *Conservation Biology, 17*(6), 1491–1499. https://doi.org/10.1111/j.1523-1739.2003.00059.x

Tsering, D., Farrington, J. D., & Norbu, K. (2006). *Human-wildlife conflict in the Chang Tang Region of Tibet: The impact of Tibetan brown bears and other wildlife on nomadic herds with recommendation for conflict mitigation*. WWF China-Tibet Program, Lhasa.

United States Fish and Wildlife Service. (2011). U. S. Fish and Wildlife Service concludes eastern cougar extinct. https://www.fws.gov/story/2011-03/us-fish-and-wildlife-service-concludes-eastern-cougar-extinct.

Van Dung, V., Giao, P. M., Chinh, N. N., Tuoc, D., & MacKinnon, J. (1994). Discovery and conservation of the Vu Quang ox in Vietnam. *Oryx, 28*, 16–21.

Van Zyll de Jong, C. G. (1986). *A systematic study of recent bison, with particular consideration of the wood bison (Bison bison athabascae Rhoads 1898)*. National Museums of Canada.

Velez-Liendo, X., & García-Rangel, S. (2017). Tremarctos ornatus (errata version published in 2018). The IUCN Red List of Threatened Species 2017: https://doi.org/10.2305/IUCN.UK.2017-3.RLTS.T22066A45034047.en.

Verkaar, E. L., Nijman, I. J., Beeke, M., Hanekamp, E., & Lenstra, J. A. (2004). Maternal and paternal lineages in cross-breeding bovine species. Has wisent a hybrid origin? *Molecular Biology and Evolution, 21*(7), 1165–1170. https://doi.org/10.1093/molbev/msh064

Wang, S. W., & Macdonald, D. W. (2006). Livestock predation by carnivores in Jigme Singye Wangchuck national park, Bhutan. *Biological Conservation, 129*(4), 558–565. https://doi.org/10.1016/j.biocon.2005.11.024

Wang, M. S., Zeng, Y., Wang, X., Nie, W. H., Wang, J. H., Su, W. T., Otecko, N. O., Xiong, Z. J., Wang, S., Qu, K. X., Yan, S. Q., Yang, M. M., Wang, W., Dong, Y., Wu, D. D., & Zhang, Y. P. (2017). Draft genome of the gayal, Bos frontalis. *GigaScience, 6*(11), 1–7. https://doi.org/10.1093/gigascience/gix094

Weber, M. (1890). Mammalia from the Malay Archipelago. I. Primates, Prosimiae, Galeopithecidae, Carnivora, Artiodactyla, Edentata, Marsupiala. *Zoologische Ergebnisse einer Reise in Niederlandisch-Ostindien, 1*, 93–114.

Węcek, K., Hartmann, S., Paijmans, J. L. A., Taron, U., Xenikoudakis, G., Cahill, J. A., Heintzman, P. D., Shapiro, B., Baryshnikov, G., Bunevich, A. N., Crees, J. J., Dobosz, R., Manaserian, N., Okarma, H., Tokarska, M., Samuel, T., Turvey, S. T., Wójcik, J. M., Żyła, W., Szymura, J. M., Hofreiter, M., & Barlow, A. (2017). Complex admixture preceded and followed the extinction of wisent in the wild. *Molecular Biology and Evolution, 34*, 598–612.

Weir, A., & Van Cuong, D. (2004). Introduction from SFNC. In J. Hardcastle, S. Cox, N. T. Dao, & A. G. Johns (Eds.), *Rediscovering the Saola - Proceedings of "Rediscovering the Saola - a status review and planning workshop"* (p. 10). WWF Indochina Programme, Hanoi, Vietnam.

Wharton, C. H. (1957). *An ecological study of the Kouprey (Novibos sauveli Urbain)*. Monographs of the Institute of Science and Technology, Monograph 5, Manila, Philippines.

Wharton, C. H. (1966). *Man, fire and wild cattle in North Cambodia*. Proceedings: 5th Tall Timbers Fire Ecology Conference 1966. https://talltimbers.org/wp-content/uploads/2018/09/23-Wharton1966_op.pdf

Wharton, C. H. (1968). Man, Fire and Wild Cattle in Southeast Asia. *Proceedings of the Annual Tall Timbers Fire Ecology Conference, 8*, 107–167.

Wheeler, P. M., Purwanto, E., Coles, T., Carlisle, B., Widyati, A., Jones, M., Winarni, N., & Dwiyahreni, A. A. (2009). State of the Lambusango Forests 2008. In *Operation Wallacea*.

Whitfield, J. (1998). A saola poses for the camera. *Nature, 396*, 410.

Whitten, A. J., Mustafa, M., & Henderson, G. S. (2002). *The ecology of Sulawesi*. Gadjah Mada University Press.

Wiener, G., Jianlin, H. & Ruijun, L. (2006). *The Yak*. FAO Regional Office for Asia and the Pacific.

Wilson, N. (2022). The cat came back: Canada's cougar comeback. *Canadian Geographic*. Retrieved from https://canadiangeographic.ca/articles/the-cat-came-back-canadas-cougar-comeback/.

Wilson, D. E., & Reeder, D. M. (2005). *Mammal species of the world*. Johns Hopkins University Press.

Wilson, G. A., Olson, W., & Strobeck, C. (2002). Reproductive success in wood bison (Bison bison athabascae) established using molecular techniques. *Canadian Journal of Zoology, 80*(9), 1537–1548.

Winker, J. (2004). Bos sauveli (On-line), Animal Diversity Web. https://animaldiversity.org/accounts/Bos_sauveli/

Winnie, J. A., Cross, P. C., & Getz, W. M. (2008). Habitat quality and heterogeneity influence distribution and behavior in African buffalo (Syncerus caffer). *Ecology, 89*(5), 1457–1468. https://doi.org/10.1890/07-0772.1

Winterbach, H. E. K. (1998). Research review: The status and distribution of Cape buffalo Syncerus caffer caffer in southern Africa. *South African Journal of Wildlife Research, 28*(3), 82–88.

Wójcik, J. M., Kawałko, A., Tokarska, M., Jaarola, M., Valenback, P., & Pertoldi, C. (2009). Post-bottleneck mtDNA diversity in a free-living population of European bison: Implications for conservation. *Journal of Zoology, 277*, 81–87.

Wolf, C., & Ripple, W. J. (2016). Prey depletion as a threat to the world's large carnivores. *Royal Society Open Science, 3*, 160252. https://doi.org/10.1098/rsos.160252

Woodroffe, R. (2000). Predators and people: Using human densities to interpret declines of large carnivores. *Animal Conservation, 3*, 165–173.

Woodroffe, R., & Sillero-Zubiri, C. (2020). Lycaon pictus (amended version of 2012 assessment). The IUCN Red List of Threatened Species 2020: https://doi.org/10.2305/IUCN.UK.2020-1.RLTS.T12436A166502262.en.

World Population Review. (2024a). Wolf Population by Country 2024. https://worldpopulationreview.com/country-rankings/wolf-population-by-country.

World Population Review. (2024b). Coyote Population by Country 2024. https://worldpopulationreview.com/country-rankings/coyote-population-by-country.

World Wildlife Fund (WWF). (2019). *Assessment of conservation of European bison in Ukraine in view of future conservation actions for the species*. World Wildlife Fund Ukraine. Retrieved from https://d2ouvy59p0dg6k.cloudfront.net/downloads/europan_bison_ukraine_final.pdf

Wozencraft, W. C. (2005a). Species Neofelis nebulosa. In D. E. Wilson & D. M. Reeder (Eds.), *Mammal species of the world: A taxonomic and geographic reference* (3rd ed., pp. 545–546). Johns Hopkins University Press.

Wozencraft, W. C. (2005b). Species Puma concolor. In D. E. Wilson & D. M. Reeder (Eds.), *Mammal species of the world: A taxonomic and geographic reference* (3rd ed.). Johns Hopkins University Press.

Wozencraft, W. C. (2005c). Species Panthera pardus. In D. E. Wilson & D. M. Reeder (Eds.), *Mammal species of the world: A taxonomic and geographic reference* (3rd ed.). Johns Hopkins University Press.

Yang, W., Liu, W., Liu, Y., Corry, R. C., et al. (2014). Cost-effective targeting of riparian buffers to achieve water quality and wildlife habitat benefits. *International Journal of River Basin Management*. [Online], *12*(1), 43–55.

Yindee, M., Vlamings, B. H., Wajjwalku, W., Techakumphu, M., Lohachit, C., Sirivaidyapong, S., Thitaram, C., Amarasinghe, A. A. A. W. K., Alexander, P. A. B. D. A., Colenbrander, B., & Lenstra, J. A. (2010). Y-chromosomal variation confirms independent domestications of swamp and river buffalo. *Animal Genetics, 41*(4), 433–435.

Young, E. (2002). Suspected foot and mouth hits endangered ox. *New Scientist*. https://www.newscientist.com/article/dn2345-suspected-foot-and-mouth-hits-endangered-ox/

Zhang, Y., Colli, L., & Barker, J. S. F. (2020). Asian water buffalo: Domestication, history and genetics. *Animal Genetics, 51*(2), 177–191. https://doi.org/10.1111/age.12911

Zielke, L., Wrage-Mönnig, N., Müller, J., & Neumann, C. (2019). Implications of spatial habitat diversity on diet selection of European bison and Przewalski's horses in a rewilding area. *Diversity, 11*, 63.

Zoo Idaho. (2016). *American Black Bear (Ursus americanus)*. Retrieved from https://zooidaho.org/animals/fact-sheets/american-black-bear/

Zoological Wildlife Foundation. (2024). *Standard Bengal Tiger*. Retrieved from https://zoologicalwildlifefoundation.com/visit/animals/bengal-tiger/

Zumeta, D. C., & Ellefson, P. V. (2000). Conserving the Biological Diversity of Forests: Program and Organizational Experiences of State Governments in the United States. *Environmental Management, 26*(4), 393–402. https://doi.org/10.1007/s0026702207

Wild Cattle and Their Carnivores

Introduction

The predatory and defense interactions between large carnivores and herbivores involve both individual tactics and larger, group-based strategies at interspecies and intraspecies levels (Campbell, 2017; Campbell & Jha, 2021). Wild cattle are powerful opponents for all large carnivores; as noted by Roth (2004), "in general, the horns of bovids are effective anti-predator deterrents." Wild herbivores also have natural herding behavior and may use this group organization to defend themselves against carnivores, exploiting their advantages in size, strong muscular bodies, and horns (Sinclair et al., 2003; Owen-Smith & Mills, 2008a, b; Owen-Smith, 2021). For example, Palmer and Packer (2021) argue that "large mammalian herbivores use a diverse array of strategies to survive predator encounters including flight, grouping, vigilance, warning signals, and fitness indicators. While anti-predator strategies appear to be driven by specific predator traits, no prior studies have rigorously evaluated whether predator hunting characteristics predict reactive anti-predator responses." Sinclair et al. (2003: 288) point to the importance of herbivore size in predation in a case study of the Serengeti in Tanzania, where "the primary cause of mortality for adults of a particular species is determined by two factors—the species diversity of both the predators and prey and the body size of that prey species relative to other prey and predator" and "threshold occurs at prey body sizes of ~150 kg, above which ungulate species have few natural predators and exhibit food limitation. Thus, biodiversity allows both predation (top-down) and resource limitation (bottom-up) to act simultaneously to affect herbivore populations" (see also Schaller, 1972; Elliot et al., 1977). Lamichhane et al. (2018) add an example of herbivore size and predation from Chitwan National Park, Nepal, where tigers and leopards accounted for more than 90% of the livestock predation, with tigers killing both large (cattle and buffalo) and medium-sized (goats, sheep, pigs) livestock, but leopards killing medium-sized livestock.

Four main issues are relevant to these interactions between wild herbivores and their predators (Tables 1, 2, 3, and 4) (Campbell, 2017; Campbell & Jha, 2021; Jha

© The Author(s), under exclusive license to Springer Nature Switzerland AG 2025

M. O. Campbell, *Cattle, Their Predators and Geomatics Research*,
https://doi.org/10.1007/978-3-031-97363-5_2

Table 1 Hunting strategies of large and medium-sized carnivores

Carnivore	Status	Hunting method	Wild cattle taken
Lion	Dominant, apex predator	Group-based, ambush	All individuals
Tiger	Dominant, apex predator	Solitary ambush	All individuals
Jaguar	Dominant, apex predator	Solitary ambush	Perhaps all individuals
Cougar	Subordinate, apex predator	Solitary ambush	Smaller individuals
Leopard	Subordinate, apex predator	Solitary ambush	Smaller individuals
Snow leopard	Dominant, apex predator	Solitary ambush	Smaller individuals
Clouded leopard	Subordinate, mesopredator	Solitary ambush	Smaller individuals
Brown bear	Dominant, apex predator	Solitary ambush	All individuals
American black bear	Subordinate, apex predator	Solitary ambush	Smaller individuals
Asiatic black bear	Subordinate, apex predator	Solitary ambush	Smaller individuals
Sloth bear	Subordinate, apex predator	Solitary ambush	Smaller individuals
Gray wolf	Dominant, apex predator	Group-based, pursuit	Perhaps all individuals
African wild dog	Dominant, apex predator	Group-based, pursuit	Smaller individuals
Dhole	Subordinate, apex predator	Group-based, pursuit	Smaller individuals
Coyote	Subordinate, mesopredator	Group-based, pursuit	Smaller individuals
Hyena	Subordinate, apex predator	Group-based, pursuit	Smaller individuals

All individuals refer to larger males to calves. Smaller individuals refer to females and/or younger animals, including calves. Although there are no wild cattle in South America, domestic water buffalo have been introduced into this region. Sources: Sheldon et al. (2009), Goswami and Friscia (2010), Campbell (2017), Campbell & Jha (2021), Jha & Campbell (2023)

& Campbell, 2023). These are the killing methods of large carnivores, including the comparative hunting strategies of large carnivores; the individual and group-based strategic defenses by wild herbivores, partly related to the sizes of predators and prey species; the similarity and differences between the habitats of large carnivores and wild herbivores; and the herbivore reactions to carnivore favored habitats, partly related to the size and power of the carnivores and herbivores involved (Campbell & Jha, 2021). Regarding the last point, as noted by Valeix et al. (2009b: 1483), "a central question in the study of predator–prey relationships is to what extent prey behaviour is determined by avoidance of predators." For the first point, the main issues concerning large carnivore hunting strategies are individual and group-based methods (Funston et al., 1998). Lang and Farine (2017: 1230) write about the social organization of animal predators, noting that "social predation—the act of hunting and feeding with others—is one of the most successful life-history traits in the animal kingdom" and "although many predators hunt and feed together, a diversity of mechanisms exist by which individuals forage socially." A framework of social

Table 2 Individual and group-based strategic defenses by wild herbivores

Cattle	Main carnivore	Defense
Gaur	Tigers, crocodiles kill adults, leopards, dholes kill calves	Killing of predators, one-to-one confrontation, frontal group defense with horns, charging, defense of young, fleeing
Banteng	Tigers, crocodiles kill adults, leopards, dholes kill calves	Killing of predators, one-to-one confrontation, frontal group defense with horns, charging, defense of young, fleeing
Water Buffalo	Tigers, crocodiles kill adults, leopards, dholes kill calves	Killing of predators, one-to-one confrontation, frontal group defense with horns, charging, defense of young, fleeing
African Buffalo	Lions, leopards, wild dogs, hyenas kill adults, jackals kill calves	Killing of predators, one-to-one confrontation, frontal group defense with horns, charging, defense of young, fleeing
American bison	Brown and black bears, cougars, wolves	One-to-one confrontation, frontal group defense with horns, charging, defense of young, fleeing
European bison	Wolves, brown bears	One-to-one confrontation, frontal group defense with horns, charging, defense of young, fleeing
Wild yak	Wolf, brown and black bears, snow leopards, lynx	One-to-one confrontation, frontal group defense with horns, charging, defense of young, fleeing

For gaur, Khaewphakdee et al. (2020) note that 45.2% of tiger kills were adults, and for banteng, 55.7% were adults, but the average weight of the banteng killed (423.9 kg) was heavier than gaur (397.9). Variar et al. (2023) note that gaur (Bos gaurus) constituted 26.33% of tiger diets in India's Western Ghats. Leopards in Cambodia may prey on adult banteng. Sources: Schaller (1967), Schaller and Junrang (1988), Oli et al. (1993), Oli (1994), Ikeda (2004), Namgail et al. (2007), Xu et al. (2007), Bhumpakphan (2008), Leslie and Schaller (2009), Tilson and Nyhus (2009), Ashokkumar et al. (2010a, b, 2011), Castelló (2016), Campbell (2017, 2018), Zangmo et al. (2018), Khaewphakdee et al. (2020), World Wildlife Fund (2020), Rahman et al., 2021, Khelkar (2022), Jha & Campbell (2021, 2022, 2023), Chaiyarat et al. (2023), Variar et al. (2023)

predation is described, comprising "five key behavioural dimensions: sociality, communication, specialization, resource sharing, and dependence" with "key commonalities and differences among species, and informative cross-dimensional correlations" (Lang & Farine, 2017: 1230).

Regarding the killing methods of large carnivores, including the comparative hunting strategies of large carnivores, there are two main variations: carnivores that hunt singly, either from pursuit or from close ambush, and those that hunt in groups, also termed pack hunters, that hunt through joint pursuit or a surrounded ambush (Campbell, 2017). Packer and Ruttan (1988: 159) point out that "When a group captures only a single prey per hunt, the expected benefits from cooperation rarely outweigh the advantages of hunting alone, since the prey must be divided between group members. Cooperation to capture the same prey can confer sufficient mutual benefit only when solitaries have low foraging efficiency; otherwise, each individual would do better to hunt separate prey. In contrast, species that capture multiple prey in a single hunt are not faced with similar disadvantages from grouping and should not be similarly constrained to hunt alone by high individual foraging efficiency." Issues concern "the theoretical effects of kinship, individual differences in hunting ability, and behavioral dominance on the evolution of cooperative hunting of single large prey" (Packer & Ruttan, 1988: 159).

Table 3 Similarity and differences between the habitats of large carnivores and wild herbivores

Cattle	Habitat	Large carnivores
Gaur	Forests, grassy glades, tropical wet evergreen, and semi-evergreen forest (SW India), tropical moist and dry deciduous forest (Central India), tropical wet evergreen and subtropical broadleaf forests and wet grasslands (NE India)	Tigers can take adults; leopards and dhole take young gaur, avoid tigers, focus on smaller prey
Banteng	Drier, open lowland deciduous forests and secondary forests, mixed deciduous forest. Less requirement for water than the gaur, but may be present near water sources, and also mineral licks, which may be "essential" (Rahman et al., 2019: 77)	Tigers, male leopards take adults; leopards, dhole take calves, avoid tigers, take small prey
Water buffalo	Nearby fresh water, waterlogged wetlands, low-lying alluvial short grasslands, grassland with swampy ecosystems, marshes, tall grassland, much less woodland or forest. More requirement for water than the gaur or banteng	Tigers take adults, leopards and dhole take calves, avoid tigers, take smaller prey
Wild Yak	Cold, treeless uplands, including alpine tundra with grasses and sedges, above 4500 m, with seasonal, especially precipitation climatic changes, moist meadows, grasslands	Wolves, brown bears, snow leopards, take young, follow smaller prey, domestic livestock
African Buffalo	Open savanna, with mixed grasses, shrubs and tree, and shrub landcover. Water sources and wallowing areas	Lions take adults, leopards, cheetahs, hyenas, wild dogs may take calves
American Bison	Grasslands and prairies mostly in valley bottoms, mid-mountain slopes near to water, and areas with burns. Includes tall-grass, mixed-grass, and short-grass prairies. Boreal forest in northern North America	Principally wolves, rarely brown, and black bears, cougars. Rarely, coyotes can kill calves
European Bison	Deciduous forest, including mixed forests. Forest habitats with low canopy cover, with few woody plants in the ground layer. Also, successively, mixed forests, then deciduous forests, alder, and coniferous forests	Rarely hunted by wolves, brown bears

Generally, large carnivores adapt and follow wild cattle into all habitats. Male leopards, but not females, may selectively prey on banteng only in Indonesia (Rostro et al., 2018). Note that in their case study, adult banteng weigh 600–800 kg, and local leopards are the only known leopard population whose main prey weighed over 500 kg. Snow leopard habitat is similar to that of the wild yak (Shrestha et al., 2018). Sheldon et al. (2009: 260) note that, concerning the coyote killing a bison calf, "the predation is, to our knowledge, the only direct and complete observation of a lone Coyote capturing and killing a Bison calf."

Sources: (Lekagul and McNeely (1977), Krasiński (1978), Karanth and Sunquist (1995a, b)), Schaller and Liu (1996), Mattson (1997), Prayurasiddhi (1997), Nowak (1999), Jędrzejewski et al. (2000), Choudhury (2002), Saari (2002), Wyman (2002), Steinmetz (2004), Xu et al. (2007), Hance (2012), Hendricks, 2013, Steinmetz et al. (2013), Moran (2014), Phillipps (2016), Steenweg et al. (2016), Liang et al. (2017), Rostro et al. (2018)), Shrestha et al. (2018), Niedziałkowska et al., 2019, Rahman et al. (2019), Schnitzler and Hermann (2019), Chaiyarat et al. (2020), Bhattacharya and Saharia (2021), Farah et al. (2021), Kusi et al., 2021, Parks Canada, 2022, Klich et al. (2023), Menon (2024), Zhao et al. (2024)

Table 4 Herbivore spatial reactions to carnivore favored habitats

Cattle	Actions
Gaur	Gaur distribution unaffected by tiger presence, tiger partly dependent on gaur presence; "declines of wild tiger populations are primarily a consequence of prey depletion driven by adverse human impacts" (Karanth et al., 2004: 4858). Leopards are usually not predators (Lamichhane et al., 2019). In some areas, gaurs are not in the tigers' diet (Biswas & Sankar, 2002), in others, a dominant prey item (Andheria et al., 2007a, b).
Banteng	"Dhole and banteng… showed a significantly negative spatial interaction" (Pudyatmoko, 2019: 191). "The extent to which leopards or banteng adjust their behavior in response to each other is unknown…" (Saisamorn et al., 2019: 78). Only male leopards in Cambodia kill banteng, "the only known leopard population whose main prey had adult weight greater than 500 k" (Rostro et al., 2018).
Water buffalo	Tigers and buffalo favor forests and mixed forests and grassland, mostly near water, but in encounters, large buffalo herds can defend themselves against tigers (Rauniyar, 2023; Bora et al., 2024).
Wild yak	Wild yaks are larger than domestic yak, and it is disputed if the wild yak is regularly predated by the predators in its range, namely the snow leopards, dhole, Himalayan black bear, and common leopard. The main predator, the snow leopard, is argued to take prey less than 180 to 250, which would be below adult wild yaks. Yaks are known to defend their young against predation (Schaller, 1977; Li et al., 2013; Suryawanshi et al., 2013; Hunter, 2015).
African Buffalo	"The preferred prey species for lions (buffalo, kudu and giraffe) avoided risky waterholes," and "buffalo avoided the hours of the day which are dangerous when the long-term and short-term risks of encountering lions were high" (Valeix et al., 2009b: 1483). In another study, "perceived predation risk seemed to play a role in the probability of drinking for all species except buffalo, roan and zebra". Also, "congregating in large herds is one of their many anti-predator adaptations. Large herds decrease the possibility of any individual being singled out by lion" (African Wildlife Foundation, 2024).
American bison	Grazes in areas ranged by wolves, bears, and cougars, with most adults able to hold their own in these habitat types: Grasslands and prairies and boreal forest in northern North America. Able to defend against predators.
European bison	Grazes in areas ranged by wolves, bears, and cougars, with most adults able to hold their own in these habitat types: Grasslands and forest habitats with low canopy cover, with few woody plants in the ground layer. Able to defend against predators.

Carnivore and herbivore social organization are important, interrelated factors in ecology (Campbell, 2017, Campbell & Jha, 2021, Jha & Campbell, 2021). Sources: Andheria et al. (2007a, b), Valeix et al. (2009b), Schaller (1977), Li et al. (2013), Suryawanshi et al. (2013), Hunter (2015), Rostro et al. (2018)), Lamichhane et al. (2019), Pudyatmoko (2019), Saisamorn et al. (2019), Rauniyar (2023), African Wildlife Foundation (2024)

Bshary et al. (2006) write that intraspecific hunting has drawn significant attention because cooperative behavior and its cognitive requirements are closely related. Consequently, behaviors that may distinguish between various levels of cognitive complexity have been the focus of species comparisons. These behaviors include deliberate communication between partners to start a joint hunt, assuming distinct, alternating, or consistent roles during a joint hunt, and the degree of food distribution following a successful hunt. Furthermore, during a hunt, individuals must play distinct roles for true coordination. When animals are able to distinguish between roles, they might take on roles that are more dangerous or have a lower likelihood of success for them personally. While some animals can act as pursuers, others can ambush their prey or obstruct its escape routes. Few predatory species are able to acquire this kind of coordination (Bshary et al., 2006).

For example, Stander (1992: 445) uses the example of group hunting among lions, where "group hunts generally involved a formation whereby some lionesses ('wings') circled prey while others ('centres') waited for prey to move towards them. Those lionesses that occupied 'wing' stalking roles frequently initiated an attack on the prey, while lionesses in 'centre' roles moved relatively small distances and most often captured prey in flight from other lionesses. Each lioness in a given pride repeatedly occupied the same position in a hunting formation. Hunts where most lionesses present occupied their preferred positions had a high probability of success" (Stander, 1992: 445). Scheel and Packer (1991: 697) also examined a case study of cooperative lion hunting and noted that "the extent of individual participation in communal hunts varied significantly." In this study, models were used to determine lion behavior within three strategies: "'refraining' (non-participation in hunts), 'conforming' (active participation in groups in which all individuals behaved similarly) and 'pursuing' (active participation in groups where individual behaviour varies)" (Scheel & Packer, 1991: 697). The results showed two main outcomes. Compared to female lions, male lions would hold back more and pursue prey less. Refraining was more common in group hunts when hunting easier prey. For example, lions were more likely to abstain when hunting less difficult warthogs than the more difficult buffalo and zebras. The findings suggest that refraining is "cheating," and that lions who employ this tactic are therefore taking advantage of their partners' hunting habits (Scheel & Packer, 1991: 697).

Group hunting by large groups of predators may also be compared to smaller groups, to determine the role of group size in hunting success, as some findings may suggest that "despite the popular view that social predators live in groups because group hunting facilitates prey capture, the apparent tendency for hunting success to peak at small group sizes suggests that the formation of large groups is unrelated to prey capture" (MacNulty et al., 2012: 75). MacNulty et al. (2014) contend that "an intriguing aspect of social foraging behaviour is that large groups are often no better at capturing prey than are small groups, a pattern that has been attributed to diminished cooperation (i.e., free riding) in large groups" suggesting that "the formation of large groups is unrelated to prey capture." The case study was of wolf predation on bison in Yellowstone National Park, compared with a similar study of wolf predation on elk. Bison is described as the "most formidable prey," on average three times more difficult to kill than elk (MacNulty et al., 2014). The findings indicated that "improvement in elk capture success levelled off at 2–6 wolves" and "bison capture success levelled off at 9–13 wolves with evidence that it continued to increase beyond 13 wolves," and "these results are consistent with the hypothesis that hunters in large groups are more cooperative when hunting more formidable prey. Improved ability to capture formidable prey could therefore promote the formation and maintenance of large predator groups, particularly among predators that specialize on such prey" (MacNulty et al., 2014).

MacNulty et al. (2012: 75) also present a study of wolves and elk in Yellowstone National Park that found that "the relationship between success and group size is indeed nonlinear and that individuals withholding effort (free riding) is why success does not increase across large group sizes." In this case, hunting success did not increase beyond four wolves "and individual performance (a measure of effort) decreased for reasons unrelated to interference from inept hunters, individual age, or size" (MacNulty et al., 2012: 75). Individual differences among the hunting wolves were discerned, as hunting performances appeared to decline faster among those wolves that might have a reason to lack strong motivation, such as having no cubs, or sharing the hunt with highly skilled hunters, or having more risky tasks. The authors note that these findings suggest that "decreasing performance was free riding and that was why success leveled off in groups with >4 wolves that had superficially appeared to be cooperating," and this provides evidence that "nonlinear trends in group hunting success reflect a switch from cooperation to free riding" and "how hunting success per se is unlikely to promote formation and maintenance of large groups" (MacNulty et al., 2012: 75).

In many cases, group cooperation in hunting yields success, especially for smaller carnivore predation on large herbivores, that would otherwise be relatively inaccessible to solitary smaller carnivores (Clements et al., 2014). For example, Pudyatmoko and Sabarno (2007): 854) cite the example of dhole pack predation on banteng, arguing that in the case study of the Baluran National Park, there was an extreme decline in banteng numbers (126 to 15 over 4 years); according to the data, the population decline was caused by high rates of predation by dholes. The best strategy to reduce the possibilities of banteng extinction is argued to be the reduction of the average size of the dhole pack, which would influence the dholes to hunt smaller prey. The authors, however, note that both banteng and dholes are listed as endangered species, and the reduction of the size of packs would entail the killing of dholes. Consequently, there had been no action to kill dholes in the case study of the Baluran National Park. In another study in Ujung Kulon National Park, Indonesia, Rahman et al. (2021: 20278) found photographic evidence of dhole predation on a young banteng and other species, which suggested that dholes in packs to predate on banteng and may separate young from adults when attacking the young. It is argued that "future research should examine the Dhole diet and interspecific relationships between dhole and banteng to gain a better understanding of the ecological impacts of endangered predators on endangered prey in UKNP" (Rahman et al., 2021: 20278).

Solitary hunting is also practiced by large carnivores, including all big cats except the lion, and bears (Campbell, 2017). Predation risk from solitary hunters may vary at the landscape and landcover levels. For example, Miller et al. (2016a, b) compare the predation patterns of two top solitary predators, tigers and leopards, in relation to livestock and human presence. Both tigers and leopards are solitary ambush predators but may vary their predation strategies according to landcover features, prey type, and human presence and proximity (Campbell & Jha, 2021). Miller et al. (2016a, b) found that tiger and leopard predation strategies and human perceptions were remarkably similar. However, whereas the risk posed by leopards was highest in areas with open vegetation, the risk posed by tigers to livestock was highest in areas with dense forests and at moderate distances from human activity. Individuals correctly predicted increased threat in areas with high observed risk values for both carnivores and correctly perceived spatial differences between tiger and leopard hunting patterns (Miller et al., 2016a, b: 1).

Miller et al. (2016a, b; 5–6) found that the tiger and leopard solitary hunting methods were associated with different landscape features, which formed different spatial manifestations of predation risk. Tigers preyed most heavily on livestock close to moderately and extremely dense forests and at intermediate distances from highways, settlements, and wilderness. There was a threshold relationship between the relative probability of a kill and the distance to roads, villages, and shrubland. The risk peaked at 1.2 km from the road, 1.1 km from the village, and 8.0 km from the shrubland. The landscape's relative predicted tiger risk levels ranged generally in the low to medium risk categories. By contrast, leopards preyed most heavily on livestock in the vicinity of shrublands, open, moderately dense forests, and away from bodies of water. Regarding leopards, most of the landscape, including human-dominated areas with villages and agricultural fields, had a very high relative risk. The authors noted that "tiger and leopard models correctly identified 63% and 85% of known validation kills, respectively, which is significantly greater than would be expected by random chance" (Miller et al. (2016a, b): 6).

Interestingly, Miller et al. (2016a, b: 7–8) found that the livestock owners perceived attack risk from tigers and leopards as remarkably similar to the modeled, observed risk from both predators. For example, owners perceived the lowest risk of predation by tigers in the village surroundings, with a gradual increase in risk in all land use categories with denser forest and less human infrastructure. These opinions described a gradient that matched the spatial pattern of modeled, observed tiger attack risk. These owners also perceived high risk of predation by leopards in all land use categories, including agricultural fields and forests, both of which were rated as at high risk of predation, and villages and agricultural field-forest edges were rated as at extremely high risk of predation (Miller et al., 2016a, b: 7–8). In evaluating the implications of such perceived and real differences in the solitary tiger and leopard hunting techniques, Miller et al. (2016a, b) conclude that tigers and leopards use similar stalking methods but different habitat ranges, with tigers mostly exploiting dense forests away from human presence and leopards foraging over a wider range of open vegetation and human-dominated areas. Tigers are termed narrow domain hunters and leopards are termed broader domain hunters. Prey animals tend to be cautious of risk from ambush predators and may alter their foraging habitat to avoid narrow domain predators, but change their habitat or activity (space or time) to avoid the more flexible, broad-domain predators (Miller et al., 2016a, b).

Solitary hunters may also vary their hunting activities in relation to their size and that of prey species, with smaller hunters targeting smaller prey than the larger predators (Campbell, 2017; Campbell & Jha, 2021). Bargali and Ahmed (2018) argue that "prey selection in large carnivore species is based on body size, tigers prefer larger livestock whereas, leopards prefer small sized livestock prey." The case study is of the areas in and around Corbett Tiger Reserve, Uttarakhand, India. The tigers mostly killed buffalos (33.5%) and the smaller cows (66%), while the leopards took mostly cows (95%). This point is supported in a study by Rasphone et al. (2022), who present a comparative study of tigers and another solitary predator, the smaller clouded leopard, which with the "longest relative canines and largest gape of any extant felid species" can "kill large prey with a powerful nape bite" (Rasphone et al., 2022; see also Rabinowitz et al., 1987; Grassman Jr. et al., 2005; Christiansen, 2006), and may be "best adapted to take large ungulate prey" (see Pocock, 1939), despite more commonly taking primates, rodents, pangolins, and small ungulates (see also Nowell & Jackson, 1996; Grassman Jr. et al. 2005; Matsuda et al., 2008; Morino, 2010; Gray et al., 2021; Sunderland-Groves et al., 2020).

Carnivores may either avoid very large herbivores or attack smaller members of the species, mostly females, juveniles, or infants (Bhumpakphan, 2008; Zangmo et al., 2018). For example, in some cases, tigers target female and young gaurs over adult male, due to the smaller size of the former animals (Schaller, 1967; Karanth & Sunquist, 1995a, b; Ashokkumar et al., 2011). Schaller (1967) noted that predation accounted for half of gaur calf mortalities, but reports on general predation range from 1.87 to 30.4 (Zangmo et al., 2018). Ashokkumar et al. (2010a, b): 31–32) contend that "carnivore studies in Nagarahole based on scat and kills has shown that among different age-sex classes of gaur preyed up on by tiger, young ones (calves and juveniles) constituted a greater proportion (58.8%), than adult females (22.6%) and adult males (14.6%). Thus, tigers selectively preyed upon young gaur and the variation in population age-structure could influence prey selection." Leopards and wild dogs prey less on gaur due to the size differences (Schaller, 1967; Karanth & Sunquist, 1995a, b; Ashokkumar et al., 2011). Ghaskadbi et al. (2022: 14029) noted that for dholes, "there was an evident avoidance of species like gaur", and "sambar was the most preferred prey of dholes followed by chital and chausinga. Gaur, nilgai and wild pig were the least preferred prey." Citing actual percentages of total prey, "overall, dholes depended predominantly on large-sized prey like sambar (61.08%) and chital (19.08%), followed by gaur (4.02%) and nilgai (4.68%)" and "strongly avoided large, challenging prey like the gaur, nilgai and wild pigs" but also hunted calves of gaur and banteng (see also Rahman et al. (2021).

Rasphone et al., 2022 cite a study of tigers and clouded leopards both taking serow, as this species was the only prey animal selected by both clouded leopards and tigers, while the other ungulate species were consumed in proportion to their presence or even avoided. Rasphone et al. (2022) note that tigers vary in the size of their prey, citing published research indicating that tigers kill mostly larger prey such as gaur, banteng, serow, and sambar (see Pakpien et al., 2017; Simcharoen et al., 2018) and other research that describes tigers killing easier, smaller prey like wild pig and northern red muntjac (see Prommakul, 2003; Rabinowitz, 1989). Here, large prey is described as heavier than 150 kg, while smaller prey is generally below 30 kg in weight (Rasphone et al., 2022). Where the tiger killed the suitably sized (110–160 kg, within the tiger's preferred prey size of 0–250 kg, see Francis, 2008, Hayward et al., 2012a, b) serow, it was considered rare and rarely recorded. It has been hypothesized that tigers favor serow because it is a solitary species with fewer defensive tools than sambars and wild boars. Therefore, compared to wild boars and sambars that live in groups, serow may be easier to track and kill. Furthermore, if tigers had adopted a hunting strategy that took advantage of the rougher and steeper terrain that serow use, they might have been easier to hunt and kill than sambar and wild boar (Rasphone et al. (2022).

Possibly surprisingly, the much smaller clouded leopard (11–23 kg) was recorded taking large prey, weighing 50 to 150 kg, including serow and wild pig (Rasphone et al., 2022). Rasphone et al. (2022) note that "serow might even have influenced the activity patterns of clouded leopards" as studies have indicated that "clouded leopard had high activity overlap with serow, compared to lower activity overlap with northern red muntjac (0.65), squirrels (0.63), wild pig (0.60), birds (0.53), and 4 macaque species (0.41–0.52)" (see also Rasphone et al., 2020) and that "felids often synchronize their activity to that of their main prey" (see also Foster et al., 2013; Nagy-Reis et al., 2019; Yang et al., 2018). Importantly, "Overall, our results support the hypothesis by Pocock (1939) and Christiansen (2006) that the elongated canines, large gape, and stocky build of the clouded leopard are adaptations to regularly take prey much larger than their own body size" (Rasphone et al., 2022).

In another similar study, leopards in Eastern Cambodia were recorded as "part of the only leopard population in the world to prey predominantly on an animal weighing more than 500 kilograms—the banteng" (Rostro et al., 2018). In this study, the larger male leopards (sometimes 30% larger than female leopards) took banteng (62% of the biomass consumed) while females took muntjac (56% of the biomass consumed) and appeared to avoid banteng, which were less than 1% of the biomass consumed. The authors note that the predation on banteng might be attributed to the local extinction of tigers, as the latter cats hunted the banteng, which was its primary prey, went extinct in the area, making leopards the new top predator. According to earlier studies, leopards avoided tiger encounters by consuming smaller prey when tigers were in the area. The largest herbivore, the banteng, which the leopards of eastern Cambodia probably used to avoid when tigers were around, became part of their prey (Rostro et al., 2018).

Leopards elsewhere may occasionally take young domestic or wild buffalo but may avoid larger prey (Campbell, 2017). For example, Hayward et al., 2006a write that leopards may take prey as large as giraffe calves and adult male eland (see also Kingdon, 2015; Scheepers & Gilchrist, 1991), but generally prefer medium and small-sized antelopes, such as impala, bushbuck, common duiker, black-fronted duiker, red forest duikers, chital, and water chevrotain, while avoiding buffalos and guar and also elephant, black and white rhinoceros, hippopotamus, Cape buffalo, giraffe, eland, plains zebra, ostrich, blue wildebeest, topi/tsessebe, baboon and hartebeest, and probably, grysbok, mountain reedbuck, roan, sable, springbok, and waterbuck (Hayward et al. (2006a). A study by Balme et al. (2017: 638) in Sabi Sand Game Reserve, South Africa, noted that "the largest kill hoisted by a leopard in our study was a juvenile African buffalo Syncerus caffer (≈166 kg)." Henschel et al. (2005: 21) noted in a study of leopards in Lopé National Park, Gabon, Central Africa, "the most important single prey species was found to be red river hog Potamochoerus porcus (Linnaeus), making up 20% of the biomass consumed, followed by forest buffalo Syncerus caffer nanus (Boddaert) and cane rat Thryonomys swinderianus (Temminck), each comprising 13% of biomass consumed."

Kittle et al. (2017: 78) also suggest that the Sri Lankan leopard may eat native water buffalo. In the stool analysis, water buffalo remains were present in the stool in expected quantities given their availability; all buffalo carcasses came from young animals or calves. Interestingly, despite the similar size of the male sambar and the adult female buffalo, the latter were much less predated than the former. Reasons for this include the gregarious nature of buffalo and their preference for open habitats, combined with effective anti-predation behavioral strategies that force leopards to prey only on calves, the most vulnerable age group (Kittle et al. (2017: 81). Miller et al. (2016a, b: 17) also cite a study of leopards and tigers in the Kanha Tiger Reserve in central India, where both big cats killed cattle and buffaloes, but while tigers killed more older cattle and buffaloes, leopards killed more young cattle and buffaloes.

Smaller, solitary predators may also adapt their hunting strategy to avoid larger, solitary predators. For example, Rostro et al. (2015) studied lions and cheetahs in the Phinda Private Game Reserve, South Africa, and found that cheetah habitat choice was determined by a trade-off between resource acquisition and lion avoidance and that the balance of this trade distance varied by scale: more open habitats with high prey densities were positively selected within home ranges, whereas more closed habitats with low prey densities were positively selected as killing sites. Cheetahs are capable of killing young buffalo (Pienaar, 1969). Killings of buffalo by cheetahs were proportional to frequency compared to other species (Hayward et al., 2006b). Hayward et al. (2006b: 615) found that the captured young buffalo can have a similar weight to the captured antelope. Cheetahs kill the most available prey at a site with a body weight of 23–56 kg, with the maximum value (mode) being 36 kg. Blesbok, impala, Thomson's and Grant's gazelles, and springbok are clearly preferred, while prey outside this range are generally avoided.

Regarding the individual and group-based strategic defenses by wild herbivores (see Table 2), partly related to the sizes of predators and prey species, research identifies several factors and issues (Saltz et al., 2023). Saltz et al. (2023) note that there is a lack of understanding of the ecological and social formative factors for herbivore group formation. The case study was of six species of mammalian grazers in Serengeti National Park, and the findings indicated that ecological factors, specifically forage quality, impact the probability of species overlap at small spatiotemporal scales. In contrast to resident species (buffaloes, hartebeests (*Alcelaphus buselaphus* Pallas, 1766), and topi (*Damaliscus lunatus* Burchell, 1824)), migratory species (gazelles Genus *Gazella* Blainville, 1816), wildebeests (Genus *Connochaetes* Hinrich Lichtenstein, 1812), and zebras (Family *Equidae* Gray, 1821)) were more likely to have heterospecific partners available at sites with higher forage quality (Saltz et al., 2023). The results of this study demonstrated that there are conditions under which mixed groups can form. An increased likelihood of mixed group formation was linked to greater heterospecific availability in gazelles, hartebeests, wildebeests, and zebras; however, these patterns were not influenced by ecological variables. According to the findings, social and ecological factors that influence group formation differ in their spatiotemporal scales and are species-specific (Saltz et al., 2023).

Regarding the similarity and differences between the habitats of large carnivores and wild herbivores, issues emerge that define the ecology of both groups of mammals, such as vegetation, surface water, topography, and human activity (Campbell, 2017). Carnivores may or may not adapt to the environments favored by prey species, but prey abundance may influence carnivore presence, and human presence and modifications may play a role (Campbell, 2017, Campbell & Jha, 2021). Although carnivores and herbivores may share similar habitats, their interrelations may result in local avoidance by herbivores and consequent adaptation by carnivores. For example, concerning Indian ecosystems, Jornburom et al. (2020) contend that "determining prey distribution provides a logistically feasible first step to estimate the extent of tiger habitat and potential viability of individual tiger populations, by providing an estimate of the land base that will support isolated populations." In this study, despite the similarities between tiger habitats and that of their main prey species, gaur, banteng, and sambar, the main parameters for the distribution of these herbivores included avoidance of village proximity, including associated activities such as subsistence poaching, bamboo cutting, and cattle grazing (see also Lynam et al. 2012). Gaur were found to favor higher elevations and evergreen forest, which generally lacked human presence (Jornburom et al. 2020). Banteng favored lower elevations, but lower elevations attracted human activities and consequently this species incurred losses. As tigers may also favor the habitat favored by these large ungulates, the presence of ungulates may determine the presence of tigers: "The presence of tigers is strongly influenced by the distribution and abundance of their major ungulate prey" (Jornburom et al. 2020).

In a case study of lions and buffalo, a similar scenario exists, as both may favor mixed savanna landcover, in some cases with water proximity, especially in the dry season, and such environmental parameters may influence their relationships (Davidson et al., 2013; Roug et al., 2020; Schooler et al., 2022). For example, Davidson et al. (2013) used a case study of Hwange National Park, Zimbabwe, "a semi-arid African savanna structured by artificial waterholes" and found that in all seasons, buffalo were the most frequently utilized prey by both male (56%) and female (33%) lions, and they also made up the largest portion of the lions' diet biomass. Throughout the year, smaller ungulates were also taken, especially by female lions, but buffalo were the species that was most heavily targeted. Lions significantly favored killing most prey within 2 km from the waterholes; these findings suggest that the presence of surface water influences lion foraging behavior by forming passive traps (Davidson et al., 2013).

Leopards, which are highly adaptive cats that range in habitats from semi-desert to evergreen rainforest, share the central African forest and mixed savannas with the forest buffalo, the only cattle species in the African forest (Henschel et al., 2005). Henschel et al. (2005) found that leopards preyed on forest buffalo as well as smaller antelopes, primates (including adult gorillas), rodents, and even leopard and golden cat remains, and all the buffalo remains detected through the study of scats were juveniles. The most important prey species were red river hogs (19.7%), forest buffalo (13.0%), cane rat (12.6%), and red duikers (9.7%) (Henschel et al., 2005: 25). The authors note that "the high proportion of forest buffalo in the diet of leopards in this study was surprising given that in most other studies where leopards and forest buffalo overlap, buffalo remains have not been identified in scat" (Henschel et al., 2005: 25). Studies cited included Hart et al. (1996), Ososky (1998), and Ray and Sunquist (2001), with Hoppe-Dominik (1984) and Fay et al. (1995) citing leopard attacks on leopards on young buffalo. It is argued that "the high proportion of buffalo preyed upon in this study could be explained by the availability of buffalo in the savanna areas in Lope," compared with the low densities that buffalo usually attain in continuous forest (Henschel et al., 2005: 25; see White, 1994; Tutin et al., 1997).

A study by Karanth and Sunquist (2000) in the forests of Nagarahole, southern India, showed that tigers, leopards, and dholes matched their hunting presence with the activities of prey species, and leopards and dholes did not match their foraging to avoid tigers. Tigers, leopards, and dholes frequented the same habitat, despite a tendency for tigers to take prey in denser vegetation than leopards, and both carnivores generally foraged near habitat features that attracted herbivores. Karanth and Sunquist (2000: 255) conclude that there was no evidence of interspecific spatial exclusion of predators due to habitat specificity or social dominance behavior. The results suggest that ecological factors such as sufficient availability of appropriately sized prey, dense cover, and high tree densities may be the main factors structuring the predator communities of tropical forests.

Regarding the herbivore reactions to carnivore-favored habitats, which is partly related to the size and power of the carnivores and herbivores involved, and the possibilities of predation, there are interspecies differences. Both carnivores and herbivores may populate similar environments, but their avoidance and predatory activities may influence temporal and spatial habitat selection, even in mixed savanna type environments, where bush, tree, and grass density are mixed but may vary spatially Campbell, 2017). A study by Valeix et al. (2009b) argues that predators, by their spatial behavior, may influence the distribution of herbivores, taking the example of ungulates (grazers including buffalo, zebra, wildebeest, roan, and sable and browsers including giraffes, kudu, and impala) and lions in Hwange National Park, Zimbabwe, with two main types of interactions: "long-term risk of predation (likelihood of lion presence calculated over 4 months) and short-term risk of predation (actual presence of lions in the vicinity in the preceding 24 hours)" (Valeix et al. (2009b: 23). The findings indicated that the long-term risk of lion predation affected the distributions of browsers but not grazers, and herbivores used more open habitat within sight of lions, "probably an effective anti-predator behavior against such an ambush predator" (Valeix et al. (2009b: 23). The difference between grazers and browsers was noted: "Whereas the distributions of all browsers across the landscape were negatively associated with the long term risk of lion presence (as assessed by our proxy), those of grazers did not" (Valeix et al. (2009b: 29). One possible explanation was that the vegetation of the case study was mostly woodland and bushland with little open grassland, with some of the latter associated with waterholes, which would leave fewer foraging areas for grazers. The grazers' antipredator activity would be centered on the sighting of lions, as general avoidance of such areas would not be possible; the browsers were less water-dependent, and buffalo may generally avoid bushland and favor open areas where lions are more visible (Valeix et al., 2009b).

Within similar foraging guilds, such as grazers, there may also be variable habitat selection, which may vary from or parallel carnivore frequented habitats, based on the size of the herbivores and the consequent threat level from carnivores (Campbell, 2017). Hopcraft et al. (2012: 201) cite a study of the role of herbivore body size in relation to predation threat and food availability among Serengeti grazers, examining the hypothesis that "predation potentially affects populations of small grazers more than large grazers, because predators have difficulty capturing and handling large prey" and "small grazers are regulated by the digestive quality of grass, while large grazers extract sufficient nutrients from low-quality forage and are regulated by its abundance instead." The study selected five variably sized (small, intermediate, and large) mammal grazers and determined their responses to gradients of predation risk and food availability (Hopcraft et al., 2012: 201). The findings indicated that the distribution of the largest grazers, African buffalos, was associated with forage availability and quality rather than predation risk, while the distribution of the smallest grazers, Thompson's gazelles, was associated with predation risk rather than forage availability and quality. The distribution of medium-sized grazers, namely, the hartebeest and topic, was associated with both high-quality forage and lower predation risk. The authors conclude that "top-down (vegetation-mediated predation risk) and bottom-up factors (biomass and nutrient content of vegetation) predictably contribute to the division of niche space for herbivores that vary in body size" (Hopcraft et al., 2012: 201).

The remainder of this chapter examines these relationships between wild cattle and large carnivores in more detail on a continental basis. Asia, Africa, North America, and Europe are cited. South America is not included in this study because, despite the introduction of domesticated water buffalo into the Americas and predation by jaguars and cougars, the domesticated water buffalo will be considered in later chapters on domesticated cattle (Azevedo et al., 2010). This develops on the introduction and examines recent research on the status of wild cattle and their relations with carnivores capable of killing adult to young cattle, and the research methods that have influenced more critical findings, including geomatics-based methods and predictive modelling for conservation biology. The aim is to compare these relationships across continents and identify trends and knowledge gaps that may be relevant to the conservation biology of wild cattle. The relevance of all findings related to wild cattle and carnivore relationships is also examined in the context of domestic cattle ecology. The vulnerability of wild cattle to carnivores is extremely relevant to the ecology of domestic cattle, as they are generally less strong and aggressive and therefore may rely heavily on human protection. Large carnivores can also be heavily protected by human actions, an issue that can be at odds with the protection of wild and domestic animals. The topics covered in this chapter provide a background for the study of domestic cattle and carnivores, as well as relevant research methods in the following chapters.

In addition to the four relations mentioned in the introduction, namely hunting strategies and methods large carnivores, the defenses by wild herbivores, the similarity and differences between the habitats of large carnivores and wild herbivores; and the herbivore reactions to carnivore favored habitats, other issues include the role of different human activities in the relationship between cattle and carnivores, and the information and research methods that support conservation actions (including geomatics, statistics, and social research), and the relevance of developments in what may be termed Big Government (increased, tax-sourced spending on conservation), Big Business (the involvement of private business finance in conservation management), Big Tech (social media platforms and associated networks) and Big Data (the development of methods to manage, otherwise difficult vast datasets, otherwise difficult) to conservation of wild cattle, carnivores and relevant ecosystems. These issues may be particularly relevant to Asia, probably the most rapidly developing continent, in terms of technology, economics, and social issues.

Asian Carnivores and Wild Cattle

Of the 12 wild cattle species found around the world, nine are found only in Asia, and all are threatened with extinction, but are vital actors in the local Asian ecosystems. The dominant predators of adult wild cattle in Asia are tigers, brown bears, and, to a lesser extent, leopards, wolves, and Asiatic black bears which may prey on calves (Campbell, 2017). Asia, extending from Asiatic Russia to Southwest Asia (sometimes termed the Middle East) to eastern Siberia southwards to Southeast Asia, has the largest number of large and medium-sized carnivore species capable of killing adult and/or juvenile cattle and calves. Tigers, leopards, snow leopards, clouded leopards, brown, black, and sloth bears, wolves, dhole, and striped hyenas may all kill calves and, in some cases, adult female or juvenile cattle (Roth, 2004; Rajaratnam et al., 2016; Khaewphakdee et al., 2020). Asia is also the only continent with more than one species of wild cattle, namely gaur, banteng, wild water buffalo, wild yak, kouprey, saola, tamaraw, and mountain and lowland anoa.

The Tiger and Wild Cattle

The study of cattle and large carnivores must begin with the tiger, which is the largest and most powerful predator, and the only carnivore that regularly kills all the wild cattle species, from the largest (gaur) to the smaller species (anoa and tamaraw). The much rarer Asiatic lion and the smaller leopard may rarely take large adult wild cattle but mostly kill younger, smaller animals. The tiger is the most common predator of the world's largest bovine species, the gaur, banteng, and wild water buffalo (Humphrey & Bain, 1990; Seidensticker & Mcdougal, 1993a, b; Hayward et al., 2012a, b). Variar et al. (2023) report seven tiger prey species in their study of the Western Ghats in India. Gaur with sambar (Rusa unicolor Kerr, 1792) each accounted for 26.33% of the tigers' diet, while chital (*Axis axis*, Erxleben, 1777) accounted for 63.15%, and together these species accounted for 95% of the total diet. As noted by Roth (2004), tigers are, after people, the only predator that takes adult water buffalo, though water buffalo are difficult even for tigers. Water buffalo may use organized formations to confront tigers, which is usually a successful defense, and water buffalo may even kill tigers with their horns. Tigers may be successful when attacking single adults or juveniles.

Khaewphakdee et al. (2020: 5152–5159) took the novel approach documenting the average size of gaur and banteng killed by tigers in Thailand, and the findings indicated that 45.2% of gaur killed were adults, while 55.7% of banteng were adults. The average weights were 423.9 kg for banteng and 397.9 kg for gaur. The study argues that "the mean weight of both prey species is 3.5-4.5 times greater than the predicted 1:1 preferred prey to predator ratio. In the absence of medium-sized prey, killing these larger animals may be especially critical for female tigers provisioning nearly independent young when male offspring are already larger than the mother" (Khaewphakdee et al., 2020: 5152). The authors further note that "because adult male gaur are 1.3 times larger than adult male banteng, and males of both species are >3.5 times larger than Hayward et al. (2012a, b) optimum 1:1 predator to prey ratio, we hypothesize that these species, and especially the large size classes, are approaching the maximum size limit of tiger prey" (Khaewphakdee et al., 2020: 5154).

The authors note that this hypothesis would be supported by the fact that male tigers may kill a lower percentage of larger adult male gaur than smaller adult banteng, while adult female tigers, which average two-thirds the weight of adult male tigers, kill a lower percentage adult males of both species compared to male tigers (Khaewphakdee et al., 2020: 5154). The findings were that gaur kills comprised adult males (15.9%), adult females (29.3%), and calves (39%), and banteng kills included adult males (29.1%) and adult females (26.6%) and calves (26.6%). The result was that although gaur males were about 1.3 times heavier than male banteng, and gaur females were 1.1 times heavier than female banteng, the average weight gaur killed (397.9 kg) was lower than the average weight of banteng killed (423.9 kg) (Khaewphakdee et al., 2020: 5156).

The capabilities of tigers in killing extremely large prey were illustrated. Adult gaur kills averaged 737.8 kg in weight, and adults accounted for 83.7% of the weight of gaur killed by tigers. Adult banteng kills averaged 652.2 kg, and adults accounted for 85.6% of the weight of banteng killed by tigers. Male tigers killed adults (gaur 48.8%, banteng 79.4%) as did female tigers (gaur 41.1%, banteng 37.8%). The results of this study arguably refuted the logical argument that tigers would kill smaller, easier to kill prey, while avoiding large adults, and that the large gaur and banteng are almost at the size limit for tiger predation. Based on this hypothesis, it would be expected that male tigers would kill fewer adult gaur than adult banteng, but the findings indicated no significant difference between male tiger kills of adult gaur and banteng. Another hypothesis would be that the smaller female tiger would kill fewer adult males than adult female gaur and banteng, but no significant relations were found. In addition, despite their smaller size, male tigers did not kill significantly more adult gaur and banteng than female tigers did. The authors note that "given that our three predictions were not supported by significant results, our hypothesis that gaur and banteng are near the upper size limit of tiger prey is not confirmed" (Khaewphakdee et al., 2020: 5156).

Prey availability is postulated as possible explanation for tiger predation on such large animals (Khaewphakdee et al., 2020). Tiger predation on gaur and banteng is higher in Southeast Asia (including Thailand) than in South Asia (including India), possibly because of the relative absence of medium-sized prey species, including the chital (*Axis axis* Erxleben, 1777), with a distribution mostly within India, which may force tigers to hunt much larger prey, which means a higher biomass of prey is taken in Southeast Asia (over 1.39 times greater on average, and larger gaur specimens are taken in Southeast Asia) (Khaewphakdee et al., 2020). Regarding prey size, Chakrabarti (2016:660) make the argument, largely opposite to the findings of Khaewphakdee et al. (2020) that "prey digestibility and utilization combined with escalating handling time and risks of killing large prey make prey larger than the predator size unprofitable for obligate carnivores," and since tigers may reach satiation before totally consuming large carcass, and larger prey may pose a greater physical risk, such felids may prefer prey of their weight or less. Hayward and Kerley (2005) counter that another possibility, as noted partially by Khaewphakdee et al. (2020), that tigers simply kill what prey is available.

A comparison of the prey of tigers and leopards and other smaller predators shows that the size of prey is a significant factor (Jha & Campbell, 2022). For example, Andheria et al. (2007a, b): 169) cite a comparison of tigers, leopards, and dholes, in Bandipur Tiger Reserve, India, and documented a high dietary overlap especially of ungulates, but "the largest ungulates, gaur and sambar, provided 73% of biomass consumed by tigers, whereas medium-sized chital *Axis axis* and wild pig *Sus scrofa* formed 65 and 83% of the biomass intake of leopards and dholes, respectively." The chital, the most abundant prey "dominated their diets (tiger=33%, leopard=39% and dhole=73%)," and the authors conclude that the "results of the study, in conjunction with earlier work, support the prediction that abundance of ungulate prey species, as well as their availability in different size classes, are both critical factors that facilitate sympatry among the three predators" (Andheria et al., 2007a, b: 169). It is also argued that regarding the taking of large herbivores, "the presence of large ungulate species in the prey assemblage is thus more important for tigers than for leopards and dholes. The tiger's larger size (120–270 kg) enables safe capture of large and potentially dangerous prey such as adult gaur and sambar and possibly explains this pattern" (Andheria et al., 2007a, b: 174).

Other studies have further enhanced the important role of large, wild cattle in the diet of tigers (Karanth & Sunquist, 1992; Narasimmarajan et al., 2014; Jha & Campbell, 2022). For example, Ramesh et al. (2012: 269) examined spatial and temporal partitioning and high dietary overlap among sympatric tigers, leopards, and dholes in Mudumalai Tiger Reserve, Western Ghats, India during 2008–2010, and found that tigers were spatially correlated with gaur and sambar, while leopards were spatially correlated with most prey species, with no avoidance of tigers, and dholes associated with chital and moderately avoided felids. Dhole presence correlated with gaur, but predation on gaur was rare due to its size, despite the social hunting method of the dhole. A similar study by Krishnakumar et al. (2022: 141) on the same predators, tigers, leopards, and dholes, in Kalakkad-Mundanthurai Tiger Reserve, India, found that in an area of limited habitat and food resources, tigers favored gaur, while leopards and dhole favored smaller prey. Similar to the studies already cited, gaurs were the main tiger prey (75% of biomass), the smaller prey species for tigers, leopards, and dhole including Indian chevrotain, sambar, and Nilgiri langur, with the last two predator having the highest dietary overlap. With tiger preying on the huge bovids, the main competition is between the leopards and dholes over chevrotain and sambar (Krishnakumar et al., 2022).

Duangchantrasiri et al. (2019: 2449) go further in predicting the presence of tigers by the presence of gaur, banteng, and sambar, in a case study of Southeast Asia, within Thailand's Western Forest Complex (WEFCOM). The findings indicated that tiger presence "was primarily influenced by the combined presence of all three large prey species, and 37% or 5,858 km^2 of the landscape was predicted to be occupied by tigers" (Duangchantrasiri et al., 2019: 2449). The authors conclude that this provided significant information on the distribution of tigers in WEFCOM and detected factors that may limit tiger distributions, which may inform attempts to expand tiger distribution and recovery in Southeast Asia. This method of predicting tiger presence with studies of larger bovids is helpful, as numerous studies have indicated a tiger preference for gaur and banteng. For example, Seidensticker and Mcdougal (1993a), b: 105) note that although tigers kill available prey including smaller prey, they may track and kill larger ungulates, thereby accessing the mammalian biomass that is supported by few animals.

In some areas where tigers coexist with gaur and other large bovids, the tigers may also favor smaller prey and kill large bovids less often. For example, Variar et al. (2023) give the example of tiger diets in the reserve forests of Begur Range, in the Malenad–Mysore Tiger Landscape (MMTL) of the Western Ghats in India. The tigers' diet by percentage comprised chital (63.15%), gaur (26.33%), and sambar (26.33%). The study detected small prey species, including Indian crested porcupine (*Hystrix indica*) and small Indian civet (*Viverricula indica*), described as examples of "opportunistic hunting behaviour and diverse prey selection by tigers" (Variar et al., 2023). Karanth et al. (2004: 4854) cite a study of tigers, using a "simple mechanistic model" for the prediction of tiger density as a function of prey density, based on "11 ecologically diverse sites across India" (Karanth et al., 2004: 4854). The results provided evidence of a functional relationship between the abundances of large carnivores and their prey under a wide range of ecological conditions. The principal ungulate prey of tigers in decreasing order of the number of sites were wild pig (11 sites), sambar (10 sites), axis deer (9 sites), gaur (7 sites), muntjac and four-horned antelope (6 sites each), nilgai (5 sites), barasingha and chinkara (2 sites each), and wild water buffalo and hog deer (1 site each).

Gaur and banteng may also live alongside tigers in certain situations without recorded predation. For example, Biswas and Sankar (2002: 411) give a case study of Pench National Park, Central India, a region with a high ungulate population, mostly chital and sambar, and large numbers of common langur. Tiger diet mainly comprised chital (47.3%), sambar (14.5%), and wild pig Sus scrofa (10.9%), with an average. The authors note that "if there is food choice, tigers seem to kill medium- and large-sized species more often. Wild pig and sambar were consumed more than their availability, whereas chital were taken in proportion to their availability" but that gaur was not detected in the tigers' diet. Reddy et al. (2004: 384) also cite a study of tiger diets in Nagarjunasagar Srisailam Tiger Reserve, India, where the commonest mammals predated were wild boar, chital, and sambar, with low mean mass of 56.3, which was lower than in many comparative studies. The authors concluded that "in tropical dry deciduous forests with low natural prey density, smaller prey species, and high livestock biomass, tigers preferentially kill smaller prey and generally avoid livestock predation" (Reddy et al., 2004: 384).

The Asian Leopard and Wild Cattle

The role of leopards in wild cattle predation is disputed, as leopards are smaller than tigers and may avoid larger bovids, and unlike lions, they are solitary hunters, and this further restricts their predation capabilities (Jha & Campbell, 2022; Baral et al., 2024). As cited in the introduction of this chapter, the Indochinese leopard in Srepok Wildlife Sanctuary (SWS) of eastern Cambodia is recorded as the only leopard recorded as regularly preying on large bovids, in this case the banteng (Rostro et al., 2018). Leopards killed 13 species, with banteng comprising 42% of the biomass consumed, was banteng, "making this the only known leopard population whose main prey had adult weight greater than 500 kg" (Rostro et al., 2018). The large wild pig comprised 22% of biomass consumed, a comparatively high figure compared with other studies (see Karanth & Sunquist, 1995a, 1995b; Ramakrishnan et al., 1999; Hayward et al. 2006a, b, c; Ghoddousi et al., 2017). Interestingly, while male leopards took mostly banteng, at 62% of biomass consumed, and muntjac less often (7% of biomass consumed), female leopards mostly consumed muntjac (56% of biomass consumed) and rarely killed banteng (under 1% of biomass consumed). The authors contend that these findings are among the first records of leopard killing and eating banteng, and as adult banteng can reach weights of 800 kg, these leopards are "the only known leopard population in the world whose main prey had adult weight greater than 500 kg" (Rostro et al., 2018). Therefore, while male leopards appear to have expanded their feeding niche to include regular hunting of banteng, one of the largest ungulate species, female leopards still occupy the typical leopard niche as predators of relatively small ungulates. As a result, the male leopard now fills the predatory niche left vacant by the tiger's extinction, that of the largest carnivore (Rostro et al., 2018).

Africa Carnivores and Wild Cattle

In Africa, there is only one species of wild cattle, the African buffalo, which lives among a large number of carnivores, the larger of these capable of killing adults, and the smaller taking juveniles and calves, these comprising lions, leopards, hyenas, African wild dogs, and to a lesser extent cheetahs and crocodiles (Pienaar, 1969; Sinclair, 1977; Mloszewski, 1983; Prins & Iason, 1989; Henschel & Skinner, 1990; Prins, 1996a, b; di Silvestre et al., 2000; Henschel et al., 2005; Hayward, 2006; Hayward, O'Brien, et al., 2006a; Hayward, Henschel, et al., 2006b; Hay et al., 2008; Wentworth et al., 2011; Prin & Prins, 2014; African Wildlife Foundation, 2024). For many predators, buffalo may be too large: "The two main factors determining whether a predator will kill a particular prey species appear to be opportunity and physical possibility. Opportunity involves sharing the same habitat and possibly being active at the same period during day and night. Relative size must be the main factor determining the possibility of killing any particular animal and must certainly prevent the adults of the larger prey species being taken by predators other than lion" (Mitchell et al., 1965: 316). Mitchell et al. (1965: 316) note that for buffalo, lions were the only significant predator in their study, and there was competition among predators for smaller prey: "the leopard with 22 prey species has a broader food range than lion with 19, wild dog with 15 and cheetah with 11. Leopard and wild dog appear to be in direct competition for the same prey. The hyaena is not a significant predator" (Mitchell et al., 1965: 316).

The African Lion and Wild Cattle

The lion is the main predator of the African buffalo, due to its size and aggressive behavior (Prins & Iason, 1989). The buffalo is considered one of the most dangerous mammals in Africa, to predators and people (Treves & Naughton-Treves, 1999; Dunham et al., 2010; IUCN SSC Antelope Specialist Group., 2019). The African Wildlife Report (2024) notes that the African buffalo is considered to be among the most dangerous mammals in Africa, responsible for about 200 human deaths annually, and also kills more lions than any other animal. Hildebrandt (2014: 40) argues that buffaloes are rarely threatened by other animals or predators because of their imposing size, deadly weapons (horns), and unique behavior. The two greatest threats to buffalo when it comes to predators are humans and lions. Hildebrandt (2014: 40) further notes that "the buffalo attacks and even if the target changes position, the buffalo will alter its course as to compensate for the change in position. This is then followed by a horning and goring action when reaching the target with clear intent to kill. Wounded buffalo are seldom distracted from the target and usually continue until either the attacker or the victim are dead (see also Mloszewski, 1983). The buffalo frequently shows six reactions to predator threats: "increased caution, selective scrutiny, auditory-olfactory probing, flight, preventative aggression, and reciprocal aggression" (Hildebrandt (2014: 40; see also Mloszewski 1983).

Mitchell et al. (1965) cite a study of the Kafue National Park, Zambia, where buffalo were the commonest prey of lions, with only one buffalo killed by a crocodile, and none by leopards, cheetahs, and wild dogs. Here buffalo comprised nearly 20% of lion preys during the wet season, and nearly 40% during the dry season. The authors note that "there is one record of an old lioness being killed by buffalo and two records of a buffalo pulled down by lions being abandoned when charged by the herd." The writers also mention that although lions seem to favor buffalo as prey because they follow buffalo herds relentlessly, hunting buffalo is risky for them. It appears that lions can survive on easier prey, such as warthogs, if they have enough cover. However, after the grass fires have destroyed much of the land, many antelope herds relocate to the open plains in search of new vegetation to graze on. Lions are forced to kill more buffalo because hunting these animals becomes much harder without sufficient cover (Mitchell et al., 1965: 312).

The importance of lion predation on buffalo was cited by Tambling et al. (2012: 1297), who use the example of the reintroduction of lions into the Addo Elephant National Park (AENP) Main Camp Section, in South Africa, and argue that buffalo responses differ from ungulates from the northern temperate regions. For the buffalo in the case study, after the lions were reintroduced, the buffalo herds merged into larger, more defensible units, with results indicating an increase in the survival rate of young buffalo against lion predation. Buffalo breeding herds favored open habitats, especially during peak lion activity in the night and in the morning. The increases in buffalo group size and habitat preference therefore counteracted initially high levels of predation by lions on young buffalo and resulted in a return of the proportion of young animals in breeding herds to pre-lion levels (Tambling et al., 2012). The authors cite their case study as varying from the majority of such predator-prey studies, which primarily focus on North American cervids, such as elk (*Cervus elaphus* Linnaeus, 1758), moose (*Alces alces* Linnaeus, 1758), wapiti (*Cervus canadensis* Erxleben, 1777), and caribou or reindeer (*Rangifer tarandus* Linnaeus, 1758), and also bovids (bison) responding to wolf (Canis lupus) reintroductions and predation threats (Fortin et al., 2005, 2009; Mao et al., 2005; Berger, 2007). In such temperate studies, elk are shown to change their dispersal, feeding, activity, and alert behaviors following the reintroduction of wolves, modifications which may affect both trophic cascades and the structure of the elk population (Tambling et al., 2012; see also Laundre et al., 2001; Ripple & Beschta, 2003; Mao et al., 2005; Berger, 2007; Creel & Winnie, 2005).

Tambling et al. (2012: 1300) found changes in the survival of juvenile buffalo after lion reintroductions, as juvenile and adult buffalo were the main buffaloes killed by the lions. The higher than average juvenile kills were attributed to the common response of buffalo herds to the approach of lions, which was to retreat, which would expose the slower moving juveniles to lion pursuit and capture (see also Funston et al., 1998). It is observed that where the herds did not flee, but faced lions, they were able to resist lion attacks (see also Mitchell et al., 1965; Sinclair, 1977; Whyte, 2004). Tambling et al. (2012: 1300) note that the defensive posture resulted from an increased percentage of juveniles in herds. Using data from the 2004 aerial census in the KNP, an analysis of the herd size–juvenile percentage relations indicated that in herds with fewer than 60 buffalo, which was the upper limit of herd size before the reintroduction of lions, there was a significantly smaller percentage of juveniles than for herds numbering between 60 and 220, which was the range of herd size after the introduction of lions. Tambling et al. (2012: 1301) attribute these numerical differences to lower reproductive rates or lower juvenile survival of lion attacks. The authors conclude that, as the current buffalo that average over 100 individuals with the threat of lion predation, "buffalo herd sizes prior to lion reintroduction were either an anomaly and herd sizes have now returned to expected for buffalo, or buffalo that do not suffer predation occur in smaller herds" (Tambling et al. (2012: 1301).

Funston et al. (1998: 1333) present a study of the Kruger National Park, where male lions, especially nonterritorial males primarily on buffalo, while female lions preyed more frequently on smaller, medium-sized ungulates, including wildebeest (*Connochaetes taurinus* Burchell, 1823), and zebra (*Equus quagga* Boddaert, 1785 or *Equus quagga burchellii* Gray, 1824). In addition, the preferred buffalo and medium-sized ungulates were located more frequently than other prey by the non-territorial males and pride females. Funston et al. (1998: 1336) note that "buffalo, warthog and small prey, however, were killed significantly more frequently by males, the kill rates differing from the female kill rates by factors of 3.2, 1.6 and 1.3, respectively," and "buffalo were the most frequently killed large prey of males, with medium-sized prey species being killed very rarely: only two wildebeest and one kudu" (Funston et al., 1998: 1336). In terms of predation selection patterns, "buffalo, especially calves and old bulls encountered away from herds, were the most frequently selected species by male lions" (Funston et al. (1998: 1342).

Funston et al. (1998: 1342) note that there were variations in the prey selection process among male lions; nonterritorial male lions caught considerably more buffalo and were less likely to be scavengers than territorial male lions; these differences were indicative of significant social differences between the two groups of male lions. Due to their slightly larger group size and freedom to choose feeding areas, nonterritorial male lions hunted the dangerous but more rewarding (in terms of quantity of meat) buffalo more frequently than territorial male lions (Funston et al., 1998: 1342). The territorial male lions are limited to foraging within their territory and need a certain amount of body mass and fitness to maintain their competitiveness in territory defense. However, males that are not territorial might gain from having greater freedom to roam around in search of buffalo, which could lead to them consuming more food than is necessary (Funston et al. (1998: 1342). Compared to the other lion group types, they encountered buffalo more frequently, indicating that they lived in areas where buffalo were expected to be present. If their coalition's size and experience permit effective buffalo hunting, the regular killing of buffalo and the ensuing increase in food intake permit quick body growth to the size required to contend against rival male lions for territorial status (Funston et al., 1998: 1342). Non-territorial males had slightly lower success rates than the other groups (57 versus 72–73%), indicating that they had initially less experience capturing buffalo (Funston et al., 1998: 1342).

Other studies have documented the role of lions in buffalo mortality. For example, Hay et al. (2008: 850) report that in the Kruger National Park, South Africa, most buffalo deaths were due to lion predation, and the "mortality hazard associated with being an adult male buffalo in a male-only 'bachelor' group was almost four times higher than for adult females in mixed herds. The mortality rates of adult males and females within mixed herds were not statistically different." In another case study of African lions in Hwange National Park, Zimbabwe, "buffalo consistently emerged as the most frequently utilised prey in all seasons by both male (56%) and female (33%) lions, contributing the most to lion dietary biomass… buffalo was the most intensively selected species throughout the year. For female lions, kudu and to a lesser extent the group 'medium Bovidae' are the most important secondary prey" (Davidson et al., 2013).

Metzger et al. (2010) took the example of the Serengeti ecosystem, noting the decline of the populations of buffaloes, elephants, and black rhinoceroses, and the possible factors for the spatial differences, examining three issues: hunting, food shortage, and natural predation. Commenting on their research model, the authors note that "population changes were best explained by illegal hunting but model fit improved with the addition of predation mortality and the effect of food supply in areas where hunting was least," here predation referring to lion predation, and "the model incorporating unequal hunting effort, survival rates resulting from drought, carrying capacity in the far east estimated separately, and lion predation provided the best fit to the census data", with lion predation estimated at 10% (Metzger et al. (2010: 3431).

Cross et al. (2009: 467) take the interesting perspective of examining lion predation, among several other factors, for the population decline of buffaloes suffering from certain diseases, namely bovine tuberculosis (BTB) in the buffalo populations in Kruger National Park, South Africa. As such, a disease may increase the buffalo's vulnerability to predation, and may reduce reproduction and general survival. The research questions were if BTB infection increases vulnerability to predation and reduces reproduction and population growth. The findings indicated that, despite increased rates of BTB from 1991, from 27 to 45% in the southern region and 4 to 28% in the central region of Kruger National Park, the buffalo population growth rate was not statistically different among regions and did not decline over time, and there was no evidence that lions preferentially killed test-positive buffalo and death rate of the positive individuals was not greater than for test-negative animals (Cross et al., 2009: 467). The authors note that with no clear impacts of BTB on buffalo vulnerability, "the demographic impacts of bovine tuberculosis in the Kruger National Park remain undetectable despite 6 years of study on known individuals and 40 years of population counts." The study detected 55 natural mortalities of radio-collared buffalo, of which all but 14 were assumed to be killed by lions, as lions or lion spoor were seen near the carcass as there was an intact radiocollar (Cross et al., 2009: 470). Importantly, the authors note that "cursorial predators" (i.e., those with long-distance running endurance) such as large canids, compared to ambush predators like lions and leopards, are more likely to target weaker prey. Furthermore, some infected individuals may be protected from disease-induced predation by species that employ herd defense (Cross et al., 2009: 473).

The African Leopard and Wild Cattle

Leopards may prey on buffalo, mostly calves (Ng, 2015). For example, Henschel et al. (2005) report in a case study of Lopé National Park, Gabon, Central Africa, that for leopards, the red river hog (*Potamochoerus porcus* Linnaeus) was the main prey species, making up 20% of the biomass consumed, and the forest buffalo (*Syncerus caffer nanus* Boddaert) and cane rat (*Thryonomys swinderianus* Temminck), each comprised 13% of biomass consumed. Korte (2008: 116) reports on African forest buffalo in the Lopé National Park, Gabon, and notes that leopards are the only large carnivore at Lopé NP and are a natural predator of the ungulate community. Forest buffalo were one of the main prey items found in leopard scat, but the indication was that leopards prey only on juveniles (Korte, 2008: 116).

The African Hyena and Wild Cattle

Hyenas eat buffalo carcasses, but there is uncertainty whether lions or the hyenas themselves killed the buffaloes (Henschel & Skinner, 1990). Henschel and Skinner (1990: 69) performed case study of hyena diets in Kruger National Park, noting that hyaenas often scavenged on the carcasses of large ungulates, with buffalo and impala being generally most common and for one group of hyenas, buffalo, kudu, and impala were scavenged more than zebra and wildebeest (Henschel & Skinner, 1990, 74). Additionally, of "the ungulates, representing 93-4% of the mass obtained by hyaenas, buffalo were the most important species, followed sequentially by wildebeest, zebra, kudu and impala," but the source of the buffalo carcasses was in some cases uncertain, as the authors admit that "buffalo were mostly scavenged as drought victims or lion kills" and possibly "the high numbers of impala and buffalo determined from aerial censuses, as well as their importance in the hyaena diet throughout the Park, implies that hyaenas frequently hunted impala and scavenged buffalo because of a high frequency of encounter" (Henschel & Skinner, 1990: 80).

American Carnivores and Wild Cattle

In North America, there is only one species of wild cattle, the American bison. This species may be killed by brown and black bears, cougars, wolves, and to a lesser extent for calves, coyotes (Herman & Willard, 1978; Chapman & Feldhamer, 1982; Carbyn & Trottier, 1987; Reynolds & Hawley, 1987; Sheldon et al., 2009). The National Park Service (2024) lists wolves and grizzly bears as the only large predators of adult bison. Hodde (2024) notes that "the only real predators of the American bison in the wild, besides humans, are wolves and bears, but both species will typically prey on the calves versus the much stronger and larger adult animals." Wuerthner (2020) notes that bison are threatened only by wolves, but the bison's large size serves as a useful defense against attacks by other predatory animals, and buffalo may be avoided when smaller prey is available, a point also noted by Carbyn et al. (1993). However, calves and older adults may be predated (Wuerthner, 2020). The relations between bison and the main North American mammal carnivores are considered below.

The Gray Wolf and Wild Cattle

Most current scientific literature lists wolves as the main predatory species for the American bison, although attacks on such a powerful animal are rare (Tallian et al., 2017). Numerous studies have documented wolves killing even adult bison, despite mentioning the difficulties involved. Such studies focus on wolf predation on bison and/or bison strategies to avoid wolves. Concerning studies of bison avoidance strategies, the main issues concern foraging and group-size variations, where the bison may with varying effectiveness either repel wolf attacks, change foraging activities according to trade-offs between better foraging grounds and predation risk, or either avoid or ignore wolf predation risk, especially when wolves shift to smaller easier prey including deer species (see for example, Fortin et al., 2009; Harvey & Fortin, 2013). Concerning studies of wolf predation preferences, the main issues concern the wolves' targeting of smaller prey species, with variable targeting of avoidance of bison, which are the target prey species in North America (Ripple et al., 2010; MacNulty et al., 2014; Tallian et al., 2017).

For the behavior of bison towards wolf predation, a study by Harvey and Fortin (2013) examines the issues of "the foraging costs of anti-predator behaviour" as well as the "the metabolic costs" and "the missed opportunity costs of not foraging elsewhere." The authors note that "spatial patterns of bison herbivory were largely driven by a search for high net energy gains and, to a lesser degree, by the spatial game with gray wolves." Bison and wolves were more likely to use the same meadows than expected randomly, which indicated active interactions between bison and wolves. This indicated that "wolves are ahead in the spatial game they play with bison" (Harvey & Fortin, 2013). Bison were influenced by the threat of wolf arrival, to the point of reduced feeding. The authors concluded that their "assessment of bison foraging in a natural setting clarifies the complex nature of plant-herbivore interactions under predation risk and reveals how spatial patterns in herbivory emerge from multi-scale landscape heterogeneity" (Harvey & Fortin, 2013).

Another study by Fortin et al. (2009: 2480) examined bison behavior and assessed the cost-benefit trade-offs and group size changes, which may be factors for ungulate foraging and predator strategies. The question was "whether habitat selection was more strongly driven by the competing demands of forage intake vs. predator avoidance behavior." The findings suggested that there were seasonal a group-size influences. Regarding predation risk, the "group-size-dependent patterns were consistent with the hypothesis that avoidance of predation risk is the strongest driver of habitat selection" (Fortin et al., 2009: 2480). It is argued that "Bison have evolved in the presence of wolves. Their behavioral response to habitat structure has thus likely been shaped by cost–benefit trade-offs of antipredator strategies" (Fortin et al., 2009: 2488). In the study area of Prince Alberta National Park, bison were considered at risk as 4 of the 20 radio-collared female bison followed from 2005 to 2008 were killed by wolves, but in general bison generally forage in open vegetation landcover, despite predation risk, and adjustments may include larger foraging groups in open areas (for cooperative defense) and smaller groups in forest (for maneuverability and detection avoidance) (Fortin et al., 2009: 2488). The authors contend that individuals in smaller groups were found to be at higher per capita risk following an encounter with wolves because of the low variation in group composition (female to calf ratio) associated with group size for the radio-collared focal females. Based on this, the predation risk hypothesis postulates that bison should perceive risk as being comparatively higher in small groups than in large groups, which would result in shorter residence times, more time spent in the forest, etc., in contrast to the food competition hypothesis, making meadow selection by individuals in smaller groups less effective (Fortin et al., 2009: 2488–2489).

Regarding populations of bison that have been reintroduced, Jung et al. (2023) contend that although wolves'ability to hunt and scavenge bison from certain long-established populations is well established, for the majority of small, reintroduced populations in the boreal forest, there is little knowledge of these ecological interactions. Predation or scavenging of reintroduced bison in the boreal forest is, in fact, rarely documented. This is a significant omission, as it is essential to comprehend how reintroduced populations integrate into regional food webs and ecosystems. The majority of the predated bison were females or adults, and some were elderly individuals, according to case studies from reintroduced bison populations in northwest Canada. The authors hypothesize that wolves may have been slow to recognize the reintroduced bison as prey and incorporate them into their diet because verified reports of wolf-bison interactions were made between 19 and 50 years after the animals' introduction (Jung et al., 2023). It is determined that the primary reason these observations are valuable is that they offer clear proof that the reintroduced bison populations are starting to be influenced by their primary predator and assimilating into local food webs—two characteristics that are essential to ecological restoration (Jung et al., 2023).

Another study of bison responses to wolf predation is that of Hernández and Laundré (2005): 215–220), which provides evidence that wolves, introduced into Yellowstone, affected ungulate species, with the largest species, bison and elk, responding with variable behavior. The hypothesis was that elk and bison would compensate by using safer forests more often and less dangerous open meadows, which would lower the quality of their diet (Hernández & Laundré, 2005: 215). The authors draw the conclusion that the data for bison did not support any predictions, in contrast to the data for elk. In certain regions, wolves primarily hunt bison. However, wolves in Yellowstone National Park primarily hunt elk, and they did not prey on bison in large numbers during the study. The anticipated changes in habitat use and a consequent drop in diet quality were most likely not brought on by the predation risk to bison (Hernández & Laundré, 2005: 219). Additionally, in response to wolves, bison females' vigilance initially increased but, for the most part, returned to its baseline levels. Additionally, this suggests that the risk of wolf predation for bison is still low and would not be a strong enough motivator to change their habitat (Hernández & Laundré, 2005: 219).

For the behavior and abilities of wolves towards bison, Smith et al. (2000: 1128) reported that in Yellowstone National Park, wolves were more successful in killing elk than bison, elk were the main prey of wolves, and elk were more plentiful than bison, and most bison kills occurred in late winter when bison were at risk due to poor health or because the bison were young or injured. Where elk were scarce or nonexistent, wolves learned to kill bison and continued to do so. The bison killed by wolves in this study were vulnerable because the only adult bull that was killed had a broken leg; the other bison were either adult females or calves (Smith et al. (2000). Furthermore, wolves attacked bison from behind when they were fleeing, which is another way they employed killing technique. Wolves quickly lost interest in bison that did not run. The bison's behavior thus dictated whether an attack took place, how long the encounter lasted, how far they pursued, and whether the encounter ended in a successful kill or a failed attempt (Smith et al., 2000: 1132).

For reintroduced bison populations in Yellowstone, Tallian et al. (2017: 1418) cite the ecological phenomenon of "prey switching," which happens when a generalist predator kills a disproportionately higher number of common prey species while ignoring a rarer species. This phenomenon is relatively unknown in situations where the predator species is opposed by dangerous prey species, such as the bison, which are wolves'preferred but dangerous prey. In the case study of Yellowstone National Park, rising bison numbers are contrasted with falling red deer populations. The authors discovered that wolves attacked and killed a disproportionately higher number of elk, a rarer but safer species, in contrast to the prey-switching hypothesis. Wolves continued to favor bison despite the fact that they were almost twice as common as elk. It was also shown that wolves grew less inclined to hunt bison as their relative abundance rose (Tallian et al., 2017: 1418. According to one theory, wolves avoided bison because of their size and strength, which required larger packs (>11 wolves) to attack smaller herds of bison (<20 bison) with calves. Wolves were increasingly consuming bison carcasses in addition to avoiding big herds (Tallian et al., 2017: 1418). In addition to avoiding large bison herds, wolves also increasingly scavenged bison carcasses (Tallian et al., 2017).

Regarding gray wolf ecology in the United States, the Yellowstone Park area has attracted several studies, partly due to the reintroduction of wolves into the area, and their interesting relations with bears and cougars, and large ungulates including bison, elk, and various species of deer (Boyce, 1993; Wright et al., 2001, 2006; White & Garrott, 2005; Varley & Boyce, 2006; Garrott et al., 2009; Ripple et al., 2010; Metz et al., 2012; MacNulty et al., 2014; Mech et al., 2015). Boyce (1993: 234) notes that although Yellowstone National Park was established in 1872, large numbers of its wildlife were killed after this, including bison, deer, moose, elk, bighorn sheep, deer, and antelope, and the carcasses were in some cases poisoned with strychnine carnivores including wolves, coyotes, and wolverines. Although wildlife protection within Yellowstone was mentioned, public pressure in favor of predator control existed (Boyce, 1993: 234). Other developments were a federal law passed by the US congress in 1914, which called for the elimination of predators from public lands, including national parks, which resulted in the killing of cougars, wolves, and coyotes in national parks, but the Endangered Species Act of 1973 offered protection to rare species, and currently the "gray wolf is listed as endangered in the contiguous 48 states except Minnesota, where it is listed as threatened" (Boyce (1993): 234). Later developments were the Northern Rocky Mountain Wolf Recovery Plan in 1980, with a revision in 1987, an aspect including the reintroduction of wolves to Yellowstone, as natural recolonization was considered improbable (Boyce (1993): 234).

With the reintroduction of wolves into Yellowstone, wolves and cougars regained their sympatric relations. Wolves and cougars are described as ecological and aggressive competitors, although wolves are generally considered dominant (Elbroch et al., 2015; Boyce, 2018). Elbroch et al. (2015) compare reintroduced wolves and resident cougars, noting that in the case of the Southern Yellowstone Ecosystem, introduced increased numbers of wolves had an impact on the sympatric resident cougars, noting that there was competition between recolonizing wolves and resident cougars, as evidenced by direct interspecific predation and shifting cougar prey selection as wolves increased in numbers, and recolonizing wolves affected resident cougars'niche in through competition, and resident cougars may have a niche more similar to the past when these cougars and wolves were sympatric in the Yellowstone Ecosystem (Elbroch et al., 2015). This result aligns with the general findings that wolves may dominate shared ecosystems with cougars (Ruth & Murphy, 2010).

In another case study of wolves and bison in Wood Buffalo National Park, Alberta, Canada, which was established in 1922 to protect wood bison (Fuller, 1962), Carbyn and Trottier (1987) noted that where single wolves encountered bison, "single wolves watched bison (23% of observations), trailed without follow-up (14%), trailed with follow-up (27%), or harassed them without making physical contact (34%). Only rarely (2% of the observations) did they attack" (Carbyn & Trottier, 1987: 2072). For wolves in packs (defined as two or more wolves), encounters "involved trailing with follow-up (26%) or harassment (48%) and rushing with physical contact (13%)." Importantly, wolves were more likely to attack bison herds with calves than herds without calves, and single wolves were "more likely than wolves in packs to attack herds of bulls only" (Carbyn & Trottier, 1987: 2072), this suggesting that "suggesting that approach of single wolves to bison represented a more random event, whereas packs deliberately formed hunting units to seek out vulnerable prey" (Carbyn & Trottier, 1987: 2078).

In another study, Carbyn and Trottier (1988: 297–302) describe wolf attacks on bison calves in Wood Buffalo National Park, noting that "Wood Buffalo National Park and adjacent Slave River Lowlands are one of a few areas where wolves are preying on free-roaming bison. This historical association of wolves and bison was first documented at the turn of the century" and "the introduced plains bison interbred with the residual wood buffalo, and the bison observed in this study were the descendants of the original wood bison and introduced plains bison." Wolves are described as "natural predators of wood buffalo," and the reintroduction of wood bison in northern Canada from 1975 contributed to increased contacts between bison and wolves, with wolf predation on bison in late spring/early summer season directed mostly toward cow/calf herds. Wolves favored herds with calves over herds without calves, but the authors noted five defensive tactics adopted by bison to protect calves: "(1) to run to the cow, (2) to run to a herd, (3) to run to the nearest bull, (4) to get out in front and center of a stampeding herd and (5) to run through water bodies" and bulls frequently adopted defensive positions to defend calves (Carbyn and Trottier (1988: 297).

Larter et al. (1994: 318–327) add a study of wolves, bison, and moose, covering the previous 20 years in the Mackenzie Bison Sanctuary, Canada, where bison numbers increased and moose numbers appeared to have decreased, and moose were double near Mink Lake Moose compared to the Mackenzie Bison Sanctuary, and wolf activity was higher in the Mackienzie Bison Sanctuary than in the Mink Lake area. The findings indicated that bison comprised a larger proportion of the wolf diet in the Mackenzie Bison Sanctuary than in the Mink Lake context, but that moose comprised a significantly greater proportion of the wolves' diet than expected according to moose presence in the two studied areas. The study suggests that wolf predation may have a significant impact on moose populations (Larter et al., 1994: 318–327).

Tallian et al. (2017) further describe the wolf kill ratios of bison and elk, noting that wolves favored attacks on bison over elk, at relative abundance ratios of one bison per 35 elk to almost two bison per elk. Out of 187 direct observations of wolf-bison interactions, less than 5% resulted in a kill. These findings agree with published research that bison were the hardest ungulates for wolves to kill in North America (Carbyn & Trottier, 1987, 1988; Carbyn et al., 1993; Hebblewhite & Pletscher, 2002; Tallian et al., 2017). Large groups of over 20 bison could repel wolves. When weighed against other factors like the bison's composition, size, and reaction to flight, the bison group's size had the biggest impact on the success of wolf capture. The best predictor of attack probability, on the other hand, was wolf group size, which was followed by bison flight response, group composition, and group size. Because they can better elicit a flight response, larger packs may be more likely to attack bison (Tallian et al., 2017; see also Garton et al., 1990; Boyce, 1993; Hebblewhite & Pletscher, 2002; Mech & Peterson, 2003; Garrott et al., 2009; Ripple et al., 2011; MacNulty et al., 2014, 2016; Geremia et al., 2014; Mech et al., 2015; White et al., 2015).

An important limitation of wolf predation is the reliance on group hunting strategies, which are variably successful (MacNulty et al., 2014). For example, MacNulty et al. (2014) note that large groups frequently do no better at capturing prey than do small groups, which is an interesting feature of social foraging behavior. This pattern has been linked to decreased cooperation in large groups, particularly when it comes to cooperation when hunting difficult-to-catch prey. The example involves wolves hunting their most formidable prey, bison, in Yellowstone National Park.

This testifies to the idea that hunting difficult prey in large groups fosters greater cooperation, with bison being particularly challenging to hunt and kill due to their three times greater resistance to harm than elk (MacNulty et al., 2014). The results showed that improvements in elk capture success leveled off at 2–6 wolves, while bison capture success peaked at 9–13 wolves, with evidence that it continued to increase beyond 13 wolves. These findings are consistent with the theory that hunters in larger groups are more cooperative when hunting more difficult prey. Therefore, increased capacity to catch large prey could facilitate the establishment and upkeep of large predator groups, especially among predators that specialize in capturing such prey (MacNulty et al., 2014).

Further comparisons have been made between wolf predation of bison and that on elk, taking Yellowstone as a case study. Ripple et al. (2010: 32) constructed a trophic cascade including wolves, bison, and elk based on the sequence: "1) wolves prey on elk, changing elk behavior and reducing elk numbers, 2) causing reduced elk herbivory and more forage available to bison, and 3) allowing higher bison densities and additional bison effects on the ecosystem." This cascade is argued to possibly induce an increase bison numbers, with important impacts on the Yellowstone ecosystem. The authors argue that this may "represent an example of an alternative top-down pathway by which predators can influence multiple trophic levels through mediating the competitive interaction between two prey species" in this cases elk and bison (Ripple et al., 2010: 31). White and Garrott (2005): 141) add that after the reintroduction of wolves into Yellowstone, "counts of northern Yellowstone elk (Cervus elaphus) decreased more (50%) than predicted (5–30%) and will likely continue to decrease given the strong preference of wolves for elk and continued high kill rates despite this substantial reduction in elk abundance" (White and Garrott (2005): 141). In this study, the wolf reintroduction had no significant effects on the other ungulate species, namely bison, bighorn sheep, moose, mule deer, and pronghorn, but wolf recovery may "eventually contribute to increased bison and pronghorn abundance by decreasing elk and coyote abundance, respectively" (White and Garrott (2005): 141). Reasons for the greater impact of wolf predation on elk over bison: When wolves attack, bison usually defend themselves by banding together instead of running away as elk do, and adult bison have a higher chance of surviving an attack. Furthermore, bison predation has likely been mostly compensatory thus far, with little effect on bison numbers, whereas wolf predation on elk in the Yellowstone region has been primarily additive, decreasing the elk population (Ripple et al. (2010: 31, 35).

The Coyote and Wild Cattle

Coyotes may also occasionally prey on bison, at least calves. For example, Sheldon et al. (2009: 260) reported the occurrence of "a single adult male Coyote (Canis latrans) kill a Bison (Bison bison) calf in Yellowstone National Park. The predation is, to our knowledge, the only direct and complete observation of a lone Coyote capturing and killing a Bison calf." Bowerman (2017) also document an adult bison protecting a calf from coyote at Yellowstone National Park. These "stunning photos show a bison defend her newborn calf from a coyote" and the coyote "attempting to chase the tiny calf and grab it, while the mother bison fights back." It is noted that "while there aren't numbers on the bison calves killed by coyotes or wolves each year, it's not uncommon for the young to become targets" (Bowerman, 2017). However, in this case, "the photos show that despite multiple attempts to isolate the calf from its mother, the coyote came up empty handed" (Bowerman, 2017).

European Carnivores and Wild Cattle

In Europe, there is only one species of wild cattle, the European bison, the largest land mammal in Europe. Okarma (1995: 345) note the rarity of this species, as free-ranging European bison were completely exterminated, except for two populations that survived in Białowieża Primeval Forest up to 1919 and the Caucasian Mountains (up to 1927). Poaching and epizootic diseases were directly responsible for the extinction of the remnant individuals from the wild. The recovery of the bison population occurred through a captive breeding program begun in the Białowieża Primeval Forest (BPF), Poland, in 1929, and reintroduction of individuals in 1952. Current bison populations occur in Poland, the Caucasus Mountains, and Ukraine (Okarma et al. (1995).

Mysterud et al. (2007: 77) note that the European bison is an "example of nearly extinct 'charismatic megafauna'" and the Białowieża Primeval Forest in Poland is one of its few refuges. The European bison has no natural predators (Pucek et al., 2003, 2004; Mysterud et al., 2007). Okarma et al. (1995: 215) cite the case of the Białowieża Primeval Forest, Poland, where unlike predation on moose ("on average, 59% of moose mortality was due to predation by large carnivore"), currently the bison does not have any natural predators, but in the nineteenth century, "when red deer were temporarily extinct from BPF, bison were preyed upon by wolves and brown bears" but "native bears were exterminated from BPF before 1880" and "at present, as in other restored populations of European bison in eastern Europe" wolves rarely hunt bison (Okarma et al. (1995: 215).

Krasińska et al. (2013) note common ancestry with the American bison, noting that they both descend from a common ancestor living in Southeast Asia, where there are currently no bison. The ancestors of the American bison came to America via a land bridge between continents that stretches across what is now the Bering Sea. The final and complete separation of the two species occurred when the ice sheets retreated again about 10,000 years ago. For the European bison, which is primarily a browser, social associations include mixed groups of up to 20 individuals, including adult cows, 2–3-year-old young animals, and calves, with adult males forming bachelor groups of two to eight animals, or foraging alone (Plumb et al., 2020). This is similar to the social behavior of the American bison, which is primarily a grazer (Lott, 1991; Bowyer et al., 2007; Aune et al., 2017; Craine, 2021). Both the European and American bison face gray wolves and brown bears, but since the extinction of the Caspian tiger (*Panthera tigris tigris* Linnaeus 1758), the Persian leopard (*Panthera pardus tulliana* Valenciennes, 1856), and lions in Europe, the European buffalo has not faced any serious big cats, which would be comparable to the North American cougar (Turner & Anton, 1997; Rossi et al., 2020). The relation between the European bison and brown bears and wolves are considered in the following.

The Gray Wolf and Wild Cattle

The gray wolf is an important predator in remnant European forests with several ungulate species (Theuerkauf & Rouys, 2008). The European bison may only be rarely killed by wolves (Haidt et al., 2018). Mysterud et al. (2007: 78) report that wolves may rarely prey on bison. Haidt et al. (2018: 9) point out that flight initiation distance (FID) for bison is among the lowest in ungulates, which may "be related to the relatively low risk of predation, because European bison are only rarely killed by wolves." Hofling (2009) points out that European bison have evolved with several predators, such as the wolf and the brown bear, and although adult bison can defend themselves against an attack from these predators, calves and subadults can be killed, usually by stamina-based pursuit and pack killing (see also Pucek et al., 2004). Citing the case of the Białowieża Primeval Forest, where bison were reintroduced, it is noted that wolves were killed to protect the bison until 1989, and wolves persist in this with possible attacks on bison occurring (see Pucek et al., 2004; Mysterud et al., 2007).

For example, Hayward et al. (2015: 2080) in a study of bison in the Białowieza forest used optimal foraging theory, which "provides a framework for studying the perceptions of wildlife to the risks inherent in their environment" and measures foraging patches for ungulates (Druce et al., 2006). It is argued that if the predation risk is high, foragers extract more nutrients from safe patches than from risky patches (Hayward et al., 2015: 2080). In addition, giving-up densities (GUDs), which are the density of food remaining within a foraging patch over a specified time, may decline to the point where the cost of foraging outweighs the benefit of foraging, due to declining food, and other issues such as predation risk. The authors used such GUDs to assess the forage selection by European bison, which "is driven by risk (wolf predation, road accident risk or human conflict/poaching/culling avoidance) or nutrient quality. European bison were reintroduced to forest habitats after their extinction in the wild at the beginning of the twentieth century" (Hayward et al., 2015: 2080). The hypothesis that bison would have higher GUDs in areas of high predation risk from wolves, as despite the rarity of fatal wolf attacks on bison in Białowieza (2.6% of bison mortalities), bison predation by wolves is more frequent in North America (Carbyn et al., 1993; Smith et al., 2000). The results of the GUD study showed that the condition of the bison was an important factor for its risk perception, bison evidently perceived open areas near human habitation as risky, the wolf avoidance hypothesis was not supported, but a human avoidance behavior was evident (Hayward et al., 2015: 2088). Bison lose their health condition during winter and consequently select food for higher nutrient content, while accepting a higher killing risk from people rather than wolves, which may be justified as wolves only rarely killed bison in the Białowieza Forest, but humans culled sickly bison, sometimes up to 17% of bison (Hayward et al., 2015).

Comparative studies of ungulates also illustrate the relationship between bison and wolves, with the general result that bison are less often killed by wolves than other ungulates. For example, Mas-Carrió et al. (2024) studied the diet composition of red deer (*Cervus elaphus*) and European bison (*Bison bonasus*) in the Bialowieza Forest, to assess the importance of increased habitat quality in terms of density of deciduous forage species, contrasted with wolf predation risk. The results indicated that the ungulate diets were spatially variable, both within the red deer and bison populations and between these species. Bison showed less diet variability and higher niche overlap within species compared to the red deer, and unlike red deer, the bison did not show a reduction of niche overlap (which would increase dissimilar diets) with increasing predation risk from wolves, which indicated that bison feeding behavior was relatively unaffected by increased wolf presence (Mas-Carrió et al., 2024). Another study by Jędrzejewski et al. (2002: 1345) examined European bison and other ungulates (moose, wild boar, red deer, roe deer) in terms of their behavioral responses to the presence of wolves and lynx in the Białowieza Forest. The findings were compared with similar wolf-ungulate relations in North America. In this study, the average weight of prey taken by wolves was 67.2 kg in the range, 2–300 kg, with the upper end of moose and bison.

Okarma (1995: 343) also examined wolf predation of ungulates in the Białowieża Primeval Forest, noting that the wolf's preferred prey are deer (cervids), especially red deer and roe deer. Moose were recorded as the main prey only in the relative absence of the smaller deer species, and wild boar are usually avoided, and bison were not prey upon (see also Jędrzejewski et al., 2000). Okarma (1995: 358) notes that European bison were not found in wolf prey. Bison are supported by supplementary feeding during winter, and in the Polish part of BPF, 80–90% of bison aggregate in winter ranges where several feeding stations are placed. Only some of the adult bulls are distributed widely (Krasiński et al. 1994). The calves were healthy and lived within large herds that protected them from wolf predation. However, a few attacks by wolves and brown bears on bison were described in the past (Okarma, 1995: 358; see Kartsov 1903). Furthermore, Okarma (1995: 369) contends that wolves do not typically hunt European bison, that any buffalo found in wolf scat may have been consumed as carrion, and that historically, wolf predation on bison was likely low in Europe due to the fact that they coexisted in mixed ungulate communities with other medium-sized and numerous ungulates that were easier for wolves to prey on than bison (Okarma (1995: 369). In contrast to the North American context, where bison are a staple diet of wolves in Wood Buffalo National Park, this low wolf predation of buffalo means that calves are disproportionately selected based on their availability. Elk and bison are the only animals found in Wood Buffalo, with bison being more common (see also Carbyn et al., 1993).

Wolves are studied for their prey choices, which usually suggest a preference for prey smaller than adult bison (Churski et al., 2021). For example, Jędrzejewski et al. (2000): 197) studied wolves and ungulates in the Polish part of Białowieża Primeval Forest and found that the main prey species was red deer (*Cervus elaphus*), with less hunted species being European bison, moose (*Alces alces*), wild boar (Sus scrofa), and roe deer (*Capreolus capreolus*), which "were hunted less than expected based on their abundance." In addition, wolves hunted young animals of all species, with the exception of deer (the smallest species), more frequently than adults (Jędrzejewski et al., 2000: 202). Churski et al. (2021) argue that the red deer is the wolf's main prey species, whereas the European bison is rarely hunted, and that based on this hypothesis, the red deer perceives the risk of wolf predation. Therefore, it is assumed that its diet would change depending on the intensity of space use by wolves, while bison would not. It was concluded that red deer had greater changes in diet composition than bison in areas of high wolf presence compared to areas of low wolf presence. This suggests that, in addition to habitat differences, the risk of predation also plays a role in shaping the diet of red deer (Churski et al., 2021). An important point was that the European bison is less preyed upon by wolves than the American bison: for the European species, "The bison does not perceive wolf presence as a risk factor, and food availability is the main factor that determines its space-use" (Churski et al., 2021).

The Brown Bear and Wild Cattle

The European bison is occasionally killed and eaten by brown bears (Okarma, 1995), as brown bears are able to kill both species of bison (Niedziałkowska et al. (2019; see also Karcov 1903; Mattson 1997). Studies of bison reactions to danger and predators have shown variable results (Jankowski et al., 2019a, b; Mas-Carrió et al., 2024). However, studies of bison have uncovered predation by brown bears. For example, Jankowski et al. (2019a, b): 33–44) take the example of bison in the Bieszczady Mountains, where a successful attack by a bear was recorded in 2006 in Baligród Forest District, southeast Poland, where there were increased attacks by bears and wolves upon bison with increased numbers of bison, and increased scavenging of bison carcasses. Increased bear and wolf attacks on bison were reported between 2000 and 2013, as 16 bison were killed by predators (Perzanowski et al., 2014). Jankowski et al. (2019a, b): 33–44) conclude that "it can be also expected that in following years, wolves and bears of Bieszczady will also supplement their diet by the biomass originating from active hunting for wisent bulls, cows, and calves, which to certain degree will influence population numbers of that species in the region." In nearby regions, there were relatively few attacks, with for example, only two bear attacks on bison reported between 1999 and 2010 in Ukraine (Khoyetskyy, 2010).

The brown bear, like the gray wolf, is studied for its prey choices. For example, a study by Niedziałkowska et al. (2019: 5) found that, regarding the ungulates preyed upon by brown bears, "selection of medium-sized (60–130 kg) species by bears was most variable, ranging from strong avoidance to clear positive selection. Only the two largest species—European bison and moose—were consistently positively selected by bears and these were eaten more often than expected from their relative abundance in the local ungulate communities." The findings were that 8.7% of the brown bear diet was of ungulate meat, especially of European bears over Asiatic bears (means of 10.5% and 6.8%, respectively). The authors note that "brown bears' preference for 10 species of ungulates increased with body mass of prey" (Niedziałkowska et al. (2019: 1). The authors write that the brown bear's preference for hunting moose was evident across the bears' entire range, and bear were responsible for 23% of all natural elk deaths on average. Possibly, this was the result of a predator-prey relationship between these two species, which evolved from the brown bears'preference for moose and their dominant role in moose mortality. This shows that even ostensibly omnivore animals, such as brown bears, can prefer a specific kind of food (Niedziałkowska et al. (2019: 1).

Conclusions

This chapter examined literature sources dealing with the relationships between wild cattle and their predators on the five wild cattle species continents. The main problems relate to the size of the cattle species, which limits the number of predator species to big cats, bears, and social, large canids. On every continent, wild cattle are threatened by human hunting and human-caused habitat loss, as are the large predators that prey on such cattle. Serious efforts have been made to conserve and even reintroduce both cattle and predators, a situation that has led to interesting interactions between cattle and predators. Comparative studies indicate intercontinental and intracontinental differences that are of significant ecological and conservation interest. Asia is the only continent with more than one species of wild cattle but has a similar number of predators capable of killing wild cattle (one feline apex predator—tiger, lion; a smaller cat—leopard; social canines—wolves, African wild dogs; and one other—African hyena, Asian brown bear). North America and Europe have markedly similar ecological settings (coniferous and deciduous forests, grasslands/prairie, effective conservation management), and similar species (brown bears, wolves, bison), but the North American predators appear to kill more bison than their European equivalents. These issues, differences, and similarities may be relevant to the ecology of domestic and semi-wild cattle, which will be considered in the succeeding chapters.

References

African Wildlife Foundation. (2024). *African Buffalo*. Retrieved from https://www.awf.org/wildlife-conservation/african-buffalo

African Wildlife Report. (2024). *Why the Cape Buffalo is the Most Dangerous of the Big Five*. https://africanwildlifereport.com/why-the-cape-buffalo-is-the-most-dangerous-of-the-big-five/

Andheria, A., Karanth, K. U., & Kumar, N. S. (2007a). Diet and prey profiles of three sympatric large carnivores in Bandipur Tiger Reserve in India. *Journal of Zoology (London), 273*, 169–175.

Andheria, A. P., Karanth, K. U., & Kumar, N. S. (2007b). Diet and prey profiles of three sympatric large carnivores in Bandipur Tiger Reserve, India. *Journal of Zoology, 273*(2), 169–175. https://doi.org/10.1111/j.1469-7998.2007.00310.x

Ashokkumar, M., Nagarajan, R., & Desai, A. (2010a). Group size and age-sex composition of Asian elephant and Gaur in Mudumalai Tiger Reserve, southern India.

Gajah, 32, 27–34. https://www.academia.edu/34188426/Group_Size_and_Age_Sex_Composition_of_Asian_Elephant_and_Gaur_in_Mudumalai_Tiger_Reserve_Southern_India
Ashokkumar, M., Nagarajan, R., & Desai, A. A. (2010b). Gaur- A wild cattle in grave peril. *Science Reporter, 47*, 33–35.
Ashokkumar, M., Swaminathan, S., Nagarajan, R., & Desai, A. A. (2011). Distribution, ecology and conservation of the Gaur. (Bos gaurus, H. Smith, 1824). In V. K. Gupta & A. K. Varna (Eds.), *Animal diversity, natural history, and conservation* (pp. 77–94).
Aune, K., Jørgensen, D., & Gates, C. (2017). Bison bison (errata version published in 2018). The IUCN Red List of Threatened Species 2017: e.T2815A123789863. https://doi.org/10.2305/IUCN.UK.2017-3.RLTS.T2815A45156541.en.
Azevedo, F., Concone, H., Pires-da Silva, A., & Verdade, L. (2010). Puma (Puma concolor) predation on a water buffalo (Bubalus bubalis). *Mammalia, 74*(4), 431–432. https://doi.org/10.1515/mamm.2010.056
Balme, G. A., Miller, J. R. B., Pitman, R. T., & Hunter, L. T. B. (2017). Caching reduces kleptoparasitism in a solitary, large felid. *Journal of Animal Ecology, 86*(3), 634–644. https://doi.org/10.1111/1365-2656.12654
Baral, K., Bhandari, S., Adhikari, B., Kunwar, R. M., Sharma, H. P., Aryal, A., & Ji, W. (2024). Prey selection by leopards (Panthera pardus fusca) in the mid-hill region of Nepal. *Ecology and Evolution, 14*(2), e10924. https://doi.org/10.1002/ece3.10924
Bargali, H. S., & Ahmed, T. (2018). Patterns of livestock depredation by tiger (Panthera tigris) and leopard (Panthera pardus) in and around Corbett Tiger Reserve, Uttarakhand, India. *PloS One, 13*(5), e0195612. https://doi.org/10.1371/journal.pone.0195612
Berger, J. (2007). Carnivore repatriation and holarctic prey: Narrowing the deficit in ecological effectiveness. *Conservation Biology: The Journal of the Society for Conservation Biology, 21*(4), 1105–1116. https://doi.org/10.1111/j.1523-1739.2007.00729.x
Bhattacharya, B. K., & Saharia, J. (2021). *Assam's wild water buffalo is treading troubled waters*. https://India.mongabay.com/2021/09/assams-wild-water-buffalo-is-treading-troubled-waters/.
Bhumpakphan, N. (2008). *Ecology and recovery of gaur in Thailand*. (Department of Forest Biology Report). Faculty of Forestry, Kasetsart University.
Biswas, S., & Sankar, K. (2002). Prey abundance and food habit of tigers (Panthera tigris tigris) in Pench National Park, Madhya Pradesh, India. *Journal of Zoology, 256*(3), 411–420. https://doi.org/10.1017/S0952836902000456
Bora, J. K., Vardhan, V., Vijh, R. K., Deshmukh, A. V., Srinivas, Y., Mungi, N. A., Goswami, S., Jhala, H. Y., Chauhan, J. S., Kumar, U., & Jhala, Y. V. (2024). Evaluating the potential for reintroducing the endangered wild water buffalo (Bubalus arnee) in Kanha National Park, Central India. *Restoration Ecology, 32*(3), e14079. https://doi.org/10.1111/rec.14079
Bowerman, M. (2017). *Incredible images show bison protect calf from coyote at Yellowstone National Park*. https://www.usatoday.com/story/news/nation-now/2017/05/30/incredible-images-show-bison-protect-calf-coyote-yellowstone-national-park/355087001/.
Bowyer, R. T., Bleich, V., Manteca, X., Whiting, J., & Stewart, K. (2007). Sociality, mate choice, and timing of mating in American bison (Bison bison): Effects of large males. *Ethology, 113*(11), 1048–1060. https://doi.org/10.1111/j.1439-0310.2007.01411.x
Boyce, M. S. (1993). Predicting the consequences of wolf recovery to ungulates in Yellowstone National Park. *Ecological issues on reintroducing wolves into Yellowstone National Park* (Ed. R. Cook), (pp. 234–269). Scientific Monograph NPS/NRYELL/NRSM-93/22, USDI National Park Service.
Boyce, M. S. (2018). Wolves for Yellowstone: Dynamics in time and space. *Journal of Mammalogy, 99*(5), 1021–1031. https://doi.org/10.1093/jmammal/gyy115
Bshary, R., Hohner, A., Ait-el-Djoudi, K., & Fricke, H. (2006). Interspecific communicative and coordinated hunting between groupers and Giant Moray Eels in the Red Sea. *PLoS Biology, 4*(12), e431. https://doi.org/10.1371/journal.pbio.0040431
Campbell, M. (2017). *Conservation biology in the 21st century: A conservation biology of large wildlife*. New York, NY: Nova Science Publishers.

Campbell, M. (2018). *Geomatics and conservation biology*. New York, NY: Nova Science Publishers.

Campbell, M., & Jha, K. K. (2021). *Critical research techniques in animal and habitat ecology: International examples*. New York, NY: Nova Science Publishers.

Carbyn, L. N., & Trottier, T. (1987). Responses of bison on their calving grounds to predation by wolves in Wood Buffalo National Park. *Canadian Journal of Zoology., 65*(8), 2072–2078. https://doi.org/10.1139/z87-317

Carbyn, L. N., & Trottier, T. (1988). Descriptions of wolf attacks on bison calves in Wood Buffalo National Park. *Arctic, 41*, 297–302. https://doi.org/10.14430/arctic1736

Carbyn, L. N., Oosenbrug, S. M., & Anions, D. W. (1993). *Wolves, Bison and the dynamics related to the peace-Athabasca Delta in Canada's wood Buffalo National Park*. Circumpolar research series number 4 Canadian Circumpolar Institute. University of Alberta.

Castelló, J. R. (2016). *"Genus Bos: Banteng". Bovids of the world: Antelopes, gazelles, cattle, goats, sheep, and relatives* (pp. 630–635). Princeton University Press.

Chaiyarat, R., Sakchan, P., Panprayun, G., Thongthip, N., & Nakbun, S. (2020). Monitoring of forage and nutrition before and after reintroduction of banteng (Bos javanicus d' Alton, 1823) to Salakphra wildlife sanctuary Thailand. *Scientific Reports, 10*, 11135. https://doi.org/10.1038/s41598-020-67942-2

Chaiyarat, R., Ingudomnukul, P., Yimphrai, N., Nakbun, S., & Youngpoy, N. (2023). The preferred habitat of reintroduced Banteng (Bos javanicus) at the Core and the edge of Salakphra wildlife sanctuary, Thailand. *Animals: An Open Access Journal from MDPI, 13*(14), 2293. https://doi.org/10.3390/ani13142293

Chakrabarti, S., Jhala, Y. V., Dutta, S., Qureshi, Q., Kadivar, R. F., & Rana, V. J. (2016). Adding constraints to predation through allometric relation of scats to consumption. *Journal of Animal Ecology, 85*, 660–670. https://doi.org/10.1111/1365-2656.12508

Chapman, J. A., & Feldhamer, G. A. (Eds.). (1982). *Wild mammals of North America*. The Johns Hopkins University Press.

Choudhury, A. (2002). Distribution and conservation of the Gaur Bos gaurus in the Indian subcontinent. *Mammal Review, 32*(3), 199–226. https://doi.org/10.1046/j.1365-2907.2002.00107.x

Christiansen, P. (2006). Sabertooth characters in the clouded leopard (Neofelis nebulosa Griffiths 1821). *Journal of Morphology, 267*, 1186–1198.

Churski, M., Spitzer, R., Coissac, É., Taberlet, P., Lescinskaite, J., Ginkel, H. A., Kuijper, D., & Cromsigt, J. P. (2021). How do forest management and wolf space-use affect diet composition of the wolf's main prey, the red deer versus a non-prey species, the European bison? *Forest Ecology and Management, 479*, 118620. https://doi.org/10.1016/j.foreco.2020.118620

Clements, H. S., Tambling, C. J., Hayward, M. W., & Kerley, G. I. H. (2014). An objective approach to determining the weight ranges of prey preferred by and accessible to the five large African carnivores. *PLoS One, 9*(7), e101054. https://doi.org/10.1371/journal.pone.0101054

Craine, J. M. (2021). Seasonal patterns of bison diet across climate gradients in North America. *Science Reports, 11*, 6829. https://doi.org/10.1038/s41598-021-86260-9

Creel, S., & Winnie, J. J. A. (2005). Responses of elk herd size to fine-scale spatial and temporal variation in the risk of predation by wolves. *Animal Behaviour, 69*, 1181–1189. https://doi.org/10.1016/j.anbehav.2004.07.022

Cross, P. C., Heisey, D. M., Bowers, J. A., Hay, C. T., Wolhuter, J., Buss, P., Hofmeyr, M., Michel, A. L., Bengis, R. G., Bird, T. L. F., Du Toit, J. T., & Getz, W. M. (2009). Disease, predation and demography: Assessing the impacts of bovine tuberculosis on African buffalo by monitoring at individual and population levels. *Journal of Applied Ecology, 46*(2), 467–475. https://doi.org/10.1111/j.1365-2664.2008.01589.x

Davidson, Z., Valeix, M., Van Kesteren, F., Loveridge, A. J., Hunt, J. E., Murindagomo, F., & Macdonald, D. W. (2013). Seasonal diet and prey preference of the African lion in a waterhole-driven semi-arid savanna. *PLoS One, 8*(2), e55182. https://doi.org/10.1371/journal.pone.0055182

di Silvestre, I., Novelli, O., & Bogliani, G. (2000). Feeding habits of the spotted hyaena in the Niokolo Koba National Park, Senegal. *African Journal of Ecology, 38*, 102–107.

Druce, D. J., Brown, J. S., Castley, J. G., Kerley, G. I. H., Kotler, B. P., Slotow, R., & Knight, M. H. (2006). Scale-dependent foraging costs: habitat use by rock hyraxes (Procavia capensis) determined using giving-up densities. Oikos, 115(3), 513–525. https://doi.org/10.1111/j.2006.0030-1299.15179.x

Duangchantrasiri, S., Jornburom, P., Jinamoy, S., Pattanavibool, A., Hines, J. E., Arnold, T. W., Fieberg, J., & Smith, J. L. D. (2019). Impact of prey occupancy and other ecological and anthropogenic factors on tiger distribution in Thailand's western forest complex. *Ecology and Evolution, 9*(5), 2449–2458. https://doi.org/10.1002/ece3.4845

Dunham, K. M., Ghiurghi, A., Cumbi, R., & Urbano, F. (2010). Human–wildlife conflict in Mozambique: A national perspective, with emphasis on wildlife attacks on humans. *Oryx, 44*(2), 185–193.

Elbroch, L. M., Lendrum, P. E., Newby, J., Quigley, H., & Thompson, D. J. (2015). Recolonizing wolves influence the realized niche of resident cougars. *Zoological Studies, 54*, e41. https://doi.org/10.1186/s40555-015-0122-y

Elliot, J. P., Cowan, I. M. T., & Holling, C. S. (1977). Prey capture of the African lion. *Canadian Journal of Zoology, 55*, 1811–1828.

Farah, S., Atkulwar, A., Nahid, R., Gadhikar, Y., & Baig, M. (2021). Recent population expansion in wild gaur (Bos gaurus gaurus) as revealed by microsatellite markers. *Mammalian Biology, 101*, 695–707. https://doi.org/10.1007/s42991-021-00145-y

Fay, J. M., Carroll, R., Kerbis Perterhans, J. C., & Harris, D. (1995). Leopard attack on and consumption of gorillas in the Central African Republic. *Journal of Human Evolution, 29*(1), 93–99. https://doi.org/10.1006/jhev.1995.1048

Fortin, D., Beyer, H. L., Boyce, M. S., Smith, D. W., Duchesne, T., & Mao, J. S. (2005). Wolves influence elk movements: Behavior shapes a trophic cascade in Yellowstone National Park. *Ecology, 86*(5), 1320–1330. https://doi.org/10.1890/04-0953

Fortin, D., Fortin, M. E., Beyer, H. L., Duchesne, T., Courant, S., & Dancose, K. (2009). Group-size-mediated habitat selection and group fusion-fission dynamics of bison under predation risk. *Ecology, 90*(9), 2480–2490. https://doi.org/10.1890/08-0345.1

Foster, V. C., Sarmento, P., Sollmann, R., Tôrres, N., Jácomo, A. T. A., Negrões, N., Fonseca, C., & Silveira, L. (2013). Jaguar and puma activity patterns and predator-prey interactions in four Brazilian biomes. *Biotropica, 45*, 373–379.

Francis, C. M. (2008). *A guide to the mammals of Southeast Asia*. Princeton University Press.

Fuller, W. A. (1962). *The biology and management of the bison of wood Buffalo National Park*. Canadian Wildlife Service Wildlife Management Bulletin, Series 1, No. 16.52.

Funston, P. J., Mills, M. G. L., Biggs, H. C., & Richardson, P. R. K. (1998). Hunting by male lions: Ecological influences and socioecological implications. *Animal Behaviour, 56*(6), 1333–1345. https://doi.org/10.1006/anbe.1998.0884

Garrott, R. A., White, P. J., & Watson, F. G. R. (Eds.). (2009). *The ecology of large mammals in central Yellowstone: Sixteen years of integrated field studies*. Academic Press/Elsevier.

Garton, E. O., Crabtree, R. L., Ackerman, B. B., & Wright, G. (1990). In J. D. Varley & W. G. Brewster (Eds.), *The potential impact of a reintroduced wolf population on the northern Yellowstone elk herd. A Report to the United States congress, vol. 2* (pp. 359–391). Yellowstone National Park, U.S., Fish, Wildlife Service, University of Wyoming, University of Idaho, Interagency Grizzly Bear Study Team, the University of Minnesota Cooperative Parks Study Unit, Wolves for Yellowstone, Research, Analysis, National Park Service, Yellowstone National Park.

Geremia, C., Wallen, R., & White, P. J. (2014). *Population dynamics and adaptive management of Yellowstone bison*. National Park Service, Yellowstone National Park.

Ghaskadbi, P., Bathla, N., Bhandari, A., Modi, S., Nigam, P., & Habib, B. (2022). Feeding ecology of the endangered Asiatic wild dogs (Cuon alpinus) across tropical forests of the central Indian landscape. *Scientific Reports, 12*, 14029. https://doi.org/10.1038/s41598-022-17906-5

Ghoddousi, A., Soofi, M., Hamidi, A., Lumetsberger, T., Egli, L., Ashayeri, S., Khorozyan, I. H., Kiabi, B., & Waltert, M. (2017). When pork is not on the menu: Assessing trophic competition between large carnivores and poachers. *Biological Conservation, 209*, 223–229. https://doi.org/10.1016/j.biocon.2017.02.032

Goswami, A., & Friscia, A. (Eds.). (2010). *Carnivoran evolution: New views on phylogeny, form and function. Cambridge studies in morphology and molecules: New paradigms in evolutionary bio*. Cambridge University Press.

Gray, T., Borah, J., Coudrat, C.N.Z., Ghimirey, Y., Giordano, A., Greenspan, E., Petersen, W., Rostro-García, S., Shariff, M., & Wai-Ming, W. (2021). Neofelis nebulosa. The IUCN Red List of Threatened Species 2021. https://doi.org/10.2305/IUCN.UK.2021-2.RLTS.T14519A198843258.en.

Grassman, L. I., Tewes, M. E., Silvy, N. J., & Kreetiyutanont, K. (2005). Ecology of three sympatric felids in a mixed evergreen forest in North-Central Thailand. *Journal of Mammalogy, 86*(1), 29–38. https://doi.org/10.1644/1545-1542(2005)086%3C0029:eotsfi%3E2.0.co;2

Haidt, A., Kamiński, T., Borowik, T., & Kowalczyk, R. (2018). Human and the beast-flight and aggressive responses of European bison to human disturbance. *PLoS One, 13*(8), e0200635. https://doi.org/10.1371/journal.pone.0200635

Hance, J. (2012). Forgotten species: The wild jungle cattle called banteng. *Mongabay*. https://news.mongabay.com/2012/01/forgotten-species-the-wild-jungle-cattle-called-banteng/.

Hart, J. A., Katembo, M., & Punga, K. (1996). Diet, prey selection and ecological relations of leopard and golden cat in the Ituri Forest, Zaire. *African Journal Ecology, 34*(4), 364–379. https://doi.org/10.1111/j.1365-2028.1996.tb00632.x

Harvey, L., & Fortin, D. (2013). Spatial heterogeneity in the strength of plant-herbivore interactions under predation risk: The tale of bison foraging in wolf country. *PLoS One, 8*(9), e73324. https://doi.org/10.1371/journal.pone.0073324

Hay, C., Cross, P., & Funston, P. (2008). Trade-offs of predation and foraging explain sexual segregation in African buffalo. *Journal of Animal Ecology, 77*, 850–858.

Hayward, M. (2006). Prey preferences of the spotted hyaena (Crocuta crocuta) and degree of dietary overlap with the lion (Panthera leo). *Journal of Zoology, 270*, 606–614.

Hayward, M. W., & Kerley, G. I. H. (2005). Prey preferences of the lion (Panthera leo). *Journal of Zoology, 267*(03), 309. https://doi.org/10.1017/s0952836905007508

Hayward, M., O'Brien, J., Hofmeyr, M., & Kerley, G. (2006a). Prey preferences of the African wild dog Lycaon pictus (Canidae: Carnivora): Ecological requirements for conservation. *Journal of Mammalogy, 87*, 1122–1131.

Hayward, M. W., Henschel, P., O'Brien, J., Hofmeyr, M., Balme, G., & Kerley, G. I. H. (2006b). Prey preferences of the leopard (Panthera pardus). *Journal of Zoology, 270*(2), 298–313. https://doi.org/10.1111/j.1469-7998.2006.00139.x

Hayward, M., Hofmeyr, M., O'Brien, J., & Kerley, G. (2006c). Prey preferences of the cheetah (Acinonyx jubatus) (Felidae: Carnivora): Morphological limitations or the need to capture rapidly consumable prey before kleptoparasites arrive? *Journal of Zoology, 270*, 615–627. https://doi.org/10.1111/j.1469-7998.2006.00184.x

Hayward, M. W., Jędrzejewski, W., & Jêdrzejewska, B. (2012a). Prey preferences of the tiger Panthera tigris. *Journal of Zoology, 288*(1), 82–83. https://doi.org/10.1111/j.1469-7998.2011.00871.x

Hayward, M. W., Jedrzejewski, W., & Jedrzejewska, B. (2012b). Prey preferences of the tiger Panthera tigris. *Journal of Zoology, 286*, 221–231.

Hayward, M. W., Ortmann, S., & Kowalczyk, R. (2015). Risk perception by endangered European bison Bison bonasus is context (condition) dependent. *Landscape Ecology, 30*, 2079–2093. https://doi.org/10.1007/s10980-015-0232-2

Hebblewhite, M., & Pletscher, D. H. (2002). Effects of elk group size on predation by wolves. *Canadian Journal of Zoology, 80*(5), 800–809. https://doi.org/10.1139/z02-059

Hendricks, K. (2013). *Bison bonasus*. European Buffalo (On-line), Animal Diversity Web. https://animaldiversity.org/accounts/Bison_bonasus/.

Henschel, J., & Skinner, J. (1990). The diet of the spotted hyaenas Crocuta crocuta in Kruger National Park. *African Journal of Ecology, 28*, 69–82.

Henschel, P., Abernethy, K., & White, L. (2005). Leopard food habits in the lope National Park, Gabon, Central Africa. *African Journal of Ecology, 43*(1), 21–28. https://doi.org/10.1111/j.1365-2028.2004.00518.x

Herman, M., & Willard, E. E. (1978). *Rocky Mountain wolf and its habitat.* U.S. Department of Agriculture, Forest Service, National Forest System Cooperative Forestry, Forestry Research, Region.

Hernández, L., & Laundré, J. W. (2005). Foraging in the 'landscape of fear' and its implications for habitat use and diet quality of elk Cervus elaphus and bison Bison bison. *Wildlife Biology, 11*, 215–220. https://doi.org/10.2981/0909-6396(2005)11[215:FITLOF]2.0.CO;2

Hildebrandt, W. R. (2014). *Management and reproduction of the African savanna buffalo (Syncerus caffer caffer).* Master of Sciences in Animal Science at the University of Stellenbosch.

Hodde, A. (2024). *American Bison.* Retrieved from https://dnr.illinois.gov/content/dam/soi/en/web/dnr/education/documents/prairieamericanbisoninfo.pdf

Hofling, A. (2009). *Do enclosure characteristics affect anti-predator behaviour in the European bison (Bison bonasus)? Master thesis.* Linköpings universitet.

Hopcraft, J. G., Anderson, T. M., Pérez-Vila, S., Mayemba, E., & Olff, H. (2012). Body size and the division of niche space: Food and predation differentially shape the distribution of Serengeti grazers. *The Journal of Animal Ecology, 81*(1), 201–213. https://doi.org/10.1111/j.1365-2656.2011.01885.x

Hoppe-Dominik, B. (1984). Etude du spectre des proies de la panthère, Panthera pardus, dans le Parc National de Taï en Côte d'Ivoire. *Mammalia, 48*(4), 477–488. https://doi.org/10.1515/mamm.1984.48.4.477

Humphrey, S. R., & Bain, J. R. (1990). Endangered animals of Thailand. Issue 6 Of Flora, & Fauna handbook. Sandhill Crane Press.

Hunter, L. (2015). *Wild cats of the world.* Bloomsbury.

Ikeda, N. (2004). Economic impacts of livestock depredation by snow leopard Uncia uncia in the Kanchenjunga conservation area, Nepal Himalaya. *Environmental Conservation, 31*(4), 322–330.

IUCN SSC Antelope Specialist Group. (2019). Syncerus caffer. The IUCN Red List of Threatened Species 2019: e.T21251A50195031. doi:10.2305/IUCN.UK.2019-1.RLTS.T21251A50195031.en.

Jankowski, W., Januszczak, M., Wołoszyn-Gałęza, A., Kaczor, S., & Perzanowski, K. (2019a). The wisent as food supply for large predators and necrophages. *European Bison Conservation Newsletter, 12*, 33–44. https://smz.waw.pl/wp-content/uploads/2019/10/04_Jankowski-W.-Januszczak-M.-Wo%C5%82oszyn-Ga%C5%82%C4%99za-A.-Kaczor-S.-Perzanowski-K.-The-wisent-as-food-supply-for-large-predators-and-necrophages.pdf

Jankowski, W., Januszczak, M., Wołoszyn-Gałęza, A., Kaczor, S., & Perzanowski, K. (2019b). The wisent as food supply for large predators and necrophages. *European Bison Conservation Newsletter, 12*, 33–44. http://ojs.wisent.org/index.php/czasopismo/article/view/32

Jędrzejewski, W., Jędrzejewska, B., Okarma, H., Schmidt, K., Zub, K., & Musiani, M. (2000). Prey selection and predation by wolves in Białowieża Primeval Forest, Poland. *Journal of Mammalogy, 81*(1), 197–212. https://doi.org/10.1644/1545-1542(2000)081<0197:PSAPBW>2.0.CO;2

Jędrzejewski, W., Schmidt, K., Theuerkauf, J., Jędrzejewska, B., Selva, N., Zub, K., & Szymura, L. (2002). Kill rates and predation by wolves on ungulate populations in Białowieża primeval Forest (Poland). *Ecology, 83*(5), 1341–1356. https://doi.org/10.2307/3071948

Jha, K. K., & Campbell, M. (2021). *Critical research techniques in animal and habitat ecology: Examples from India.* New York, NY: Nova Science Publishers.

Jha, K. K., & Campbell, M. (2022). *Dynamics and Interrelations between Nature, Science and Society.* New York, NY: Nova Science Publishers.

Jha, K. K., & Campbell, M. (2023). Vultures of India. New York, NY: Nova Science Publishers.

Jornburom, P., Duangchantrasiri, S., Jinamoy, S., Pattanavibool, A., Hines, J. E., Arnold, T. W., Fieberg, J., & Smith, J. L. D. (2020). Habitat use by tiger prey in Thailand's western forest complex: What will it take to fill a half-full tiger landscape? *Journal for Nature Conservation, 58*, 125896. https://doi.org/10.1016/j.jnc.2020.125896

Jung, T. S., Larter, N. C., Lewis, C. J., Thacker, C., & Taylor, S. D. (2023). Wolf (Canis lupus) predation and scavenging of reintroduced bison (Bison bison): A hallmark of ecological restoration to boreal food webs. *European Journal of Wildlife Research, 69*, 50. https://doi.org/10.1007/s10344-023-01676-0

Karanth, K. U., & Sunquist, M. E. (1992). Population structure, density and biomass of large herbivores in the tropical forests of Nagarahole, India. *Journal of Tropical Ecology, 8*(1), 21–35. https://doi.org/10.1017/S0266467400006040

Karanth, K. U., & Sunquist, M. E. (1995a). Prey selection by Tiger, leopard and dhole in tropical forests. *Journal of Animal Ecology, 64*(4), 439–450. https://doi.org/10.2307/5647

Karanth, K. U., & Sunquist, M. E. (1995b). Prey selection by tiger, leopard and dhole in tropical forests. *Journal of Animal Ecology, 64*, 439–450. https://doi.org/10.1016/j.biocon.2012.12.016

Karanth, K. U., & Sunquist, M. E. (2000). Behavioural correlates of predation by tiger (Panthera tigris), leopard (Panthera pardus) and dhole (Cuon alpinus) in Nagarahole, India. *Journal of Zoology, 250*(2), 255–265. https://doi.org/10.1111/j.1469-7998.2000.tb01076.x

Karanth, K. U., Nichols, J. D., Kumar, N. S., Link, W. A., & Hines, J. E. (2004). Tigers and their prey: Predicting carnivore densities from prey abundance. *Proceedings of the National Academy of Sciences of the United States of America, 101*(14), 4854–4858. https://doi.org/10.1073/pnas.0306210101

Kartsov G. (1903). *Belovezha Primeval forest. Its historical description, contemporary game management, and monarchial hunts in the Forest*. A. Marks, St. Petersburg: 1–414.

Khaewphakdee, S., Simcharoen, A., Duangchantrasiri, S., Chimchome, V., Simcharoen, S., & Smith, J. L. D. (2020). Weights of gaur (Bos gaurus) and banteng (Bos javanicus) killed by tigers in Thailand. *Ecology and Evolution, 10*(11), 5152–5159. https://doi.org/10.1002/ece3.6268

Khelkar, P. P. (2022). Rare footage of Royal Bengal tiger killing an Indian bison gaur. *India Today.*. https://www.indiatoday.in/india/story/footage-royal-bengal-tiger-indian-bison-gaur-watch-1927841-2022-03-21

Khoyetskyy, P. (2010). Causes of European bison death in the western part of Ukraine. *European Bison Conservation Newsletter, 3*, 25–32.

Kingdon, J. (2015). Cheetah Acinonyx jubatus. In *The Kingdon field guide to African mammals* (2nd ed., pp. 403–404). Bloomsbury.

Kittle, A., Watson, A., & Fernando, S. (2017). The ecology and behaviour of a protected area Sri Lankan leopard (Panthera pardus kotiya) population. *Tropical Ecology, 58*(1), 71–86.

Klich, D., Stereńczak, K., Lisiewicz, M., Sobczuk, M., Nieszała, A., & Olech, W. (2023). An assessment of the habitat preferences of European bison with airborne laser scanning data in forest ecosystem. *Scientific Reports, 13*, 17987. (2023). https://doi.org/10.1038/s41598-023-45280-3

Korte, L. M. (2008). Habitat selection at two spatial scales and diurnal activity patterns of adult female Forest Buffalo. *Journal of Mammalogy, 89*(1), 115–125. https://doi.org/10.1644/06-MAMM-A-423.1

Krasińska, M., Krasiński, Z., & Tokarska, M. (2013). *European bison: The nature monograph*. Springer.

Krasiński, Z. (1978). Bisoniana LXVII. Dynamics and structure of the European bison population in the Białowieża Primeval Forest. *Acta Theriologica, 23*, 3–48. Retrieved from http://rcin.org.pl/ibs/Content/10494/PDF/BI002_11474_Cz-40-2_Acta-T23-nr1-3-48_o.pdf

Krasiński, Z. A., Krasińska, M., & Leniec, H. (1994). The European bison in the KnyszynfForest. Parki Nar Rez Przyr, 13, 107–114.

Krishnakumar, B. M., Nagarajan, R., & Selvan, K. M. (2022). Diet composition and prey preference of Tiger, leopard, and dhole in Kalakkad-Mundanthurai Tiger Reserve, southern Western Ghats, India. *Mammal Study, 47*(3), 141–153. https://doi.org/10.3106/ms2020-0058

Kusi, N., Manandhar, P., Senn, H., Joshi, J., Ghazali, M., Hengaju, K. D., Suwal, S. P., Lama, T. L., Poudyal, L. P., Thapa, M., & Werhahn, G. (2021). Phylogeographical analysis shows the need to protect the wild yaks' last refuge in Nepal. *Ecology and Evolution, 11*(12), 8310–8318. https://doi.org/10.1002/ece3.7660

Lamichhane, B. R., Persoon, G. A., Leirs, H., Poudel, S., Subedi, N., Pokheral, C. P., Bhattarai, S., Thapaliya, B. P., & de Iongh, H. H. (2018). Spatio-temporal patterns of attacks on human and economic losses from wildlife in Chitwan national park, Nepal. *PloS one, 13*(4), e0195373. https://doi.org/10.1371/journal.pone.0195373

Lamichhane, B. R., Leirs, H., Persoon, G. A., Subedi, N., Dhakal, M., Oli, B. N., Reynaert, S., Sluydts, V., Pokheral, C. P., Poudyal, L. P., Malla, S., & de Iongh, H. H. (2019). Factors associated with co-occurrence of large carnivores in a human-dominated landscape. *Biodiversity and Conservation, 28*, 1473–1491. https://doi.org/10.1007/s10531-019-01737-4

Lang, S. D. J., & Farine, D. R. (2017). A multidimensional framework for studying social predation strategies. *Nature Ecology, & Evolution, 1*, 1230–1239. https://doi.org/10.1038/s41559-017-0245-0

Larter, N. C., Sinclair, A. R., & Gates, C. C. (1994). The response of predators to an erupting bison, Bison bison athabascae, population. *Canadian field-naturalist. Ottawa ON, 108*(3), 318–327.

Laundre, J. W., Hernandez, L., & Altendorf, K. B. (2001). Wolves, elk, and bison: Reestablishing the "landscape of fear" in Yellowstone National Park, USA. *Canadian Journal of Zoology, 79*, 1401–1409. https://doi.org/10.1139/z01-094

Lekagul, B., & McNeely, J. A. (1977). *Mammals of Thailand.* Association for the Conservation of Wildlife.

Leslie, D. M., & Schaller, G. B. (2009). Bos grunniens and Bos mutus (Artiodactyla: Bovidae). *Mammalian Species, 836*, 1–17. https://doi.org/10.1644/836.1

Li, J., Yin, H., Wang, D., Jiagong, Z., & Lu, Z. (2013). Human-snow leopard conflicts in the Sanjiangyuan region of the Tibetan plateau. *Biological Conservation, 166*, 118–123. https://doi.org/10.1016/j.biocon.2013.06.024

Liang, X., Kang, A., & Pettorelli, N. (2017). Understanding habitat selection of the vulnerable wild yak Bos Mutus on the Tibetan plateau. *Oryx, 51*(2), 361–369. https://doi.org/10.1017/S0030605315001155

Lott, D. F. (1991). American bison socioecology. *Applied Animal Behaviour Science, 29*(1–4), 135–145. https://doi.org/10.1016/0168-1591(91)90242-P

Lynam, A. J., Tantipisanuh, N., Chutipong, W., Ngoprasert, D., Baker, M. C., Cutter, P., et al. (2012). Comparative sensitivity to environmental variation and human disturbance of Asian tapirs (Tapirus indicus) and other wild ungulates in Thailand. *Integrative Zoology, 7*(4), 389–399.

MacNulty, D. R., Smith, D. W., Mech, L. D., Vucetich, J. A., & Packer, C. (2012). Nonlinear effects of group size on the success of wolves hunting elk. *Behavioral Ecology, 23*, 75–82. https://doi.org/10.1093/beheco/arr159

MacNulty, D. R., Tallian, A., Stahler, D. R., & Smith, D. W. (2014). Influence of group size on the success of wolves hunting bison. *PLoS One, 9*(11), e112884. https://doi.org/10.1371/journal.pone.0112884

MacNulty, D. R., Stahler, D. R., Wyman, C. T., Ruprecht, J., & Smith, D. W. (2016). The challenge of understanding northern Yellowstone elk dynamics after wolf reintroduction. In D. W. Smith, D. R. Stahler, D. R. MacNulty, & S. Haas (Eds.), *Yellowstone science* (pp. 25–33). Yellowstone National Park, Mammoth.

Mao, J. S., Boyce, M. S., Smith, D. W., Singer, F. J., Vales, D. J., Vore, J. M., & Merrill, E. H. (2005). Habitat selection by elk before and after wolf reintroduction in Yellowstone National Park. *The Journal of Wildlife Management, 69*(4), 1691–1707. https://doi.org/10.2193/0022-541X(2005)69[1691:HSBEBA]2.0.CO;2

Mas-Carrió, E., Churski, M., Kuijper, D., & Fumagalli, L. (2024). Niche overlap across landscape variability in summer between two large herbivores using eDNA metabarcoding. *PLoS One, 19*(2), e0279672. https://doi.org/10.1371/journal.pone.0279672

Matsuda, I., Tuuga, A., & Higashi, S. (2008). Clouded leopard (Neofelis diardi) predation on proboscis monkeys (Nasalis larvatus) in Sabah, Malaysia. *Primates; Journal of Primatology, 49*(3), 227–231. https://doi.org/10.1007/s10329-008-0085-2

Mattson, D. J. (1997). Use of ungulates by Yellowstone grizzly bears Ursus arctos. *Biological Conservation, 81*, 161–177. https://doi.org/10.1016/S0006-3207(96)00142-5

Mech, L. D., & Peterson, R. O. (2003). Wolf-prey relations. In L. D. Mech & L. Boitani (Eds.), *Wolves: Behavior, ecology and conservation* (pp. 131–160). The University of Chicago Press.

Mech, L. D., Smith, D. W., & MacNulty, D. R. (2015). *Wolves on the Hunt: The behavior of wolves hunting wild prey*. University of Chicago Press.

Menon, A. (2024). *More wild water buffaloes in central India to fix population decline and ecological degradation*. Retrieved from https://india.mongabay.com/2024/02/more-wild-water-buffaloes-in-central-india-to-fix-population-decline-and-ecological-degradation/

Metz, M. C., Smith, D. W., Vucetich, J. A., Stahler, D. R., & Peterson, R. O. (2012). Seasonal patterns of predation for gray wolves in the multi-prey system of Yellowstone National Park. *The Journal of Animal Ecology, 81*(3), 553–563. https://doi.org/10.1111/j.1365-2656.2011.01945.x

Metzger, K. L., Sinclair, A. R. E., Hilborn, R., Grant, J., Hopcraft, C., & Mduma, S. A. R. (2010). Evaluating the protection of wildlife in parks: The case of African buffalo in Serengeti. *Biodiversity and Conservation, 19*, 3431–3444. https://doi.org/10.1007/s10531-010-9904-z

Miller, J. R., Jhala, Y. V., & Schmitz, O. J. (2016a). Human perceptions Mirror realities of carnivore attack risk for livestock: Implications for mitigating human-carnivore conflict. *PLoS One, 11*(9), e0162685. https://doi.org/10.1371/journal.pone.0162685

Miller, J. R. B., Jhala, Y. V., & Jena, J. (2016b). Livestock losses and hotspots of attack from tigers and leopards in Kanha Tiger Reserve, Central India. *Regional Environmental Change, 16*(Suppl 1), 17–29. https://doi.org/10.1007/s10113-015-0871-5

Mitchell, B. L., Shenton, J. B., & Uys, J. C. M. (1965). Predation on large mammals in the Kafue National Park, Zambia. *Zoologica Africana, 1*, 297–318.

Mloszewski, M. J. (1983). *The behavior and ecology of the African Buffalo*. Cambridge University Press.

Moran, M. D. (2014). Bison grazing increases arthropod abundance and diversity in a tallgrass prairie. *Environmental Entomology, 43*(5), 1174–1184. https://doi.org/10.1603/EN14013

Morino L. (2010). Clouded leopard predation on a wild juvenile siamang. *Folia Primatologica; International Journal of Primatology, 81*(6), 362–368. https://doi.org/10.1159/000324303

Mysterud, A., Barton, K., Jedrzejewska, B., Krasinski, Z., Niedziałkowska, M., Kamler, J., Yoccoz, N., & Stenseth, N. (2007). Population ecology and conservation of endangered megafauna: The case of European bison in Białowieża primeval Forest, Poland. *Animal Conservation, 10*(1), 77–87. https://doi.org/10.1111/j.1469-1795.2006.00075.x

Nagy-Reis, M. B., Iwakami, V. H. S., Esevo, C. A., & Setz, E. Z. F. (2019). Temporal and dietary segregation in a neotropical small-felid assemblage and its relation to prey activity. *Mammalian Biology, 95*, 1–8.

Namgail, T., Fox, J. L., & Bhatnagar, Y. V. (2007). Carnivore-caused livestock mortality in trans-Himalaya. *Environmental Management, 39*, 490–496.

Narasimmarajan, K., Mahato, S., & Parida, A. (2014). Population density and biomass of the wild prey species in a tropical deciduous forest, Central India. *TAPROBANICA: The Journal of Asian Biodiversity, 6*(1), 1–6. https://doi.org/10.4038/tapro.v6i1.7053

National Park Service. (2024). Bison. Retrieved from https://www.nps.gov/yell/learn/nature/bison.htm.

Ng, C. (2015). *Syncerus caffer*. African buffalo (on-line), Animal Diversity Web. https://animaldiversity.org/accounts/Syncerus_caffer/

Niedziałkowska, M., Hayward, M. W., Borowik, T., Jędrzejewski, W., & Jędrzejewska, B. (2019). A meta-analysis of ungulate predation and prey selection by the brown bear Ursus arctos in Eurasia. *Mammal Research, 64*, 1–9. https://doi.org/10.1007/s13364-018-0396-7

Nowak, R. (1999). *Walker's mammals of the world* (6th ed.). The John Hopkins University Press.

Nowell, K. & Jackson P. (Eds.) (1996). *Wild cats: Status survey and conservation action plan*. Gland: IUCN/SSC Cat Specialist Group.

Okarma, H. (1995). The trophic ecology of wolves and their predatory role in ungulate communities of forest ecosystems in Europe. *Acta Theriologica, 40*, 335–338.

Okarma, H., Jędrzejewska, B., Jędrzejewski, W., Krasiński, Z. A., & Miłkowski, L. (1995). The roles of predation, snow cover, acorn crop, and man-related factors on ungulate mortality in

Białowieża Primeval Forest, Poland. *Acta Theriologica, 40*, 197–217. https://rcin.org.pl/ibs/dlibra/publication/28323/edition/12344/content

Oli, M. K. (1994). Snow leopards and blue sheep in Nepal: Densities and predator: Prey ratio. *Journal of Mammalogy, 75*(4), 998–1004.

Oli, M. K., Taylor, I. R., & Rogers, D. M. (1993). Diet of the snow leopard (Panthera uncia) in the Annapurna conservation area, Nepal. *Journal of Zoology, 231*(3), 365–370.

Ososky, J. J. (1998). *Diet of leopards and golden cats in Ndoki Park, republic of Congo*. MSc thesis,. Northern Illinois University.

Owen-Smith, N. (2021). The big mammal menagerie: Herbivores, carnivores and their ecosystem impacts. In *Only in Africa: The ecology of human evolution* (pp. 141–246). Cambridge University Press.

Owen-Smith, N., & Mills, M. G. (2008a). Shifting prey selection generates contrasting herbivore dynamics within a large-mammal predator-prey web. *Ecology, 89*(4), 1120–1133. https://doi.org/10.1890/07-0970.1

Owen-Smith, N., & Mills, M. G. (2008b). Predator-prey size relationships in an African large-mammal food web. *The Journal of Animal Ecology, 77*(1), 173–183. https://doi.org/10.1111/j.1365-2656.2007.01314.x

Packer, C., & Ruttan, L. M. (1988). The evolution of cooperative hunting. *The American Naturalist, 132*, 159–198.

Pakpien, S., Simcharoen, A., Duangchantrasiri, S., Chimchome, V., Pongpattanurak, N., & Smith, J. L. D. (2017). Ecological correlates at kill sites influence tiger (Panthera tigris) hunting success in Huai Kha Khaeng Wildlife Sanctuary, Thailand. *Tropical Conservation Science, 10*, 1–7.

Palmer, M. S., & Packer, C. (2021). Reactive anti-predator behavioral strategy shaped by predator characteristics. *PLoS One, 16*(8), e0256147. https://doi.org/10.1371/journal.pone.0256147

Parks Canada. (2022). Plains bison facts. https://parks.canada.ca/pn-np/sk/princealbert/nature/bisons/faits-facts.

Perzanowski, K., Welz, M., Januszczak, M., Kaczor, S., & Wołoszyn-Gałęza, A. (2014). Mortality cases of wisents in Bieszczady Mountains. *European Bison Conservation Newsletter, 7*, 57–62.

Phillipps, Q. (2016). *Banteng and water Buffalo. Phillipps' field guide to the mammals of Borneo: Sabah, Sarawak, Brunei, and Kalimantan*. Princeton University Press.

Pienaar, U. D. V. (1969). Predator-prey relationships amongst the larger mammals of the Kruger National Park. *Koedoe, 12*(1), 108–175. https://doi.org/10.4102/koedoe.v12i1.753

Plumb, G., Kowalczyk, R., & Hernandez-Blanco, J. A. (2020). Bison bonasus. The IUCN Red List of Threatened Species 2020: e.T2814A45156279. https://doi.org/10.2305/IUCN.UK.2020-3.RLTS.T2814A45156279.en.

Pocock, R. I. (1939). *The fauna of British India. Mammalia, I. Primates and carnivore*. Taylor and Francis.

Prayurasiddhi, T. (1997). *The ecological separation of gaur (Bos gaurus) and banteng (Bos javanicus) in Huai Kha Khaeng wildlife sanctuary, Thailand*. PhD thesis,. University of Minnesota.

Prin, T., & Prins, H. H. (2014). African buffalo Syncerus caffer (Sparrman, 1779). In *Ecology, evolution and behaviour of wild cattle: Implications for conservation* (pp. 326–372). Cambridge University Press.

Prins, H. (1996a). *Ecology and behaviour of the African Buffalo*. Chapman, & Hall.

Prins, H. H. T. (1996b). *Ecology and behaviour of the African Buffalo. Chapman, & Hall Wildlife Ecology and behaviour series (WEBS)*. Chapman, & Hall.

Prins, H., & Iason, G. (1989). Dangerous lions and nonchalant buffalo. *Behaviour, 108*, 262–296.

Prommakul, P. (2003). Habitat utilization and prey of the tiger (Panthera tigris [Linnaeus]) in the eastern Thung Yai Naresaun wildlife sanctuary. *Journal of Wildlife in Thailand, 11*, 1–12.

Pucek, Z. (Ed.); Pucek, Z., Belousova, I. P., Krasińska, M., Krasiński Z. A., & Olech, W. (comps.) (2004). European bison. Status survey and conservation action plan. IUCN/SSC Bison Specialist Group.

Pucek, Z., Belousova, I., Krasinski, Z., Krasinska, M., & Olech, W. (2003). European bison (Bison bonasus) current state of the species and an action plan for its conservation. *Convention on the Conservation of European Wildlife and Natural Habitats, 20*, 28–29.

Pudyatmoko, S. (2019). Spatiotemporal inter-predator and predator–prey interactions of mammalian species in a tropical savanna and deciduous forest in Indonesia. *Mammal Research, 64*, 191–202. https://doi.org/10.1007/s13364-018-0391-z

Pudyatmoko, S., Djuwantoko, & Sabarno, Y. (2007). Evidence of Banteng (Bos javanicus) decline in Baluran National Park, Indonesia. *Journal of Biological Sciences, 7*, 854–859. https://doi.org/10.3923/jbs.2007.854.859

Rabinowitz, A. (1989). The density and behavior of large cats in a dry tropical forest mosaic in Huai Kha Khaeng wildlife sanctuary, Thailand. *Natural History Bulletin of the Siam Society, 37*, 235–251.

Rabinowitz, A., Andau, P., & Chai, P. P. K. (1987). The clouded leopard in Malaysian Borneo. *Oryx, 21*, 107–111.

Rahman, D. A., Herliansyah, R., Rianti, P., Rahmat, U. M., Firdaus, A. Y., & Syamsudin, M. (2019). Ecology and conservation of the endangered Banteng (Bos javanicus) in Indonesia tropical lowland Forest. *HAYATI Journal of Biosciences, 26*(2), 68. https://doi.org/10.4308/hjb.26.2.68

Rahman, D. A., Syamsudin, M., Firdaus, A. Y., Afriandi, H. T., & Anggodo. (2021). Photographic record of dholes predating on a young Banteng in southwestern Java, Indonesia. *Journal of Threatened Taxa, 13*(14), 20278–20283. https://doi.org/10.11609/jott.7498.13.14.20278-20283

Rajaratnam, R., Vernes, K., & Sangay, T. (2016). A Review of livestock predation by large carnivores in the Himalayan Kingdom of Bhutan. In F. Angelici (Ed.), *Problematic wildlife*. Springer. https://doi.org/10.1007/978-3-319-22246-2_8

Ramakrishnan, U., Coss, R. G., & Pelkey, N. W. (1999). Tiger decline caused by the reduction of large ungulate prey: Evidence from a study of leopard diets in southern India. *Biological Conservation, 89*(2), 113–120. https://doi.org/10.1016/S0006-3207(98)00159-1

Ramesh, T., Kalle, R., Sankar, K., & Qureshi, Q. (2012). Spatio-temporal partitioning among large carnivores in relation to major prey species in Western Ghats. *Journal of Zoology, 287*(4), 269–275. https://doi.org/10.1111/j.1469-7998.2012.00908.x

Rasphone, A., Kamler, J. F., & Macdonald, D. W. (2020). Temporal partitioning by felids, dholes, and their potential prey in northern Laos. *Mammal Research, 65*, 679–689.

Rasphone, A., Bousa, A., Vongkhamheng, C., Kamler, J. F., Johnson, A., & Macdonald, D. W. (2022). Diet and prey selection of clouded leopards and tigers in Laos. *Ecology and Evolution, 12*(7), e9067. https://doi.org/10.1002/ece3.9067

Rauniyar, T. (2023). Nepal fails in ambitious attempt to revive wild water buffalo population. https://dialogue.earth/en/nature/nepal-fails-in-ambitious-attempt-to-revive-wild-water-buffalo-populations/.

Ray, J., & Sunquist, M. (2001). Trophic relations in a community of African rainforest carnivores. *Oecologia, 127*(3), 395–408. https://doi.org/10.1007/s004420000604

Reddy, H. S., Srinivasulu, C., & Thulsi Rao, K. (2004). Prey selection by the Indian tiger (Panthera tigris tigris) in Nagarjunasagar Srisailam Tiger Reserve, India. *Mammalian Biology, 69*, 384–391. https://doi.org/10.1078/1616-5047-00160

Reynolds, H. W., & Hawley, A. W. L. (1987). Bison ecology in relation to agricultural development in the Slave River lowlands, Northwest Territories. Occasional paper no. 63. Canadian wildlife service.

Ripple, W. J., & Beschta, R. L. (2003). Wolf reintroduction, predation risk, and cottonwood recovery in Yellowstone National Park. *Forest Ecology and Management, 184*, 299–313. https://doi.org/10.1016/S0378-1127(03)00154-3

Ripple, W. J., Painter, L. E., Beschta, R. L., & Gates, C. C. (2010). Wolves, elk, bison, and secondary trophic cascades in Yellowstone National Park. *The Open Environmental Research Journal, 3*(1). https://doi.org/10.2174/1874213001003030031

Ripple, W. J., Painter, L. E., Beschta, R. L., & Gates, C. C. (2011). Wolves, elk, bison, and secondary trophic cascades in Yellowstone National Park. *The Open Environmental Research Journal, 3*(1), 31. https://doi.org/10.2174/1874213001003030031

Rossi, L., Scuzzarella, C. M., & Angelici, F. M. (2020). Extinct or perhaps surviving relict populations of big cats: Their controversial stories and implications for conservation. In F. Angelici & L. Rossi (Eds.), *Problematic wildlife II*. Springer. https://doi.org/10.1007/978-3-030-42335-3_12

Rostro-García, S., Kamler, J. F., & Hunter, L. T. (2015). To kill, stay or flee: The effects of lions and landscape factors on habitat and kill site selection of cheetahs in South Africa. *PLoS One, 10*(2), e0117743. https://doi.org/10.1371/journal.pone.0117743

Rostro-García, S., Kamler, J. F., Crouthers, R., Sopheak, K., Prum, S., In, V., Pin, C., Caragiulo, A., & Macdonald, D. W. (2018). An adaptable but threatened big cat: Density, diet and prey selection of the Indochinese leopard (Panthera pardus delacouri) in eastern Cambodia. *Royal Society Open Science, 5*(2), 171187. https://doi.org/10.1098/rsos.171187

Roth, J. (2004). *Bubalus bubalis* wild water buffalo (on-line), Animal Diversity Web. https://animaldiversity.org/accounts/Bubalus_bubalis/.

Roug, A., Muse, E. A., Clifford, D. L., Larsen, R., Paul, G., Mathayo, D., Mpanduji, D., Mazet, J. A. K., Kazwala, R., Kiwango, H., & Smith, W. (2020). Seasonal movements and habitat use of African buffalo in Ruaha National Park, Tanzania. *BMC ecology, 20*(1), 6. https://doi.org/10.1186/s12898-020-0274-4

Ruth, T. K., & Murphy, K. (2010). Competition with other carnivores for prey. In M. Hornocker & S. Negri (Eds.), *Cougar: Ecology and conservation* (pp. 163–172). University of Chicago Press.

Saari, J. (2002). *Bos javanicus*. Banteng (On-line), Animal Diversity Web. Accessed 18 Mar 2024 at https://animaldiversity.org/accounts/Bos_javanicus/

Saisamorn, A., Duengkae, P., Pattanavibool, A., Duangchantrasiri, S., Simcharoen, A., & Smith, J. L. (2019). Spatial and temporal analysis of leopards (Panthera pardus), their prey and tigers (Panthera tigris) in Huai Kha Khaeng wildlife sanctuary, Thailand. *Folia Oecologica, 46*(2), 73–82. https://doi.org/10.2478/foecol-2019-0010

Saltz, J. B., Palmer, M. S., & Beaudrot, L. (2023). Identifying the social context of single- and mixed-species group formation in large African herbivores. *Philosophical Transactions of the Royal Society of London. Series B, Biological Sciences, 378*(1878), 20220105. https://doi.org/10.1098/rstb.2022.0105

Schaller, G. (1967). *The deer and the Tiger: A study of wildlife in India*. University of Chicago Press.

Schaller, G. B. (1972). *The Serengeti lion: A study of predator-prey relations*. University. Chicago Press.

Schaller, G. B. (1977). *Mountain monarchs. Wild sheep and goats of the Himalaya*. University of Chicago Press.

Schaller, G. B., & Junrang, R. (1988). Effects of a snowstorm on Tibetan antelope. *Journal of Mammalogy, 69*(3), 631–634.

Schaller, G. B., & Liu, W. (1996). Distribution, status, and conservation of wild yak Bos grunniens. *Biological Conservation, 76*(1), 1–8. https://doi.org/10.1016/0006-3207(96)85972-6

Scheel, D., & Packer, C. (1991). Group hunting behaviour of lions: A search for cooperation. *Animal Behaviour, 41*(4), 697–709. https://doi.org/10.1016/S0003-3472(05)80907-8

Scheepers, J. L., & Gilchrist, D. (1991). Leopard predation on giraffe calves in the Etosha National Park. *Madoqua, 18*(1), 49. https://journals.co.za/doi/pdf/10.10520/AJA10115498_346

Schnitzler, A., & Hermann, L. (2019). Chronological distribution of the tiger Panthera tigris and the Asiatic lion Panthera leo persica in their common range in Asia. *Mammal Review, 49*(4), 340–353. https://doi.org/10.1111/mam.12166

Schooler, S. L., Finnegan, S. P., Fowler, N. L., Kellner, K. F., Lutto, A. L., Parchizadeh, J., van den Bosch, M., Zubiria Perez, A., Masinde, L. M., Mwampeta, S. B., Boone, H. M., Gantchoff, M. G., Hill, J. E., Kautz, T. M., Wehr, N. H., Fyumagwa, R., & Belant, J. L. (2022). Factors influencing lion movements and habitat use in the western Serengeti ecosystem, Tanzania. *Scientific Reports, 12*(1), 18890. https://doi.org/10.1038/s41598-022-22053-y

Seidensticker, J., & Mcdougal, C. (1993a). Tiger predatory behaviour, ecology and conservation. In N. Dunstone & M. L. Gorman (Eds.), *Mammals as predators: The proceedings of a symposium held by the Zoological Society of London and the mammal society: London, 22nd and 23rd November 1991*. Oxford Academic. https://doi.org/10.1093/oso/9780198540670.003.000

Seidensticker, J., & McDougal, C. (1993b). Tiger predatory behaviour, ecology and conservation. In N. Dunstone & M. L. Gorman (Eds.), *Mammals as predators: The proceedings of a symposium held by the Zoological Society of London and the mammal society, 105–125* (p. 65). The Zoological Society of London. https://doi.org/10.1093/oso/9780198540670.003.0006

Sheldon, J. W., Reed, G., Burnett, A. C., Li, K., & Crabtree, R. L. (2009). Coyote, Canis latrans, predation on a bison, Bison bison, calf in Yellowstone National Park. *Canadian Field Naturalist, 123*(3), 260–261. https://doi.org/10.2478/foecol-2019-0010

Shrestha, B., Aihartza, J., & Kindlmann, P. (2018). Diet and prey selection by snow leopards in the Nepalese Himalayas. *PLoS One, 13*(12), e0206310. https://doi.org/10.1371/journal.pone.0206310

Simcharoen, A., Simcharoen, S., Duangchantrasiri, S., Bump, J., & Smith, J. L. D. (2018). Tiger and leopard diets in western Thailand: Evidence for overlap and potential consequences. *Food Webs, 15*, e00085.

Sinclair, A. R. E. (1977). *The African buffalo: A study of resource limitation of populations*. University of Chicago Press.

Sinclair, A. R., Mduma, S., & Brashares, J. S. (2003). Patterns of predation in a diverse predator-prey system. *Nature, 425*(6955), 288–290. https://doi.org/10.1038/nature01934

Smith, A. T., & Xie, Y. (2008). *A guide to the mammals of China*. Princeton University Press.

Stander, P. E. (1992). Cooperative hunting in lions: The role of the individual. *Behavioral Ecology and Sociobiology, 29*, 445–454. https://doi.org/10.1007/BF00170175

Steenweg, R., Hebblewhite, M., Gummer, D., Low, B., & Hunt, B. (2016). Assessing potential habitat and carrying capacity for reintroduction of plains bison (Bison bison bison) in Banff National Park. *PLoS One, 11*(2), e0150065. https://doi.org/10.1371/journal.pone.0150065

Steinmetz, R. (2004). Gaur (Bos gaurus) and Banteng (B. Javanicus) in the lowland forest mosaic of Xe Pian protected area, Lao PDR: Abundance, habitat use, and conservation. *Mammalia, 68*(2–3), 141–157. https://doi.org/10.1515/mamm.2004.015

Steinmetz, R., Seuaturien, N., & Chutipong, W. (2013). Tigers, leopards, and dholes in a half-empty forest: Assessing species interactions in a guild of threatened carnivores. *Biological Conservation, 163*, 68–78. https://doi.org/10.1016/j.biocon.2012.12.016

Sunderland-Groves, J. L., Tandang, M. V., Patispathika, F. H., Marzec, A., Knox, A., Nurcahyo, A., Husson, S. J., & Sihite, J. (2020). Suspected Sunda clouded leopard (Neofelis diardi) predation attempts on two reintroduced Bornean orangutans (Pongo pygmaeus wurmbii) in Bukit Batikap Protection Forest, Central Kalimantan, Indonesia. *Primates, 62*(1), 41–49. https://doi.org/10.1007/s10329-020-00842-1

Suryawanshi, K. R., Bhatnagar, Y. V., Redpath, S., & Mishra, C. (2013). People, predators, and perceptions: Patterns of livestock depredation by snow leopards and wolves. *Journal of Applied Ecology, 50*(3), 550–560. https://doi.org/10.1111/1365-2664.12061

Tallian, A., Smith, D. W., Stahler, D. R., Metz, M. C., Wallen, R. L., Geremia, C., Ruprecht, J., Wyman, C. T., & MacNulty, D. R. (2017). Predator foraging response to a resurgent dangerous prey. *Functional Ecology, 31*(7), 1418–1429. https://doi.org/10.1111/1365-2435.12866

Tambling, C. J., Druce, D. J., Hayward, M. W., Castley, J. G., Adendorff, J., & Kerley, G. I. (2012). Spatial and temporal changes in group dynamics and range use enable anti-predator responses in African buffalo. *Ecology, 93*(6), 1297–1304. https://doi.org/10.1890/11-1770.1

Theuerkauf, J., & Rouys, S. (2008). Habitat selection by ungulates in relation to predation risk by wolves and humans in the Białowieża Forest, Poland. *Forest Ecology and Management, 256*(6), 1325–1332. https://doi.org/10.1016/j.foreco.2008.06.030

Tilson, R., & Nyhus, P. J. (2009). *Tigers of the world: The science, politics and conservation of Panthera tigris*. Academic.

Treves, A., & Naughton-Treves, L. (1999). Risk and opportunity for humans coexisting with large carnivores. *Journal of Human Evolution, 36*(3), 275–282.

Turner, A., & Anton, M. (1997). *The big cats and their fossil relatives*. Columbia University Press.

Tutin, C. E. G., White, L. J. T., & Mackanga-Missandzou, A. (1997). The use by rain forest mammals of natural forest fragments inan equatorial African savanna. *Conservation Biology, 11*, 1190–1203. https://doi.org/10.1046/j.1523-1739.1997.96211.x

Variar, A. S., Anoop, N. R., Komire, S., Vinayan, P. A., Sujin, N. S., Raj, A., & Prasadan, P. K. (2023). Prey selection by the Indian tiger (Panthera tigris tigris) outside protected areas in India's Western Ghats: Implications for conservation. *Food Webs, 34*, e00268. https://doi.org/10.1016/j.fooweb.2022.e00268

Varley, N., & Boyce, M. S. (2006). Adaptive management for reintroductions: Updating a wolf recovery model for Yellowstone National Park. *Ecological Modelling, 193*(3–4), 315–339. https://doi.org/10.1016/j.ecolmodel.2005.09.001

Wentworth, J., Tambling, C., & Kerley, G. (2011). Evidence for prey selection by spotted hyaena in the eastern cape, South Africa. *Acta Theriologica, 56*, 389–392.

White, L. J. T. (1994). Biomass of rain forest mammals in the lope´ reserve, Gabon. *Journal of Animal Ecology, 63*(3), 449–512. https://doi.org/10.2307/5217

White, P. J., & Garrott, R. A. (2005). Yellowstone's ungulates after wolves—Expectations, realizations, and predictions. *Biological Conservation, 125*(2), 141–152. https://doi.org/10.1016/j.biocon.2005.01.048

White, P. J., Wallen, R. L., & Hallac, D. (2015). *Yellowstone bison—Conserving an American icon in modern society*. Yellowstone Association, Yellowstone National Park.

Whyte, I. J. (2004). *Census results for elephant and buffalo in the Kruger National Park between 1997 and 2004*. Scientific report 03/04. Internal Report. Kruger National Park, South African National Parks, Pretoria, South Africa.

Wright, G. J., Peterson, R. O., Smith, D. W., & Lemke, T. O. (2001). Selection of northern Yellowstone elk by gray wolves and hunters. *The Journal of Wildlife Management, 70*, 1070–1078. https://doi.org/10.2193/0022-41X(2006)70[1070:SONYEB]2.0.CO;2

Wright, G. J., Peterson, R. O., Smith, D. W., & Lemke, T. O. (2006). Selection of northern Yellowstone elk by Gray wolves and hunters. *The Journal of Wildlife Management, 70*(4), 1070–1078. https://doi.org/10.2193/0022-541X(2006)70[1070:SONYEB]2.0.CO;2

Wuerthner. (2020). *Bison ecology, ecological influence, behavior, and decline*. Retrieved from https://www.thewildlifenews.com/2020/09/02/bison-ecology-ecological-influence-behavior-and-decline/

Wyman, T. (2002). Grizzly bear predation on a bull bison in Yellowstone National Park. *Ursus, 13*, 375–377. http://www.jstor.org/stable/3873220

Xu, A. C., Jiang, Z. G., Li, C. W., Guo, J. X., Wu, G. S., & Cai, P. (2007). Summer food habits of brown bears in Kekexili nature reserve, Qinghai-Tibetan plateau, China (vol 17, pg 132, 2006). *Ursus, 18*(1), 132–137.

Yang, H., Zhao, X., Han, B., Wang, T., Mou, P., Ge, J., & Feng, L. (2018). Spatiotemporal patterns of Amur leopards in Northeast China: Influence of tigers, prey, and humans. *Mammalian Biology, 92*, 120–128.

Zangmo, P., Gurung, D. B., Letro, L., & Wangmo, S. (2018). Distribution, abundance and occupancy of gaur (Bos gaurus Smith) in the Royal Manas National Park, Bhutan. *Bhutan Journal of Natural Resources & Development, 5*(1), 1–12. https://doi.org/10.17102/cnr.2018.01

Zhao, H., Wang, N., Cheng, H., Wang, Y., Liu, X., Qiao, B., & Zhao, L. (2024). Mapping conservation priorities for wild yak (Bos mutus) habitats on the Tibetan plateau, China. *The Science of the Total Environment, 914*, 169803. https://doi.org/10.1016/j.scitotenv.2023.169803

Cattle Ancestry, Domestication and Carnivores

Introduction

Domesticated cattle (species *Bos taurus* Linnaeus 1758, genus *Bos* L. 1758; subfamily Bovinae J. E. Gray, 1821), are large, bovid ungulates, currently distributed across five continents (as defined by the United Nations Geoscheme, by the United Nations Statistics Division, UNSD, 2024). Due to their enormous global population, and ubiquitous existence and use in human life spaces, domesticated cattle define the various wild cattle species and overlap with these other species through semi-domesticated breeds such as the domesticated yak (from the Wild yak), domesticated water buffalo (from the wild water buffalo), and the gayal (from the wild gaur) (Simoons, 1984; Jacques et al., 2021). The origins of these domesticated cattle are frequently debated and even disputed (Arbuckle & Kassebaum, 2021:10). Many authorities dispute two possible strands: cattle either emerged from one source or several sources. For example, Arbuckle and Kassebaum (2021: 10) argue that early morphologically domestic cattle and pre-domestic cattle farming most likely originated in multiple Southwest Asian regions rather than in a single location.

The origins of cattle domestication are traced to the extinct aurochs (*Bos primigenius primigenius* (Bojanus, 1827)), mostly of the southwestern Asian region. Zeyland et al. (2013: 265) cite the "complete mitochondrial genome of wild aurochs reconstructed from ancient DNA" and argue that the extinct aurochs, which have been assessed as the ancestor of domestic cattle, were among the largest wild animals inhabiting Europe, Asia, and North Africa. This species gradually became extinct, with the extinction process climaxing in Poland in 1627, where the last recorded aurochs (a female) died. However, some aspects of the history of cattle domestication and the distribution of aurochs genetic material in modern cattle breeds are uncertain. In their study of a Polish case study "the mtDNA sequence of the analyzed PWA (Polish Wild Aurochs) individual belongs to haplogroup P" (Zeyland et al., 2013: 265).

Regarding the global domestication of cattle and the impact of ancestral genes, Decker et al. (2014) note that despite the importance of cattle for human

© The Author(s), under exclusive license to Springer Nature Switzerland AG 2025

M. O. Campbell, *Cattle, Their Predators and Geomatics Research*,
https://doi.org/10.1007/978-3-031-97363-5_3

development, there is little information available about the history of cattle populations and breeds, especially for American, African, and Asian breeds. The authors state that "cattle migration, movement and trading followed by admixture have been important forces in shaping modern bovine genomic variation" (Decker et al., 2014). Using an autosomal single nucleotide polymorphism marker-based study on 134 domesticated cattle breeds, the findings indicate that the three major groups of Asian indicine, Eurasian taurine, and African taurine were consistently detected. Features detected or inferred included indicine introgression into Africa, wild African aurochs ancestry into African taurine, causing a divergence of the latter from Eurasian taurine, Eurasian taurine/indicine hybridization in Asia, and the effect of species other than *Bos taurus taurus* and *B. t. indicus* in Asian cattle (Decker et al., 2014). Other findings were the introgression of African taurine into Iberian and Italian cattle, the Iberian source of American Criollo cattle (rather than directly from Africa, with African ancestry from Iberian ancestors), and indicine introgression into American cattle in the Americas (rather than in Europe).

The general consensus is that there are three original clusters: taurine cattle, domesticated from wild aurochs (*Bos primigenius* Bojanus 1827) (Table 1) about 10,500 years BP, in the region of central Anatolia to Western Iran; zebu cattle, domesticated from the Indian form of aurochs between 7000 and 6000 BP by Mesopotamian peoples in the region of modern Pakistan; and possibly an African cluster, sometimes termed sanga cattle, variously named *Bos africanus* (Kerr, 1792), *Bos sanga* (Fitzinger, 1860), or *Bos taurus africanus* (Blench & MacDonald, 2006; Edwards et al., 2007; Ajmone-Marsan et al., 2010; Mwai et al., 2015). Felius et al. (2014: 705) note that "several bovine species have been domesticated but taurine cattle (*Bos taurus*) and zebu (*Bos indicus*, account for almost all cattle. Both descend from the wild aurochs (*Bos primigenius*), which at the end of the last glacial period (12,000 BP) was endemic over most of Asia, Europe, North Africa and the once green Sahara. This huge and reputedly fierce species has been extinct since 1627, when the last animal died in Poland. Only few contemporary pictures of aurochs exist, but skeletal remains allow reconstructing its morphology. The subspecies *B.p. primigenius* in Southwest Asia and *B.p. namadicus* in India were the ancestors of taurine and zebu cattle, respectively." Regarding the species and genera of the Bovini tribe, Verkaar et al. (2004: 1165) also argue that "reconstructions of their

Table 1 The physical characteristics of the aurochs

Physical attribute	Quantification or description
Length (cm)	<300
Shoulder height (cm)	135–180
Weight (kg)	<2200
Coat color	Dark brown to black
Horns (cm)	70–107
Habitat	Open land cover for grazing
Domestication	Evidence of smaller size
Predators	Wolves, bears, big cats for calves

Sources: Kysely (2008), Goderie et al. (2015), and Petruzzello (2024)

phylogeny are currently incomplete" but "mitochondrial DNA indicates that four distinct maternal lineages diverged after an early split in buffalo species," these being taurine cattle and zebu, bison, American bison and yak, and banteng, gaur, and gayal, respectively. Xia et al. (2023) and other authors studying the global distribution and adaptive evolution of domestic cattle from a genomic perspective list several types of inter- and intraspecies introgressions that suggest adaptation and complex lineages (Table 2). They point out that modern cattle belong to seven major mitochondrial haplogroups and that taurine and indicine cattle, respectively, are descended from two genetically distinct and geographically separate aurochs ancestors from western and southern Asia, as supported by molecular evidence for uniparental and autosomal markers. Furthermore, "at least" eight major autosomal ancestral groups exist (Xia et al., 2023) (Table 3).

The domestication process is justified by the beneficial impact of cattle domestication on ancient human livelihoods, as human societies became increasingly structured and sophisticated, and cattle provided meat, hides, milk, horns for glue, weapons, and transport. Sahu et al. (2020: 50) note that "cattle are the most popular animal after dogs that are closely associated with humans." Arbuckle and Kassebaum (2021: 10) describe cattle as among "the most significant animal partners in human history, and the origins of cattle management as well as domestic cattle have been the focus of scholarly interest for decades" (see also older literature, such as Peake & Fleure, 1927; Reed, 1960). Decker et al. (2014) add that "the domestication and development of cattle has considerably impacted human societies…." Ajmone-Marsan et al. (2010: 148) also write that "since their domestication in the Neolithic, cattle have belonged to our cultural heritage. The reconstruction of their history is an active field of research that contributes to our understanding of human history." Stuart (2014) adds that "as the most common type of large, domesticated ungulate, cattle play an integral role in supporting the world's human population, whether it is through food (meat, dairy products), the production of leather, dung for fuel or manure or as draft animals for agricultural use (mainly in poorer and less-developed countries)."

An important issue concerns the extent of genetic change in cattle after domestication. Most studies use genome-based research, including assessment of Y-DNA, mtDNA, and autosomal DNA to examine evolutionary trends with domestication, which may be due to environmental changes. For example, Xia et al. (2023) note

Table 2 Seven major mitochondrial haplogroups and others

Breeding group	Mitochondrial haplogroups	Classifications
Taurine	T1, T2, T3, T4, T5	Major mitochondrial haplogroups
Indicine	I1, I2	
Rare aurochs introgressions	E, R, P, Q, and C	Rare mitochondrial haplogroups
Taurine	Y1, Y2a, Y2b	Y chromosome haplogroups
Indicine	Y3a, Y3b	

Chen et al. (2010), Lenstra et al. (2014), Xia et al. (2019a), Xia et al. (2023), Zhang et al. (2020), Xia et al. (2021), Cubric-Curik et al. (2021), Xia et al. (2019b), Cao et al. (2019), Edwards et al. (2007), and Pérez-Pardal et al. (2018)

Table 3 The eight main autosomal ancestral groups

Cattle group	Environmental range description
African taurine cattle	Tropical forest and savanna, humid and sub-humid environments, tsetse fly-infested, in west, central, and east Africa (Gautier et al., 2009; Kim et al., 2017)
East Asian taurine cattle	Temperate and frigid, hypoxic environments, some of with alleles from yak introgression, Northeast Asia and the Qinghai-Tibetan Plateau (Chen et al., 2018a; Wu et al., 2018)
Eurasian taurine cattle	Semiarid, frigid, temperate, subtropical open grassland and mixed vegetation in Central Asia (Chen et al., 2021; Kantanen et al., 2009)
European taurine cattle	Temperate climates, mixed forest and grassland, some with alleles from European aurochs, ancestors of most breeds (Achilli et al., 2008; Park et al., 2015).
Indian-Pakistani indicine cattle	Tropical forest, grassland, and mixed vegetation in hot and semiarid regions, with some mixture with wild species including aurochs, gaur, and water buffalo (Chen et al., 2010)
African indicine cattle	Tropical, semiarid savanna in East and Central Africa, with some mixture with African taurine and South Asian indicine breeds (Bahbahani et al., 2017; Kim et al., 2017, 2020)
Diversified East Asian indicine cattle	Tropical, hot-humid mixed vegetation, with alleles from other wild and/or domestic Asian bovine species (Chen et al., 2018a; Sinding et al., 2021)
Indonesian breeds	Tropical humid forest, with a mixture of indicine, banteng, and/or Bali cattle ancestries (Mohamad et al., 2009; Sudrajad et al., 2020)

Sources: Felius et al. (2011), Chen et al. (2018a, 2020), Utsunomiya et al. (2019), Verdugo et al. (2019), Felius et al. (2014), Gao et al. (2017), Ajmone-Marsan et al. (2010), Gautier et al. (2009), Kim et al. (2017), Wu et al. (2018), Chen et al. (2021), Kantanen et al. (2009), Achilli et al. (2008), Park et al. (2015), Chen et al. (2010), Bahbahani et al. (2017), Kim et al. (2017, 2020), Sinding et al. (2021), Mohamad et al. (2009), and Sudrajad et al. (2020)

that there are domestic cattle throughout the world that live in environments that are constantly changing and unpredictable. A range of local environmental and management conditions, such as UV radiation, disease, and stable feeding systems, have enabled them to adapt to various selection pressures to which they have been subjected. Different phenotypic and genetic characteristics between today's cattle breeds/populations are a consequence of this selection pressure. The original aurochs lineage, described below, is no longer strongly represented in these cattle. Recent studies on cattle ancestry have been based on autosomal ancestral groups, mitochondrial haplogroups, and Y-chromosomal haplogroups (Table 2). Cattle may also have adapted to environmental change (Table 4).

Lenstra et al. (2014) argue that mitochondrial DNA (mtDNA) analysis has recently become critical for molecular genetic studies of the diversity and history of cattle populations and ancestry. This meta-analysis of bovine mtDNA sequences from both current and archaeological samples provides a global picture of the geographical differentiation of maternal lineages shaped by the global distribution of cattle. Ancient DNA studies contribute significantly to historical reconstruction by providing geographical and historical information. Examples of research advances based on this technique include the determination of two domestication events, for taurine (*Bos taurus*) and zebu (*Bos indicus*), the introduction of taurine mtDNA in

Table 4 Cattle adaptation to cold, hot, and high-altitude environments

Environment	Biological adaptation
Adaptation to a cold environment	Yakut cattle range to the Arctic circle and have adapted to frigid winters (<−50 °C). Yakut cattle have a breed-specific mutation in an evolutionarily conserved NRAP gene adapting to cold temperatures. NRAP amino acid change may slow down metabolism and enhance their heart function to supply blood during the winter periods
Adaptation to a hot environment	Indicine cattle have several phenotypes that reflect adaptations to tropical environments, including their hump, large ears, and excess skin. These features adapt the body temperature to the balance between metabolic heat production and loss
Adaptation to high altitudes	Cattle physiological adaptations to hypoxic environments. EPAS1 is a gene for adaption to cold altitudes. HMGA2 is another such gene and ADH7 gene relates to shortness in Tibetan cattle (Wang et al., 2021)

Xia et al. (2023), Buggiotti et al. (2021), Weldenegodguad et al. (2019), Truszkowska et al. (2017), Berman (2011), Utsunomiya et al. (2019), Li et al. (2012), Wang et al. (2021), Beall et al. (2010), Simonson et al. (2010), Yi et al. (2010), Kader et al. (2015)

African zebu, and the development of the taurine (T) and indicine (I) haplogroups among the mtDNA haplotypes and the rarer P, Q, and R haplogroups (Lenstra et al., 2014).

These findings largely correlated with archeological findings regarding the beginning of agricultural development, and the role of cattle domestication in these events. Although all the evidence based on genetic and archeological study indicate at least some origins from southwest Asia, there is also evidence of other clusters of domestication in Europe, South and East Asia, and Africa, which are currently under study for their relative and associated contributions to global cattle domestication and agricultural development (Lenstra et al., 2014).

Global studies of cattle domestication also involve migration of people and cattle between the clusters of initial domestication. In terms of behavioral change between wild ancestral and domesticated cattle, Palmeira et al. (2008: 118) note that compared to similar-sized wild prey species, domestic livestock are relatively easy to kill and exhibit almost none of their ancestors' anti-predator behavior (see also Linnell et al., 1999). Sahu et al. (2020: 51) noted that despite a wealth of research on cattle, a review of the literature reveals that, in the majority of research papers and projects, the study of cattle behavior was not one of the primary goals of the research; rather, the majority of documents concentrated solely on foraging and reproductive behavior from an economic perspective.

Kluever et al. (2008: 321) used the "vigilance rates" (the percentage of foraging time the ungulate spends scanning for predators) and write that despite the fact that "vigilant behavior in wild ungulates is critical to guard against predation" there are few studies that have "examined vigilant behavior in domesticated ungulates." The authors concluded that domesticated "cattle exhibit vigilance at lower levels compared to wild ungulates, but this behavior appears to be at least partially an anti-predatory behavior. Our findings provide support that predators can influence cattle behavior." The authors further note that "adult female cattle in our study site exhibited lower baseline vigilance rates (3%) than studies with female wild ungulates while actively foraging" with examples cited of American bison (9.6–18.9%, see Laundre et al., 2001), African buffalo (4.2–5.6%, see Underwood, 1982), female elk (12–30%, see Laundre et al., 2001), and giraffe (18–25%, see Cameron & Du Toit, 2005) (Kluever et al., 2008: 325). Kluever et al. (2008: 325) argue that "the reason for lower vigilance rates in cattle is unknown but is likely due to the process of domestication, although it is also possible that species in the genus Bos have inherently lower vigilance. Unfortunately, this is a difficult issue to resolve because cattle's wild ancestor, the auroch (Bos primigenius Bojanus), is extinct," and there are few studies of this issue on other Bos wild cattle such as the gaur, banteng, kouprey, and wild yak (see also Van Vuure, 2005).

A similar study found that cattle brain sizes decline with domestication (Balcarcel et al., 2021). According to the study's findings, domestic taurine cattle's brains are smaller than those of their aurochs, their wild ancestors. However, breed-specific differences in brain size mean that some domestic cattle have much smaller brains than others. The level of selection for docility is linked to variations in husbandry practices among various groups (Balcarcel et al., 2021). The authors discover a distinct pattern where a gradient in brain shrinkage corresponds to a gradient in behavioral selection when they investigate whether variations in brain size are related to the degree of human contact. Compared to dairy breeds, which are heavily bred for docility, bullfighting cattle, which are bred for fighting and aggressive temperaments, have significantly larger brains. The findings supported a key tenet of animal domestication theory: variations in brain size appear to be explained by the interaction of basic aspects of the home environment, such as the choice of docility, the lack of predators, and human provision of resources (Balcarcel et al., 2021).

The rest of this chapter will examine the evolution of domestic cattle from ancestral forms, taking a geographical perspective, noting the source regions of these cattle and the spread of domestication in southwest Asia, south and east Asia, Africa, Europe, and the Americas. The hypothesis of cattle domestication in southwest Asia and southern Asia is followed. An examination is also made of possible source areas in Africa. In some cases, domestication reduced the body size and horns of cattle breeds, thereby increasing cattle vulnerability to predation, despite the declining numbers of carnivores.

Southwest Asian Cattle

Although there is a general consensus that cattle originate from the general region of southwest Asia, the more detailed dynamics around the precise movements and flows are frequently disputed. This section will examine the origins, physical attributes, and ecology of the aurochs and compare this lineage with other evolving strands of cattle descent in southwest Asia. The extinct aurochs (*Bos primigenius* Bojanus 1827) (Table 1), which ranged across Eurasia and North Africa are generally regarded as the first ancestral species for domestic cattle (Arbuckle & Kassebaum, 2021: see also Duerst, 1908; Zeuner, 1963; Balasse et al., 1997). It is also noted by recent genomic research findings that there are two original ancestral clusters for domestic cattle: the taurine cattle of Neolithic SW Asia, and zebu cattle (*Bos indicus*, Linnaeus 1758), from a South Asian origin, with both groups increasingly mixing (Verdugo et al., 2019; Arbuckle & Kassebaum, 2021). Several studies have documented the origins in the southwest Asian region. Achilli et al. (2008: R157) note that by "sequencing entire mitochondrial genomes from modern cattle that extinct wild aurochsen from Europe occasionally transmitted their mitochondrial DNA (mtDNA) to domesticated taurine breeds. However, the vast majority of mtDNAs belong either to haplogroup I (B. indicus) or T (B. taurus)… MtDNAs of haplotype T fall into subclades whose ages support a single Neolithic domestication event for B. taurus in the Near East, 9–11 thousand years ago (kya)."

The current consensus is that taurine cattle were domesticated around 10,000 years BP from southwest Asian wild aurochs, with possible spread to Europe and elsewhere, but variable arguments and hypotheses have been stated. Troy et al. (2001:1088) point out that many of the major domestic animal species in Europe have restricted wild ancestor ranges, suggesting that these species originated in Anatolia or the Fertile Crescent further east. Nonetheless, it is unknown if the wild ox (Bos primigenius) was domesticated in Europe to contribute locally to the agricultural economy despite its widespread distribution. Janák et al. (2024) point out that although Europe was not the main hub for the domestication of cattle, its subsequent growth and spread from the Middle East produced a complicated pattern of breed diversity in cattle. It is suggested that historical bovine diversity is currently at the crossroads of two prominent genetic research trajectories: the first is based on archaeogenetics, which reconstructs the genetic profiles of ancient populations backward using genetic data from archaeological evidence; the second is based on genetic data collected from living cattle, which is made possible by the cattle's significant economic role (Janák et al., 2024). The study looked at genetic, genomic, and archaeogenetic information regarding the history of cattle in Central Europe (Janák et al., 2024).

Research on the domestication of cattle and associated livestock has increasingly used more advanced techniques, including deoxyribonucleic acid (DNA). For example, Ajmone-Marsan et al. (2010: 148–157) have noted that the reconstruction of livestock farming's earlier history is based on archaeological discoveries, pictorial representations, and DNA analyses. Currently, analyses of ancient and modern cattle samples using DNA markers for autosomal, paternal, or maternal inheritance complement archeological data. In addition, the majority of molecular research to date has focused on nuclear microsatellites and the mtDNA control region. FAO supports conservation efforts in order to reduce fragmentation of research efforts and facilitate global data integration. The authors anticipate that single nucleotide polymorphism (SNP) high-throughput genotyping, Y-chromosomal haplotypes, complete mtDNA sequences, and analysis of ancient DNA, and eventually genome sequencing will all become more important (Ajmone-Marsan et al., 2010: 148–157).

Frantz et al. (2020:449) contend that the sequencing of complete ancient genomes has been made possible by recent methodological advancements like enhanced techniques for extraction of ancient DNA and next-generation sequencing. The process by which animals established domestic relationships with humans and encountered novel selection pressures has been reconstructed with the aid of these genomes. MacHugh (2017: 329) notes that paleogenomics is a new field that emerged from the revolution in ancient DNA research in the last 10 years brought about by the availability of increasingly powerful DNA sequencing and genomics ancillary technologies. Advances in science have contributed to the understanding of the origins and distribution of livestock and domestic animals in the Upper Paleolithic and Neolithic, as well as the genetic basis of animal domestication. These methods were especially illuminating because they demonstrated substantial interbreeding between early domestic populations and their wild relatives as well as high-resolution patterns of artificial and natural selection.

The results of these increasingly sophisticated research methods have been that increasingly complex assessments have been made concerning the ecology and distribution of the aurochs, their genetic relationship with taurine cattle, and the time sequence of these dynamics. While some studies affirm "the traditional narrative that taurine cattle domestication occurred 8500 BC in the Euphrates valley" (see Arbuckle & Kassebaum, 2021:10), recent complex findings have questioned the precise genetic relationship between the aurochs and taurine cattle. For example, Arbuckle and Kassebaum (2021:12) make their own proposition: "Despite the success of the Euphrates-EPPNB cattle domestication narrative, we argue that the history of cattle domestication is more complex. Zooarchaeological evidence from the Tigris drainage in southeastern Turkey is particularly important showing an alternative history of cattle domestication in a neighboring region."

Beja-Pereira et al. (2006: 8113) note that current genetic and archaeological evidence supports at least two main sites of domestication in India and in the Near East, where zebu and the taurine breeds emerged independently. This hypothesis suggests that the present-day European breeds are descended from cattle domesticated in the Near East and then spread during the spread of herding and farming lifestyles. However, using "previously undescribed genetic evidence in contrast with this view, based on mtDNA sequences from five Italian aurochsen dated between 7,000 and 17,000 years B.P. and >1,000 modern cattle from 51 breeds" the findings presented "do not support the currently accepted hypothesis of a single Neolithic origin in the Near East" (Beja-Pereira et al., 2006: 8116). The data support the theory that the origins of European cattle are diverse, and it is argued that the domestication and spread of livestock were undoubtedly more complex than previously believed. Domesticated Middle Eastern breeds that were brought to Europe during the Neolithic era most likely mixed, at least in some areas, with local fauna and cattle introduced from Africa by sea (Beja-Pereira et al., 2006: 8116–8117).

Ajmone-Marsan et al. (2010: 149) describe the position of the southwest Asian origin, from a genetic viewpoint, noting that taurine cattle have the highest diversity in southwest Asia, with four different mtDNA haplogroups (T, T1, T2, and T3). One of these, T3, which is the dominating haplogroup in European cattle, has also been found in ancient cattle DNA from the European Neolithicum. Northern and central European aurochs carry a different P haplotype and thus are excluded as maternal ancestors of European cattle. In contrast, southwestern Asian aurochs carried the mitochondria now found in taurine domestic cattle, which is consistent with the notion that taurine cattle originate from the Fertile Crescent." Scheu et al. (2008:1257–1264) cite an example from Northern Germany, which indicates that domestication of cattle was not derived from local wild breeds, but from southwestern Asia: "The data therefore suggests that the beginning of livestock farming in northern Germany did not precede the Neolithic transition period but was related to it. The only Neolithic Rosenhof house gave an early radiocarbon date of 4000 ± 50 cal B.P. BC and marks the beginning of the Neolithic in Northern Europe." The spread of cattle breeds within Europe after domestication is also documented (MacHugh et al., 1999).

One identified complexity concerns the reliability of the genetic background. For example, Bonfiglio et al. (2012) argue for possible secondary sources for domestic cattle. The argument is that although many genetic studies on modern cattle have established a common origin for the taurine breeds in the Near East, during the Neolithic transition at about 10,000 years BP, the possibility of the identification of rare mitochondrial DNA (mtDNA) haplogroups like P, Q, and R has sparked discussions about independent and/or secondary domestication events. Haplogroup T1 has been the focus of numerous studies because of its geographic distribution, which suggests a potential independent domestication event in Africa and a genetic contribution from African cattle to the development of Iberian and Creole cattle. Additionally, the origin and structure of haplogroup T1 can be effectively studied using whole mitochondrial genome sequence analysis, which raises the resolution of phylogeographic studies (Bonfiglio et al., 2012). Furthermore, up to eight founding T1 haplotypes provide evidence for the T1 subhaplogroups Near Eastern origin. But it is also possible that a sub-haplogroup (T1d) arrived from the Middle East and began in domesticated populations in North Africa (Bonfiglio et al., 2012).

Bollongino et al. (2012): 2101) note that "archaeozoological and genetic data indicate that taurine cattle were first domesticated from local wild ox (aurochs) in the Near East some 10,500 years ago. However, while modern mitochondrial DNA (mtDNA) variation indicates early Holocene founding event(s), a lack of ancient DNA data from the region of origin, variation in mutation rate estimates, and limited application of appropriate inference methodologies have resulted in uncertainty on the number of animals first domesticated." Possibly, only 80 female aurochs may have been domesticated at the beginning. Such a low number implies that the process was constrained by the challenge of sustainably managing and breeding the wild ancestors of domestic cattle, which is consistent with archaeological data showing that initial domestication occurred in a restricted area (Bollongino et al., 2012: 2101).

Schibler et al. (2014) support this position pointing out that most European taurine cattle genetic and archaeozoological evidence support an original domestication around 11,000 BP in the southwest Asia from a small number of aurochs with the mitochondrial DNA T macro-haplogroup. Gene flow between wild European aurochs of P haplogroup and domestic cattle of T haplogroup coexisting over thousands of years may have been "sporadic." The case study is of DNA evidence for the mixing of wild stock into a domestic cattle herd from a Neolithic lake-dwelling in Switzerland. The example is of a female skeleton, dated to over 5360 BP, associated to the Horgen culture. This had a mtDNA P haplotype variant of the European aurochs, and the authors argue that "this is either a single event or, based on osteological characteristics of the Horgen cattle, a rare instance of intentional breeding with female aurochs" (Schibler et al., 2014).

Upadhyay et al. (2017a, b) note that the southwest Asian domestication was preceded by a dispersal through migration of herders to Europe. The possible gene flow from wild aurochs still present at the time of this early dispersion is uncertain, and the extant primitive cattle populations may have some similarities to aurochs-like features and characteristics. The study used genome-wide single nucleotide polymorphisms, to analyze the history and interrelationships of a sample ancient aurochs and European cattle populations. The findings indicated historic variations for northwestern and southern European cattle, a zebu gene flow in the Balkan and Italian Podolic cattle, which supported the view of wide gene flow between aurochs and European cattle populations, with impacts from isolation and variable breeding which have created differences in the European cattle. The authors conclude that the genomic composition of European cattle populations is influenced by several factors, including gene flow between domestic and wild-cattle populations, human migrations, distance, and crossbreeding (Upadhyay et al., 2017a, b).

Further possibilities concerning the aurochs ancestry are offered by Pitt et al. (2018: 123–136), who note that there are limitations of current knowledge of ancestral populations, but the domestication of cattle may have occurred in two or three events, creating the taurine (*Bos taurus*) and indicine (*Bos indicus*) species that share the aurochs (*Bos primigenius*) as the common ancestor. These authors acknowledge the general position that indicine and taurine cattle were domesticated in the Indus Valley and Fertile Crescent, respectively, but they also suggest a proposed "additional domestication event for taurine in the Western Desert of Egypt" (Pitt et al., 2018: 123–136). Using medium density Illumina Bovine SNP array data for over 3000 individuals, for taurine and indicine populations, the findings suggested only two episodes of domestication, "that the additional genetic variation component usually detected in African taurine cattle may be explained by hybridization with local aurochs in Africa after the domestication of taurine cattle in the Fertile Crescent" (Pitt et al., 2018: 123).

The relation between the aurochs and later breeds is also highlighted by Park et al. (2015) who note that domestication of the wild aurochs gave rise to the taxa, *B. taurus* and *B. indicus*. It is argued that some aspects of this domestication are still unclear, these being the phylogenetic status of aurochs, the gene flow from aurochs into early domestic populations, and the genomic regions of selection processes during and after domestication. The case study was of whole-genome sequencing data generated from a 6750-year-old British aurochs bone and genome sequence data from 81 additional cattle and genome-wide single nucleotide polymorphism data from 1225 modern cattle. The findings indicated that phylogenomic analyses point to the aurochs as an outgroup to the domestic *B. taurus* lineage, thus indicating the possible Near Eastern origin of European cattle. Some of the traditional British and Irish breeds have closer genetic links with the aurochs specimen than some European cattle populations. This supports a localized gene flow from aurochs into the ancestors of modern British and Irish cattle, perhaps through management by early British herders. It is concluded that the study presents critical information on the origins and functional evolution of modern cattle, showing complexity in the interface between early European domestic populations and wild aurochs.

Edwards et al. (2007: 1377) present a similar study, using mitochondrial DNA analysis, arguing for a Near Eastern Neolithic origin for domestic cattle and the need for evidence on the domestication of European aurochs, which may not be strongly linked to the domesticated breeds. The authors note that although the aurochs are considered the "wild progenitor of modern cattle," it is "unclear whether European aurochs contributed to this process" (Edwards et al., 2007: 1377). The study is based on "high-confidence mitochondrial DNA sequences from 59 archaeological skeletal finds, which were attributed to wild European cattle populations based on their chronological date and/or morphology. All pre-Neolithic aurochs belonged to the previously designated P haplogroup, indicating that this represents the Late Glacial Central European signature" and "one new and highly divergent haplotype in a Neolithic aurochs sample from Germany, which points to greater variability during the Pleistocene" (Edwards et al., 2007: 1377). As previously shown, the majority of ancient and contemporary European domestic cattle carry haplotypes previously designated T. These findings imply that there is little connection between aurochs and native cattle, as evidenced by the Neolithic and Bronze Age samples of European aurochs that morphologically carried mitochondrial DNA of the P haplotype, suggesting a continuity of late glacial and early Holocene aurochs populations in Europe. When coupled with the recent finding of a T haplotype at a very early Neolithic site in Syria, this offers strong support for the hypothesis that the Middle East is home to graceful domestic populations. Since the Early Neolithic, P-haplotypes, who carried robust native aurochs populations, have colonized the primarily T-haplotype-bearing aurochs populations in Europe. The conclusion is that introgression was rare and domestic cattle were presumably kept apart from wild aurochs during the coexistence period (Edwards et al., 2007: 1377).

Hristov et al. (2017: 383–391) cite a study of cattle origins, noting that "two main hypotheses regarding the early history of cattle domestication, from their ancestor—the aurochs, have been discussed: multipoint domestication centers or single point origin and subsequent worldwide dissemination." The study examines molecular data for the Balkan aurochs and the Neolithic/Chalcolithic cattle populations. The findings "did not show different genetic profile of wild and domestic populations. All haplotypes were found to belong to the basic macro-haplogroup T. The majority of specimens (n = 14) were defined to form a new Balkan-specific T6 haplogroup. Only two of the ancient samples analyzed were assigned to the T3 haplotype predominating in Europe" (Hristov et al., 2017: 383–391). The authors argue that their study moves in two analytical directions, namely "local independent domestication processes in Neolithic South-Eastern Europe" and a "single point domestication in the Middle East and subsequent cattle dissemination in Europe" (Hristov et al., 2017: 383–391). The conclusion is that the findings do not "exclude the possibility for independent domestication events followed by a second wave of parallel dissemination of cattle herds via the Mediterranean route" (Hristov et al., 2017: 383–391).

Götherström et al. (2005: 2345–2350) argue that the domestication of cattle, one of the main pillars of European Neolithization, is believed to have originated in the Middle East, where aurochs were domesticated. According to mitochondrial DNA (mtDNA) data, there is a distinct difference between ancient British aurochs and contemporary European cattle. Nonetheless, the Y chromosome haplotypes of cattle breeds from northern Europe resemble ancient European aurochs more than they do modern cattle breeds from the Middle East and southern Europe. In addition, there is a significant north-south gradient in the frequency of two distinct Y-chromosome haplotypes in contemporary European cattle breeds. Of the 21 European aurochs, or early domestic cattle, that date from the period between 9500 and 1000 BC, 20 have the northern haplotype. This suggests that modern breeds in central and northern Europe bear a paternal imprint from their local hybridization with male aurochs. It might have been challenging to prevent or deliberate for aurochs bulls and domestic cows to mate in secret in order to increase the quality of the breeding stock. As demonstrated by mtDNA, the data indicates that domestic cattle origins may be much more complex than previously believed, indicating that they are not limited to a few geographic regions (Götherström et al., 2005: 2345).

Scheu et al. (2015) argue that "cattle domestication started in the 9(th) millennium BC in Southwest Asia. Domesticated cattle were then introduced into Europe during the Neolithic transition" but "the scarcity of palaeogenetic data from the first European domesticated cattle still inhibits the accurate reconstruction of their early demography." The study used mitochondrial DNA from ancient and modern domesticated cattle from Europe, Western Anatolia, and Iran to assess the Neolithic dispersal process and the role of the local European aurochs population in cattle domestication. The findings suggest that a female founder population from the Middle East had an impact, that there was no major barrier to the spread of cattle from the Middle East and Anatolia to Europe, and that there was evidence of gene flow between the European and Middle Eastern/Anatolian cattle populations during the early Neolithic period, but that this flow decreased later. According to the authors, this is the most likely scenario. Cattle were domesticated in the Middle East in a geographically restricted manner, and they migrated to Europe during the Neolithic transition without undergoing substantial maternal interbreeding with the native wild population. Intercontinental trade relationships between Neolithic farmers are suggested by evidence of gene flow between cattle populations from Southwest Asia and Europe during the early stages of the European Neolithic.

Arbuckle and Kassebaum (2021: 12) argue for a more complex narrative of cattle origins than commonly promoted. Compared with the common hypothesis of a domestication of taurine cattle period around the tenth to ninth millennium BC, "instead, dramatic changes in cattle phenotype including body and horn size are evident in a wide arc including the Upper Euphrates, Upper Tigris, and southern Levant almost a thousand years later (eighth millennium BC)." It is argued that commonly assessed origins of taurine cattle, the Euphrates Jordan Valley, and central Anatolia areas "do not represent the beginning of close relationships between humans and aurochs which extend temporally in both directions… that a long history of pre-domestic cattle management preceded the appearance of 'domestic 'cattle, whose slow reproductive rates, combined with early herders 'learning by doing, 'and an apparent preference for the 'aurochs aesthetic 'likely made it necessary for herders to draw from local aurochs populations thereby inhibiting the appearance of domestic phenotypes" (Arbuckle & Kassebaum, 2021: 17).

In terms of the relation between human action and cattle domestication, Arbuckle and Kassebaum (2021): 10) consider the "traditional narrative," where the northern Syrian upper Euphrates valley has been referred to as the "hearth" of domesticated taurine cattle. This process is dated to the middle of the ninth millennium BC and is connected to the PPNB (Pre-Pottery Neolithic B), an early Neolithic culture that flourished in Mesopotamia and the Eastern Mediterranean between 10,800 and 8500 years BP (Arbuckle & Kassebaum, 2021: 11). It is stated that aurochs and human cattle management are appropriate in this area (see also Helmer & Gourichon, 2008). Helmer et al. (2005) found evidence of the potential emergence of domesticated cattle at the Dja'de al-Mughara site in northern Syria, 10,500 BP. These cattle are identified by the smaller size of the male cattle, which may be the result of human selection favoring smaller, nonaggressive males. These allegedly domesticated cattle have since been documented in Iran, the Caucasus, and the eastern Mediterranean (Vigne et al., 2011; Arbuckle, 2013; Marom & Bar-Oz, 2009, 2013; Arbuckle et al., 2016).

Arbuckle et al., 2016 argue that the appearance of domestic cattle in the Eastern Fertile Crescent (EFC) region of SW Asia "was a minor component of animal economies in the EFC in the late Pleistocene and early Holocene but increased dramatically in the sixth millennium BC." Importantly, research suggests that "small-sized Bos, likely representing domesticates, appear suddenly in the region without any transitional forms in the early to mid sixth millennium BC. This suggests that domestic cattle were imported into the EFC, possibly associated with the spread of the Halaf archaeological culture, several millennia after they first appear in the neighboring northern Levant." The authors note that although "the domestication of cattle (Bos taurus) from wild aurochs (Bos primigenius) was one of the most important consequences of the Neolithic period in Southwest Asia… very little is known about the timing and mechanisms of the appearance of domestic cattle in the eastern wing of the Fertile Crescent" and due to lack of critical research, "the timing and mechanisms for the appearance of domestic cattle in this core area of the Neolithic Revolution remain poorly understood" (Arbuckle et al. (2016).

South and East Asian Cattle

The development of cattle breeds to the south and east of the aurochs area of southwest Asia primarily concerns the evolution of *Bos indicus*, generally termed as indicine cattle or zebu (Utsunomiya et al., 2019). An important issue concerns how the indicine cattle evolved from, link with, or differ from their possible aurochs ancestors and the methods used to study these issues. These methods have developed from simple archeological techniques to recent DNA and genomic-based methods. As noted by Ajmone-Marsan et al. (2010): 149) in a study which described how molecular-genetic research complements archeological research, these studies were based on "a molecular assessment of the divergence of taurine cattle, the common type of cattle in the temperate climate zone, and the humped zebu, which better tolerates hot and arid environments and is preferred in tropical climates. This divergence, which is suggested by mtDNA sequencing", may precede the domestication of these species, and separate domestications of *Bos primigenius taurus* in the Fertile Crescent and *Bos primigenius indicus* in the Indus Valley with temporal estimates for the divergence of taurine and zebu cattle at "330,000 years or even two million years" (Ajmone-Marsan et al., 2010: 149). These authors further note that using MtDNA, there is increased knowledge of the sites of domestications, and "a hotspot of zebu mtDNA diversity in the Indian peninsula relative to the surrounding regions indicates domestication of B. indicus in India, followed by migrations toward southwest Asia, China, and southeast Asia" (Ajmone-Marsan et al., 2010: 149).

For South Asia, Felius et al. (2014) note that zebu domestication occurred around 2000 years after the taurine domestication, in the Indus Valley, with evidence dated at 8000 BP. It is argued that the zebu first spread about 4500 BC, when Central Asian invaders entered the Indus Valley by way of the northern passes, and residents relocated with their livestock eastward into the Ganges Valley, and some cattle were transported east to Bengal and Bihar around 3500 BC. Evidence of zebu presence has also been detected on the Deccan Plateau in central India. Zebu cattle also entered China from the south or southwest after 3000 BP and mixed with the taurine cattle from the north creating a mixed zebu-taurine mtDNA and Y-chromosomal DNA, and mixing occurred with bibovine cattle (banteng, gaur, or gayal). Zebu also reached Indonesia over 1000 years BP with evidence of maternal banteng ancestry, possibly indicating zebu were interbred with domesticated banteng. There is also evidence of a westward migration of zebu, in the Mesopotamian region after 6500 BP and to Egypt and East Africa about 4000 BP (Felius et al., 2014).

Pérez-Pardal et al. (2018) located all the zebu Y-chromosome haplotypes within three lineages: Y3A, which is the most widely distributed; Y3B, which is detected only in West Africa; and Y3C found in South and Northeast India. The authors note that the three zebu-specific lineages have estimated divergence times that may predate domestication. De novo domestication of genetically distinct paternal lineages or a more intricate process involving gene flow between wild and domestic animals are both supported by coalescent demographic models. The recent cattle introductions in East Africa are argued to be consistent with the restriction of Y3A haplotypes to this region. In addition, the existence of Y3B haplotypes in West Africa that are unique to the region raises the possibility that these haplotypes may represent the oldest zebu lineage that was brought to Africa about 3000 years BP, which was later replaced elsewhere (Pérez-Pardal et al., 2018).

There is evidence of a westward movement of cattle from South Asia to southwest Asia. For example, Xia et al. (2023) argue that cattle moved from the Indus Valley to the Middle East about 4200 years ago due to climate change, which changed their genetic makeup as they adapted to a dry environment (Verdugo et al., 2019). The ancestral taurine cattle rapidly adapted to frigid and cold temperate climates, in places such as Siberia and the Qinghai-Tibetan Plateau, as a result of their eastward migration, which brought them to the northern part of East Asia (Felius et al., 2014). Cattle from Indochina and southern China also migrated there approximately 3000 years ago. These migrations produced several admixtures, with populations of intermediate taurine-indicine populations. Indicine cattle came into contact with taurine populations in the southwest and east of Asia, which created multiple hybrid taurine populations.

For East Asia, several studies have applied advanced techniques to examine the genetic character of domestic cattle in China and Mongolia, including possible ancestral links with southwest Asia, genetic introgression from taurine and indicine cattle, and the geographical demarcation of such genetic influences from southwest and South Asia. For example, Chen et al. (2018a), b) used an analysis of the genomes of 49 modern breeds and eight East Asian ancient samples and found that that "East Asian cattle populations are mainly composed of three distinct ancestries, including an earlier East Asian taurine ancestry that reached China at least ~3.9 kya, a later introduced Eurasian taurine ancestry, and a novel Chinese indicine ancestry that diverged from Indian indicine approximately 36.6–49.6 kya." Another crucial result was that "historic introgression events… helped domestic cattle from southern China and the Tibetan Plateau achieve rapid adaptation by acquiring ~2.93% and ~ 1.22% of their genomes from banteng and yak, respectively" (Chen et al., 2018a, b). In addition, the findings aligned with previous archaeological evidence that "suggested that B. taurus may have been imported into East Asia from West Asia during the late Neolithic period (between 5000 and 4000 YBP) and B. indicus may have dispersed from the Indian subcontinent to East Asia at a later stage, from 3500 to 2500 YBP, which led the hybridisation between taurine and indicine cattle in Central China" and studies that indicated evidence of multiple introgressions within the genus *Bos*, including banteng introgression into Chinese Hainan cattle and bovine introgression into Mongolian yak. The authors conclude that "Asian cattle have three distinct ancestral lineages: East Asian taurine; Eurasian taurine; and Chinese indicine" (Chen et al., 2018a, b).

Three issues concern the links between East Asian aurochs and indicine cattle, and the ingression of yak and banteng into indicine cattle. Xia et al. (2023) researched on yak introgression into Tibetan taurine cattle and noted that yak may have inhabited the Tibetan Plateau for millions of years and adapted to environmental conditions, including enlarged lungs and hearts. Domestic taurine cattle were introduced to the Tibetan Plateau by humans later, only a few thousand years ago, and were not adapted to the Tibetan Plateau and suffered from pulmonary hypertension during early period. Yak introgression into Tibetan cattle genomes partially facilitated cattle adaptation to high altitude (Xia et al., 2023). Several adaptive introgressed genes have been identified that are associated with the response-to-hypoxia pathway. Cited adaptive genes include COPS5, IL1A, IL1B, MMP3, EGLN1, EGLN2, HIF3α, RYR2, and SDHD (Xia et al., 2023; see also Chen et al., 2018a; Wu et al., 2018).

Regarding the introgression of banteng-like sequences into East Asian indicine cattle, Xia et al. (2023) note that banteng provide a genetic match but not a clear source of East Asian indicine cattle, and these findings suggest that more wild *Bos* diversity must be sequenced to fully explain their evolution (Sinding et al., 2021). The authors also note the inference "that between 2.38 and 3.84% of the southeastern Chinese indicine genome is of banteng ancestry" and the introgression genes include *T2R12*, *TAS2R9*, and *TAS2R6*" and some introgressed genes conducive to the local adaptation of East Asian indicine cattle to the hot and humid tropical climate have been detected" including genes that enable adaptation to hot climates, including heat-shock protein (HSP) genes, such as HSPA1A, HSPB8, HSPA8, HSPA4, HSPB2, and HSF2, and genes for hair cell differentiation and blood circulation, such as ATOH, GNA14, VPS13, and KIF2B (Chen et al., 2018a).

For the links between East Asian aurochs and indicine cattle, the East Asian aurochs of the C haplogroup occurred in northern China during the Holocene and coexisted with early domestic cattle, perhaps contributing to modern East Asian cattle, which is currently unclear (Xia et al., 2023; see also Brunson et al., 2016; Cai et al., 2018; Zhang et al., 2013). The authors note that there was an accelerated introgression of indicine cattle in southwest and central Asia around 4200 years BP, which may have been associated with climate change that led to the increased use of arid-adapted indicine bulls for breeding purposes (Xia et al., 2023; see also Verdugo et al., 2019).

The domestication of the zebu indicine cattle is argued to be highly expansive: Utsunomiya et al. (2019: 557) note that "Zebu cattle were domesticated in the Indus Valley 8000 years before present (YBP). From the domestication site, they expanded to Africa, East Asia, southwestern Asia and Europe between 4000 and 1300 YBP, intercrossing with B. taurus to form clinal variations of zebu ancestry across the landmass of Afro-Eurasia." The authors further note that "the zebu genome is characterized by two mitochondrial haplogroups (I1 and I2), one Y chromosome haplogroup (Y3) and three major autosomal ancestral groups (Indian-Pakistani, African and Chinese). Phenotypically, zebu animals are recognized by their hump, large ears and excess skin. They are rustic, resilient to parasites and capable of bearing the hot and humid climates of the tropics."

However, the precise trajectory of the zebu cattle evolution is discussed and disputed. For example, Chen et al. (2010: 1) note that archaeological evidence suggests that the domestication of zebu cattle within the Indian subcontinent and the "exact geographic origins and phylogenetic history of zebu cattle is uncertain." In addition, "although investigations of mitochondrial DNA (mtDNA) sequence variation have confirmed the independent domestic origins of B. taurus and B. indicus cattle from genetically divergent wild aurochs progenitors and have shed much light on the ancestry of B. taurus cattle, similar studies involving zebu mtDNA are limited" (Chen et al., 2010: 2). The authors used 844 zebu mitochondrial DNA (mtDNA) sequences from 19 Asiatic countries identifying 2 distinct mitochondrial haplogroups (I1 and I2). It is argued that the increase in nucleotide diversity ($P < 0.001$) for both the I1 and I2 haplogroups within the northern part of the Indian subcontinent is consistent with an origin for all domestic zebu in this area (Chen et al., 2010: 1). The two haplogroups, I1 and I2, differed in the spatial configuration in South Asia. For haplogroup I2, a diverse pattern was identified that prevented clear identification of the place of origin for this zebu maternal lineage. For haplogroup I1, the genetic diversity was highest within the Indus Valley among the three possible domestication clusters of the Indus Valley, Ganges, and South India, thus suggesting that the Indus Valley is the most probable origin for the I1 haplogroup and an important location of zebu domestication (Chen et al., 2010: 1).

Felius et al. (2014) note that between 5000 and 4000 BP, taurine cattle began to spread and reached as far east as northern China or Mongolia. Analyses of cattle remains from five archaeological sites in northern China, dating from 4500 to 2300 BP, using mtDNA support this hypothesis. The East Asian taurine cattle, including the North Siberian Yakut, display mtDNA haplotypes of the T2, T3, and T4 haplogroups. A founder effect during eastward expansion is most likely how the T4 haplogroup, a subtype of the common haplogroup T3, developed. This T3 haplogroup is only found in East Asia (Felius et al., 2014).

A major strand of research focuses on the interbreeding of indicine cattle with wild breeds. Zhang et al. (2020: 637–657) note that taurine and indicine cattle and yak had significant genetic interactions. Taurine and zebu cattle are distributed globally, but their populations have experienced bottlenecks that have caused Y-chromosomal and mitochondrial DNA to significantly diverge phylogenetically. The isolation caused by distance, the various production objectives, the development of popular breeds, and the great diversity of European breeds are what define them. Cattle hybridization occurred with other cattle species as well as with taurine and zebu cattle due to their overlapping and vast ranges. Tibetan cattle and yaks have exchanged gene variants, the Chinese gayal carries zebu mitochondrial DNA, some Indonesian zebu are descended from crosses between zebu bulls and banteng cows, and approximately 5% of American bison contain taurine mtDNA. There is a possibility that environmental adaptation was aided by introgression, according to genomic analysis.

Chen et al. (2023a, b) argue that the "adaptive genetic changes of indicine cattle following their dispersal into East Asia from the Indian subcontinent have remained poorly documented." This study examines the possible admixture of indicine cattle with the wild species banteng and/or gaur. The study used "high-quality whole-genome sequencing data from 354 indicine cattle of 57 breeds/populations, including major indicine phylogeographic groups worldwide" and detected "introgression from other bovine species" (Chen et al., 2023a; b). According to Asian agroecologies, genomic regions containing genes related to morphology, immunity, and heat tolerance changed. Genetic factors were present for adaptation to tropical humid and semiarid environments. According to the findings, localized banteng and/or gaur introgression helped East Asian indica cattle adapt quickly and successfully to hot, humid climates (Chen et al., 2023a, b).

Qi et al. (2010: 242–252) argue that the hybridization between yak and taurine taurus or indicine cattle has been common across the yak range, with important gene flow. It is argued that some 3000 years ago, ancient nomadic peoples are thought to have started the hybridization of cattle and yak. At relatively high altitudes, beef bulls are frequently utilized in crossbreeding with yak cows, while reciprocal crossing is done at lower altitudes within their range. The ancient historical records attest to the hybridization of cattle and yak. It is thought that the first instances of yak and native cattle hybridization in China happened in the Yin Dynasty (about 1100 BC). The study examined cattle admixture on domestic yak, with samples from China, Bhutan, Nepal, India, Pakistan, Kyrgyzstan, Mongolia, and Russia, based on mitochondrial DNA and autosomal microsatellite loci. The findings indicated that the interspecies admixture was low but variable, with "higher levels of admixture in the Qinghai-Tibetan Plateau and Mongolian and Russian regions, and lower levels in the Himalayan and Pamir Plateau region" (Qi et al., 2010: 242). These findings "show that the subsequent gene flow between yak and cattle took place and has affected contemporary genetic make-up of domestic yak" (Qi et al., 2010: 242).

Chen et al. (2023a, b) also explore the issue of yak and cattle hybridization, noting that "domestic yak, cattle, and their hybrids are fundamental to herder survival at high altitudes on the Tibetan Plateau. However, little is known about their history. Bos remains are uncommon in this region, and ancient domestic yak have not been securely identified." The study used "zooarchaeological analyses of 193 Bos specimens and sequenced five nuclear genomes from recently excavated assemblages at Bangga" (Chen et al., 2023a, b). The findings indicated that cattle outnumbered yak and ancient mitochondrial DNA and nuclear genome sequences detected taurine cattle and supported domestic yak and yak-cattle hybridization about 2500 years BP. The study suggested that the use of diverse Bos species and hybrids increased cattle and herder adaptation to the Tibetan plateau environmental conditions. Ancient cattle and yak at Bangga were related to "contemporary cattle livestock, indicating associations of cattle and yak husbandry on the plateau" (Chen et al., 2023a, b). Xuebin et al. (2005: 117) examined a similar issue but cover Mongolian, Russian, and Chinese yak populations from the Qinghai-Tibetan Plateau, which is considered the center of yak domestication, using 15 microsatellite loci for genetic variation within populations and between populations. The findings suggested that yak populations in Mongolia and Russia were genetically highly diverse within populations, less between populations, and a low, although significant, genetic differentiation between populations. The authors conclude that "both phylogenetic and principal component analyses support a close genetic relationship between the Gobi Altai, south Gobi and north Hangai populations, and between the Hovsgol and Buryatia populations respectively. Our results indicate that these yak populations should be considered as distinct genetic entities in respect of conservation and breeding programmes" (Xuebin et al., 2005: 117).

Wu et al. (2018: 1139) examine interspecies introgression of *Bos* species at a broader level, including taurine cattle, zebu, gayal, gaur, banteng, yak, wisent, and bison. The findings suggested that the gayal is an independent species or subspecies, while there has been introgression between domestic cattle to yak and from yak to Tibetan cattle. The authors suggest that the *Bos* genus evolved as several genetically related species with associated evolutionary trajectories (Wu et al., 2018: 1139). The example is cited of the introgression from the yak to the cattle, in which introgression may have adaptive significance, exemplified by the introgression of EGLN1 alleles from yak to cattle that lowers the hemoglobin and red blood cell concentration, which is an adaptive response to hypoxia-induced polycythemia (Wu et al., 2018: 1143) and "genes in the response-to-hypoxia pathway (for example, EGLN1, EGLN2 and HIF3a) have been introgressed from yak to Tibetan cattle, probably facilitating their adaptation to high altitude" (Wu et al., 2018: 1139).

In addition, for East Asia, Brunson et al. (2023) note that the genetic history of cattle in East Asia remains understudied: Genome-wide analyses of 23 ancient Mongolian cattle show that East Asian aurochs and ancient East Asian taurine cattle are closely related, but neither are closely related to modern East Asian breeds. The study found "binary variation in aurochs diet throughout the early Neolithic, and genomic evidence shows millennia of sustained male-dominated introgression" with association between ancient Mongolian aurochs and the European Hereford breed. These results suggested the possibility of human management of aurochs in Northeast Asia before and during the original beginning of taurine cattle herding. Brunson et al. (2023) note that interbreeding of East Asian aurochs and cattle suggests management but left no link with modern eastern breeds, and many paleogenomic studies of such domestication focus mostly on Western Eurasian origins and ignore several indigenous domestications of eastern bovines such as yaks.

Brunson et al. (2023) note that despite current genetic evidence suggesting that modern cattle lineages derive from two aurochs domestication episodes: in the Near East by 10,000 years ago that originated taurine cattle (*Bos taurus*), and in South Asia by 8000 years ago that originated indicine/zebu cattle (*Bos indicus*) other domestication events were possible with significant interspecies mixing in Asia, including taurine-indicine hybrids and mixing of these with East Asian bovines such as yak, gayal/mithun, and banteng, with taurine-yak hybrids occurring in Central Mongolia. Brunson et al. (2023) further note that there are important questions, such as: "how are East Asian aurochs populations related to East Asian domestic cattle? Were aurochs domesticated in East Asia? And did aurochs and domestic cattle interbreed when taurine cattle were first adopted in East Asia?"

The current evidence suggests that Mongolian aurochs are genetically close to other aurochs and taurine cattle but are distinct from indicine cattle. The Neolithic Mongolian aurochs are part of the East Asian mitochondrial haplogroup C identified exclusively in East Asian aurochs from 10,000 to 4000 years BP. The Mongolian aurochs group differed from African, European, and the Middle Eastern aurochs, although part of the clade included sequenced aurochs and taurine cattle (Brunson et al., 2023). Human management of cattle may have been important, as there is evidence of gene flow between male aurochs and female taurine cattle, the extinction of the aurochs in China during the Zhou Dynasty (1046–256 BCE) (see also Tang, 2020), aurochs ancestry during the Xiongnu and Turkic periods, with interbreeding and that aurochs domestication occurred before the arrival of taurine cattle to Northeast Asia around 5000 years BP, followed by intention interbreeding the two populations (Brunson et al., 2023).

For Indonesian cattle, Sudrajad et al. (2020) argue that the domestication studies covered genetic diversity, up to the genome level, "but little information exists regarding the history of domestication of cattle, which remains unclear. Their study "genotyped seven Indonesian cattle breeds using an Illumina BovineSNP50 Bead Chip to provide insight into their domestication and demographic history in a worldwide population context" and the findings "indicated the presence of hybrid cattle, with Bos javanicus and *Bos indicus* ancestries being most prevalent, as well as purebred cattle" and "the breeds were interconnected through several migration events" (Sudrajad et al., 2020).

Mohamad et al. (2009) argue that "two bovine species contribute to the Indonesian livestock, zebu (Bos indicus) and banteng (Bos javanicus), respectively. Although male hybrid offspring of these species is not fertile, Indonesian cattle breeds are supposed to be of mixed species origin. However, this has not been documented and is so far only supported by preliminary molecular analysis." The study used mitochondrial, Y-chromosomal and microsatellite DNA, and the findings indicated a banteng introgression of 10–16% in the Indonesian zebu indicine cattle. In addition, East-Javanese Madura and Galekan cattle had higher levels of autosomal banteng introgression (20–30%) and "combine a zebu paternal lineage with a predominant (Madura) or even complete (Galekan) maternal banteng origin… In contrast, we did not find evidence for zebu introgression in five populations of the Bali cattle, a domestic form of the banteng" (Mohamad et al., 2009).

For Japanese cattle, Noda et al., 2018 pose the question "the mtDNA haplogroup P of modern Asian cattle: A genetic legacy of Asian aurochs?" Noting that "the mtDNA haplogroups of most modern cattle belong to haplogroups T and I, several additional haplogroups (P, Q, R, C and E) have been identified in modern cattle and aurochs. Haplogroup P was the most common haplogroup in European aurochs" the authors argue that currently "it has been identified in only three of >3,000 submitted haplotypes of modern Asian cattle" (Noda et al. (2018)). In a case study of the complete mtDNA D-loop region of Japanese Shorthorn cattle and representative bovine mtDNA sequences, the haplotype P of Japanese Shorthorn cattle was studied with data from European aurochs and three modern Asian cattle sequences. The authors conclude that the mtDNA haplogroup P in Japanese Shorthorn cattle was detected with a high frequency, and the phylogenetic networks had two main clusters, Pa for haplogroup P in European aurochs and Pc in the modern Asian cattle (Noda et al. (2018)). The findings showed that "no shared haplotypes are observed between the European aurochs and the modern Asian cattle. This finding suggests the possibility of local and secondary introgression events of haplogroup P in northeast Asian cattle and will contribute to a better understanding of its origin and genetic diversity" (Noda et al. (2018)).

Mannen et al. (2020) also present a study of Japanese cattle, arguing that studies of mitochondrial DNA (mtDNA) variation have shown that worldwide domestic cattle have only a few major haplogroups. Two, T and I, are common and largely define *Bos taurus* and *Bos indicus*, respectively, while the other three, P, Q, and R, are found only in taurine breeds. In addition, concerning Haplogroup P, it is noted that this haplogroup is typical of the extinct European aurochs, "while intriguingly modern P mtDNAs have only been found in northeast Asian cattle. These Asian P mtDNAs are extremely rare with the exception of the Japanese Shorthorn breed, where they reach a frequency of 45.9%" (Mannen et al., 2020). The findings indicated that there was a "post-glacial expansion of aurochs carrying haplogroup P from Europe to Asia" and "all Asian P mtDNAs belong to a single sub-haplogroup (P1a), so far never detected in either European or Asian aurochs remains, which was incorporated into domestic cattle of continental northeastern Asia possibly ~3700 years ago" and (haplogroup P1a mtDNAs found in the Japanese Shorthorn breed probably reached Japan about 650 years ago from Mongolia/Russia, in agreement with historical evidence" (Mannen et al., 2020).

For Chinese cattle, Xia et al. (2019a: 70) note that there is a high genetic diversity, and the paternal origins require more research. Cai et al. (2018: 72) present a study indicating that "DNA confirms that aurochs were much more abundantly available in Neolithic China, many aurochs remains may have been mistakenly identified as Bison exiguous, all aurochs belong to Haplogroup C with no direct contribution to modern cattle." The study reports that "ancient DNA analysis of 24 archaeological bovid remains recovered from large Neolithic (6300 BP to 5000 calBP) pit and ditch features at Houtaomuga, Northeast China, identified 23 of these samples as aurochs." These DNA-based identifications contrast with the morphological analysis of the remains, which identified them as Bison exiguous (Xia et al., 2019a, b: 70). The abundance of aurochs remains at this site contradicts the general assumption that this species was not present in large numbers in Neolithic China. It also suggests archaeologists need to "revise the notion that wild aurochs played an insignificant role in the lifeways of Neolithic peoples in China" (Xia et al., 2019a, b: 70). Notably, the findings also indicated that phylogenetic analyses of the mitochondrial DNA (mtDNA) D-loop showed the aurochs belonged to a "unique haplogroup (Haplogroup C) that is indigenous to East Asia and made no direct contribution to modern domesticated cattle Bos taurus. Moreover, temporal changes in haplotype frequencies were observed among the identified aurochs, suggesting population fluctuations potentially caused by human hunting activities occurred among Chinese aurochs during the Neolithic" (Xia et al., 2019a, b: 70). Importantly, the study also identified "one sample (HT31) radiocarbon dated to ca. 5500–5300 calBP as Bos taurus, making it one of the earliest known taurine cattle specimens in China. HT31's location in Northeast China and early date points to the existence of another entrance for domesticated cattle into China, the Northeast China Route via the Mongolian Steppe" (Xia et al., 2019a: 70).

Xia et al. (2019b: 64) present a study based on the polymorphism of Y-SNPs and Y-STRs of 24 cattle breeds/populations and identified. 14 Y-chromosome haplotypes that represented taurine haplotypes generally found in northern and central China, and detected mainly in northern and central China, and the indicine haplotype was found mainly in southern China. The authors concluded that multidimensional scaling analysis differentiated Tibetan cattle population structures from the other three groups of cattle, and the majority of the genetic variation was indicated by the genetic differences among groups. The findings indicated that Chinese cattle have high paternal diversity and possibly some of the original lineages from the domestication center in the Near East without strong admixture from cattle with Y1 haplotypes (Xia et al., 2019b: 64). Chen et al. (2018a), b, 246) present a similar study of the population structure and ancestry of Qinchuan cattle. The findings indicated that the Qinchuan cattle had a dominant *Bos taurus* ancestry, with some *Bos indicus* ancestry, and less of a *Bos javanicus* ancestry. The study also found a similar admixture pattern in Qinchuan and Turkish cattle breeds, which suggests similar degrees of zebu introgression.

Another Y chromosome-based study was presented by Li et al. (2013a: 6633), which aimed to determine "the Y chromosome genetic diversity and paternal origin of Chinese cattle… using a recently discovered USP9Y marker that could distinguish between taurine and indicine cattle more efficiently." Regionally, the taurine Y2 haplogroup dominated in northern China (91.4%) and the indicine Y3 dominated in southern China (81.2%). Central China is an admixture zone with Y2 predominating overall (72.0%). The authors conclude that these results suggest that Chinese cattle have two paternal origins, one from *B. taurus* (Y2) and the other from *B. indicus* (Y3). The Y1 haplogroup may be from the imported beef cattle breeds in Western countries. The distributions of the Y2 and Y3 haplogroup frequencies showed a pattern of male indicine introgression from southern to northern China, and male taurine introgression from northern to southern China (Li et al., 2013a, b: 6633).

Li et al. (2013b: 446) conducted another study on the paternal origin and genetic diversity using Y-SNPs (a single-nucleotide polymorphism on the Y chromosome, used for paternal genealogical DNA testing) and Y-STRs (Y-chromosomal short tandem repeat polymorphisms for paternal lineages). The results showed that the taurine "Y2 dominates in northern China (91.4%), and Y3 dominates in southern China (90.8%). Central China is an admixture zone, although Y2 predominates overall (72.0%). The geographical distributions of the Y2 and Y3 haplogroup frequencies revealed a pattern of male indicine introgression from south to north China. The three Y haplogroups were further classified into one Y1 haplotype, five Y2 haplotypes and one Y3 haplotype in Chinese native bulls" (Li et al., 2013b: 446). The authors conclude that because of the association between taurine and indicine types, Chinese cattle have a wide distribution of genetic diversity, and the Y haplotype distribution of Chinese cattle shows a geographical structure that is aligned with mtDNA findings and historical and geographical information (Li et al., 2013b: 446).

This contrast between taurine and indicine cattle ancestry in China is supported by Gao et al. (2017), who note that indigenous Chinese cattle have both taurine and indicine origins and occupy several environments. Their findings showed an inconsistent distribution of taurine and indicine cattle ancestries, which varied from 10%

indicine cattle in the north and more than 90% in the south and southwest China, the introgression of banteng and gayal into southern Chinese cattle, and inconsistent yak genetic impacts around the Tibetan area (Gao et al., 2017). The findings are consistent with other research findings on mitochondrial DNA analysis, and the distribution of taurine and indicine mitochondrial and Y-chromosomal DNA diversity align with the autosomal cline. The geographic distribution of the genomic admixture of different bovine species may be result of "the impact of prehistoric immigrations, gene flow, major rivers acting as genetic barriers, local breeding objectives and environmental adaptation" (Gao et al., 2017). The authors conclude that the whole-genome scan for genetic differentiation and association analyses with both environmental and morphological covariables similar to those of previous research findings and detect several genes associated with adaptation (Gao et al., 2017).

Cao et al. (2019: 412) also conducted a study that referenced whole-genome resequencing analyses which showed four paternal lineages (European, Eurasian taurine, East Asian taurine, and Chinese indicine) in East Asian cattle, represented by haplogroups Y1, Y2a, Y2b, and Y3a, and aimed "to present a survey of the distribution of these Y-haplogroup data in a broad panel of Chinese cattle" (Cao et al., 2019: 412). The findings indicated that Y2a is the dominant haplogroup in northern China and exists in some central and eastern breeds and in southern China. In contrast, Y2b was found to be dominant in Tibetan cattle and occurring in some northern and central breeds. The indicine Y-chromosomes occurred mostly in southern China. The Y1 was scattered and possibly represented the introgression of European bulls. The authors concluded that the Y-SNP markers combined with Y-specific markers may inform research on the recent and past history of the cattle paternal lineages (Cao et al., 2019: 412).

Cai et al. (2007: 175) used phylogenetic analysis of Chinese cattle using mitochondrial gene polymorphism. The results of this mitochondrial-based study were similar to those that used Y chromosomes, as the "zebu mtDNA dominated in the southern breeds (63.3-100%), while the taurine dominated in the northern breeds (81.8-100%)." In addition, "six cattle breeds from the central area of China exhibited intermediate frequencies of zebu mtDNA (25-71.4%)" and "this polymorphism revealed a declining south-to-north gradient of female zebu introgression and a geographical hybrid zone of Bos taurus and Bos indicus in China" (Cai et al., 2007: 175). The authors also report that the studied cattle had higher nucleotide (0.00923) and haplotype diversity (0.848) than those reported in other published research, and the taurine lineage showed higher nucleotide diversity (0.00330) and haplotype diversity (0.746) than the zebu lineage (0.00136; 0.661). Six cattle breeds from the central area of China also showed intermediate frequencies of zebu mtDNA (25–71.4%) (Cai et al., 2007: 175).

Xia et al. (2019a, b: 70) conducted complete mitochondrial DNA D-loop sequences of a sample of cattle in China to assess their maternal lineage diversity. The findings showed five main haplogroups (T1a, T2, T3, T4, and T5) for *Bos taurus* and two haplogroups (I1 and I2) for *Bos indicus*. In addition, in Asia the T1a distribution clustered mainly in northeast China, Korea, and Japan, and haplogroups

T2, T3, and T4 clustered in Chinese cattle, while T5 was intermittent in Mongolian and Pingwu cattle. I1 was widely distributed, but I2 was restricted to southwestern China (Yunnan-Guizhou Plateau and the Tibet Autonomous Region) and Xinjiang Uygur Autonomous Region (Xia et al., 2019a, b: 70). The authors comment that these findings "are consistent with their complex historical background, migration route and ecological environment" (Xia et al., 2019a, b: 70).

Another important topic concerns the text to which indicine cattle have adapted to tropical environments. For example, Berman (2011: 2147) poses the question: "Are adaptations present to support dairy cattle productivity in warm climates?" The study argues that dairy cattle health is seriously impacted by environmental heat stress in warmer weather. Evidence suggests that cattle bred in warm climates such as zebu and sanga are adapted to warmer climates. Cattle management may acknowledge these adaptations for increasing cattle productivity. Zebu and sanga breeds may also be more resistant to ticks and tick-transmitted diseases (Berman (2011: 2147). It is argued that the position that higher sweat rates in zebu-descended breeds are due to morphological differences in sweat glands between breeds is unclear, and that coat characteristics and their responses to seasonal heat, particularly in temperate breeds, are also unclear (Berman, 2011: 2147). Tropical breeds have "lower maintenance requirements and milk yields, and limited response to improved feeding and management" which are useful characteristics for contexts of scarcity, seasonality of food sources and temperatures are high. The authors conclude that generally, the adaptations found in tropical breeds were not based on increased heat dissipation capacity, but the resistance reduced climate-induced strain by reduced milk production. Zebu cattle also minimally exhibit this negative relationship between reproductive efficiency and milk yield. The authors argue that "this association, coupled with limited feed intake, acting over millennia, probably created the selection pressure for a low milk production in these breeds" (Berman (2011: 2147).

A similar study by Hansen (2004:82) notes that zebu cattle have evolved genes supporting physiological and cellular thermotolerance during their evolution from taurine cattle breeds, and zebu cattle are more adept at controlling their body temperature in response to heat stress than taurine cattle with European ancestry. In addition, exposure to high temperatures has fewer negative effects on zebu cells than on European cells. The superior body temperature regulation is because of lower metabolic rates and increased heat loss capacity. Zebu cattle, compared to European breeds, have larger sweat glands and less tissue resistance to heat flow from the core to the skin, and some features of the hair coat reduce solar radiation absorption and may increase heat loss through conduction and convection. Due to genetic adaptations, zebu embryos are less susceptible to developmental stunting from high temperatures than those of European breeds. Crossbreeding systems employing the zebu genotype to produce cattle for beef and dairy production in hot climates have been tried, but their success has been limited by other undesirable genetic traits of these cattle. A different strategy would be to avoid undesirable genes and introduce specific zebu cattle thermotolerance genes into European breeds (Hansen, 2004:82).

Asian Cattle also adapted to cold climates, in northern Asian parts of China and Russia. For example, Huang et al. (2023) argue that animals' genomes are shaped by cold climates and must have a sufficient amount of genetic variation to enable them to adapt to temperature changes. However, little is known about the pattern of genome-wide adaptations to cold environments in cattle. In order to distinguish between divergent selection signatures, particularly between the heat-adapted (20.2 °C to 24.73 °C) and cold-adapted (6.24 °C to 10.3 °C) Chinese native cattle breeds, their study used whole genome resequencing and SNP genotyping data for Chinese and other cattle (Huang et al., 2023). The authors noted that the genomic analyses showed candidate genes (e.g., UQCR11, DNAJC18, EGR1, and STING1), which were functionally associated with thermogenesis and energy metabolism, and the adaptive loci of cattle exposed to cold temperatures were characterized (Huang et al., 2023).

Yan et al. (2022: 275) sequenced the genomes of cattle from cold and warm areas in China, with the findings indicating that there were two genetic clusters, based on northern and southern groups. Cold tolerance was measured by genomic scans of cold-tolerant breeds, and the identification of potential candidate genes in the thermogenesis-related pathways. Yan et al. (2022: 275) found "a substitution of PRDM16 (p. P779L) in northern cattle, which maintains brown adipocyte formation by boosting thermogenesis-related gene expression, indicating a vital role of this gene in cold tolerance." These results showed an example of genetic variation and reverse mutation in domestic cattle as a result of environmental temperatures. For the cold-tolerant lineages, the average habitat temperature was 2–6 °C, and for the cold-intolerant cattle lineages the average habitat temperature was 20–25 °C. It is noted that "through characterization of population history and selective sweeps, we identified PRDM16 as a candidate gene under selection, which is responsible for the modification of BAT function and underpins cold-tolerance in northern cattle" (Yan et al., 2022: 276). The conclusions were that "on the one hand, well-functioning PRDM16 is required for northern cattle to resist extreme cold, and on the other hand, functional inactivation of PRDM16 impairs beige adipocyte formation, which is beneficial for the environmental adaptability of southern cattle," and these results provided information on adaptive genetic variations in cattle in different temperature regions (Yan et al., 2022: 282).

Liu et al. (2023: 4803) contend that in cattle, prolonged exposure to cold environments may result in reduced development and decline and even death. The authors cite the G allele and the WNK1 gene as possible markers for cold tolerance in Chinese cattle, as their findings indicated that G allele frequencies increased from northern to southern China, with an opposite trend in the frequencies of the A allele. Consequently, the WNK1 gene is surmised to be a candidate gene marker associated with cold tolerance. WNK1 is described as "a member of With-no-lysine kinases (WNKs), widely expressed in animal organs and tissues. WNK1 and WNK4 are expressed in adipose tissue, and WNK4 promotes adipogenesis. WNK1 does not directly affect adipogenesis but has been shown to promote WNK4 expression in several tissues or organs." The study was based on northern group cattle, southern group cattle, central group cattle, and special group cattle (Tibetan cattle) in China (Liu et al., 2023: 4803).

Mei et al. (2021: 959) present a study of Mongolian cattle, ranging in the northern Chinese region of Inner Mongolia, and Minnan cattle, from South China, which represent *Bos taurus* and *Bos indicus*, respectively. The former breed is more well adapted to cold climates, and the latter "is known for its tolerance to muggy weather" (Mei et al., 2021: 959). The findings indicated the two breeds differed and linkage disequilibrium analysis showed that both have experienced variable and intense natural or artificial selection. Potential selected genes (PSRs) were taken from the two samples of cattle, which with the analyzed copy number variation (CNV)-related genes indicated possible associations with phenotypic characteristics, including attributes such as environmental adaptability, feed efficiency, and meat/milk production (Mei et al., 2021: 959).

Wang et al. (2021: 1491) researched on high (≥1800 m) and low altitude (<1500 m) cattle lineages, using high-density genotype data to identify the influence of domestic yak and banteng on the Chinese indicine breeds, which contributed to the latter's divergence from the zebu, which originated in India. The findings showed "evidence for introgression, admixture, and migration in most of the Chinese breeds. Selection signature analyses between high-altitude and low altitude adapted breeds revealed candidate genes and pathways that are putatively involved in hypoxia adaptation" (Wang et al., 2021: 1491). The results suggested "that the up-regulation of ACSS2 expression in the liver promotes the metabolic adaptation of cells to hypoxia via the hypoxia-inducible factor pathway" and "high altitude adaptation involved the introgression of alleles from high-altitude adapted yaks" into Chinese *Bos taurus taurus* before their breed formation and selection, and in addition, the adaptation to cold, high altitude environments was supported by admixture and introgression with adapted cattle (Wang et al., 2021: 1491).

European Cattle

The history of ancient European cattle is linked to the ancestral southwest Asian aurochs, the European aurochs, and the taurine cattle (Götherström et al., 2005; Edwards et al., 2007; Felius et al., 2014). The southwest Asian aurochs are generally suggested as at least associated with the ancestry of the European taurine cattle. For example, Troy et al. (2001: 1088) contend that the restricted wild ancestor ranges of numerous significant native European species indicate their origins in the Fertile Crescent or Anatolia, which are located farther east. This study used mitochondrial DNA and found that the DNA of modern cattle contains four star-like clusters of haplotypes, one of which predominates in Europe and is frequent in the Near East. The authors conclude that this suggests a Near-Eastern origin for European cattle (Troy et al., 2001: 1088). Pellecchia et al. (2007: 1175) also describe cattle and human migrations from the eastern Mediterranean to ancient Etruria (Tuscany, Central Italy), supported by mtDNA of *Bos taurus* breeds, which have "a strikingly high level of mtDNA variation, which is found neither in the rest of Italy nor in

Europe. The Tuscan bovines are genetically closer to Near Eastern than to European gene pools and this Eastern genetic signature is paralleled in modern human populations from Tuscany, which are genetically close to Anatolian and Middle Eastern ones." Felius et al. (2014) note that the first evidence of domesticated possibly taurine cattle in Europe is from the Pre-Pottery Neolithic at Argissa-Magula, in Thessaly, Greece, dated 8500 BP. There may have been two migration routes from southwest Asia, along the Mediterranean coasts and along the Danube River. Domestic cattle reached Central Europe approximately 7500 BP through the Danube route and Northern Europe about 1000 years later (Felius et al., 2014).

The role of European aurochs in this ancestry is frequently debated (Götherström et al., 2005). European aurochs have been recorded as present during the domestication cultures in Europe (Hofman-Kamińska et al., 2019; de Carvalho et al., 2022; Janák et al., 2024). Edwards et al. (2007: 1377) argue that although the aurochs are generally considered the wild progenitor of modern cattle, the contribution of the European aurochs is unclear. In this study, mitochondrial DNA sequences were used to assess the status of this ancestry, and the results suggested that all the pre-Neolithic aurochs were from the previously designated P haplogroup, which indicated a representation of the Late Glacial Central European signature and "one new and highly divergent haplotype in a Neolithic aurochs sample from Germany, which points to greater variability during the Pleistocene" (Edwards et al., 2007: 1377). Furthermore, the Neolithic and Bronze Age samples that were classified with confidence as European aurochs using morphological criteria all carry P haplotype mitochondrial DNA, suggesting continuity of Late Glacial and Early Holocene aurochs populations in Europe. The authors argue that published research suggests that most ancient and modern European domestic cattle carry haplotypes that were previously designated T. "This position, in combination with our new finding of a T haplotype in a very Early Neolithic site in Syria, lends persuasive support to a scenario whereby gracile Near Eastern domestic populations, carrying predominantly T haplotypes, replaced P haplotype-carrying robust autochthonous aurochs populations in Europe, from the Early Neolithic onward." While these breeds coexisted, current evidence suggests that domestic cattle separated from wild aurochs, with minimal introgression (Edwards et al., 2007: 1377).

Götherström et al. (2005: 2345) contribute to this argument, stating that "rather than originating from a few geographical areas only, as indicated by mtDNA, our data suggest that the origin of domestic cattle may be far more complex than previously thought."

Their argument is that mitochondrial DNA (mtDNA) data suggest a clear separation between modern European cattle and the ancient British aurochsen. The results, however, suggest that the Y chromosome haplotypes of north European breeds are more similar to haplotypes from ancient specimens of European aurochsen than to current southern European and southwest Asian cattle breeds. There is also a clear north-south gradient across Europe among modern cattle breeds in the frequencies of two distinct Y chromosome haplotypes; the northern haplotype is found "in 20 out of 21 European aurochsen or early domestic cattle dated 9500–1000 BC" (Götherström et al., 2005: 2345). These findings suggest that there was some local hybridization between local male aurochsen and cattle, which may have contributed to a paternal imprint on the genetics of current central and northern European breeds. The authors note that there may have been some undetected mating or deliberate breeding between aurochs bulls and domestic cows (Götherström et al., 2005: 2345).

The cattle ancestry and the genetic associations with aurochs in Europe are also supported by Lari et al. (2011), who notes that for the "complete mitochondrial genome of an 11,450-year-old aurochsen (Bos primigenius) from Central Italy" also argue that while the origin of current European breeds in Europe is contested, the current archaeological and early genetic evidence suggests a single southwestern Asian/eastern Mediterranean (Near Eastern) origin and a later diffusion with herding and farming. However, some recent genetic data suggest some local domestication events or some local, aurochs introgression. The authors argue that some Italian aurochsen were genetically similar to modern cattle breeds, but "highly divergent from the North-Central European aurochsen" and using "complete mitochondrial genome sequences … suggest that previously identified sub-groups within the widespread modern cattle mitochondrial T clade are polyphyletic, and they support the hypothesis that modern European breeds have multiple geographic origins" (Lari et al., 2011).

The European aurochs have also been studied for their size relative to domestic cattle. For example, Wright and Viner-Daniels (2015: 8) present a study identifying a south-north body size cline and a west-east body size cline for European aurochs. It is argued that "domestic cattle are smaller than their wild ancestors, but there is also a degree of overlap between the two species, which means that distinguishing them can be problematic. However, previous analyses of aurochs morphology have generally been patchy, and do not provide a picture of aurochs variation across Europe according to environment, climate and geography." Some studies by zooarchaeologists may use unsuitable comparative biometrical data from geographical areas and time periods to identify remains from their local study areas, which may not be generalizable to larger areas (Wright & Viner-Daniels, 2015: 8). Regarding the south-north cline, it is possible it is due to Bergmann's Rule (1847) that animals in colder climates are larger than those of warmer climates, to facilitate retention of body heat. Wright and Viner-Daniels (2015: 18) note that "the combination of the south-north and west-east patterns could ultimately reflect an overall north-east/south-west cline in aurochs body size, which can be investigated more effectively with the addition of more data from eastern areas of Europe." In addition, the most Italian species were smaller than the British specimens, and the southern European populations were smaller northern European populations. However, "after domestication this pattern can no longer be identified, but there is tentative evidence for a west-east cline in body size, as German and Polish samples consistently contain the largest individuals" (Wright & Viner-Daniels, 2015: 18).

Bonfiglio et al. (2010) add to this argument, noting that when the domestic taurine cattle slowly migrated from the southwest Asian "Fertile Crescent" to Europe, the local wild aurochsen were still present and numerous, as they had for millennia, which could have facilitated the inbreeding of local aurochs and domestic cattle.

Significantly, this scenario suggests conditions for interbreeding and creation of secondary domestication clusters with local aurochs populations (Bonfiglio et al., 2010). The findings using mitochondrial genome analyses suggested that "not all modern taurine mtDNAs belong to the shallow macro-haplogroup T of Near Eastern origin, as demonstrated by the detection of three branches (P, Q and R) radiating prior to the T node in the bovine phylogeny. These uncommon haplogroups represent excellent tools to evaluate if sporadic interbreeding or even additional events of cattle domestication occurred" and "revealed major similarities between haplogroups Q and T" (Bonfiglio et al., 2010). The authors conclude Q may represent a different branch lineage that was domesticated in Southwest Asia alongside the T subhaplogroup. Haplogroup R, on the other hand, was limited to Italy, suggesting that it originated in the European aurochs. Haplogroup R may have been obtained by interbreeding between domestic cattle and aurochs, or secondary domestication of aurochs in Italy.

Mona et al. (2010) introduce another complexity: using DNA fragments from the mitochondrial hypervariable region of samples from Italy, the findings suggested that Italian aurochsen were different from northern/central European aurochsen but genetically similar to modern bovine breeds and "the genetic variation pattern in both Italian and northern/central European aurochsen is compatible with demographic stability after the last glaciation" (Mona et al., 2010). The authors concluded that different aurochsen groups "may have occurred in Italy and northern/central Europe after the last glaciation, and the Italian and Fertile Crescent aurochsen may have shared some mtDNA sequences among themselves and with modern breeds, and there was some genetic homogeneity between aurochs populations in Southern Europe and the Middle East (southwest Asia), and the post-glacial recolonization of northern and central Europe advanced from eastern, and not southern refugia" (Mona et al., 2010).

Gravlund et al. (2012: 103) cite evidence of lower admixture levels between local European aurochs and cattle. The findings of this study, based on extracted DNA from 39 Danish aurochs specimens dated to 9830–2865 14 Cyr BP, the "first study of genetic variation of Danish aurochs" suggested that "aurochs from southern Scandinavia display a star-shaped population genetic structure," and this suggests that some ancestral haplotypes have undergone local and recent diversification; these may have originated from the ancestral population in Western Europe before migrating northward during glacial retreat. These analyses do not support scenarios that suggest multiple invasions by genetically distinct aurochs. Instead, the findings point to a single, ongoing northward migration (Gravlund et al., 2012: 103). The authors also point out that their data imply—albeit with scant evidence—that aurochs in Northwestern Europe experienced a population expansion that started soon after the glacial ice receded from Denmark and remained stable in size until the inevitable population decline that had to happen before the species went extinct. The lack of haplotypes resembling contemporary domestic cattle in their aurochs sample implies that, if any, introgression between these species was likely restricted (Gravlund et al., 2012: 103).

Cubric-Curik et al. (2021: 663) performed a mitogenome analysis of cattle including a sample from southeast Europe, which was an important agricultural region during the European Neolithization, with high cattle biodiversity. The results indicated the dominant T3 haplogroup, and other rarer haplogroups including Q, T5, T4, T2, and T1, and "maternal line in Austrian Murbodner cattle that possess surviving aurochs' mitochondria haplotype P1 that diverged prior to the Neolithization process" (Cubric-Curik et al., 2021: 663). The authors note that this is "convincing evidence for rare female-mediated adaptive introgression of wild aurochs into domesticated cattle in Europe" with "a more precise estimated divergence time than previously available" (Cubric-Curik et al., 2021: 663). The authors conclude that, unlike some other studies, their results "support the existence of interactions between aurochs and cattle during domestication and dispersal of cattle in the past, contribute to the conservation of maternal cattle diversity and enable functional analyses of the surviving aurochs P1 mitogenome" (Cubric-Curik et al., 2021: 663). In addition, discovery of surviving aurochs' mitochondria provides strong evidence for gene flow between European wild aurochs and domesticated cattle and supports the protection of already endangered maternal cattle diversity. In addition, Murbodner cattle with the P1 haplotype may be useful for aurochs "de-extinction programmes and for research on the expression of mitonuclear interacting gene complexes" (Cubric-Curik et al., 2021: 663).

European taurine cattle have also been compared with Asian taurine cattle. Chen et al. (2021) argue that although European taurine and East Asian taurine are both clades in *Bos taurus*, their genomic differences are currently unclear. The study focused on Mongolian (East Asian taurine cattle, distributed in Inner Mongolia Autonomous Region of China, Mongolia, and East Siberian Regions of Russia) cattle genomes and compared them with 10 breeds in other nations, including Hanwoo (Korean) and Mishima (Japanese) breeds. The authors argue that published research suggests that East Asian taurine breeds originated/migrated from Mongolia, therefore "it is of importance to investigate the phylogenetic relationship between Mongolian and other East Asian taurine cattle at the whole-genome level" (Chen et al., 2021: see Kantanen, 2010; Lee et al., 2014). Chen et al. (2010) note that in contrast to the other breeds, the European taurine typically exhibit rapid growth and large body size and "the functional enrichment analysis of candidate genes under positive selection revealed that metabolic pathways were significantly overrepresented in European taurine… These results imply that putatively selected genes might contribute to the characteristics of rapid growth and large body size in European taurine" (Chen et al., 2021).

The varied ancestry of European cattle has also been investigated in regard to indicine cattle. Kantanen et al. (2009: 404) note that "maternally inherited mitochondrial DNA (mtDNA) has been used extensively to determine origin and diversity of taurine cattle (Bos taurus) but global surveys of paternally inherited Y-chromosome diversity are lacking." Their study examined mtDNA information on less studied Eurasian breeds and Y-chromosomal microsatellite data. The findings indicated that mitochondrial haplogroup T3 was the most prevalent and that T4 was found exclusively in Siberian Yakut cattle. According to the mtDNA data, there may have been taurine/zebu cattle hybrids in the Central Asian and Ukrainian regions. It appears that southern and southeast European races have since been affected by this zebu influence (Kantanen et al., 2009: 404). Additionally, compared to Middle Eastern and Anatolian races, the frequency of the most prevalent Y-chromosomal microsatellite haplotype in the Eurasian sample set was higher. It was discovered that the taurine Y-chromosomal microsatellite haplotypes were arranged in a network based on the Y haplogroups Y1 and Y2. These findings may contest the widely held belief that local European hybridization of cattle with male aurochs is the source of Y1 (Kantanen et al., 2009: 404).

Studies have also been conducted on how European cattle breeds have adapted to cold climates. As an illustration, Buggiotti et al. (2021: 3093) compared the Kholmogory as a typical taurine cattle breed and studied the evolutionary history of the world's northernmost native cattle breeds from Russia. On the other hand, Yakut cattle, who split off from European taurine cattle approximately 5000 years ago, have a large number of ancestors and some genetic variants that allow them to adapt to the harsh Arctic conditions. The findings pointed to a number of shared genetic pathways in both breeds that are connected to their ability to adapt to extreme weather. However, the genes impacted by these pathways' selection were essentially distinct. At least 16 species of hibernating/cold-adapted mammals from six different phylogenetic orders share a recent convergent change in amino acid residues, which is exemplified by a breed-specific missense mutation in the highly conserved NRAP gene found in Yakut cattle. This points to a fast fixation in a single isolated population of cattle exposed to a severe environment and a convergent evolutionary event along the mammalian family tree (Buggiotti et al., 2021: 3093).

In addition, Weldenegodguad et al. (2019) observe that the northernmost regions on Earth for cattle are Northern Finland and the Sakha Republic in the Russian Federation. Whole-genome sequencing was used in the study to characterize the genetic makeup of three uncommon native breeds of cattle that are native to these northern Eurasian regions: Yakutian, Western, and Eastern Finncattle (Weldenegodguad et al., 2019). The findings demonstrated that there were two major prehistoric population sizes among the ancestral aurochs of Eurasian taurine cattle, and these populations were linked to significant climatic changes. Weldenegodguad et al. (2019) discovered that, compared to modern European taurine breeds, modern Yakutian cattle displayed a higher level of within-population variation in terms of the number of SNPs and nucleotide diversity. This finding is inconsistent with the findings of marker-based cattle breed diversity studies, suggesting assortment bias in earlier analyses (Weldenegodguad et al., 2019). The results of this study indicated a number of genomic regions under selection that may have contributed to the adaptation to the northern and subarctic environments, including genes involved in disease resistance, sensory perception, cold adaptation, and growth. They also supported two theories: either the effective population size of the ancestral Asiatic taurine cattle may have been higher than that of the European cattle, or they indicate the hybrid origins of the Yakutian cattle ancestries and possibly the lack of intensive artificial selection (Weldenegodguad et al., 2019).

In a similar study, Yurchenko et al. (2018) investigated the genomes of nine Russian native cattle breeds that have adapted to extremely cold environments. The authors argue that in comparison with breeds of European and Asian origins "we found known and novel candidate genes that could potentially be related to domestication, economically important traits and environmental adaptations in cattle" (Yurchenko et al., 2018). The genomes of the Russian cattle breeds contained specific areas that appear to be under selective pressure, housing genes that could be linked to adaptations to extreme environments (like AQP5, RAD50, and RETREG1). The research uncovered genetic markers indicative of selective sweeps close to crucial genes influencing economically relevant attributes such as milk yield (such as DGAT1, ABCG2), body growth (such as XKR4), and reproductive capacity (such as CSF2). The researchers propose that these candidate genes should be considered for further study to enhance existing breeds and aid in the development of commercial breeds better adapted to similar climatic conditions (Yurchenko et al., 2018).

African Cattle

The presence of cattle in Africa has been dated to 7000–3500 BP in Egypt, Libya, and the Sahara (Felius et al., 2014). It is argued that dairy pastoralists arrived in the prehistoric, more vegetated Sahara around 7000 BP, and created rock carvings possibly the ancestors of modern West African cattle. The Sahara then became much drier around 4000 BP and the pastoralists moved out of the area. Before zebu arrived in Northeast Africa, remains of both long- and short-horned cattle were detected at several locations dating from 5600 to 3000 BP (see below). Cattle herding moved southwards starting in 2500 BP. Along the North African Mediterranean coast and later the West African Atlantic coast, taurine cattle most likely also spread westward from Egypt. Felius et al. (2014) also note that taurine cattle may have moved from Egypt westward along the North African coast and then along the West African coast.

There is debate and a lack of clarity surrounding the genetic background of pastoralism in Africa, for example, Hanotte et al. (2002: 336). Kim et al. (2020: 1099) note that "cattle pastoralism plays a central role in human livelihoods in Africa. However, the genetic history of its success remains unknown." Mwai et al. (2015): 911) argue that "at least 150 indigenous African cattle breeds have been named, but the majority of African cattle populations remain largely uncharacterized. As cattle breeds and populations in Africa adapted to various local environmental conditions, they acquired unique features. We know now that the history of African cattle was particularly complex and while several of its episodes remain debated, there is no doubt that African cattle population evolved dramatically over time. Today, we find a mosaic of genetically diverse population from the purest Bos taurus to the nearly pure Bos indicus." Gebrehiwot et al. (2020) note that "Humpless Bos taurus cattle are one of the earliest domestic cattle in Africa, followed by the arrival of humped Bos indicus cattle. The diverse indigenous cattle breeds of Africa are derived from these migrations, with most appearing to be hybrids between Bos taurus and Bos indicus… African indigenous cattle breeds have a large genetic diversity and are either pure African Bos taurus or admixtures of African Bos taurus and Bos indicus."

Most studies suggest that the ancestry of modern African taurine cattle is most likely derived from native aurochs, which were brought to Africa 6800 years ago with the introduction of taurine cattle from southwest Asia. Male Indian cattle were first introduced into Africa around 4000 years BP, and the introduction increased 1300 years BP. Cattle from the original taurine lineage were mostly replaced by indicine cattle around 200 years BP, because they were less prone to rinderpest than the taurine cattle. One significant taurine-indicine admixture event has been dated to between 750 and 1050 years ago, or roughly 150 generations ago. African indicine populations with varying taurine/indicine genome compositions that are better suited to arid climates are the consequence of male-mediated introgression of indicines into local taurine herds (Xia et al., 2023; see also (Ajmone-Marsan et al., 2010; Felius et al., 2014; Pitt et al., 2019; Verdugo et al., 2019; Kim et al., 2020).

Hanotte et al. (2002: 336) examined the genetic signatures of cattle and found evidence of genetic influence from Europe and the Middle East. Kim et al. (2020: 1099), used a whole-genome sequence analysis of indigenous African cattle representative of the main cattle breeds and found "a major taurine × indicine cattle admixture event dated to circa 750–1050 year. ago, which has shaped the genome of today's cattle in the Horn of Africa" and taurine or indicine ancestry, including "immune-, heat-tolerance- and reproduction-related genes" and "one highly divergent locus in African taurine cattle, which is putatively linked to trypanotolerance and present in crossbred cattle living in trypanosomosis-infested areas" (Kim et al., 2020: 1099). The authors conclude that "a combination of past taurine and recent indicine admixture-derived genetic resources is at the root of the present success of African pastoralism" (Kim et al., 2020: 1099).

Kim et al. (2020: 1099) comment that "cattle pastoralism plays a central role in human livelihood in Africa. However, the genetic history of its success remains unknown." Whole-genome sequencing analysis was used in the study on African cattle from 16 breeds that are representative of the main cattle groups. The study found that the current genome of cattle in the Horn of Africa was shaped by a significant taurine-indicine admixture event that occurred between 750 and 1050 years ago, and 16 loci in crossbred animals with an excess of taurine or indicine ancestry that are linked to adaptations to the African environment, including the immune system, heat tolerance, and reproduction genes. In addition, in crossbred cattle ranging in trypanosomosis-affected areas, there was a divergent locus in taurine cattle, possibly related to trypanotolerance. The findings showed that African pastoralism is supported by a combination of older taurine genetic resources and more recent indicine admixtures (Kim et al., 2020: 1099).

Kim et al. (2023) argue that "present-day African cattle retain a unique genetic profile composed of a mixture of the Bos taurus and Bos indicus populations introduced into the continent at different time periods. However, details of the admixture history and the exact origins of the source populations remain obscure." In their study, a significant contribution (up to ~20%) was detected "from a basal taurine lineage, which might represent the now-extinct African aurochs" and "the indicine ancestry of African cattle, although most closely related to so-far sampled North Indian indicine breeds, has a small amount of additional genetic affinity to Southeast Asian indicine breeds"; the findings supported "the hypothesis of aurochs introgression into African taurine…" and the "origin of indicine ancestry in Africa might be different indicine populations than the ones found in North India today" (Kim et al., 2023).

Kim et al. (2017) describe a genetic variation study of African cattle based on genome sequencing of five native populations and comparison with commercial taurine breed genomes. The findings demonstrated that the African zebu and sanga cattle exhibited the highest genetic diversity. Additionally, selection signatures for traits that are environmentally adaptive were found, specifically those affecting anemia, feeding behavior in trypanotolerant N'Dama, coat color, and other characteristics, and regulate the growth of horns and resistance to heat and ticks, especially in zebu breeds (Kim et al., 2017). The findings, according to the authors, are revealed at a genome-wide level, emphasizing the distinctive adaptive diversity of African cattle and the prospects for long-term increases in livestock productivity on the continent (Kim et al., 2017).

Paguem et al. (2020) examined the adaptive traits of African indigenous taurine cattle which are "shaped by husbandry management, regional climate and exposure to endemic pathogens," noting that these breeds may be "less productive with respect to milk and meat production" partly and possibly due to "small size, traditional beliefs, husbandry practices, limited feed resources, disease burden and lack of sustained breeding for trait improvement" (Paguem et al., 2020). The Namchi taurine cattle breed is resistant to trypanosome infection and exposure to tick infestation, but the introduced zebu cattle are commoner, and less resistant to local pathogens. The study used a whole genome sequencing approach and sequenced the genomes of cattle breeds from Cameroon. The studied breeds were the "Namchi, an endangered trypano-tolerant taurine breed, the Kapsiki, an indigenous trypano-susceptible taurine breed, and three Zebu (Bos indicus indicus) breeds: Ngaoundere Gudali, White Fulani and Red Fulani" (Paguem et al., 2020). The study compared the single nucleotide polymorphisms (SNPs) obtained from analysis using the genome ARS-UCD1.2 with reference genomes of European taurine Holstein, the Asian indicine Brahman, and African trypanotolerant N'Dama breeds. The study identified genetic signatures to assess cattle-breed-specific phenotypes including trypanotolerance and thermotolerance in the Namchi and Kapsiki breeds, respectively. The authors note that phylogenetic analysis clustered the taurine breeds Kapsiki, N'Dama, and Holstein together (excepting the Namchi breed) and the indicine White and Red Fulani, Gudali, and Brahman separately. It is argued that "the deviating result for Namchi indicates a hybrid status of the selected animal with a recent introgression of Zebu genes into its genome" (Paguem et al., 2020).

An important issue concerns the association between the zebu and sanga cattle in Africa. For example, Frisch et al. (1997: 77) present "new evidence supporting the classification of southern African sangas as Bos taurus and East African zebus as 'taurindicus'," based on karyotype, DNA markers, and protein polymorphisms. The findings suggested that "southern African sanga breeds have a submetacentric Y chromosome typical of Bos taurus. Frequencies of four DNA markers support the hypothesis that the Tuli, a southern African sanga, had taurine ancestors and the Boran had both taurine and indicine ancestors. Frequencies for several protein polymorphisms strongly suggest that southern African sangas have more in common with taurine than with indicine breeds, while East African zebus are an admixture of African taurine and Asian indicine breeds" (Frisch et al., 1997: 77). Lenstra (2020) note that "complete mtDNA sequences… show that the African T1 and East-Asian T4 haplotypes are closely related to mtDNA variants from Southwest Asia" (Lenstra et al., 2014).

Pitt et al. (2018: 123) note that because of their recent divergence and the relatively recent gene flow between them, African taurine cattle and Asian indicine cattle share high levels of genetic variation. It is also conceivable that there could be unidirectional or bidirectional migration events between European taurine and Asian indicine cattle; however, more research is needed to understand the intricate human-induced dispersal patterns of domestic cattle. Furthermore, a number of factors, including a more recent divergence in indicine cattle or less gene flow between Africa and Europe than between Africa and Asia, may account for the significant difference between the African and Asian indicine groups and the European and African taurine populations. An alternative explanation could be that African taurine populations are more affected by genetic drift than Eurasian taurine or indicine populations (Pitt et al., 2018: 123). The authors conclude that their study confirmed that in contrast to American hybrids, which showed a greater influence from taurine breeds, African hybrids were formed with a greater contribution from African indicine genetic origin due to the large differentiation between African and Eurasian taurine and the high levels of admixture in African indicine cattle from both Asian indicine and African taurine cattle (Pitt et al., 2018: 133).

The study by Hanotte et al. (2000: 387) examines the sub-Saharan geographical distribution and the frequency of an indicine and a taurine Y-specific allele among African cattle breeds. The authors argue that "the taurine allele is probably the oldest one on the continent. However, the taurine and the indicine alleles were present in 291 males (30%), and 693 males (70%), respectively" and "the taurine allele is the commonest only among the sanga breeds of the southern African region and the trypanotolerant taurine breeds of West Africa." In addition, the "sanga cattle from the Lake Victoria region and the Kuri cattle of Lake Chad, cattle populations surrounded by zebu breeds were, surprisingly, completely devoid of the indicine allele. Human migration, phenotypic preferences by the pastoralists, adaptation to specific habitats and to specific diseases are the main factors explaining the present-day distribution of the alleles in sub-Saharan Africa" (Hanotte et al., 2000: 387).

Ibeagha-Awemu et al. (2004:673) studied *Bos indicus* and *Bos taurus* cattle breeds in Cameroon and Nigeria using genotype information on 28 markers, with the findings showing that "the level of the Indian zebu genetic admixture proportions in the African zebus was higher than the African taurine and European taurine admixture proportions, and ranged from 58.1% to 74.0%. The African taurine breed, Muturu was free of Indian zebu genes while its counter Namchi was highly introgressed (30.2%)." Through phylogenic reconstruction and principal component analysis, close associations were detected between the zebu breeds in Cameroon and Nigeria, and there was a large genetic divergence between the African taurine, European taurine and Indian indicine, and "a central position for the African zebus" (Ibeagha-Awemu et al., 2004:673).

In a similar study, Bahbahani et al. (2017) examined the genetic background of the East African Shorthorn Zebu (EASZ) cattle, which are an "ancient hybrid between Asian zebu × African taurine cattle preferred by local farmers due to their adaptability to the African environment." Although the authors admit that "the genetic controls of these adaptabilities are not clearly understood yet," the findings suggested an "even admixed autosomal zebu × taurine genomic structure in the population..." In addition, genome-wide SNP analysis of zebu × taurine admixed cattle populations from Uganda and Nigeria showed some shared genetic background between KEASZ and Uganda cattle and across the KEASZ, Uganda, and Nigeria cattle. The authors conclude that "these regions intersect with genes and QTL associated with reproduction and environmental stress (e.g., immunity and heat stress) suggesting that the genome of the zebu × taurine admixed cattle has been uniquely selected to maximize hybrid fitness both in terms of reproduction and survivability" (Bahbahani et al., 2017).

An important issue for African cattle domestication concerns trypanotolerance. Gautier et al. (2009) also refer to the environmental impacts of cattle introductions into West Africa, arguing that "the recent introduction of cattle in West Africa after past waves of migration from remote centres of domestication has imposed dramatic changes in their environmental conditions, in particular through exposure to new pathogens." The adaptation of West African cattle shows an adaptation to West African conditions. The authors studied these adaptations at the whole genome level in a data set including 36,320 SNPs genotyped in nine West African cattle populations (Gautier et al., 2009). The authors conclude that "the main possible underlying selective pressures may be related to climatic conditions but also to the host response to pathogens such as Trypanosoma species. Overall, these results might open the way towards the identification of important variants involved in adaptation to tropical conditions and in particular to resistance to tropical infectious diseases" (Gautier et al., 2009).

Gifford-Gonzalez and Hanotte (2011) note that "the endemic tsetse fly transmits trypanosomes to hoofed animals. Indigenous wild bovids are trypanotolerant, the result of their long co-evolution with the infection, but domestic livestock develop acute symptoms that often result in death. Not all infected animals die, but trypanosomes persist in recovered, asymptomatic animals, and may resurge if helminthic infestations, low plane of nutrition, or late stages of pregnancy cause loss of condition." Some breeds in the Lake Chad basin areas, "unlike the straight-backed breeds of West African closed habitats, these breeds do not display appreciable trypanotolerance" (Gifford-Gonzalez & Hanotte, 2011). Other breeds, including West African shorthorns, are trypanotolerant. It is argued that "the genetics of trypanotolerance in cattle has been intensively explored. It is under complex genetic control involving several genes and gene networks with responses to parasitaemia and to anaemia—a major consequence of trypanosomiasis infection—under distinct genetic controls.

Both taurine and indicine genetic factors contribute to trypanotolerance in crossbred populations, suggesting that at least some general mechanism of tolerance to blood parasite infection exists among tropical domestic bovines" (Gifford-Gonzalez & Hanotte, 2011). Other breeds of cattle exhibit variable levels of trypanotolerance. The authors note that research has examined "the introgression of indicine Y-chromosome heritage" in African cattle, evidence of taurine African cattle from the eastern Saharan region dispersing into West Africa, and evolving into trypanotolerant taurine cattle, and dispersion into East and Southern Africa, the Horn of Africa as the original entry point of zebu indicine cattle into Africa, and mitochondrial and Y-chromosome-based evidence of South Asian zebu bulls crossbreeding with African taurine cows, producing sanga breeds, which expanded from East Africa across the continent (Gifford-Gonzalez & Hanotte, 2011).

Noyes et al. (2011: 9304) add that trypanosomiasis is a "major constraint on cattle productivity in sub-Saharan Africa. Some African Bos taurus breeds are highly tolerant of infection, but the potentially more productive Bos indicus zebu breeds are much more susceptible. Zebu cattle are well adapted for plowing and haulage, and increasing their tolerance of trypanosomiasis could have a major impact on crop cultivation as well as dairy and beef production." In addition, "some African Bos taurus cattle breeds, such as N'Dama, are tolerant of infection with T. congolense, remaining apparently healthy despite the presence of parasites. This capacity to remain productive while harboring potentially lethal trypanosome infections is known as trypanotolerance and is an evolutionary adaptation between the host and pathogen. N'Dama cattle are native to West Africa, where they were introduced about 3000 years ago, whereas the introduction of zebu (Bos indicus) cattle to the region is recent, probably only intensifying after a rinderpest epidemic just over a century ago." The study examined selected genes within the quantitative trait loci (QTL) which had been found to regulate cattle response to infection. The two breeds examined were the trypanotolerant N'Dama and the less tolerant Boran cattle after both had been infected with *Trypanosoma congolense*. The results identified a previously undescribed polymorphism (multiple possible traits in a gene) in the trypanotolerance QTL which may be involved in the response to infection. The authors concluded that their "This combination of genetic, in vitro analysis and expression data has provided unique insights into the differences between susceptible Boran and tolerant N'Dama cattle in their response to infection with T. congolense. We see clear evidence of differential innate responses early in the infection, apparently mediated by sequence differences in previously identified QTL regions" (Noyes et al., 2011:9308).

Paling and Dwinger (1993: 60) note that the "trypanotolerant cattle, like the N'Dama can survive and produce in tsetse affected areas without interventions. This taurine breed has been indigenous to Africa for approximately 7000 years and forms presently about 6% of the bovine population of Africa… experimental infections showed conclusively the superior resistance to the effects of infection of the N'Dama cattle when compared to zebu cattle and have confirmed that trypanotolerance is innate in N'Dama cattle." Murray et al. (1990: 369) added that the taurine N'Dama and West African Shorthorn breeds of West and Central Africa have a "significant degree of innate resistance to trypanosomiasis and several other important infectious diseases. These cattle are extremely well adapted to the environment and are now recognised as having considerable production potential. The ability to resist the development of anaemia in the face of infection, as assessed by packed red cell volume percent (PCV), has been shown to be correlated with the capacity to be productive, thereby identifying regulation of PCV as a key trait of trypanotolerance" (Murray et al., 1990: 369).

Tick disease resistance has been mentioned for zebu cattle and compared with trypanotolerance. For example, Mattioli et al. (1998: 57) presented a case study based on N'Dama, Gobra zebu, and Gobra x N'Dama crossbreeds, which were "exposed to field-tick infestations and natural high tsetse challenge over more than one year." The findings indicated that "N'Dama cattle exhibited a superior capacity to control trypanosome infections and limit tick burdens… N'Dama cattle showed lower trypanosome infection rates, higher mean PCV value, lower requirement for trypanocide treatments and lower tick load than crossbred and Gobra cattle" than the other breeds and cross breeds (Mattioli et al., 1998: 57). It is noted that "no death occurred in N'Dama cattle maintained under high surveillance of trypanosome infection, while approximately 8% of crossbreds and 50% of Gobras died of trypanosomosis" while in another group, all Gobra and over 70% of crossbred cattle and less than 16% of the N'Dama died. In both groups, differences in mortality were significantly higher ($P < 0.01$) in Gobras than in N'Damas. The authors concluded that "cattle of the three breeds suffered from trypanosomosis and that trypanosome infections affected tick susceptibility. However, N'Dama showed a superior ability to limit both the pathological effects of trypanosomosis and the level of tick infestations" and therefore the "multi-disease resistant N'Dama cattle is recommended for the low-input traditional African farming systems in areas where trypanosomosis, ticks and tick-borne diseases are constraints to livestock production" (Mattioli et al., 1998: 57).

Mattioli and Dempfle (1995: 63) also present a study suggesting that "although resistance is mainly associated with zebu (Bos indicus) breeds, studies carried out in The Gambia show a higher resistance to ticks and tick-borne diseases in N'dama (Bos taurus) than in Gobra (B. indicus) cattle. Tick resistance in N'dama breed appears to be effective against those species with long hypostome, such as Amblyomma variegatum and Hyalomma spp." Mattioli et al. (1995: 95) studied the Gambian N'Dama and Gobra zebu cattle for naturally occurring tick infection, using an assessment of the serological prevalence of *Anaplasma marginale* antigen and *Babesia bigemina* antibody. The results showed significantly higher numbers of *Amblyomma variegatum* and *Hyalomma* species in Gobra zebu than on N'Dama cattle. *A. marginale* antigen was significantly lower in the N'Dama breed, with similar levels of the serological prevalence of the *B. bigemina* antibody in both N'Dama and Gobra cattle. Another result was "liveweight gain was significantly higher in N'Dama in comparison with Gobra cattle during the period of abundance of A. variegatum and Hyalomma species in adults. The significantly lower burdens of A. variegatum and Hyalomma spp. ticks in Gambian N'Dama in comparison with Gobra zebu cattle confirms previous findings. It also shows a more deleterious effect of ticks and tick-borne diseases on animal production and health in Gobra zebu than in N'Dama cattle. The presence of a tick resistant trait in N'Dama cattle is suggested" (Mattioli et al., 1995: 95).

Mattioli et al. (2000: 263) added a similar study of immunogenetic influences on tick resistance in African cattle, based on the N'Dama and trypanosusceptible Gobra zebu cattle, arguing that "resistance to tick attack and tick-borne micro-organisms (TBMs) varies among different breeds of cattle" and "generally, zebu cattle possess a higher resistance to ticks and TBMs than European cattle. The host's immune system would appear to be the single most important factor that regulates this resistance." The study examined the effector immune mechanisms that determine resistance against ticks and TBMs in the studied cattle. The findings suggested that the cellular immune response was more effective than the humoral immunity in regulating resistance to ticks and TBMs. There were also similarities between the immune mechanisms of the N'Dama when infected with trypanosomes and when subjected to tick bites and TBMs. The authors conclude that "in the N'Dama breed, resistance against ticks per se also has a genetic basis. Therefore, the N'Dama appears to be a unique breed in that it exhibits resistance to several parasitic diseases and/or infections, including helminths, when compared to other cattle breeds in West Africa" and therefore "the multi-parasite resistant traits of the N'Dama breed should be exploited in those areas where trypanosomosis, ticks and tick-borne diseases constrain animal production" (Mattioli et al., 2000: 263).

North and South American Cattle

There are no indigenous domestic cattle species in the Americas, but various breeds were introduced with European colonization. Ajmone-Marsan et al. (2010: 154) note that after the discovery and conquest of America by the Europeans, livestock farming accompanied human emigration. This movement of cattle occurred along four main routes, namely from Spain and North Africa to the Caribbean, from Portugal to Brazil, from Northern Europe with North America and Australia and India to Brazil, the first three mostly of taurine cattle and the last mostly of zebu cattle (Ajmone-Marsan et al., 2010: 154). The first livestock came with the second voyage of Columbus, who left Cádiz in 1493 and went to Santo Domingo. The first cattle in continental America arrived in what is now Colombia in 1524. Cattle were eventually brought to Peru, Chile, Paraguay, Argentina, Uruguay, and Venezuela over the course of the following 50 years (Ajmone-Marsan et al., 2010: 155).

Regarding the Caribbean area, McTavish et al. (2013:1398) note that "domesticated cattle were introduced to the Caribbean in 1493 by Christopher Columbus, and between 1493 and 1512, Spanish colonists brought additional cattle in subsequent expeditions" introducing cattle to southern North America and northern South America (see also Rouse, 1977). It is also noted that Christopher Columbus brought the first cattle in the Americas from the Canary Islands to the Caribbean Island of Hispaniola on his second voyage across the Atlantic in 1493, and Spanish colonists continued to import cattle until then. Cattle from the Canary Islands descended from animals of Portuguese and Spanish origin introduced 20 years earlier by early Spanish explorers. Therefore, these cattle probably had some ancestry with North African cattle breeds and may therefore have contained an indicative genetic component via an earlier gene flow from Africa to the Iberian Peninsula. Imported cattle multiplied rapidly in the Caribbean, and by 1512 the importation of cattle by ship was no longer necessary (Barragy, 2003).

McTavish et al. (2013:1398) note that feral Spanish, which were the ancestors of modern American cattle are more adapted than European cattle to the American environment, including drought and tick-borne diseases (Figueroa et al., 1992). Decker et al. (2009) found New World cattle to be related to European taurine cattle; however, indicine cattle were introduced to North America through Jamaica by the 1860s and later to Brazil. McTavish et al. (2013: 1399) note that "now there are "naturalized" Brazilian indicine (Nelore) and indicine/taurine hybrid (Canchim) breeds. In some samples of Spanish-derived breeds from South America, mtDNA haplogroups and a Y chromosome microsatellite marker suggest indicine introgression in New World cattle. In particular, recent male-mediated introgression of indicine alleles into taurine breeds appears common in Brazil" and "introgression in New World Cattle breeds suggest that European–African admixture (which results in greater apparent divergence) may have driven the apparent sister group relationship between Texas Longhorns and all other European taurine cattle" (McTavish et al., 2013: 1399: see also Hoyt (1982); Giovambattista et al., 2000; Decker et al., 2009; Ginja et al., 2010; Mirol et al., 2003).

Genetic studies on American cattle have made important findings, such as some breeds being more thermotolerant. For example, Olson et al. (2003: 80) detected a "major gene (designated as the slick hair gene), dominant in mode of inheritance, that is responsible for producing a very short, sleek hair coat. Cattle with slick hair were observed to maintain lower rectal temperatures (RT). The gene is found in Senepol cattle and criollo (Spanish origin) breeds in Central and South America. This gene is also found in a Venezuelan composite breed, the Carora, formed from the Brown Swiss and a Venezuelan criollo breed." In addition, "the effect of the slick hair gene on RT depended on the degree of heat stress and appeared to be affected by age and/or lactation status. The decreased RT observed for slick-haired crossbred calves compared to normal-haired contemporaries ranged from 0.18 to 0.4 degrees C. An even larger decrease in RT (0.61 degrees C; $P < 0.01$) was observed in lactating Carora x Holstein F1 crossbred cows, even though it did not appear that these cows were under severe heat stress." The authors concluded that the improved thermotolerance of crossbred calves due to their slick hair coats did not result in increased weaning weight, possibly because both the slick hair and normal-haired calves were suckled by slick haired dams. The slick haired calves grew faster immediately after weaning, and their growth in the cooler months of the year was not significantly affected by reduced hair growth. In the Carora x Holstein cross-bred cows, the smooth coat had a positive effect on milk production under dry, tropical conditions (Olson et al., 2003: 80).

Domestication and Wild Cattle Attacks

An important aspect of domestication is the reduction or elimination of "wild" behavior in the domesticated species (Price, 1984). Price (1999: 245) notes that "animal domestication is best viewed as a process, more specifically, the process by which captive animals adapt to man and the environment he provides. Since domestication implies change, it is expected that the phenotype of the domesticated animal will differ from the phenotype of its wild counterparts. Adaptation to the captive environment is achieved through genetic changes occurring over generations, and environmental stimulation and experiences during an animal's lifetime (Price, 1984). In this sense, domestication can be viewed as both an evolutionary process and a developmental phenomenon." Ferreira et al. (2023) note that "animal domestication leads to diverse behavioral, physiological, and neurocognitive changes in domesticated species compared to their wild relatives," but "the widely held belief that domesticated species are inherently less 'intelligent' (i.e., have lower cognitive performance) than their wild counterparts requires further investigation..." and among the studied research, "30% of these studies showed superior cognitive abilities in wild animals, while another 30% highlighted superior cognitive abilities in domesticated animals. The remaining 40% of studies found similar cognitive performance between the two groups. Therefore, the question regarding the presumed intelligence of wild animals and the diminished cognitive ability of domesticated animals remains unresolved" (Ferreira et al., 2023).

For cattle, given their size and evident strength, domestication must imply a less aggressive approach toward people. Wild cattle are noted for attacks on humans, with some even fatal attacks (Pickles, 1987; Norwood et al., 2000; Burke, 2020). For example, for the European bison, Haidt et al. (2018) "investigated flight initiation distance (FID) and aggressive responses of the European bison–the largest terrestrial mammal of Europe–to human disturbance in the Białowieża Forest." Although "aggression was observed in only 0.4% of approach attempts" and FID was "influenced by habitat, sex, and supplementary feeding intensity" two important findings were that "females showed greater timidity than males, FID was lower in forest than in open habitats, and supplementary feeding caused a drop in FID" and "in 84.5% of all documented aggression cases, bison attacks were provoked by humans approaching too close to the bison or by deliberate scaring them" (Haidt et al., 2018). In addition, "goring was observed in 22.7% of all aggression cases and no fatalities were recorded," and the authors concluded that "bison are not dangerous animals and only manifest aggression in response to prolonged disturbance at close ranges" (Haidt et al., 2018).

For American bison, Osborne (2022) cites a case of a 25-year-old woman gored by a bison in Yellowstone National Park, United States, when "the woman came within ten feet of the bison as it neared a boardwalk at Black Sand Basin—just north of Old Faithful—and the animal tossed her ten feet in the air…The woman suffered a "puncture wound and other injuries" and was transported to Eastern Idaho Regional Medical Center, per the park. Further, "fifty-six people were injured and two died from bison attack in Yellowstone between 1978 and 1992, while another

25 were injured between 2000 and 2015. The vast majority of these attacks were the result of visitors getting too close, often for a photograph." Dura (2023) reported that "a bison severely injured a Minnesota woman on Saturday in Theodore Roosevelt National Park" and she suffered "significant injuries to her abdomen and foot." In another incident "a bison charged and gored a 47-year-old Phoenix woman in Yellowstone National Park. She sustained significant injuries to her chest and abdomen and was taken by helicopter to an Idaho Falls hospital." Shaw (2017) reported that a 65-year-old man was charged by a buffalo in Theodore Roosevelt National Park, suffering a torn thigh, bruises, and unconsciousness. In addition, in 2015, five people were gored by bison in Yellowstone Park. The author concludes that "a bison is not a cow. Domestic cattle have been bred for placid grazing and packing on pounds. Bison, by contrast, are tough-as-nails beasts with hardcore Ice Age roots, designed to withstand the blizzards, droughts, wolves, and grizzlies of the North American grasslands."

For African buffalo, the attack rate on humans has been compared and equated with that of lions (Dunham et al., 2010), and as noted by Mukamuri et al. (2023: 14) "encounters with buffalo, especially bachelor males, when local residents move around their natural habitat (e.g. to fetch water) can end in death. Although they are sometimes referred to as 'widow maker' or 'black death', buffalo 'only' kill an estimated 200 people each year in Africa, much fewer than crocodiles, hippopotamus and elephants." Geleta et al. (2019): 18) note that African buffaloes "come into conflict with humans mostly due to crop raiding and injuries or death to humans when wounded during poaching." Furstenburg et al. (2023: 361) notes that the African buffalo "is the most dangerous of all African game species, especially if wounded or solitary." For the gaur, Choudhury (2002: 213) notes that in one case study "between September 1981 and January 1997 at least 12 people have been killed and 50 injured by such straying animals in north Bengal; domestic cattle and goats have also been killed or injured and houses and plantations damaged."

For domesticated cattle, numerous attacks are reported. For example, Byard, 2024 argues that "lethal episodes arising from interactions with cattle may be the result of a response of the animals to simple handling and herding, or from reactions to deliberate provocation or goading intended to incite aggressive behavior for public entertainment purposes" and "cattle cause significant numbers of deaths globally and are considered the most dangerous large animal in Britain. Behavior may be unpredictable even in apparently docile domesticated animals, and attacks may be by a single animal or a herd and result in injuries from kicking, head/butting/charging, stomping, goring, and crushing" (Byard, 2024). Serious injuries may include "fractures of the spine or skull with cerebral contusions and lacerations associated with subarachnoid, subdural, and extradural hemorrhages, flail chest, hemo- and pneumothoraces, and organ disruption, intestinal perforations, splenic rupture, perineal and vaginal tears, urethral lacerations and avulsions, and bladder and rectal perforations, and complete and partial transections and lacerations" (Byard (2024).

Domestic water buffalo have been cited in many attacks on humans, in the media, and in scientific studies. For example, the British Broadcasting Corporation (BBC, 2020) reported that "a man has died after a water buffalo attacked three people at a commercial property. The 57-year-old man was pronounced dead at the scene, after police officers were called to the address at Gwehelog, near Usk, Monmouthshire. A man, 19, was also critically injured and taken to Cardiff's University Hospital of Wales by air ambulance. A woman, 22, suffered a serious leg injury and is being treated at Newport's Royal Gwent Hospital… the buffalo always appeared to be quite quiet, and that the owners had been seen stroking them." Shennan (2020) commenting on this incident cited communications describing water buffalo as "tame, friendly and curious animals," but in addition "to best avoid being attacked by a water buffalo, you should keep clear of any fields or farms where they are grazing. The incident that killed the 57-year-old man took place on a farm in the area that keeps a herd of water buffalo. Water buffalo body language is hard to understand, and generally the animals don't display any changes in their behaviour before charging, so it's best to keep away entirely."

There have been several other domestic water buffalo attacks on humans. For example, Byard (2017:934) reports that "a 72-year-old man was trampled by a water buffalo while attempting to move the animal within an enclosed area at an abattoir. At autopsy, there were numerous injuries involving the chest, head, neck, and left upper arm. Blunt force injury had resulted in multiple rib fractures with a flail chest and fracture/dislocation of the mid-cervical vertebrae caused either by physical crushing by the animal against the walls of the enclosure or by stomping. There was also evidence of crush asphyxia with bilateral conjunctival hemorrhages and petechial hemorrhages. In addition, there was a deep degloving injury of the upper left arm compatible with goring by one of the buffalo horns." Bakkannavar et al. (2010: 102) argue that "attacks on humans by domestic animals causing fatal injuries are not uncommon in rural areas of India. But injuries due to buffalo gore are rarely observed in villages and are different from other casualties like stab injuries, road fatalities, etc." The authors add that "most water buffaloes are not trained to be driven. Instead, the herdsman must walk alongside or ahead of them. They then instinctively follow. Usually, buffaloes are timid and lazy animals and are not dangerous to people, provided they are left alone. When provoked, handlers and bystanders occasionally sustain injuries from these animals." In this case, "a 46 year old male was attacked by a domestic water buffalo in a village at the foothills of Western Ghats of Karnataka State, India. The incident occurred while he was leading them back to their shelter after the days work in the paddy field. He was brought dead" (Bakkannavar et al., 2010).

Domestic bulls can often attack people. Martínez Hernández et al. (2021: 446) argue that "the presence of wild bulls in different countries around the world, and their behaviour in the various environments, highlights the global importance of having a knowledge about possible attacks by these animals and consequent injuries in humans." The Spanish province of Castellon (Valencian Community, Spain) has a large number of bull-related activities worldwide, with over 5000 popular events annually, and in the period between January 1978 and December 2019 there were 572 hospital admissions based on bull horn injuries, with most injuries to the lower extremities, perineum, and abdomen, and the "most frequently injured organs were the intestine and liver" and the "most frequent complications were skin devitalisation, infection and post-operative eventration. The recorded mortality was 0.87%" (Martínez Hernández et al., 2021: 446).

Maity et al. (2022) note that bull attacks cause a significant number of injuries and have become an important public health problem among children and adults, especially in countries such as India. These injuries may be provoked, as in bullfighting, or unprovoked, as in domestic settings, usually in rural areas where livestock are kept. In addition, injuries from bull and cow horns are a health problem in countries which are more dependent on livestock farming. For example, in rural India, men of younger working age are most commonly injured, followed by middle-aged women. The authors note that animal attack injuries represent a less studied niche of surgical issues. The attack often affects the chest and abdomen, but areas such as the perineum, skull, spine, and extremities can also be affected. The injuries can be blunt or penetrating depending on the type of impact. The complications of these wounds are due to multiple wound routes, muscle tears, evisceration of internal organs, and possible wound infection (Maity et al., 2022). Wasadikar et al. (1997: 3) reported 50 cases of bull horn injuries at a Rural Medical College over a 5-year period in India. Kulkarni et al. (2016: 201) also noted that "bull gore injuries are among the most common accidents in rural India, where people make their living rearing the livestock. The most common site of injury in bull gore cases is the abdomen and perineal region… Perineal injuries are associated with delayed complications like anovaginal fistula, urethrorectal fistula, loss of sphincter function due to injury to the anal canal." In addition, "chest injuries are in the form of multiple rib fractures and penetrating injuries involving lungs" (Kulkarni et al., 2016: 203).

Conclusions

The domestication of cattle is assessed by the series of domestication events, some disputed on their importance and effects, across southwest Asia to East and South Asia, and influenced possibly secondary events in Europe, North Asia, and Africa. The justification of such domestication is clearly its role in the development of human civilization. The development of genetic-based studies has clearly assumed a prominent or even indispensable role in the study of cattle domestication. The ancestry of domestic cattle is nevertheless disputed, with numerous disputes concerning the genetic, ecological, and temporal variations. Consensus has however been achieved concerning the aurochs ancestry and relations with taurine and indicine cattle. The next chapter will examine the relation between the ancestral cattle and their contemporary carnivores. The ecology of cattle, which evolved from the ecology of wild cattle, is less understood. However, this relationship is becoming increasingly important as livestock predators, charismatic large carnivores, are increasingly protected repopulation preexisting habitats. Therefore, there is significant conflict between agriculturists, for whom cattle are a vital resource, and environmentalists and conservationists, for whom the protection of large carnivores is a global concern and an existential right. Such conflicts are increasing with growing human populations and exploitation of limited land resources, the increasing polarization of global and domestic geopolitics, and increasing interest group participation through the global media. This chapter provides a background to these issues by examining the process of cattle domestication from the ancient breeds and the slow process by which domestic breeds spread over the human-inhabited world. Current research suggests that cattle breeds are descended from the wild aurochs in southwest Asia, which later interbreed with *Bos taurus* (taurine cattle) in western Eurasia, and *Bos indicus* (zebu indicine cattle) in south, southeastern, and east Asia, and later Africa. Later inbreeding has been documented between the auroch/domesticated cattle breeds and other wild species, such as gaur, banteng, water buffalo, and yak in Asia, European aurochs and bison in Europe, and to a lesser extent American bison in North America. It is concluded that knowledge of the problems of cattle domestication is critical to the current scenarios of political, economic, socio-cultural, and environmental cooperation and conflict that enable and constrain more effective sustainable development.

References

Achilli, A., Olivieri, A., Pellecchia, M., Uboldi, C., Colli, L., Al-Zahery, N., Accetturo, M., Pala, M., Hooshiar Kashani, B., Perego, U. A., Battaglia, V., Fornarino, S., Kalamati, J., Houshmand, M., Negrini, R., Semino, O., Richards, M., Macaulay, V., Ferretti, L., Bandelt, H. J., et al. (2008). Mitochondrial genomes of extinct aurochs survive in domestic cattle. *Current Biology: CB, 18*(4), R157–R158. https://doi.org/10.1016/j.cub.2008.01.019

Ajmone-Marsan, P., Garcia, J. F., & Lenstra, J. A. (2010). On the origin of cattle: How aurochs became cattle and colonized the world. *Evolutionary Anthropology: Issues, News, and Reviews, 19*(4), 148–157. https://doi.org/10.1002/evan.20267

Arbuckle, B. S. (2013). The late adoption of cattle and pig husbandry in Neolithic Central Turkey. *Journal of Archaeological Science, 40*, 1805–1815.

Arbuckle, B. S., & Kassebaum, T. M. (2021). Management and domestication of cattle (Bos taurus) in Neolithic Southwest Asia. *Animal Frontiers: The Review Magazine of Animal Agriculture, 11*(3), 10–19. https://doi.org/10.1093/af/vfab015

Arbuckle, B. S., Price, M. D., Hongo, H., & Öksüz, B. (2016). Documenting the initial appearance of domestic cattle in the eastern fertile crescent (northern Iraq and western Iran). *Journal of Archaeological Science, 72*, 1–9. https://doi.org/10.1016/j.jas.2016.05.008

Bahbahani, H., Tijjani, A., Mukasa, C., Wragg, D., Almathen, F., Nash, O., Akpa, G. N., Mbole-Kariuki, M., Malla, S., Woolhouse, M., Sonstegard, T., Van Tassell, C., Blythe, M., Huson, H., & Hanotte, O. (2017). Signatures of selection for environmental adaptation and zebu × taurine hybrid fitness in East African Shorthorn Zebu. *Frontiers in Genetics, 8*, 68. https://doi.org/10.3389/fgene.2017.00068

Bakkannavar, S. M., Monteiro, F. N., Bhagavath, P., & Pradeep Kumar, G. (2010). Death by attack from a domestic buffalo. *Journal of Forensic and Legal Medicine, 17*(2), 102–104. https://doi.org/10.1016/j.jflm.2009.09.008

Balasse, M., Bocherens, H., Tresset, A., Mariotti, A., & Vigne, J.-D. (1997). Emergence of dairy production in the Neolithic? Contribution of isotopic analysis of cattle archaeological bones. *Earth and Planetary Science, 325*(12), 1005–1010.

Balcarcel, A. M., Veitschegger, K., Clauss, M., & Sánchez-Villagra, M. R. (2021). Intensive human contact correlates with smaller brains: Differential brain size reduction in cattle types. *Proceedings. Biological sciences, 288*(1952), 20210813. https://doi.org/10.1098/rspb.2021.0813

Barragy, T. J. (2003). *Gathering Texas gold*. Cayo Del Grullo Press.

Beja-Pereira, A., Caramelli, D., Lalueza-Fox, C., Vernesi, C., Ferrand, N., Casoli, A., Goyache, F., Royo, L. J., Conti, S., Lari, M., Martini, A., Ouragh, L., Magid, A., Atash, A., Zsolnai, A., Boscato, P., Triantaphylidis, C., Ploumi, K., Sineo, L., Mallegni, F., et al. (2006). The origin of European cattle: Evidence from modern and ancient DNA. *Proceedings of the National Academy of Sciences of the United States of America, 103*(21), 8113–8118. https://doi.org/10.1073/pnas.0509210103

Bergmann, C. (1847). Ueber die Verhältnisse der Wärmeökonomie der Thiere zu ihrer Grösse. *Göttinger Studien, 1*, 595–708.

Berman, A. (2011). Invited review: Are adaptations present to support dairy cattle productivity in warm climates? *Journal of Dairy Science, 94*(5), 2147–2158. https://doi.org/10.3168/jds.2010-3962

Blench, R. M., & MacDonald, K. C. (2006). *The origins and development of African livestock: Archaeology, genetics, linguistics and ethnography*. Routledge.

Bollongino, R., Burger, J., Powell, A., Mashkour, M., Vigne, J. D., & Thomas, M. G. (2012). Modern taurine cattle descended from small number of near-eastern founders. *Molecular Biology and Evolution, 29*(9), 2101–2104. https://doi.org/10.1093/molbev/mss092

Bonfiglio, S., Achilli, A., Olivieri, A., Negrini, R., Colli, L., Liotta, L., Ajmone-Marsan, P., Torroni, A., & Ferretti, L. (2010). The enigmatic origin of bovine mtDNA haplogroup R: Sporadic interbreeding or an independent event of Bos primigenius domestication in Italy? *PLoS One, 5*(12), e15760. https://doi.org/10.1371/journal.pone.0015760

Bonfiglio, S., Ginja, C., De Gaetano, A., Achilli, A., Olivieri, A., Colli, L., Tesfaye, K., Agha, S. H., Gama, L. T., Cattonaro, F., Penedo, M. C., Ajmone-Marsan, P., Torroni, A., & Ferretti, L. (2012). Origin and spread of Bos taurus: New clues from mitochondrial genomes belonging to haplogroup T1. *PLoS One, 7*(6), e38601. https://doi.org/10.1371/journal.pone.0038601

British Broadcasting Corporation. (2020). *Water buffalo attack leaves one dead and two hurt in Usk*. https://www.bbc.com/news/uk-wales-52552766.

Brunson, K., Witt, K. E., Monge, S., Williams, S., Peede, D., Odsuren, D., Bukhchuluun, D., Cameron, A., Szpak, P., Amartuvshin, C., Honeychurch, W., Wright, J., Pleuger, S., Erdene, M., Tumen, D., Rogers, L., Khatanbaatar, D., Batdalai, B., Galdan, G., & Janz, L. (2023). Ancient Mongolian aurochs genomes reveal sustained introgression and management in East Asia. bioRxiv: The preprint server for biology, 2023.08.10.552443. https://doi.org/10.1101/2023.08.10.552443.

Buggiotti, L., Yurchenko, A. A., Yudin, N. S., Vander Jagt, C. J., Vorobieva, N. V., Kusliy, M. A., Vasiliev, S. K., Rodionov, A. N., Boronetskaya, O. I., Zinovieva, N. A., Graphodatsky, A. S., Daetwyler, H. D., & Larkin, D. M. (2021). Demographic history, adaptation, and NRAP convergent evolution at amino acid residue 100 in the world northernmost cattle from Siberia. *Molecular Biology and Evolution, 38*(8), 3093–3110. https://doi.org/10.1093/molbev/msab078

Burke, M. (2020). *Bison attacks woman taking photos of herd at South Dakota state park*. Retrieved from https://www.nbcnews.com/news/us-news/bison-attacks-woman-taking-photos-herd-south-dakota-state-park-n1236730.

Byard, R. W. (2017). Causes and mechanisms of death in fatal water Buffalo attacks. *Journal of Forensic Sciences, 62*(4), 934–936. https://doi.org/10.1111/1556-4029.13358

Byard, R. W. (2024). Death and injuries caused by cattle: A forensic overview. *Forensic Science, Medicine, and Pathology, 21*, 401. https://doi.org/10.1007/s12024-024-00786-8

Cai, X., Chen, H., Lei, C.Z., Wang, S., Xue, K. & Zhang, B. (2007). mtDNA diversity and genetic lineages of eighteen cattle breeds from Bos taurus and Bos indicus in China. *Genetica 131*, 175–183.

Cai, D., Zhang, N., Zhu, S., Chen, Q., Wang, L., Zhao, X., Ma, X., Royle, T. C. A., Zhou, H., & Yang, D. Y. (2018). Ancient DNA reveals evidence of abundant aurochs (*Bos primigenius*) in Neolithic Northeast China. *Journal of Archaeological Science, 98*, 72–80. https://doi.org/10.1016/j.jas.2018.08.003

Cao, Y., Xia, X., Hou, J., Chen, N., Zhao, X., Chen, S., Dang, R., Huang, Y., Chen, H., & Lei, C. (2019). Y-chromosomal haplogroup distributions in Chinese cattle. Animal Genetics, 50(4), 412–413. https://doi.org/10.1111/age.12804

Cameron, E. Z., & Du Toit, J. T. (2005). Social influences on vigilance behaviour in giraffes, Giraffa camelopardalis. *Animal Behaviour, 69*, 1337–1344.

Chen, S., Lin, B. Z., Baig, M., Mitra, B., Lopes, R. J., Santos, A. M., Magee, D. A., Azevedo, M., Tarroso, P., Sasazaki, S., Ostrowski, S., Mahgoub, O., Chaudhuri, T. K., Zhang, Y. P., Costa, V., Royo, L. J., Goyache, F., Luikart, G., Boivin, N., Fuller, D. Q., et al. (2010). Zebu cattle are an exclusive legacy of the South Asia neolithic. *Molecular Biology and Evolution, 27*(1), 1–6. https://doi.org/10.1093/molbev/msp213

Chen, N., Cai, Y., Chen, Q., Li, R., Wang, K., Huang, Y., Hu, S., Huang, S., Zhang, H., Zheng, Z., Song, W., Ma, Z., Ma, Y., Dang, R., Zhang, Z., Xu, L., Jia, Y., Liu, S., Yue, X., Deng, W., et al. (2018a). Whole-genome resequencing reveals world-wide ancestry and adaptive introgression events of domesticated cattle in East Asia. *Nature Communications, 9*(1), 2337. https://doi.org/10.1038/s41467-018-04737-0

Chen, N., Huang, J., Zulfiqar, A., Li, R., Xi, Y., Zhang, M., Dang, R., Lan, X., Chen, H., Ma, Y., & Lei, C. (2018b). Population structure and ancestry of Qinchuan cattle. *Animal Genetics, 49*(3), 246–248. https://doi.org/10.1111/age.12658

Chen, Q., Shen, J., Hanif, Q., Chen, N., Huang, Y., Dang, R., Lan, X., Chen, H., & Lei, C. (2021). Whole genome analyses revealed genomic difference between European taurine and east Asian taurine. *Journal of Animal Breeding and Genetics = Zeitschrift fur Tierzuchtung und Zuchtungsbiologie, 138*(1), 56–68. https://doi.org/10.1111/jbg.12501

Chen, N., Xia, X., Hanif, Q., Zhang, F., Dang, R., Huang, B., Lyu, Y., Luo, X., Zhang, H., Yan, H., Wang, S., Wang, F., Chen, J., Guan, X., Liu, Y., Li, S., Jin, L., Wang, P., Sun, L., Zhang, J., et al. (2023a). Global genetic diversity, introgression, and evolutionary adaptation of indicine cattle revealed by whole genome sequencing. *Nature Communications, 14*(1), 7803. https://doi.org/10.1038/s41467-023-43626-z

Chen, N., Zhang, Z., Hou, J., Chen, J., Gao, X., Tang, L., Wangdue, S., Zhang, X., Sinding, M. S., Liu, X., Han, J., Lü, H., Lei, C., Marshall, F., & Liu, X. (2023b). Evidence for early domestic yak, taurine cattle, and their hybrids on the Tibetan Plateau. Science. *Advances, 9*(50), eadi6857. https://doi.org/10.1126/sciadv.adi6857

Choudhury, A. (2002). Distribution and conservation of the Gaur Bos gaurus in the Indian subcontinent. *Mammal Review, 32*(3), 199–226. https://doi.org/10.1046/j.1365-2907.2002.00107.x

Cubric-Curik, V., Novosel, D., Brajkovic, V., Rota Stabelli, O., Krebs, S., Sölkner, J., Šalamon, D., Ristov, S., Berger, B., Trivizaki, S., Bizelis, I., Ferenčaković, M., Rothammer, S., Kunz, E., Simčič, M., Dovč, P., Bunevski, G., Bytyqi, H., Marković, B., Brka, M., et al. (2021). Large-scale mitogenome sequencing reveals consecutive expansions of domestic taurine cattle and supports sporadic aurochs introgression. *Evolutionary Applications, 15*(4), 663–678. https://doi.org/10.1111/eva.13315

de Carvalho, C. N., Muñiz, F., Cáceres, L. M., Belaústegui, Z., Rodríguez-Vidal, J., Belo, J., Moreira, N., Cachão, M., Cunha, P. P., Figueiredo, S., Galán, J. M., Zhang, Y., Gómez, P., Toscano, A., Ruiz, F., Ramírez-Cruzado, S., Giles-Guzmán, F., Finlayson, G., Finlayson, S., & Finlayson, C. (2022). Aurochs roamed along the SW coast of Andalusia (Spain) during late Pleistocene. *Scientific Reports, 12*(1), 9911. https://doi.org/10.1038/s41598-022-14137-6

Decker, J. E., Pires, J. C., Conant, G. C., McKay, S. D., Heaton, M. P., Chen, K., Cooper, A., Vilkki, J., Seabury, C. M., Caetano, A. R., Johnson, G. S., Brenneman, R. A., Hanotte, O., Eggert, L. S., Wiener, P., Kim, J. J., Kim, K. S., Sonstegard, T. S., Van Tassell, C. P., Neibergs, H. L., et al. (2009). Resolving the evolution of extant and extinct ruminants with high-throughput phylogenomics. *Proceedings of the National Academy of Sciences of the United States of America, 106*(44), 18644–18649. https://doi.org/10.1073/pnas.0904691106

Decker, J. E., McKay, S. D., Rolf, M. M., Kim, J., Molina Alcalá, A., Sonstegard, T. S., Hanotte, O., Götherström, A., Seabury, C. M., Praharani, L., Babar, M. E., de Almeida, C., Regitano, L., Yildiz, M. A., Heaton, M. P., Liu, W. S., Lei, C. Z., Reecy, J. M., Saif-Ur-Rehman, M., Schnabel, R. D., & Taylor, J. F. (2014). Worldwide patterns of ancestry, divergence, and admixture in domesticated cattle. *PLoS Genetics, 10*(3), e1004254. https://doi.org/10.1371/journal.pgen.1004254

Duerst, J. U. (1908). Animal remains from the excavations at Anau, and the horse of Anau and its relation to the races of domestic horses. In R. Pumpelly (Ed.), *Explorations in Turkestan: Expedition of 1904*. Publication 73, Vol. II (pp. 341–494). Carnegie institution of Washington.

Dunham, K. M., Ghiurghi, A., Cumbi, R., & Urbano, F. (2010). Human–wildlife conflict in Mozambique: A national perspective, with emphasis on wildlife attacks on humans. *Oryx, 44*(2), 185–193. https://doi.org/10.1017/S003060530999086X

Dura, J. (2023). *Bison attack visitors in North Dakota and Wyoming national parks*. The Associated Press.

Edwards, C. J., Bollongino, R., Scheu, A., Chamberlain, A., Tresset, A., Vigne, J. D., Baird, J. F., Larson, G., Ho, S. Y., Heupink, T. H., Shapiro, B., Freeman, A. R., Thomas, M. G., Arbogast, R. M., Arndt, B., Bartosiewicz, L., Benecke, N., Budja, M., Chaix, L., Choyke, A. M., et al. (2007). Mitochondrial DNA analysis shows a near eastern Neolithic origin for domestic cattle and no indication of domestication of European aurochs. *Proceedings. Biological Sciences, 274*(1616), 1377–1385. https://doi.org/10.1098/rspb.2007.0020

Felius, M., Beerling, M.-L., Buchanan, D. S., Theunissen, B., Koolmees, P. A., & Lenstra, J. A. (2014). On the history of cattle genetic resources. *Diversity, 6*(4), 705–750. https://doi.org/10.3390/d6040705

Felius, M., Koolmees, P. A., Theunissen, B., European Cattle Genetic Diversity Consortium, & Lenstra, J. A. (2011). On the Breeds of Cattle—Historic and Current Classifications. *Diversity, 3*(4), 660–692. https://doi.org/10.3390/d3040660

Ferreira, V. H. B., Lansade, L., Calandreau, L., Cunha, F., & Jensen, P. (2023). Are domesticated animals dumber than their wild relatives? A comprehensive review on the domestication effects on animal cognitive performance. *Neuroscience and Biobehavioral Reviews, 154*, 105407. https://doi.org/10.1016/j.neubiorev.2023.105407

Figueroa, J. V., Chieves, L. P., Johnson, G. S., & Buening, G. M. (1992). Detection of Babesia bigemina-infected carriers by polymerase chain reaction amplification. *Journal of Clinical Microbiology, 30*(10), 2576–2582. https://doi.org/10.1128/jcm.30.10.2576-2582.1992

Frantz, L. A. F., Bradley, D. G., Larson, G., & Orlando, L. (2020). Animal domestication in the era of ancient genomics. *Nature Reviews. Genetics, 21*(8), 449–460. https://doi.org/10.1038/s41576-020-0225-0

Frisch, J. E., Drinkwater, R., Harrison, B., & Johnson, S. (1997). Classification of the southern African sanga and east African shorthorned zebu. *Animal Genetics, 28*(2), 77–83. https://doi.org/10.1111/j.1365-2052.1997.00088.x

Furstenburg, D., Gandiwa, P. T., Oberem, P., & Oberem, P. (2023). African Buffalo production systems. In A. Caron, D. Cornélis, P. Chardonnet, & H. H. T. Prins (Eds.), *Ecology and Management of the African Buffalo* (pp. 355–381). Cambridge University Press.

Gao, Y., Gautier, M., Ding, X., Zhang, H., Wang, Y., Wang, X., Faruque, M. O., Li, J., Ye, S., Gou, X., Han, J., Lenstra, J. A., & Zhang, Y. (2017). Species composition and environmental adaptation of indigenous Chinese cattle. *Scientific Reports, 7*(1), 16196. https://doi.org/10.1038/s41598-017-16438-7

Gautier, M., Flori, L., Riebler, A., Jaffrézic, F., Laloé, D., Gut, I., Moazami-Goudarzi, K., & Foulley, J. L. (2009). A whole genome Bayesian scan for adaptive genetic divergence in west African cattle. *BMC Genomics, 10*, 550. https://doi.org/10.1186/1471-2164-10-550

Gebrehiwot, N. Z., Strucken, E. M., Aliloo, H., Marshall, K., & Gibson, J. P. (2020). The patterns of admixture, divergence, and ancestry of African cattle populations determined from genome-wide SNP data. *BMC Genomics, 21*(1), 869. https://doi.org/10.1186/s12864-020-07270-x

Geleta, M., Jebessa, H., & Bekele, A. (2019). Human-Buffalo conflict around Jorgo-Wato protected Forest, Western Ethiopia. *Global Veterinaria, 21*, 17–23.

Gifford-Gonzalez, D., & Hanotte, O. (2011). Domesticating animals in Africa: Implications of genetic and archaeological findings. *Journal of World Prehistory, 24*, 1–23. https://doi.org/10.1007/s10963-010-9042-2

Ginja, C., Penedo, M. C., Melucci, L., Quiroz, J., Martínez López, O. R., Revidatti, M. A., Martínez-Martínez, A., Delgado, J. V., & Gama, L. T. (2010). Origins and genetic diversity of

New World Creole cattle: Inferences from mitochondrial and Y chromosome polymorphisms. *Animal Genetics, 41*(2), 128–141. https://doi.org/10.1111/j.1365-2052.2009.01976.x

Giovambattista, G., Ripoli, M. V., De Luca, J. C., Mirol, P. M., Lirón, J. P., & Dulout, F. N. (2000). Male-mediated introgression of Bos indicus genes into Argentine and Bolivian creole cattle breeds. *Animal Genetics, 31*(5), 302–305. https://doi.org/10.1046/j.1365-2052.2000.00658.x

Goderie, R., Lenstra, J. A., Upadhyay, M., Crooijmans, R., & Linnartz, L. (2015). *Aurochs genetics, a cornerstone of European biodiversity*. Retrieved from https://rewildingeurope.com/wp-content/uploads/2016/01/Aurochs-genetics_summary_final.pdf.

Götherström, A., Anderung, C., Hellborg, L., Elburg, R., Smith, C., Bradley, D. G., & Ellegren, H. (2005). Cattle domestication in the near east was followed by hybridization with aurochs bulls in Europe. *Proceedings. Biological sciences, 272*(1579), 2345–2350. https://doi.org/10.1098/rspb.2005.3243

Gravlund, P., Aaris-Sørensen, K., Hofreiter, M., Meyer, M., Bollback, J. P., & Noe-Nygaard, N. (2012). Ancient DNA extracted from Danish aurochs (Bos primigenius): Genetic diversity and preservation. *Annals of anatomy = Anatomischer Anzeiger: official organ of the Anatomische Gesellschaft, 194*(1), 103–111. https://doi.org/10.1016/j.aanat.2011.10.011

Haidt, A., Kamiński, T., Borowik, T., & Kowalczyk, R. (2018). Human and the beast-flight and aggressive responses of European bison to human disturbance. *PLoS One, 13*(8), e0200635. https://doi.org/10.1371/journal.pone.0200635

Hanotte, O., Tawah, C. L., Bradley, D. G., Okomo, M., Verjee, Y., Ochieng, J., & Rege, J. E. (2000). Geographic distribution and frequency of a taurine Bos taurus and an indicine Bos indicus Y specific allele amongst sub-saharan African cattle breeds. *Molecular Ecology, 9*(4), 387–396. https://doi.org/10.1046/j.1365-294x.2000.00858.x

Hanotte, O., Bradley, D. G., Ochieng, J. W., Verjee, Y., Hill, E. W., & Rege, J. E. (2002). African pastoralism: Genetic imprints of origins and migrations. *Science New York, N.Y., 296*(5566), 336–339. https://doi.org/10.1126/science.1069878

Hansen, P. J. (2004). Physiological and cellular adaptations of zebu cattle to thermal stress. *Animal Reproduction Science, 82-83*, 349–360. https://doi.org/10.1016/j.anireprosci.2004.04.011

Helmer, D., & Gourichon, L. (2008). Premières données sur les modalités de subsistance à Tell Aswad (Syrie, PPNB Moyen et Récent, Néolithique Céramique Ancien)—fouilles 2001–2005. In L. Gourichon, H. Buitenhuis, & A. Choyke (Eds.), *Archaeozoology of the near east* (Vol. 8, pp. 119–151). Maison de l'Orient et de la Méditerranée.

Helmer, D., Gourichon, L., Monchot, H., Peters, J., & Sana, S. M. (2005). Identifying early domestic cattle from pre-pottery Neolithic sites on the Euphrates using sexual dimorphism. In J.-D. Vigne, J. Peters, & D. Helmer (Eds.), *The first steps of animal domestication: New archaeological approaches* (pp. 86–95. Proceedings of the 9th ICAZ conference, Durham 2002). Oxford.

Hofman-Kamińska, E., Bocherens, H., Drucker, D. G., Fyfe, R. M., Gumiński, W., Makowiecki, D., Pacher, M., Piličiauskienė, G., Samojlik, T., Woodbridge, J., & Kowalczyk, R. (2019). Adapt or die-response of large herbivores to environmental changes in Europe during the Holocene. *Global Change Biology, 25*(9), 2915–2930. https://doi.org/10.1111/gcb.14733

Hoyt, A. M. (1982). *History of Texas longhorns*. Texas Longhorn J 1982, 1–48. http://doublehelixranch.com/History.html.

Hristov, P., Spassov, N., Iliev, N., & Radoslavov, G. (2017). An independent event of Neolithic cattle domestication on the South-Eastern Balkans: Evidence from prehistoric aurochs and cattle populations. *Mitochondrial DNA. Part A, DNA mapping, Sequencing, and Analysis, 28*(3), 383–391. https://doi.org/10.3109/19401736.2015.1127361

Huang, N., Zhao, L., Wang, J., Jiang, Q., Ju, Z., Wang, X., Yang, C., Gao, Y., Wei, X., Zhang, Y., Xiao, Y., Liu, W., Lu, S., & Huang, J. (2023). Signatures of selection in indigenous Chinese cattle genomes reveal adaptive genes and genetic variations to cold climate. *Journal of Animal Science, 101*, skad006. https://doi.org/10.1093/jas/skad006

Ibeagha-Awemu, E. M., Jann, O. C., Weimann, C., & Erhardt, G. (2004). Genetic diversity, introgression and relationships among west/central African cattle breeds. *Genetics, Selection, Evolution: GSE, 36*(6), 673–690. https://doi.org/10.1186/1297-9686-36-6-673

Jacques, G., d'Alpoim Guedes, J., & Zhang, S. (2021). Yak domestication: A review of linguistic, archaeological, and genetic evidence. *Ethnobiology Letters., 12*(1), 103–114. https://doi.org/10.14237/ebl.12.1.2021.1755

Janák, V., Novák, K., & Kyselý, R. (2024). Late history of cattle breeds in Central Europe in light of genetic and archaeogenetic sources-overview, thoughts, and perspectives. *Animals, 14*(4), 645. https://doi.org/10.3390/ani14040645

Kantanen, J. (2010). *The origin and genetic diversity of native Yakutian cattle as revealed by DNA-marker analysis* (Good to eat, good to live with: Nomads and animals in northern Eurasia and Africa, northeast Asian study series 11) (pp. 195–201). Center for Northeast Asian Studies, Tohoku University.

Kantanen, J., Edwards, C. J., Bradley, D. G., Viinalass, H., Thessler, S., Ivanova, Z., Kiselyova, T., Cinkulov, M., Popov, R., Stojanović, S., Ammosov, I., & Vilkki, J. (2009). Maternal and paternal genealogy of Eurasian taurine cattle (Bos taurus). *Heredity, 103*(5), 404–415. https://doi.org/10.1038/hdy.2009.68

Kim, J., Hanotte, O., Mwai, O. A., Dessie, T., Bashir, S., Diallo, B., Agaba, M., Kim, K., Kwak, W., Sung, S., Seo, M., Jeong, H., Kwon, T., Taye, M., Song, K. D., Lim, D., Cho, S., Lee, H. J., Yoon, D., Oh, S. J., et al. (2017). The genome landscape of indigenous African cattle. *Genome Biology, 18*(1), 34. https://doi.org/10.1186/s13059-017-1153-y

Kim, K., Kwon, T., Dessie, T., Yoo, D., Mwai, O. A., Jang, J., Sung, S., Lee, S., Salim, B., Jung, J., Jeong, H., Tarekegn, G. M., Tijjani, A., Lim, D., Cho, S., Oh, S. J., Lee, H. K., Kim, J., Jeong, C., Kemp, S., et al. (2020). The mosaic genome of indigenous African cattle as a unique genetic resource for African pastoralism. *Nature Genetics, 52*(10), 1099–1110. https://doi.org/10.1038/s41588-020-0694-2

Kim, K., Kim, D., Hanotte, O., Lee, C., Kim, H., & Jeong, C. (2023). Inference of admixture origins in indigenous African cattle. *Molecular Biology and Evolution, 40*(12), msad257. https://doi.org/10.1093/molbev/msad257

Kluever, B. M., Breck, S. W., Howery, L. D., Krausman, P. R., & Bergman, D. L. (2008). Vigilance in cattle: The influence of predation, social interactions, and environmental factors. *Rangeland Ecology & Management, 61*(3), 321–328. https://doi.org/10.2111/07-087.1

Kulkarni, S. R., Biradar, S. B., Nagur, B. K., Reddy, M., & Savsaviya, J. K. (2016). Bull horn injuries in rural area: A case series. *International Journal of Scientific Studies, 3*(10), 201–203.

Kysely, R. (2008). Aurochs and potential crossbreeding with domestic cattle in Central Europe in the Eneolithic period. A metric analysis of bones from the archaeological site of Kutná Hora-Denemark (Czech Republic). *Anthropozoologica., 43*(2), 7–37.

Lari, M., Rizzi, E., Mona, S., Corti, G., Catalano, G., Chen, K., Vernesi, C., Larson, G., Boscato, P., De Bellis, G., Cooper, A., Caramelli, D., & Bertorelle, G. (2011). The complete mitochondrial genome of an 11,450-year-old aurochsen (Bos primigenius) from Central Italy. *BMC Evolutionary Biology, 11*, 32. https://doi.org/10.1186/1471-2148-11-32

Laundre, J. W., Hernandez, L., & Altendorf, K. B. (2001). Wolves, elk and bison: Reestablishing the "landscape of fear" in Yellowstone National Park, U.S.A. *Canadian Journal of Zoology, 79*, 1401–1409.

Lee, S. H., Park, B. H., Sharma, A., Dang, C. G., Lee, S. S., Choi, T. J., Choy, Y. H., Kim, H. C., Jeon, K. J., Kim, S. D., Yeon, S. H., Park, S. B., & Kang, H. S. (2014). Hanwoo cattle: Origin, domestication, breeding strategies and genomic selection. *Journal of Animal Science and Technology, 56*, 2. https://doi.org/10.1186/2055-0391-56-2

Lenstra, J. A. (2020). Cattle: Domestication and development. In C. Smith (Ed.), *Encyclopedia of global archaeology*. Springer. https://doi.org/10.1007/978-3-030-30018-0_2201

Lenstra, J. A., Ajmone-Marsan, P., Beja-Pereira, A., Bollongino, R., Bradley, D. G., Colli, L., De Gaetano, A., Edwards, C. J., Felius, M., Ferretti, L., Ginja, C., Hristov, P., Kantanen, J., Lirón, J. P., Magee, D. A., Negrini, R., & Radoslavov, G. A. (2014). Meta-analysis of mitochondrial DNA reveals several population bottlenecks during worldwide migrations of cattle. *Diversity, 6*(1), 178–187. https://doi.org/10.3390/d6010178

Li, R., Xie, W. M., Chang, Z. H., Wang, S. Q., Dang, R. H., Lan, X. Y., Chen, H., & Lei, C. Z. (2013a). Y chromosome diversity and paternal origin of Chinese cattle. *Molecular Biology Reports, 40*(12), 6633–6636. https://doi.org/10.1007/s11033-013-2777-y

Li, R., Zhang, X. M., Campana, M. G., Huang, J. P., Chang, Z. H., Qi, X. B., Shi, H., Su, B., Zhang, R. F., Lan, X. Y., Chen, H., & Lei, C. Z. (2013b). Paternal origins of Chinese cattle. *Animal Genetics, 44*(4), 446–449. https://doi.org/10.1111/age.12022

Linnell, J. D. C., Odden, J., Smith, M. E., Aanes, R., & Swenson, J. E. (1999). Large carnivores that kill livestock: Do "problem individuals" really exist? *Wildlife Society Bulletin, 27*, 698–705.

Liu, X., Chen, J., Xu, X., Liu, J., Zhang, J., Cheng, H., Ahmed, Z., Huang, B., & Lei, C. (2023). A missense mutation of the WNK1 gene affects cold tolerance in Chinese domestic cattle. *Animal Biotechnology, 34*(9), 4803–4808. https://doi.org/10.1080/10495398.2023.2196316

MacHugh, D. E., Troy, C. S., McCormick, F., Olsaker, I., Eythórsdóttir, E., & Bradley, D. G. (1999). Early medieval cattle remains from a Scandinavian settlement in Dublin: Genetic analysis and comparison with extant breeds. *Philosophical Transactions of the Royal Society of London. Series B, Biological Sciences, 354*(1379), 99–109. https://doi.org/10.1098/rstb.1999.0363

MacHugh, D. E., Larson, G., & Orlando, L. (2017). Taming the past: Ancient DNA and the study of animal domestication. *Annual Review of Animal Biosciences, 5*, 329–351. https://doi.org/10.1146/annurev-animal-022516-022747

Maity, S., Puranik, A., Baskaran, S., Sairam, M. V., & Kompally, P. (2022). Patterns and Management of Unprovoked Bull Attack Injuries: A retrospective case series. *Cureus, 14*(12), e33075. https://doi.org/10.7759/cureus.33075

Mannen, H., Yonezawa, T., Murata, K., Noda, A., Kawaguchi, F., Sasazaki, S., Olivieri, A., Achilli, A., & Torroni, A. (2020). Cattle mitogenome variation reveals a post-glacial expansion of haplogroup P and an early incorporation into northeast Asian domestic herds. *Scientific Reports, 10*(1), 20842. https://doi.org/10.1038/s41598-020-78040-8

Marom, N., & Bar-Oz, G. (2009). 'Man made oases': Neolithic patterns of wild ungulate exploitation and their consequences for the domestication of pigs and cattle. *Before Farming, 1*, 1–12. https://doi.org/10.1371/journal.pone.0055958

Marom, N., & Bar-Oz, G. (2013). The prey pathway: A regional history of cattle (Bos taurus) and pig (Sus scrofa) domestication in the northern Jordan Valley, Israel. *PLoS One, 8*(2), e55958. https://doi.org/10.1371/journal.pone.0055958

Martínez Hernández, A., Martínez Ramos, D., García Moreno, M. V., Abdlekader Mohamed, N., López Loscos, E., Aliaga Hilario, E., Queralt Martín, R., Chordá Palomero, R., & Laguna Sastre, J. M. (2021). Bull horn injuries. A 40-year retrospective study with 572 patients. *American Journal of Surgery, 222*(2), 446–452. https://doi.org/10.1016/j.amjsurg.2020.11.031

Mattioli, R. C., & Dempfle, L. (1995). Recent acquisitions on tick and tick-borne disease resistance in N'dama (Bos taurus) and Gobra zebu (Bos indicus) cattle. *Parassitologia, 37*(1), 63–67.

Mattioli, R. C., Bah, M., Kora, S., Cassama, M., & Clifford, D. J. (1995). Susceptibility to different tick genera in Gambian N'Dama and Gobra zebu cattle exposed to naturally occurring tick infestations. *Tropical Animal Health and Production, 27*(2), 95–105. https://doi.org/10.1007/BF02236320

Mattioli, R. C., Jaitner, J., Clifford, D. J., Pandey, V. S., & Verhulst, A. (1998). Trypanosome infections and tick infestations: Susceptibility in N'Dama, Gobra zebu and Gobra x N'Dama crossbred cattle exposed to natural challenge and maintained under high and low surveillance of trypanosome infections. *Acta Tropica, 71*(1), 57–71. https://doi.org/10.1016/s0001-706x(98)00051-5

Mattioli, R. C., Pandey, V. S., Murray, M., & Fitzpatrick, J. L. (2000). Immunogenetic influences on tick resistance in African cattle with particular reference to trypanotolerant N'Dama (Bos taurus) and trypanosusceptible Gobra zebu (Bos indicus) cattle. *Acta Tropica, 75*(3), 263–277. https://doi.org/10.1016/s0001-706x(00)00063-2

McTavish, E. J., Decker, J. E., Schnabel, R. D., Taylor, J. F., & Hillis, D. M. (2013). New World cattle show ancestry from multiple independent domestication events. *Proceedings of the National Academy of Sciences of the United States of America, 110*(15), E1398–E1406. https://doi.org/10.1073/pnas.1303367110

Mei, C., Gui, L., Hong, J., Raza, S. H. A., Aorigele, C., Tian, W., Garcia, M., Xin, Y., Yang, W., Zhang, S., & Zan, L. (2021). Insights into adaption and growth evolution: A comparative genomics study on two distinct cattle breeds from northern and southern China. *Molecular Therapy. Nucleic acids, 23*, 959–967. https://doi.org/10.1016/j.omtn.2020.12.028

Mirol, P. M., Giovambattista, G., Lirón, J. P., & Dulout, F. N. (2003). African and European mitochondrial haplotypes in South American Creole cattle. *Heredity, 91*(3), 248–254. https://doi.org/10.1038/sj.hdy.6800312

Mohamad, K., Olsson, M., van Tol, H. T., Mikko, S., Vlamings, B. H., Andersson, G., Rodríguez-Martínez, H., Purwantara, B., Paling, R. W., Colenbrander, B., & Lenstra, J. A. (2009). On the origin of Indonesian cattle. *PLoS One, 4*(5), e5490. https://doi.org/10.1371/journal.pone.0005490

Mona, S., Catalano, G., Lari, M., Larson, G., Boscato, P., Casoli, A., Sineo, L., Di Patti, C., Pecchioli, E., Caramelli, D., & Bertorelle, G. (2010). Population dynamic of the extinct European aurochs: Genetic evidence of a north-south differentiation pattern and no evidence of post-glacial expansion. *BMC Evolutionary Biology, 10*, 83. https://doi.org/10.1186/1471-2148-10-83

Mukamuri, B., Wichatitsky, E. G., Gandiwa, E., Perrotton, A., Kupika, O. L., & Monin, L. (2023). African Buffalo and the human societies in Africa: Social values and interaction outcomes. In A. Caron, D. Cornélis, P. Chardonnet, & H. H. T. Prins (Eds.), *Ecology and Management of the African Buffalo* (pp. 1–20). Cambridge University Press.

Murray, M., Trail, J. C., & D'Ieteren, G. D. (1990). Trypanotolerance in cattle and prospects for the control of trypanosomiasis by selective breeding. *Revue scientifique et technique (International Office of Epizootics), 9*(2), 369–386. https://doi.org/10.20506/rst.9.2.506

Mwai, O., Hanotte, O., Kwon, Y. J., & Cho, S. (2015). African indigenous cattle: Unique genetic resources in a rapidly changing world. *Asian-Australasian Journal of Animal Sciences, 28*(7), 911–921. https://doi.org/10.5713/ajas.15.0002R

Noda, A., Yonesaka, R., Sasazaki, S., & Mannen, H. (2018). The mtDNA haplogroup P of modern Asian cattle: A genetic legacy of Asian aurochs? *PLoS One, 13*(1), e0190937. https://doi.org/10.1371/journal.pone.0190937

Norwood, S., McAuley, C., Vallina, V. L., Fernandez, L. G., McLarty, J. W., & Goodfried, G. (2000). Mechanisms and patterns of injuries related to large animals. *The Journal of Trauma, 48*(4), 740–744. https://doi.org/10.1097/00005373-200004000-00025

Noyes, H., Brass, A., Obara, I., Anderson, S., Archibald, A. L., Bradley, D. G., Fisher, P., Freeman, A., Gibson, J., Gicheru, M., Hall, L., Hanotte, O., Hulme, H., McKeever, D., Murray, C., Oh, S. J., Tate, C., Smith, K., Tapio, M., Wambugu, J., et al. (2011). Genetic and expression analysis of cattle identifies candidate genes in pathways responding to Trypanosoma congolense infection. *Proceedings of the National Academy of Sciences of the United States of America, 108*(22), 9304–9309. https://doi.org/10.1073/pnas.1013486108

Olson, T. A., Lucena, C., Chase, C. C., Jr., & Hammond, A. C. (2003). Evidence of a major gene influencing hair length and heat tolerance in Bos taurus cattle. *Journal of Animal Science, 81*(1), 80–90. https://doi.org/10.2527/2003.81180x

Osborne, M. (2022). *Woman Gored by a Bison in Yellowstone National Park*. Smithsonian Magazine.

Paguem, A., Abanda, B., Achukwi, M. D., Baskaran, P., Czemmel, S., Renz, A., & Eisenbarth, A. (2020). Whole genome characterization of autochthonous Bos taurus brachyceros and introduced Bos indicus indicus cattle breeds in Cameroon regarding their adaptive phenotypic traits and pathogen resistance. *BMC Genetics, 21*(1), 64. https://doi.org/10.1186/s12863-020-00869-9

Paling, R. W., & Dwinger, R. H. (1993). Potential of trypanotolerance as a contribution to sustainable livestock production in tsetse affected Africa. *The Veterinary Quarterly, 15*(2), 60–67. https://doi.org/10.1080/01652176.1993.9694374

Palmeira, F. B. L., Crawshaw, P. G., Jr., Haddad, C. M., Ferraz, K. M. P. M. B., & Verdade, L. M. (2008). Cattle depredation by puma (Puma concolor) and jaguar (Panthera onca) in Central-Western Brazil. *Biological Conservation, 141*(1), 118–125. https://doi.org/10.1016/j.biocon.2007.09.015

Park, S. D., Magee, D. A., McGettigan, P. A., Teasdale, M. D., Edwards, C. J., Lohan, A. J., Murphy, A., Braud, M., Donoghue, M. T., Liu, Y., Chamberlain, A. T., Rue-Albrecht, K., Schroeder, S., Spillane, C., Tai, S., Bradley, D. G., Sonstegard, T. S., Loftus, B. J., & MacHugh, D. E. (2015). Genome sequencing of the extinct Eurasian wild aurochs, Bos primigenius, illuminates the phylogeography and evolution of cattle. *Genome Biology, 16*, 234. https://doi.org/10.1186/s13059-015-0790-2

Peake, H. J., & Fleure, H. J. (1927). *Peasants and potters*. Oxford University Press.

Pellecchia, M., Negrini, R., Colli, L., Patrini, M., Milanesi, E., Achilli, A., Bertorelle, G., Cavalli-Sforza, L. L., Piazza, A., Torroni, A., & Ajmone-Marsan, P. (2007). The mystery of Etruscan origins: Novel clues from Bos taurus mitochondrial DNA. *Proceedings. Biological sciences, 274*(1614), 1175–1179. https://doi.org/10.1098/rspb.2006.0258

Pérez-Pardal, L., Sánchez-Gracia, A., Álvarez, I., Traoré, A., Ferraz, J. B. S., Fernández, I., Costa, V., Chen, S., Tapio, M., Cantet, R. J. C., Patel, A., Meadow, R. H., Marshall, F. B., Beja-Pereira, A., & Goyache, F. (2018). Legacies of domestication, trade and herder mobility shape extant male zebu cattle diversity in South Asia and Africa. *Scientific Reports, 8*(1), 18027. https://doi.org/10.1038/s41598-018-36444-7

Petruzzello, M. (2024, June 21). Aurochs. Encyclopedia Britannica. https://www.britannica.com/animal/aurochs

Pickles, G. (1987). Injuries by wild animals in the African bush. *Journal of the Royal Army Medical Corps, 133*(3), 159–160. https://doi.org/10.1136/jramc-133-03-09

Pitt, D., Sevane, N., Nicolazzi, E. L., MacHugh, D. E., Park, S. D. E., Colli, L., Martinez, R., Bruford, M. W., & Orozco-terWengel, P. (2018). Domestication of cattle: Two or three events? *Evolutionary Applications, 12*(1), 123–136. https://doi.org/10.1111/eva.12674

Price, E. O. (1984). Behavioral aspects of animal domestication. *The Quarterly Review of Biology, 59*(1), 1–32. https://doi.org/10.1086/413673

Price, E. O. (1999). Behavioral development in animals undergoing domestication. *Applied Animal Behaviour Science, 65*(3), 245–271. https://doi.org/10.1016/S0168-1591(99)00087-8

Qi, X. B., Jianlin, H., Wang, G., Rege, J. E., & Hanotte, O. (2010). Assessment of cattle genetic introgression into domestic yak populations using mitochondrial and microsatellite DNA markers. *Animal Genetics, 41*(3), 242–252. https://doi.org/10.1111/j.1365-2052.2009.01989.x

Reed, C. A. (1960). A review of the archaeological evidence on animal domestication in the prehistoric near east. In R. J. Braidwood & B. Howe (Eds.), *Prehistoric investigations in Iraqi Kurdistan, studies in ancient oriental civilization 31: 119–146*. Oriental Institute of the University of Chicago SAOC.

Rouse, J. E. (1977). *The criollo: Spanish cattle in the Americas* (pp. xvi+–303).

Sahu, B. K., Parganiha, A., & Pati, A. K. (2020). Behavior and foraging ecology of cattle: A review. *Journal of Veterinary Behavior, 40*, 50–74. https://doi.org/10.1016/j.jveb.2020.08.004

Scheu, A., Hartz, S., Schmölcke, U., Tresset, A., Burger, J., & Bollongino, R. (2008). Ancient DNA provides no evidence for independent domestication of cattle in Mesolithic Rosenhof, northern Germany. *Journal of Archaeological Science, 35*(5), 1257–1264. https://doi.org/10.1016/j.jas.2007.08.012

Scheu, A., Powell, A., Bollongino, R., Vigne, J. D., Tresset, A., Çakırlar, C., Benecke, N., & Burger, J. (2015). The genetic prehistory of domesticated cattle from their origin to the spread across Europe. *BMC Genetics, 16*, 54. https://doi.org/10.1186/s12863-015-0203-2

Schibler, J., Elsner, J., & Schlumbaum, A. (2014). Incorporation of aurochs into a cattle herd in Neolithic Europe: Single event or breeding? *Scientific Reports, 4*, 5798. https://doi.org/10.1038/srep05798

Shaw, E. (2017). *Recent bison attacks are a reminder to keep your distance*. Retrieved from https://www.earthtouchnews.com/natural-world/animal-behaviour/recent-bison-attacks-are-a-reminder-to-keep-your-distance/

Shennan, R. (2020). *This is why there are water buffalo in the UK, and how to avoid an attack after man is killed in Wales*. https://www.blackpoolgazette.co.uk/read-this/this-is-why-there-are-water-buffalo-in-the-uk-and-how-to-avoid-an-attack-after-man-is-killed-in-wales-2844456.

Simoons, F. J. (1984). Gayal or mithan. In I. L. Mason (Ed.), *Evolution of domesticated animals* (pp. 34–38). Longman.

Sinding, M.-H. S., Ciucani, M. M., Jazmín Ramos-Madrigal, Carmagnini, A., Rasmussen, J. A., Feng, S., Chen, G., Vieira, F. G., Mattiangeli, V., Ganjoo, R. K., Larson, G., Sicheritz-Pontén, T., Petersen, B., Frantz, L., Gilbert, T. P., & Bradley, D. G. (2021). Kouprey (Bos sauveli) genomes unveil polytomic origin of wild Asian Bos. *IScience, 24*(11), 103226–103226. https://doi.org/10.1016/j.isci.2021.103226

Stuart, S. (2014). Foreword. In M. Melletti & J. Burton (Eds.), *Ecology, evolution and behaviour of wild cattle implications for conservation*. Cambridge University Press.

Sudrajad, P., Subiharta, S., Adinata, Y., Lathifah, A., Lee, J. H., Lenstra, J. A., & Lee, S. H. (2020). An insight into the evolutionary history of Indonesian cattle assessed by whole genome data analysis. *PLoS One, 15*(11), e0241038. https://doi.org/10.1371/journal.pone.0241038

Tang, Z. (2020). Recognition of historical and geographical distribution of Bos Primigenius in northern China. *Agricultural Archaeology, 3*, 7–12.

Troy, C. S., MacHugh, D. E., Bailey, J. F., Magee, D. A., Loftus, R. T., Cunningham, P., Chamberlain, A. T., Sykes, B. C., & Bradley, D. G. (2001). Genetic evidence for near-eastern origins of European cattle. *Nature, 410*(6832), 1088–1091. https://doi.org/10.1038/35074088

Underwood, R. (1982). Vigilance behavior in grazing African antelopes. *Behaviour, 79*, 81–107.

United Nations Statistics Division (UNSD). (2024). *Methodology*. https://unstats.un.org/unsd/methodology/m49/#ftn13.

Upadhyay, M. R., Chen, W., Lenstra, J. A., Goderie, C. R. J., MacHugh, D. E., Park, S. D. E., Magee, D. A., Matassino, D., Ciani, F., Megens, H. J., van Arendonk, J. A. M., Groenen, M. A. M., & European Cattle Genetic Diversity Consortium, & RPMA Crooijmans. (2017a). Genetic origin, admixture and population history of aurochs (Bos primigenius) and primitive European cattle. *Heredity, 119*(6), 469. https://doi.org/10.1038/hdy.2017.59

Upadhyay, M. R., Chen, W., Lenstra, J. A., Goderie, C. R., MacHugh, D. E., Park, S. D., Magee, D. A., Matassino, D., Ciani, F., Megens, H. J., van Arendonk, J. A., Groenen, M. A., & European Cattle Genetic Diversity Consortium, & RPMA Crooijmans. (2017b). Genetic origin, admixture and population history of aurochs (Bos primigenius) and primitive European cattle. *Heredity, 118*(2), 169–176. https://doi.org/10.1038/hdy.2016.79

Utsunomiya, Y. T., Milanesi, M., Fortes, M. R. S., Porto-Neto, L. R., Utsunomiya, A. T. H., Silva, M. V. G. B., Garcia, J. F., & Ajmone-Marsan, P. (2019). Genomic clues of the evolutionary history of Bos indicus cattle. *Animal Genetics, 50*(6), 557–568. https://doi.org/10.1111/age.12836

Van Vuure, C. (2005). *Retracing the aurochs. History, morphology and ecology of an extinct wild ox*. Pensoft.

Verdugo, M. P., Mullin, V. E., Scheu, A., Mattiangeli, V., Daly, K. G., Maisano Delser, P., Hare, A. J., Burger, J., Collins, M. J., Kehati, R., Hesse, P., Fulton, D., Sauer, E. W., Mohaseb, F. A., Davoudi, H., Khazaeli, R., Lhuillier, J., Rapin, C., Ebrahimi, S., Khasanov, M., et al. (2019). Ancient cattle genomics, origins, and rapid turnover in the Fertile Crescent. *Science (New York, N.Y.), 365*(6449), 173–176. https://doi.org/10.1126/science.aav1002

Verkaar, E. L., Nijman, I. J., Beeke, M., Hanekamp, E., & Lenstra, J. A. (2004). Maternal and paternal lineages in cross-breeding bovine species. Has wisent a hybrid origin? *Molecular Biology and Evolution, 21*(7), 1165–1170. https://doi.org/10.1093/molbev/msh064

Vigne, J.-D., Carrere, I., Briois, F., & Guilaine, J. (2011). The early process of mammal domestication in the near east. *Current Anthropology, 52*, S255–S271.

Wang, X., Ju, Z., Jiang, Q., Zhong, J., Liu, C., Wang, J., Hoff, J. L., Schnabel, R. D., Zhao, H., Gao, Y., Liu, W., Wang, L., Gao, Y., Yang, C., Hou, M., Huang, N., Regitano, L. C. A., Porto-Neto, L. R., Decker, J. E., Taylor, J. F., et al. (2021). Introgression, admixture, and selection facilitate genetic adaptation to high-altitude environments in cattle. *Genomics, 113*(3), 1491–1503. https://doi.org/10.1016/j.ygeno.2021.03.023

Wasadikar, P. P., Paunikar, R. G., & Deshmukh, S. B. (1997). Bull horn injuries in rural India. *Journal of the Indian Medical Association, 95*(1), 3–16.

Weldenegodguad, M., Popov, R., Pokharel, K., Ammosov, I., Ming, Y., Ivanova, Z., & Kantanen, J. (2019). Whole-genome sequencing of three native cattle breeds originating from the northernmost cattle farming regions. *Frontiers in Genetics, 9*, 728. https://doi.org/10.3389/fgene.2018.00728

Wright, E., & Viner-Daniels, S. (2015). Geographical variation in the size and shape of the European aurochs (Bos primigenius). *Journal of Archaeological Science, 54*, 8–22. https://doi.org/10.1016/j.jas.2014.11.021

Wu, D. D., Ding, X. D., Wang, S., Wójcik, J. M., Zhang, Y., Tokarska, M., Li, Y., Wang, M. S., Faruque, O., Nielsen, R., Zhang, Q., & Zhang, Y. P. (2018). Pervasive introgression facilitated domestication and adaptation in the Bos species complex. *Nature Ecology & Evolution, 2*(7), 1139–1145. https://doi.org/10.1038/s41559-018-0562-y

Xia, X., Qu, K., Zhang, G., Jia, Y., Ma, Z., Zhao, X., Huang, Y., Chen, H., Huang, B., & Lei, C. (2019a). Comprehensive analysis of the mitochondrial DNA diversity in Chinese cattle. *Animal Genetics, 50*(1), 70–73. https://doi.org/10.1111/age.12749

Xia, X., Yao, Y., Li, C., Zhang, F., Qu, K., Chen, H., Huang, B., & Lei, C. (2019b). Genetic diversity of Chinese cattle revealed by Y-SNP and Y-STR markers. *Animal Genetics, 50*(1), 64–69. https://doi.org/10.1111/age.12742

Xia, X., Zhang, S., Zhang, H., Zhang, Z., Chen, N., Li, Z., Sun, H., Liu, X., Lyu, S., Wang, X., Li, Z., Yang, P., Xu, J., Ding, X., Shi, Q., Wang, E., Ru, B., Xu, Z., Lei, C., Chen, H., et al. (2021). Assessing genomic diversity and signatures of selection in Jiaxian red cattle using whole-genome sequencing data. *BMC Genomics, 22*(1), 43. https://doi.org/10.1186/s12864-020-07340-0

Xia, X., Qu, K., Wang, Y., Sinding, M. S., Wang, F., Hanif, Q., Ahmed, Z., Lenstra, J. A., Han, J., Lei, C., & Chen, N. (2023). Global dispersal and adaptive evolution of domestic cattle: A genomic perspective. *Stress Biology, 3*(1), 8. https://doi.org/10.1007/s44154-023-00085-2

Xuebin, Q., Jianlin, H., Lkhagva, B., Chekarova, I., Badamdorj, D., Rege, J. E., & Hanotte, O. (2005). Genetic diversity and differentiation of Mongolian and Russian yak populations. *Journal of Animal Breeding and Genetics = Zeitschrift fur Tierzuchtung und Zuchtungsbiologie, 122*(2), 117–126. https://doi.org/10.1111/j.1439-0388.2004.00497.x

Yan, C. L., Lin, J., Huang, Y. Y., Gao, Q. S., Piao, Z. Y., Yuan, S. L., Chen, L., Ren, X., Ye, R. C., Dong, M., Zhang, H. L., Zhou, H. Q., Jiang, X. X., Jin, W. Z., Zhou, X. M., & Yan, C. G. (2022). Population genomics reveals that natural variation in PRDM16 contributes to cold tolerance in domestic cattle. *Zoological Research, 43*(2), 275–284. https://doi.org/10.24272/j.issn.2095-8137.2021.360

Yurchenko, A. A., Daetwyler, H. D., Yudin, N., Schnabel, R. D., Vander Jagt, C. J., Soloshenko, V., Lhasaranov, B., Popov, R., Taylor, J. F., & Larkin, D. M. (2018). Scans for signatures of selection in Russian cattle breed genomes reveal new candidate genes for environmental adaptation and acclimation. *Scientific Reports, 8*(1), 12984. https://doi.org/10.1038/s41598-018-31304-w

Zeuner, F. (1963). The history of the domestication of cattle. In *Man and cattle: Proceedings of a symposium on domestication*. Royal Anthropological Society of Great Britain and Ireland. Occasional paper: 9–15.

Zeyland, J., Wolko, L., Bocianowski, J., Szalata, M., Słomski, R., Dzieduszycki, A. M., Ryba, M., Przystałowska, H., & Lipiński, D. (2013). Complete mitochondrial genome of wild aurochs (Bos primigenius) reconstructed from ancient DNA. *Polish Journal of Veterinary Sciences, 16*(2), 265–273. https://doi.org/10.2478/pjvs-2013-0037

Zhang, K., Lenstra, J. A., Zhang, S., Liu, W., & Liu, J. (2020). Evolution and domestication of the Bovini species. *Animal Genetics, 51*(5), 637–657. https://doi.org/10.1111/age.12974

Cattle Ancestry and Ancient Carnivores

Introduction

As most records have found that the aurochs were much larger than modern cattle, despite some possible size overlaps, physical characteristics of cattle, size, resistance to predators, and intelligence may be assessed through examination of changes during the period of domestication (Linnell et al., 1999; Kluever et al., 2008; Balcarcel et al., 2021). Van Vuure notes that while adult aurochs were probably not vulnerable to predators due to their enormous size and strength, calves of aurochs would have been susceptible to attack by brown bears and gray wolves. Prehistoric lions, tigers, and spotted hyenas were likely predators. The aurochs were described in historical accounts as being agile despite their size and capable of great aggression when agitated, but otherwise fearless toward people. Predation possibilities may also have been related to the variable sizes of adult aurochs. For example, de Carvalho et al. (2022) note that aurochs peaked in size during the early Late Pleistocene interglacial period. The size of the preserved metapodial bones increased from the Middle Pleistocene to the early Late Pleistocene, and it decreased from the Late Pleistocene to the early Holocene. The aurochs' decline in size was caused by the Last Glacial. Furthermore, because these animals were prey to a number of large carnivores during the early Late Pleistocene, the increase in body size during this time may have been caused by predation pressure. Predation pressure and husbandry have been blamed for the Holocene increase in aurochs' size found in tracks (de Carvalho et al., 2022). The main predators across the range of the aurochs are presented in Table 1, and extinct species are compared with related, current species in terms of species association and ecology (Table 2).

McTavish et al. (2013: 1398) note that "archeological and genetic research has shown that modern cattle breeds are descended from multiple independent domestication events of the wild aurochs (Bos primigenius) ~10,000 years ago. Two primary areas of domestication in the Middle East/Europe and the Indian subcontinent resulted in taurine and indicine lines of cattle, respectively." The aurochs and their immediate descendants were among the most powerful mammals in their

© The Author(s), under exclusive license to Springer Nature Switzerland AG 2025

M. O. Campbell, *Cattle, Their Predators and Geomatics Research*,
https://doi.org/10.1007/978-3-031-97363-5_4

Table 1 Late Pleistocene/Early Holocene and current predators and their largest prey

Late Pleistocene species	Current closest species	Largest prey species of current species
Caspian tiger (*Panthera tigris* Linnaeus, 1758)	Siberian tiger (*Panthera tigris* Linnaeus, 1758)	Ussuri brown bear (*Ursus arctos lasiotus* Gray, 1867) Ussuri black bear (*Ursus thibetanus ussuricus* Heude, 1901) Moose (*Alces alces* Linnaeus, 1758)
Cave lion (*Panthera spelaea* Goldfuss, 1810)	African/Asiatic lion (*Panthera leo* Linnaeus, 1758)	African buffalo (*Syncerus caffer* Sparrman, 1779) Common eland (*Taurotragus oryx* Pallas, 1766) Giraffe (Genus *Giraffa*, Linnaeus, 1758)
Cave leopard (*Panthera pardus spelaea* Bächler, 1936)	African/Asiatic leopard (*Panthera pardus* Linnaeus, 1758)	Common eland (*Taurotragus oryx* Pallas, 1766) Blue wildebeest (*Connochaetes taurinus* Burchell, 1823) Medium-sized antelopes (*Bos* Linnaeus, 1758)
Cave bear (*Ursus spelaeus* Rosenmüller, 1794)	Eurasian brown bear (*Ursus arctos* Linnaeus, 1758)	European bison (*Bison bonasus*, Linnaeus, 1758) Moose (*Alces alces* Linnaeus, 1758) Reindeer (*Rangifer tarandus* Linnaeus, 1758)
Cave hyena (*Crocuta crocuta spelaea* Goldfuss, 1823)	Spotted hyena (*Crocuta crocuta* Erxleben, 1777)	Giraffe (Genus *Giraffa*, Linnaeus, 1758) African buffalo (*Syncerus caffer* Sparrman, 1779) Blue wildebeest (*Connochaetes taurinus* Burchell, 1823)
Pleistocene wolf (*Canis lupus spelaeus* Goldfuss, 1823)	Gray wolf (*Canis lupus* Linnaeus, 1758)	European bison (*Bison bonasus*, Linnaeus, 1758) Moose (*Alces alces* Linnaeus, 1758) Reindeer (*Rangifer tarandus* Linnaeus, 1758)

Sources: Kruuk (1972), Schaller (1976), Estes (1992), and Sinclair et al. (2003)

ecosystems, superseded only by elephants, rhinoceroses, and possible hippopotamuses (Kysely, 2008; Chrószcz et al., 2011; Stokstad, 2015). Assessments of the aurochs' ecological role may be based on hypothetical comparisons with current mammals of comparative size, social organization, and ferocity, especially concerning the loss of competitive ability against carnivores and lower intelligence with domestication (Balcarcel et al., 2021). The main issues concern the environments of the aurochs through the domestication period up to the recent status of the taurine and indicine cattle, including the favored habitats, sympatric predatory species, and relevant human actions (Bollongino et al., 2012; Decker et al., 2014; Stokstad, 2015; Upadhyay et al., 2017). Current research on these topics is based on the assessments of the size and possible power of the studied species, compared to modern examples, fossil evidence of species ranges, archaeological evidence of

Table 2 Comparison of modern large herbivore species with aurochs

Modern species	Modern species weight (kg)	Aurochs/cattle type	Aurochs weight (kg)
Giraffe	1192–828	Large adult male	1000
African buffalo	870–425	Adult females, smaller males, large male taurine, indicine cattle	900–800
Eurasian bison	920–424		
Moose	595–490		
Irish elk (extinct)	600–450		
Eland	942–300		
Ussuri brown bear	>350 kg	Smaller females, juveniles/calves, taurine and indicine breeds	<700
Ussuri black bear	132–167		
Blue wildebeest	410–140		
Reindeer	80–182		
Medium-sized antelopes	<100		

Note that Wright and Viner-Daniels (2015) argue for a north-south cline for aurochs, with larger specimens from the north: "A south-north body size cline is detected for aurochs from Marine Isotope Stage 9 and during the early Holocene" and "Tentative evidence for a west-east body size cline is also presented during more recent periods." Sources: Heptner and Naumov (1967), Franzmann (1981), Skinner and Smithers (1990), Kingdon (1997, 2015), Semenov (2014), Estes (2024), and Jeong et al. (2021). Note that the Irish elk (*Megaloceros giganteus*, Blumenbach, 1799) is now extinct (Lister & Stuart, 2019)

human actions, and the extinction of the predatory species (Palmeira et al., 2008; Upadhyay et al., 2017).

Three critical issues regarding the aurochs' ability to deal with predators compared to ancient and modern domestic cattle concern possible positive correlations between increased cattle domestication, decreasing size, declining intelligence, and reduced predator defense abilities. For example, concerning the decline of intelligence with domestication, Balcarcel et al. (2021) note that domestic taurine cattle "have smaller brains than their wild ancestor, Bos primigenius (aurochs), but… brain size varies significantly by breed, with some having much smaller brains than others. Differences in husbandry practices between several breed categories align with a range of human engagement, which also aligns with the degree of selection for docility" (Balcarcel et al., 2021). Balcarcel et al. (2021) examined the relationship between variations in brain size and the degree of human contact. They discovered a distinct pattern whereby a gradient in behavioral selection corresponds with a gradient in brain shrinkage. Dairy breeds, heavily selected for docility, have much smaller brains than bullfighting cattle, which are bred for fighting and aggressive temperaments. The findings support a central tenet of animal domestication theory: variations in brain size appear to be explained by the interaction of three basic aspects of the home environment: lack of predators, docility preference, and human provision of resources (Balcarcel et al., 2021).

Regarding the decline of predator resistance in domesticated cattle, Linnell et al. (1999: 698) argue that compared to wild prey species of comparable size, domestic livestock are relatively easy to kill and exhibit almost none of the anti-predator behavior of their ancestors (see also Palmeira et al., 2008). Klüver et al. (2008: 321) used vigilance rates (the proportion of foraging time that the ungulate spends searching for predators) and write that few studies have examined vigilant behavior in both domesticated and wild ungulates, despite the fact that it is essential for protecting ungulates from predators. The findings indicate that while domesticated cattle appear less vigilant than wild ungulates, this behavior seems to be at least partially aimed at thwarting predators. The authors further note that "adult female cattle in our study site exhibited lower baseline vigilance rates (3%) than studies with female wild ungulates while actively foraging" with examples cited of American bison (9.6%–18.9%, see Laundré et al., 2001), African buffalo (4.2%–5.6%, see Underwood, 1982), female elk (12%–30%, see Laundré et al., 2001), and giraffe (18%–25%, see Cameron & Du Toit, 2005; Kluever et al., 2008: 325). Kluever et al. (2008: 325) argue that while it is possible that species in the genus *Bos* are inherently less alert, the domestication process is most likely the cause of decreased alertness. Unfortunately, there are few studies on this issue in other Bos wild cattle, such as gaur, banteng, kouprey, and wild yak, and the aurochs are extinct, making the problem challenging to solve, and there are few studies of this issue on other Bos wild cattle such as the gaur, banteng, kouprey, and wild yak (see also Van Vuure, 2005).

Regarding the declining size of cattle with domestication, Mulville (2024) notes that "the aurochs was far larger than most modern domestic cattle with a shoulder height of 2 m and weighing in the region of 1000 kg. To compare, large, domesticated cattle today are about 1.5 m tall." Chrószcz et al. (2011) describe the aurochs as 30% larger than domestic cattle. Ajmone-Marsan et al. (2010: 148) note that compared to aurochs, neolithic cattle were smaller, and their size continued to decline until the Middle Ages. Cattle simultaneously developed a wide range of observable phenotypes and adapted to habitats ranging from lush pastures to deserts. Short-horned and even hornless (polled) cattle became more common because of the need for and distaste for large horns for fighting in a farm setting. Although domestic cattle are smaller than their wild counterparts, Wright and Viner-Daniels (2015: 8) note that there is some overlap between the two species, making it difficult to tell them apart. Furthermore, throughout the Pleistocene and Holocene, aurochs in Europe showed morphological variation. In the early Holocene, aurochs from southern Europe were smaller in size than those from northern regions. This pattern is no longer discernible after domestication, although the largest individuals are consistently found in German and Polish samples, suggesting a west-east decline in body size (Wright & Viner-Daniels, 2015: 18).

Manning et al. (2015) note that "osteometric measurements from fossil remains dating between c. 5600 and 1500 BCE reveals a substantial reduction in body mass of 33% in Neolithic central European domestic cattle," and they reject "the hypothesis that it was caused by an increase in the proportion of smaller adult females in the population. Instead, we find some support for the hypothesis that the size decrease was driven by a demographic shift towards smaller newborns from subadult breeding as a result of intensifying meat production strategies during the

Neolithic" (Manning et al., 2015). Regarding the role of aurochs in this decline in size, the authors note that because male aurochs were larger than domestic bulls, introgression may be the cause of the opposite trend, which is a rise in size. We would expect to see a gradual retardation of the rate of size increase until the size remained approximately constant over time (subject to random drift), but not a decrease in size, even if introgression was initially common and became less common over time. Moreover, current aDNA research supports the importation of cattle stock from the Near East and offers scant proof of the genetic contribution of native aurochs to the domestic gene pool (Manning et al., 2015).

Edwards et al. (2007: 1377) note that the decline in size during domestication in Europe may have occurred through migrations of smaller breeds from southwest Asia. Importantly, research has indicated that "most ancient and modern European domestic cattle carry haplotypes previously designated T. This, in combination with our new finding of a T haplotype in a very Early Neolithic site in Syria, lends persuasive support to a scenario whereby gracile Near Eastern domestic populations, carrying predominantly T haplotypes, replaced P haplotype-carrying robust autochthonous aurochs populations in Europe, from the Early Neolithic onward. During the period of coexistence, it appears that domestic cattle were kept separate from wild aurochs and introgression was extremely rare"; this refers to cross breeding between domesticated taurine cattle, and European aurochs carry the P haplogroup (Edwards et al., 2007: 1377).

Elayadeth-Meethal et al. (2018: 10608) describe the evolution of heat tolerance in taurine and indicine cattle as a factor for size reduction: with the results showing that standard-sized cattle breeds "can acclimatize in the short term to higher temperatures but reach their tolerance limit under prevailing tropical conditions" and dwarf-sized cattle breeds are adapted to the warmer climate. Therefore, "increased hemoglobin concentration, reduced cellular size, and smaller body size enhance thermal tolerance… The results illustrate mechanistic ways of dwarfing, body size-dependent tolerance, and differential fitness in a large mammal species under harsh field conditions, providing a background for comparing similar populations during global climate change" (Elayadeth-Meethal et al., 2018: 10608).

The rest of this chapter examines the relationship between the ancestors of modern cattle, namely, the aurochs, and early breeds of taurine and indicine cattle, and their predators, considering issues such as the size, intelligence and defensive ability of the cattle, and the hunting strategies of the predators. The main predators for these early cattle were tigers in eastern Europe and most of Asia, lions in southern and western Asia, and wolves and bears across Eurasia. Smaller predators, such as leopards and smaller bears, may have been able to kill calves. The evidence indicates that cattle predation became easier with the reduced strength of domestic cattle compared with the wild ancestors, but human interventions have affected the predators, in some cases to the point of extinction.

Scimitar Cats, Aurochs, and Ancestral Cattle

The scimitar cats were contemporary with the aurochs in Eurasia. Jiangzuo et al. (2023) describes "the Plio-Pleistocene sabretoothed cat Homotherium represents a unique member of Felidae, with special hunting and social behaviours." Antón et al. (2014: 259) describe the genus *Homotherium* as "a genus of lion-sized, highly successful Plio-Pleistocene sabre-toothed felids" and as "the last member of the long-lived machairodontine tribe Homotherini." The genus *Homotherium* has been located in fossil form in Eurasia, Africa, and North America, and recently, South America, with the Eurasian type species being *Homotherium latidens* (Owen, 1846). Barnett et al. (2020: 5018) describe *Homotherium* as "a genus of large-bodied scimitar-toothed cats, morphologically distinct from any extant felid species, that went extinct at the end of the Pleistocene. They possessed large, saber-form serrated canine teeth, powerful forelimbs, a sloping back, and an enlarged optic bulb, all of which were key characteristics for predation on Pleistocene megafauna." Jiangzuo et al. (2023) notes that "the iconic sabretooth Homotherium is thought to have hunted cooperatively, but the origin of this behaviour and correlated morphological adaptations are largely unexplored"; research the case of "the most primitive species of Amphimachairodus (Amphimachairodus hezhengensis sp. nov.), a member of Machairodontini basal to Homotherium, from the Linxia Basin, northeastern border of the Tibetan Plateau (9.8–8.7 Ma)"; and argue that based on the fossil, the "the long snout, laterally oriented and posteriorly located orbit of Amphimachairodus suggest a better ability to observe the surrounding environment, rather than targeting single prey, pointing to an adaptation to the open environment or social behavior."

Indications of hunting behavior were derived genetically, as "comparative genomic analyses found signatures of positive selection in several genes, including those involved in vision, cognitive function, and energy consumption, putatively consistent with diurnal activity, well-developed social behavior, and cursorial hunting" (Barnett et al., 2020: 5018). There is also evidence of adaptability: "fossil evidence suggests that Homotherium succeeded in expanding its distribution despite potential competition with other sympatric large cats, including lions (Panthera leo and the now-extinct cave lion P. spelaea), leopards (P. pardus) and the now-extinct dirk-toothed cat Megantereon cultridens across Eurasia and Africa, tigers (P. tigris) in Southeast Asia, and jaguars (P. onca), American lions (P. atrox), and other now-extinct saber-toothed cats (Smilodon spp.) in the Americas" (Barnett et al., 2020: 5020).

Domínguez-Rodrigo et al. (2022) argue that the evidence of their study suggests that lion-sized sabretooths "probably hunted and most certainly consumed effectively the carcasses of animals ranging in size from peccaries to juvenile mammoths." Andersson et al. (2011) examine the possibility that scimitar cats killed large prey. Their findings suggest that although "sabretoothed carnivorans are widely interpreted as hunters of larger and more powerful prey than that of their present-day non-sabretoothed relatives," it is evident that "the extended gape of sabretooths only results in considerable increase in bite depth when biting into prey with a radius of less than ~10 cm. For sabretooths, this size-reversed functional advantage suggests predation on species within a similar size range to those attacked by present-day carnivorans, rather than 'megaherbivores' as previously believed."

Tigers, Aurochs, and Ancestral Cattle

The tiger would be one of the main predators or aurochs, perhaps even for adult cattle. The area of the original aurochs populations would be principally inhabited by the currently extinct Caspian tiger (southwest Asia, southwestern Russia, central Asia), which is closely related to the Siberian/Amur tiger (both *Panthera tigris tigris*, the Caspian tiger formerly *Panthera tigris virgata*, the Siberian *Panthera tigris altaica*) (Goodrich et al., 2022). Driscoll et al. (2009) suggest that the Caspian tiger migrated from Eastern China about 10,000 years BP, and "flourished in Central Asian riverine forest systems in a range disjunct from that of other tigers but was driven to extinction in 1970 prior to a modern molecular evaluation. For over a century naturalists puzzled over the taxonomic validity, placement, and biogeographic origin of this enigmatic animal." Regarding the relationship between the Caspian and Siberian tigers, Driscoll et al. (2009) used aDNA from museum collections and found that "Caspian tigers carry a major mtDNA haplotype differing by only a single nucleotide from the monomorphic haplotype found across all contemporary Amur tigers" and that "phylogeographic analysis with extant tiger subspecies suggests that less than 10,000 years ago the Caspian/Amur tiger ancestor colonized Central Asia via the Gansu Corridor (Silk Road) from eastern China then subsequently traversed Siberia eastward to establish the Amur tiger in the Russian Far East" (see also Dybas, 2010). The decline of the Caspian tiger populations was due to human hunting and human-induced habitat conversion, agricultural expansion, prey species depletion, and killing in response to livestock depletion (Kitchener & Dugmore, 2000; Driscoll et al., 2009; Dybas, 2010; Cooper et al., 2016) (Figs. 1 and 2).

Regarding prey species, one way to guess or determine the prey of the Caspian tiger could be to examine the prey of the modern tigers in similar environments. For modern tigers, the prey are mostly smaller mammals, which might be similar in size to aurochs calves. For example, Miller et al. (2013: 849) note that in their study of Amur tigers in the Russian Far East, "of the total observed kills, 27.9% were wild boar, 24.3% were red deer, and 23.4% were roe deer. Among known wild boar kills, 25.8% were adults, 67.7% were juveniles and piglets, and 6.5% could not be accurately classified into an age class. Red deer kills consisted of 70.4% adults and 29.6% juveniles and calves. Among roe deer kills, 53.9% were adults, 26.9% were juveniles, and 19.2% could not be accurately classified. Juvenile ungulates comprised 18 (50.0%) of 36 observed summer large ungulate kills and 17 (51.5%) of 33 winter kills. Overall, wild ungulate species represented 90.1% of all tiger kills, with non-ungulate or domestic prey comprising the remaining kills." These results differed from other studies, "which showed red deer and wild boar comprised 63%–92% of all kills from 6 sites across tiger range in Russia. In contrast, only 52.3% of our confirmed kills were red deer or wild boar, and only 48.8% of

Fig. 1 Caspian tiger–public domain

predicted kills were classified as large prey" (Miller et al., 2013: 852; see also Miquelle et al., 1999).

The Caspian tiger might be capable of larger aurochs, such as females or juveniles, inferred from the capability of South Asian and Southeast Asian tigers to kill large wild cattle such as gaur and banteng. For example, Khaewphakdee et al. (2020: 5152) in a study of tiger predation on wild cattle in Thailand found that "the average weight of banteng killed was 423.9 kg, which was similar to the 397.9 kg average weight for gaur." Ahrestani (2018: 43) notes that gaur "calves are preyed on by leopards, dhole and tigers, although adults are preyed on primarily by tigers. Leopards and tigers typically kill calves and females by asphyxiation using bites to their throats… Males, however, have a thick neck and often a dewlap, which makes killing them by asphyxiation difficult. Tigers have been known to overcome this by biting the hock, thereby hamstringing and bringing down male B. gaurus… the remains of B. gaurus were regularly found in tiger feces within Eravikulam National Park, India… analysis of scats of the primary predators in Mudumalai Tiger Reserve, India, showed that B. gaurus made up 0.19% of dhole diet and about 5% of tiger diet and adult female and adult male B. gaurus comprised 23% and 15% of tiger kills, respectively, in Nagarahole, India" (Ahrestani, 2018: 43).

Fig. 2 Amur tiger–Permission Chris Unger-unsplash

Lions, Aurochs, and Ancestral Cattle

Concerning lions during the late Pleistocene, just before the domestication of the aurochs, there were two main species, the cave lion (*Panthera spelaea* Goldfuss, 1810, or *Panthera leo spelaea* M. Boule & L. De Villeneuve, 1927) and the modern lion (*Panthera leo* Linnaeus, 1758). The latter includes the subspecies or haplogroup named the Asiatic lion. Regarding the prey of the cave lion, Reumer et al. (2003: 260) note that "the common large felid from the late Pleistocene of Europe is the cave lion (Panthera leo spelaea), which is supposed to have preyed upon nearly all of the available larger mammals in its vicinity, such as horse, deer, bison, aurochs, yak, musk ox, sheep, goat, saiga antelope, juvenile mammoth, and juvenile rhinoceros."

Hemmer (2011: 201) gives a comprehensive description of the evolution and ecology of the cave lion, noting that some studies view the cave lion as a subspecies of the species *Panthera leo* (e.g., *Panthera leo spelaea*, *Panthera leo leo*, *Panthera leo atrox*) this termed the monospecific concept, or separate species within the genus *Panthera* (*Panthera leo*, *Panthera atrox*, *Panthera spelaea*, or *Leo leo*, *Leo atrox*, *Leo spelaea*). Hemmer (2011: 201) also notes that "the speciation of the lion obviously began later than 1.7 Myr BP in Africa" and "the first appearance of the cave lion line in Europe is represented by the early *P. leo fossilis*, dated to the Early Middle Pleistocene MIS 15" (marine isotope stage). In addition, "the Late Middle Pleistocene was the time of transition to the later *P. leo spelaea* and, at its end, of the first dispersal of the cave lion to North America. At the end of the Late Pleistocene, the genetic variability of cave lion populations seems to have been drastically reduced (Hemmer, 2011: 201). Cave lions became extinct in the 12th millennium BP in Eurasia and in the 11th millennium BP in North America" (Hemmer, 2011: 201). It is also noted that "the Late Pleistocene cave lion had a considerably stouter stature than recent lions, with an estimated body-mass range of between 140 and 400 kg… Prey focus mass is estimated for males at 1000 kg (usual range 100–2000 kg) and for females at 500 kg (usual range 50–1000 kg). All large mammals of the mammoth steppe, including young mammoth, were likely to be the regular prey of cave lions… Only two competitor species could have had any significant impact on the cave lion—namely, the lesser scimitar cat (*Homotherium latidens* Owen, 1846) and Palaeolithic man" (Hemmer, 2011: 201).

Stanton et al. (2020) note that the "cave lion is an extinct felid that was widespread across the Holarctic throughout the Late Pleistocene. Its closest extant relative is the lion (*Panthera leo)*, but the timing of the divergence between these two taxa, as well as their taxonomic ranking are contentious." These authors present a study that indicates that "the mitochondrial DNA divergence between cave lions and lions to be 1.85 million years (95% 0.52–2.91 Mya)" and that "cave lions existed as at least two subspecies during the Pleistocene, and that lions and cave lions were distinct species" (Stanton et al., 2020). de Manuel et al. (2020: 10927) present a similar study indicating a divergence between the cave lion and the modern lion: "cave and modern lions shared an ancestor ca. 500,000 years ago and that the 2 lineages likely did not hybridize following their divergence. Within modern lions, we found 2 main lineages that diverged ca. 70,000 years ago, with clear evidence of subsequent gene flow." The Asiatic lion was probably sympatric with the aurochs and their immediate descendants. Broggini et al. (2024) note that the Asiatic lion *Pantheras leo* "is not a subspecies, but rather a haplogroup of the northern *P.l. leo* that exited Africa at least ~31 kya. The progenitor to all lions existed ~1.2 Mya, possibly in SE Africa, and later exited Africa and split into the two cave lion lineages ~175 kya. Species demography is correlated to major climactic events."

Current research suggests that the physical size of lions declined through the Pleistocene (Marciszak & Gornig, 2024). Marciszak and Gornig (2024) cite extensive literature sources suggesting that "the first lions that entered Eurasia were among the largest felids that ever existed" and "some of the specimens… were truly giants of gargantuan proportions, having no equivalent among modern felids. Their size, body weight and structure were more reminiscent of the largest, modern ursids like brown bear Ursus arctos Linnaeus, 1758 and polar Ursus maritimus Phipps, 1774 than of any modern felid. In general perception, the image of the Pleistocene lion Panthera spelaea (Goldfuss, 1810) brings to mind an image of a huge, massive and maneless animal, considerably larger than the extant lion Panthera leo (Linnaeus,

1758)." In addition, "in the evolution of *Panthera spelaea* is an evolutionary trend of decreasing body size" as of the three subspecies, the trend is largest and oldest to youngest and smallest (with exceptions), namely from *Panthera spelaea fossilis*, to the more recent and smaller *Panthera spelaea intermedia*, to the yet smaller and more recent *Panthera spelaea spelaea* (Marciszak & Gornig, 2024). Compared with rivals, such as the scimitar cat, the authors note that "the cave lion was considered to be extremely massive"…"the weight of P. spelaea was 160%–180% of the body mass of P. leo of similar linear size"… "Late Pleistocene P. s. spelaea, although still large, is distinctly smaller than P. s. fossilis" and "the enormous size (body mass 400–500 kg) and probably social life guaranteed P. s. fossilis a leading position among other carnivores. By virtue of its impressive posture, it is likely that the early lion was able to displace any European carnivore from its rightful kill. Even Homotherium, despite its fearsome sabres, does not stand a chance to defend its kill against the cow-sized rival."

Lions inhabited areas where the aurochs originated or where common, based on modern records. For example, Németh et al. (2017: 42) give the example of the Carpathian Basin, where lion remains of Holocene age are found in the Hungarian area, and the "wide geographical distribution of the localities indicates that lions were widely distributed in the central part of the Carpathian Basin… Most of the lion records are dated to the Middle and Late Copper Age, and no Bronze Age or younger lion remains are known from the Carpathian Basin." This area was a habitat of the aurochs and the European bison, as Németh et al. (2017: 45) note that "vast territories in the Carpathian Basin were once covered by open landscapes: dry grasslands (steppe) and forest steppe. The importance of large herbivores in grassland ecosystems is clear, and their role was filled by large domestic animals for a long time in the Carpathian Basin… The aurochs and the European bison, usually linked to temperate deciduous forests, are thought to be 'refugee species' that originally inhabited mosaic landscapes of grassland and woodland habitats" and "this is concordant with the wood-pasture hypothesis…that the presence of large herbivores maintained an open landscape composed of a mosaic of groves, closed-canopy woodlands, open parkland, and regenerating scrub in the primeval Holocene woodlands of the European lowlands."

Diedrich (2014) writing about the Late Pleistocene apex predators in Europe, noted that "Late Pleistocene spotted hyena Crocuta crocuta spelaea (Goldfuss, 1823) and steppe lion Panthera leo spelaea (Goldfuss, 1810) were top predators in Central Europe." In addition, "the larger Late Pleistocene felids focussed on cervids (reindeers specialization during the high glacial = Last Glacial Maximum – LGM), on bovids (steppe bison/aurochs), and possibly on saiga antelope and on the cave bear, hunting deep in caves during their hibernations and targeting cubs" (Diedrich, 2014). Bocherens et al. (2011: 250) note that the study of the prey preferences of extinct predators may be based on a comparison with current carnivores, and the traces of the fossil remains of their prey. For the cave lion and comparisons with contemporary species, the main comparison is with the modern lion: "in the case of the cave lion, fortunately there is a very close modern relative abundantly studied for its dietary ecology and for competition with other large predators. Since modern lions consume essentially large ungulates such as zebras and large antelopes but are unable to tackle adults of very large species such as elephants and rhinoceros, it is generally assumed that cave lions should have had a similar prey choice shifted to larger sized prey, as the average size of cave lions was larger than that of modern lions (see also Hemmer, 2004). This leads to the assumption that cave lions were primarily consuming horse and bison, as well as reindeer, muskoxen and young individuals of mammoths and woolly rhinoceros." Of course, comparisons between modern and cave lions may be tenuous: "Although the habitats used by modern lions range from desert to closed woodland, none of the modern lion populations dwells under ecological conditions even remotely comparable to the steppe-tundra of the late Pleistocene, making the ecological analogy between modern and fossil lions very difficult."

For the traces of prey items, Bocherens et al. (2011: 250) note that "approaches such as stable isotopic tracking may provide such direct quantitative evidence at an individual level." Research evidence suggests "possible antagonism between cave lions and cave hyenas. Here again, it is very difficult to infer quantitatively this type of ecological interaction between predators in the past with such evidence, while stable isotope approaches allow to test dietary competition between predators by comparing their signatures: different isotopic signatures indicate that the average source of proteins was different for both predators." The authors conclude that concerning cave lion preferences "some individuals strongly oriented on reindeer and others possibly on juvenile cave bears. In the post Last Glacial Maximum sites, the isotopic data point to a preference on reindeer for cave lion. It appears that cave lion was the most reindeer oriented large predator during the late Pleistocene in Northwestern Europe, and that this prey preference was still strongly established until the extirpation of the species, which seems therefore to coincide chronologically with the extirpation of reindeer notably in the region" (Bocherens et al., 2011: 259).

Bocherens et al. (2011: 249) used bone collagen isotopic signatures in the Belgian Ardennes and the Swabian Jura covering 40,000 and 25,000 years ago and the Late-glacial of the northwestern Alp foreland and the Paris Basin, with the findings suggesting "an individualistic prey choice for cave lions, with some individuals more oriented on reindeer and others on young cave bears." Notably, "the most recent western European cave lions seem to have been consuming mainly reindeer until the local extirpation of this prey species, which coincides chronologically with their own extinction. This restricted prey choice may be involved in the extinction of this large predator in Western Europe" (Bocherens et al., 2011: 249).

Leopards, Aurochs, and Ancestral Cattle

Leopard ancestry and ancestral forms require comprehensive study. Leopards were distributed through Eastern Europe and Western and Central Asia during the Late Pleistocene (Németh et al., 2017; Paijmans et al., 2018). Paijmans et al. (2018) argue that "resolving the historical biogeography of the leopard is a complex issue, because patterns inferred from fossils and from molecular data lack congruence.

Fossil evidence supports an African origin and suggests that leopards were already present in Eurasia during the Early Pleistocene. Analysis of DNA sequences, however, suggests a more recent, Middle Pleistocene shared ancestry of Asian and African leopards." The authors further note that the difficulties of determining the ancestry of leopards have led researchers "to propose a two-stage hypothesis of leopard dispersal out of Africa: an initial Early Pleistocene colonisation of Asia and a subsequent replacement by a second colonisation wave during the Middle Pleistocene" (Paijmans et al., 2018).

The leopards of the Late Pleistocene have been classified as the cave leopard (*Panthera pardus spelaea*) (Paijmans et al., 2018). Diedrich (2013: 167) writes that "the final European Late Pleistocene 'Ice Age leopard' *Panthera pardus spelaea* (Bächler, 1936) is validated taxonomically. The skull shows heavy signs of sexual dimorphism with closest cranial characters to the Caucasian *Panthera pardus ciscaucasica* (Persian leopard). Late Pleistocene leopards were distributed northernmost, up to Southern England with the youngest stratigraphic records by skeletons and cave art in the MIS 2/3 (about 32,000–26,000 BP)." Paijmans et al. (2018) add that *Panthera pardus spelaea* "likely went extinct before the Last Glacial Maximum (LGM), around 25,000 years ago. The disappearance of the European leopard populations coincides with the extinction of many other Pleistocene carnivores, such as the cave bear, cave hyena and Homotherium." Németh et al. (2017: 43) argue that "the Holocene distribution patterns of leopards and lions are remarkably similar to each other, in agreement with the present-day distribution of the two big cats: where lions occur one can find leopards as well," although the leopard has the "widest habitat tolerance of all Old World felids." In addition, "assuming that the distribution pattern, ecology and biogeography of the two species were similar during the entire Holocene, the leopard's range would have included the Carpathian Basin via the Balkans" (Németh et al., 2017:43; Paijmans et al., 2018).

Diedrich (2013: 167) reports that "the last Ice Age glacial leopard habitat was the mountain/alpine boreal forest (not mammoth steppe lowland), where those hunted even larger prey such as alpine game (Ibex, Chamois)." An assessment of prey taken by the cave leopard could be based on assessments on prey preferences of modern leopards. For example, Hayward (2006: 298) notes that "Leopards Panthera pardus have a catholic diet and are generally thought to prey on medium-sized ungulates" and "leopards preferentially prey upon species within a weight range of 10–40 kg. Regression plots suggest that the most preferred mass of leopard prey is 25 kg, whereas the mean body mass of significantly preferred prey is 23 kg. Leopards prefer prey within this body mass range, which occur in small herds, in dense habitat and afford the hunter minimal risk of injury during capture. Consequently, impala, bushbuck and common duiker are significantly preferred, with chital likely to also be preferred with a larger sample size from Asian sites." From these results, it may be inferred that very young aurochs and older taurine and indicine could be targeted by leopards. Current records show that leopards attack adult domestic cattle (Babrgir et al., 2017: Consolee et al., 2020; Naha et al., 2020; Kilikar, 2024).

Bears, Aurochs, and Ancestral Cattle

The cave bear is determined to have become extinct before the domestication of cattle 11,000 to 9000 year BP. For example, Rafferty (2023) pointed out that European cave deposits contain the remains of over 100,000 cave bears. Between 28,000 and 27,000 years ago, the cave bear went extinct through a slow and gradual process. The largest cave bears were about the same size as the North American Kodiak bears (*Ursus arctos middendorffi*) and polar bears (*U. maritimus*), with weights ranging from 400 to 1000 kg. The cave bear is surmised to be mostly vegetarian based on its very large head and prominent teeth. In terms of the relationship between the cave bear and the brown bear, Barlow et al. (2018: 1563) using "multifold coverage genomic data" note that "cave bears hybridized with brown bears (Ursus arctos) during the Pleistocene" and "segments of cave bear DNA still persist in the genomes of living brown bears, with cave bears contributing 0.9%–2.4% of the genomes of all brown bears investigated. Our results show that even though extinction is typically considered as absolute, following admixture, fragments of the gene pool of extinct species can survive for tens of thousands of years in the genomes of extant recipient species" (Barlow et al., 2018: 1563).

Fortes et al. (2016: 4907) examined the "complete mitochondrial genomes of extinct Late Pleistocene cave bears and middle Holocene brown bears that each inhabited multiple geographically proximate caves in northern Spain." Differences between cave bears and brown bears were detected, in that for cave bears "each cave almost exclusively contains a unique lineage of closely related haplotypes. This remarkable pattern suggests extreme fidelity to their birth site in cave bears, best described as homing behaviour, and that cave bears formed stable maternal social groups at least for hibernation" (Fortes et al., 2016: 4907). However, for brown bears there was no "strong association of mitochondrial lineage and cave, suggesting that these two closely related species differed in aspects of their behaviour and sociality" (Fortes et al., 2016: 4907). The authors argue that "this difference is likely to have contributed to cave bear extinction, which occurred at a time in which competition for caves between bears and humans was likely intense and the ability to rapidly colonize new hibernation sites would have been crucial for the survival of a species so dependent on caves for hibernation as cave bears" (Fortes et al., 2016: 4907).

Hofreiter et al. (2002: 1244) used the genetic diversity, population history, and relationship among different cave bear populations, aged from about 26,500 to at least 49,000 years and compared these results with that of brown bears. The findings suggested that "about 49,000 years ago, the mtDNA diversity among cave bears was about 1.8-fold lower than the current species-wide diversity of brown bears (Ursus arctos). However, the current brown bear mtDNA gene pool consists of three clades, and cave bear mtDNA diversity is similar to the diversity observed within each of these clades." The findings also demonstrate that, in terms of mtDNA, geographically distinct populations of the high alpine cave bear form were polyphyletic. This implies that large size evolved independently at least twice and that small body size may have been an inherited characteristic of cave bears (Hofreiter et al., 2002: 1244).

During the period of significant aurochs habitation in southwest and central Asia and Europe, brown bear populations may have been more common than currently evident, with significant migrations between population clusters (Saarma et al., 2007; Murtskhvaladze et al., 2010). For example, Albrecht et al. (2017) note that humans and climate change were factors for the "Holocene decline of the largest living terrestrial carnivore, the brown bear, on a continental scale." It is argued that "despite the broad climatic niche of the brown bear, increasing winter temperatures contributed substantially to its Holocene decline—both directly by reducing the species' reproductive rate and indirectly by facilitating human land use. The first local extinctions occurred during the Mid-Holocene warming period, but the rise of the Roman Empire 2000 years ago marked the onset of large-scale extinctions, followed by increasingly rapid range loss and fragmentation" (Albrecht et al., 2017).

The brown bear would be capable of killing juvenile or perhaps even smaller, adult female aurochs, which may be surmised by the predation on similarly sized moose. For example, Niedziałkowska et al. (2019: 1) note that in a study of brown bear ungulate predation in Europe and Asia, "ungulate meat makes up, on average, 8.7% of brown bear diets, with European bears feeding on ungulates more (mean 10.5%) than Asiatic bears (6.8%)" the factor for the lower figure in Asia probably due to the edible Siberian pine (*Pinus* sp.). Importantly, "brown bears' preference for 10 species of ungulates increased with body mass of prey. The bear significantly preferred preying upon moose Alces alces throughout its range. Bears were the most important predator of moose and caused, on average, 23% of total natural mortality in moose populations" (Niedziałkowska et al., 2019: 1).

Hyenas, Aurochs, and Ancestral Cattle

Rivals et al. (2022) note that "Extinct hyenas of the genus Crocuta were typical representatives of the Late Pleistocene mammoth fauna of Eurasia. The geographical distribution of these animals was vast, stretching from the British Isles to the Pacific coast" with the study based on *Crocuta spelaea* and *Crocuta ultima ussurica*. Diedrich and Žák (2006: 237) using a study of fossil remains note that the main animals preyed upon by Upper Pleistocene hyenas were Przewalski's horse (*Equus ferus przewalskii* I. S. Polyakov, 1881) and the woolly rhinoceros (*Coelodonta antiquitatis*, Blumenbach, 1799), with other species including steppe bison (*Bison priscus*, Bojanus, 1825), reindeer (*Rangifer tarandus* Linnaeus, 1758), red deer (*Cervus elaphus* Linnaeus, 1758), Irish deer (*Megaloceros giganteus*, Blumenbach, 1799), European wild ass (*Equus hydruntinus* Regalia, 1907), chamois (*Rupicapra rupicapra* Linnaeus, 1758), Alpine ibex (*Capra ibex*, Linnaeus, 1758), cave bear (*Ursus spelaeus*, Rosenmüller, 1794), cave or steppe lion (*Panthera spelaea*, Goldfuss, 1810), wolves (*Canis lupus*, Linnaeus, 1758), and rarely wolverine (*Gulo gulo* Linnaeus, 1758) and mammoth (*Mammuthus*, Brookes, 1828) (Diedrich & Žák, 2006: 237). Diedrich and Žák (2006: 238) further note that the foraging behavior of the Upper Pleistocene hyena may be studied by using the bone accumulations and by actuopaleontological comparisons to the ecological behaviour of the African spotted hyena (Crocuta crocuta crocuta Erxleben) based on the den and prey material deposit sites. Bocherens et al. (2011: 250) note that "some predators are also closely related to the ones formerly sharing the same environment with cave lions, such as cave hyenas, even if behavioural differences are to be expected between modern and Pleistocene hyenas, the latter being most probably a more active predator than modern hyenas." In addition, "the isotopic signatures and therefore dietary choice of cave lions did not overlap with those of cave hyenas, indicating competitive exclusion between the large predators" (Bocherens et al., 2011: 249).

Another study by Diedrich (2016: 66) of bones in a hyena cave in the Czech Republic, Bohemian Mountains had mostly similar as at two other Czech Republic sites, mainly of Przewalski's horse (*Equus ferus przewalskii*, I. S. Polyakov, 1881) (50%), which included pregnant females being hunted in spring/early summer. There were also smaller numbers of bones of fewer remains of reindeer (*Rangifer tarandus*, Linnaeus, 1758) (15%), woolly rhinoceros (*Coelodonta antiquitatis*, Blumenbach, 1799) (4%), Bison priscus (2%), alpine chamois (*Rupicapra rupicapra*, Linnaeus, 1758) (2%) or Alpine ibex (*Capra ibex*, Linnaeus, 1758) (1%) and no mammoth remains, which "correlate to hyena den bone assemblages in middle high elevated mountain regions of northern/central Europe. The rareness of cave bears in the Bohemian Mountains, on which hyenas specialized to feed additionally in European mountain regions, explains predominant predation on horses" (Diedrich, 2016: 66).

Marra et al. (2004: 784) used taphonomic analysis (the study of how organisms are fossilized) of two sites of *Crocuta crocuta spelaea* prey remains in the S. Teodoro Cave in northeastern Sicily and the cave of Bois Roche in southwestern France. Noted differences concerned the remains of the dominant presence of large herbivores (mainly bovids and equids) in the French cave, and smaller mammals (cervids) in the Sicilian cave. Other caves have been used as dens by hyena populations. At Bois Roche the prey assemblage is dominated by large herbivores (mainly bovids and equids) with a smaller representation of medium-sized animals (mainly cervids, including *Cervus elaphus siciliae* and *Equus hydruntinus*), despite the strong similarities between the two sites in terms of "the proportional representation of hyenas and other carnivores, in kinds of damage to the bone surfaces and in the presence of coprolites" (fossilized feces) (Marra et al., 2004: 784).

The authors note that the reasons for these differences are environmental: "while all of the Bois Roche fauna is widely distributed in continental Western Europe, the macromammal assemblage of S. Teodoro belongs to an insular faunal complex which is composed of non-endemic taxa (*Equus hydruntinus*, *Sus scrofa*, Canis cf. lupus, *Vulpes vulpes*, *Crocuta crocuta spelaea*) associated with some endemic taxa of reduced size (*Elephas mnaidriensis*, *Cervus elaphus siciliae, Bos primigenius siciliae*)" (Marra et al., 2004:784). In addition, "humans were present in the vicinity of the Bois Roche cave, but they are completely absent from San Teodoro, although the possibility of human presence in the island would seem to be suggested by the occurrence of non-endemic faunas derived from the Italian continent" (Marra et al., 2004: 784). Regarding the prey species, for San Teodoro "the endemic red deer (*Cervus elaphus siciliae*) is the most common herbivore, followed by much lower frequencies of the small horse Equus hydruntinus, the endemic elephant *Elephas mnaidriensis*, the endemic auroch *Bos primigenius siciliae* and the wild boar *Sus scrofa*" (Marra et al., 2004:789). In addition, the "the auroch is represented by few remains, in particular teeth and hard bones" (Marra et al., 2004:791).

For a similar study of spotted hyenas, Stiner (2004: 771) notes that "in Late Pleistocene Italy, spotted hyenas competed in certain niche dimensions with wolves and in other dimensions with Paleolithic humans." In addition, "spotted hyenas of the Italian peninsula consumed essentially the same ungulate species as Paleolithic humans did, and both of these predators depended heavily on bone marrow. Wolves tended to consume more hillside-adapted ungulates in the same area and period, suggesting some spatial separation of the three predators according to topography" and "there is complete overlap in the prey age groups commonly harvested by spotted hyenas and wolves, but pronounced differences between this pair of predators and Paleolithic humans of the Middle and Upper Paleolithic periods; from the narrow perspective of ungulate exploitation, Middle and Upper Paleolithic humans were quite similar." Importantly, the "variation in prey age selection by spotted hyenas in Late Pleistocene Italy is indistinguishable from patterns documented for modern African spotted hyenas. They are cursorial predators with a well defined tendency to produce 'attritional' or U-shaped mortality patterns while hunting common prey species, whether feeding on red deer, aurochs, and equids in Italy or antelopes and zebra in sub-Saharan Africa; primeadult prey are notably under-represented relative to the structure of a living population" (Stiner, 2004: 773). The authors add that "the main prey of spotted hyenas, wolves, and humans in Late Pleistocene Italy were red deer (*Cervus elaphus*), aurochs (Bos primigenius), horse (Equus caballus), roe deer (Capreolus capreolus), fallow deer (*Dama dama*), wild boar (Sus scrofa), and ibex (Capra ibex)" (Stiner, 2004: 773).

Wolves, Aurochs, and Ancestral Cattle

The gray and white wolves and their ancestral species and subspecies were important aspects of Eurasian ecologies for thousands of years before cattle domestication. For example, Diedrich (2022) notes Eurasian gray and white wolf ancestors went through "800,000 years evolution, adaptation, pathologies and European dog origins." It is documented that the ancestral wolves included "the oldest known wolf" which appeared about 800,000 years BP during the Marine Isotope Stage 21 in Eurasia, represented by the "unspecialized short-legged old Mammoth steppe wolf Canis lupus bohemica nov. spec." From this species, interglacial *Canis lupus mosbachensis* (Soergel, 1925) short-legged Mosbach gray wolf subspecies emerged about 600,000–420,000 years BP during MIS 15–11. Later lineages in the late

Middle Pleistocene were the southern interglacial gray and northern glacial white wolves in Eurasia after 320,000 BP (MIS 8), including the white wolf *Canis lupus spelaeus* (Goldfuss, 1823) and the warmer climate gray wolf *Canis lupus brevis*, the Greenland *Canis lupus arctos*, and Siberian *Canis lupus albus* (Kerr, 1792).

Loog et al. (2020: 1596) note that "grey wolves (Canis lupus) are one of the few large terrestrial carnivores that have maintained a wide geographical distribution across the Northern Hemisphere throughout the Pleistocene and Holocene. Recent genetic studies have suggested that, despite this continuous presence, major demographic changes occurred in wolf populations between the Late Pleistocene and early Holocene, and that extant wolves trace their ancestry to a single Late Pleistocene population." The findings suggest that the ancestry of contemporary wolf populations is based on an expansion from Beringia at the end of the Last Glacial Maximum, due to Late Pleistocene ecological changes in the Northern Hemisphere. The authors conclude that their "study provides direct ancient genetic evidence that long-range migration has played an important role in the population history of a large carnivore and provides insight into how wolves survived the wave of megafaunal extinctions at the end of the last glaciation" (Loog et al., 2020: 1596).

Wolves were among the predominant predators of western Eurasia, during the time of the aurochs until more recent historic times (Hindrikson et al., 2017). Van Vuure (2005) noted that "the main predator of the aurochs in Europe was the wolf." Flower et al. (2021) note that the wolf "has been a major keystone predator in the Palaearctic since the late Middle Pleistocene"; using an isotope case study of British fossil wolves, the findings indicated "clear patterns of changing wolf prey choice through time. Notwithstanding issues of collagen preservation obscuring some dietary choices in the oldest samples, both small and large prey (hare, horse) were taken by wolves in the MIS 7c-a interglacial, large prey only (reindeer, bison) during MIS 5a and a broader range of large prey items (horse, woolly rhinoceros, bison) during MIS 3." In addition, "where wolves and spotted hyaenas co-existed, they occupied the same dietary niches, and the former was not outcompeted by the latter" (Flower et al., 2021).

There was also a possibility of dire wolf presence in the aurochs range. For example, Lu et al. (2020: 87) report "the first record of the dire wolf fossil in Eurasia… from a Late Pleistocene underwater sand mine site near the city of Harbin in Northeastern China. Other mammalian fossils from the same site suggest that this dire wolf coexisted with the typical Mammuthus-Coelodonta fauna of Eurasia." Consistent with other records of dire wolves, the specimen was "much larger than gray wolves and other large canids from the same region" (Lu et al., 2020: 87). The authors note that "when the body mass spectra of medium and large-bodied carnivores from Asia and North America were compared, the Asian dire wolf occupies a position nearly overlapping the spotted hyena (Crocuta crocuta ultima). During the Late Pleistocene, C. crocuta ultima was one of the most widely distributed and dominant carnivores in Asia. Competition from hyenas may have kept dire wolf Asian populations at very low levels leading to their rarity as fossils rare in Asia." The originality of the finding was based on the fact that previous findings of dire wolf remains were from North America (Lu et al., 2020: 87). It is argued that "the dire wolf was one of the successful top predators in North America during the Pleistocene. It is best known from the southern part of North America, and it even immigrated to South America. Fossils of dire wolves are very rare north of 42° North latitude in North America. That distribution supported the belief that the low temperatures and ice sheets in the higher latitudes of North American formed an insurmountable barrier for the dispersal of dire wolves" (Lu et al., 2020: 87).

Conclusions

The evidence presented in this review chapter supports the view that the aurochs and later taurine and indicine cattle shared complex environments with numerous predators, with probable successful predation on at least smaller aurochs individuals, evidenced by aurochs and predator fossil parts in caves, and the ability of ancient and current predators to kill prey species approximating juvenile or even female aurochs, especially as some authorities dispute successful predation on large male aurochs. The assumption is that successful predation on aurochs implies the ability to kill the smaller taurine and indicine cattle, especially with the evidence that domestication reduces large ungulate anti-predator behavior. The role of domestication in the possible reduction of the defensive capabilities of cattle is clearly important, although both cattle and predator species appear to decline in size through the Holocene time span. Ancestral species related to the current tigers in Eurasia, Asiatic and African lions, Asiatic leopards, Eurasian brown bears, hyenas, and wolves all appear to have been larger than the current species. The changing relations between cattle and predators of declining size and the increased role of human action may provide a background for current cattle carnivores.

References

Ahrestani, F. S. (2018). Bos frontalis and Bos gaurus (Artiodactyla: Bovidae). *Mammalian Species, 50*(959), 34–50. https://doi.org/10.1093/mspecies/sey004

Ajmone-Marsan, P., Garcia, J., & Lenstra, J. A. (2010). On the origin of cattle: How aurochs became cattle and colonized the world. *Evolutionary Anthropology, 19*(4), 148–157. https://doi.org/10.1002/evan.20267

Albrecht, J., Bartoń, K. A., Selva, N., Sommer, R. S., Swenson, J. E., & Bischof, R. (2017). Humans and climate change drove the Holocene decline of the brown bear. *Scientific Reports, 7*(1), 10399. https://doi.org/10.1038/s41598-017-10772-6

Andersson, K., Norman, D., & Werdelin, L. (2011). Sabretoothed carnivores and the killing of large prey. *PLoS One, 6*(10), e24971. https://doi.org/10.1371/journal.pone.0024971

Antón, M., Salesa, M. J., Galobart, À., & Tseng, Z. J. (2014). The Plio-Pleistocene scimitar-toothed felid genus Homotherium Fabrini, 1890 (Machairodontinae, Homotherini): Diversity, palaeogeography and taxonomic implications. *Quaternary Science Reviews, 96*, 259–268. https://doi.org/10.1016/j.quascirev.2013.11.022

Babrgir, S., Farhadinia, M. S., & Moqanaki, E. M. (2017). Socio-economic consequences of cattle predation by the endangered Persian leopard Panthera pardus saxicolor in a Caucasian conflict hotspot, northern Iran. *Oryx, 51*, 124–130. https://doi.org/10.1017/S0030605315000903

Balcarcel, A. M., Veitschegger, K., Clauss, M., & Sánchez-Villagra, M. R. (2021). Intensive human contact correlates with smaller brains: Differential brain size reduction in cattle types. *Proceedings. Biological sciences, 288*(1952), 20210813. https://doi.org/10.1098/rspb.2021.0813

Barlow, A., Cahill, J. A., Hartmann, S., Theunert, C., Xenikoudakis, G., Fortes, G. G., Paijmans, J. L. A., Rabeder, G., Frischauf, C., Grandal-d'Anglade, A., García-Vázquez, A., Murtskhvaladze, M., Saarma, U., Anijalg, P., Skrbinšek, T., Bertorelle, G., Gasparian, B., Bar-Oz, G., Pinhasi, R., Slatkin, M., et al. (2018). Partial genomic survival of cave bears in living brown bears. *Nature Ecology & Evolution, 2*(10), 1563–1570. https://doi.org/10.1038/s41559-018-0654-8

Barnett, R., Westbury, M. V., Sandoval-Velasco, M., Vieira, F. G., Jeon, S., Zazula, G., Martin, M. D., Ho, S. Y. W., Mather, N., Gopalakrishnan, S., Ramos-Madrigal, J., de Manuel, M., Zepeda-Mendoza, M. L., Antunes, A., Baez, A. C., De Cahsan, B., Larson, G., O'Brien, S. J., Eizirik, E., Johnson, W. E., et al. (2020). Genomic adaptations and evolutionary history of the extinct scimitar-toothed cat, Homotherium latidens. *Current Biology: CB, 30*(24), 5018–5025. e5. https://doi.org/10.1016/j.cub.2020.09.051

Bocherens, H., Drucker, D. G., Bonjean, D., Bridault, A., Conard, N. J., Cupillard, C., Germonpré, M., Höneisen, M., Münzel, S., Napierala, H., Patou-Mathis, M., Stephan, E., Uerpmann, H., & Ziegler, R. (2011). Isotopic evidence for dietary ecology of cave lion (Panthera spelaea) in North-Western Europe: Prey choice, competition and implications for extinction. *Quaternary International, 245*, 249–261. https://doi.org/10.1016/j.quaint.2011.02.023

Bollongino, R., Burger, J., Powell, A., Mashkour, M., Vigne, J. D., & Thomas, M. G. (2012). Modern taurine cattle descended from small number of near-eastern founders. *Molecular Biology and Evolution, 29*(9), 2101–2104. https://doi.org/10.1093/molbev/mss092

Broggini, C., Cavallini, M., Vanetti, I., Abell, J., Binelli, G., & Lombardo, G. (2024). From caves to the savannah, the mitogenome history of modern lions (Panthera leo) and their ancestors. *International Journal of Molecular Sciences, 25*(10), 5193. https://doi.org/10.3390/ijms25105193

Cameron, E. Z., & Du Toit, J. T. (2005). Social influences on vigilance behaviour in giraffes, Giraffa camelopardalis. *Animal Behaviour, 69*, 1337–1344.

Chrószcz, A., Janeczek, M., & Miklikova, Z. (2011). The role of European aurochs (Bos Prigeminius Bojanus, 1827) in early cattle breeding in context of aurochs remains from celtic settlement Liptovska Mara (North Slovakia). *Veterinarija ir Zootechnika, 55*, 3–7.

Consolee, K. T., Gao, C., Vitekere, K., Li, C., Yan, H., & Jiang, G. (2020). Human-leopard conflict: An emerging issue of North China leopard conservation in tieqiaoshan provincial nature reserve in Shanxi Province, China. *Animals: An Open Access Journal from MDPI, 10*(6), 996. https://doi.org/10.3390/ani10060996

Cooper, D. M., Dugmore, A. J., Gittings, B. M., Scharf, A. K., Wilting, A., & Kitchener, A. C. (2016). Predicted Pleistocene–Holocene range shifts of the tiger (Panthera tigris). *Diversity and Distribution, 22*(11), 1199–1211.

de Carvalho, C. N., Muñiz, F., Cáceres, L. M., Belaústegui, Z., Rodríguez-Vidal, J., Belo, J., Moreira, N., Cachão, M., Cunha, P. P., Figueiredo, S., Galán, J. M., Zhang, Y., Gómez, P., Toscano, A., Ruiz, F., Ramírez-Cruzado, S., Giles-Guzmán, F., Finlayson, G., Finlayson, S., & Finlayson, C. (2022). Aurochs roamed along the SW coast of Andalusia (Spain) during Late Pleistocene. *Scientific Reports, 12*(1), 9911. https://doi.org/10.1038/s41598-022-14137-6

de Manuel, M., Barnett, R., Sandoval-Velasco, M., Yamaguchi, N., Garrett Vieira, F., Zepeda Mendoza, M. L., Liu, S., Martin, M. D., Sinding, M. S., Mak, S. S. T., Carøe, C., Liu, S., Guo, C., Zheng, J., Zazula, G., Baryshnikov, G., Eizirik, E., Koepfli, K. P., Johnson, W. E., Antunes, A., et al. (2020). The evolutionary history of extinct and living lions. *Proceedings of the National Academy of Sciences of the United States of America, 117*(20), 10927–10934. https://doi.org/10.1073/pnas.1919423117

Decker, J. E., McKay, S. D., Rolf, M. M., Kim, J., Molina Alcalá, A., Sonstegard, T. S., et al. (2014). Worldwide patterns of ancestry, divergence, and admixture in domesticated cattle. *PLoS Genetics, 10*(3), e1004254.

Diedrich, C. G. (2013). Late Pleistocene leopards across Europe–northernmost European German population, highest elevated records in the Swiss Alps, complete skeletons in the Bosnia Herzegowina Dinarids and comparison to the Ice Age cave art. *Quaternary Science Reviews, 76*, 167–193. https://doi.org/10.1016/j.quascirev.2013.05.009

Diedrich, C. G. (2014). Palaeopopulations of Late Pleistocene top predators in Europe: Ice Age spotted hyenas and steppe lions in battle and competition about prey. *Paleontology Journal*. https://doi.org/10.1155/2014/106203

Diedrich, C. G. (2016). Late Pleistocene hyena skeleton remains of a communal/prey depot cave den in the Bohemian Mountains (Czech Republic) – Its osteology, taphonomy and palaeoecology. *Acta Zoologica, 98*(1), 66–93. https://doi.org/10.1111/azo.12152

Diedrich, C. G. (2022). Eurasian Grey and White wolf ancestors—800,000 years evolution, adaptation, pathologies and European dog origins. *Acta Zoologica, 105*(1), 25–37. https://doi.org/10.1111/azo.12451

Diedrich, C. G., & Žák, K. (2006). Prey deposits and den sites of the Upper Pleistocene hyena Crocuta crocuta spelaea (Goldfuss, 1823) in horizontal and vertical caves of the Bohemian Karst (Czech Republic). *Bulletin of Geosciences, 81*(4), 237–276. (25 figures). Czech Geological Survey, Prague.

Domínguez-Rodrigo, M., Egeland, C. P., Cobo-Sánchez, L., Baquedano, E., & Hulbert, R. C., Jr. (2022). Sabertooth carcass consumption behavior and the dynamics of Pleistocene large carnivoran guilds. *Scientific Reports, 12*(1), 6045. https://doi.org/10.1038/s41598-022-09480-7

Driscoll, C. A., Yamaguchi, N., Bar-Gal, G. K., Roca, A. L., Luo, S., Macdonald, D. W., & O'Brien, S. J. (2009). Mitochondrial phylogeography illuminates the origin of the extinct Caspian tiger and its relationship to the Amur tiger. *PLoS One, 4*(1), e4125. https://doi.org/10.1371/journal.pone.0004125

Dybas, C. L. (2010). The once and future tiger. *Bioscience, 60*(11), 872–877. https://doi.org/10.1525/bio.2010.60.11.3

Edwards, C. J., Bollongino, R., Scheu, A., Chamberlain, A., Tresset, A., Vigne, J. D., Baird, J. F., Larson, G., Ho, S. Y., Heupink, T. H., Shapiro, B., Freeman, A. R., Thomas, M. G., Arbogast, R. M., Arndt, B., Bartosiewicz, L., Benecke, N., Budja, M., Chaix, L., Choyke, A. M., et al. (2007). Mitochondrial DNA analysis shows a Near Eastern Neolithic origin for domestic cattle and no indication of domestication of European aurochs. *Proceedings. Biological sciences, 274*(1616), 1377–1385. https://doi.org/10.1098/rspb.2007.0020

Elayadeth-Meethal, M., Thazhathu Veettil, A., Maloney, S. K., Hawkins, N., Misselbrook, T. H., Sejian, V., Rivero, M. J., & Lee, M. R. F. (2018). Size does matter: Parallel evolution of adaptive thermal tolerance and body size facilitates adaptation to climate change in domestic cattle. *Ecology and Evolution, 8*(21), 10608–10620. https://doi.org/10.1002/ece3.4550

Estes, R. (1992). *The behavior guide to African mammals: Including hoofed mammals, carnivores, primates*. University of California Press.

Estes, R. (2024, August 7). *Cape buffalo*. Encyclopedia Britannica. https://www.britannica.com/animal/Cape-buffalo

Flower, L., Schreve, D., & Lamb, A. (2021). Nature of the beast? Complex drivers of prey choice, competition and resilience in Pleistocene wolves (Canis lupus L., 1754). *Quaternary Science Reviews, 272*, 107212. https://doi.org/10.1016/j.quascirev.2021.107212

Fortes, G. G., Grandal-d'Anglade, A., Kolbe, B., Fernandes, D., Meleg, I. N., García-Vázquez, A., Pinto-Llona, A. C., Constantin, S., de Torres, T. J., Ortiz, J. E., Frischauf, C., Rabeder, G., Hofreiter, M., & Barlow, A. (2016). Ancient DNA reveals differences in behaviour and sociality between brown bears and extinct cave bears. *Molecular Ecology, 25*(19), 4907–4918. https://doi.org/10.1111/mec.13800

Franzmann, A. W. (1981). Alces alces. *Mammalian Species, 154*, 1–7. The American Society of Mammalogists.

Goodrich, J., Wibisono, H., Miquelle, D., Lynam, A. J., Sanderson, E., Chapman, S., Gray, T. N. E., Chanchani, P., & Harihar, A. (2022). Panthera tigris. The IUCN red list of threatened species 2022. https://doi.org/10.2305/IUCN.UK.2022-1.RLTS.T15955A214862019.en

Hayward, M. W., Henschel, P., O'Brien, J., Hofmeyr, M., Balme, G., & Kerley, G. I. H. (2006). Prey preferences of the leopard (Panthera pardus). *Journal of Zoology, 270*(2), 298–313. https://doi.org/10.1111/j.1469-7998.2006.00139.x

Hemmer, H. (2004). Notes on the ecological role of European cats (Mammalia: Felidae) of the last two million years. Paleontología. In E. Baquedano & S. Rubio Jara (Eds.), *Zona Arqueológica 4. Miscelánea en homenaje a Emiliano Aguirre* (Vol. II, pp. 214–232). Museo Arqueológico Regional.

Hemmer, H. (2011). The story of the cave lion – Panthera leo spelaea (Goldfuss, 1810) – A review. *Quaternaire, Supplement, 4*, 201–208.

Heptner, V. G., & Naumov, M. P. (1967). *Mammals of the Soviet Union 2*. Vysshaya Shkola Publishers.

Hindrikson, M., Remm, J., Pilot, M., Godinho, R., Stronen, A. V., Baltrūnaité, L., Czarnomska, S. D., Leonard, J. A., Randi, E., Nowak, C., Åkesson, M., López-Bao, J. V., Álvares, F., Llaneza, L., Echegaray, J., Vilà, C., Ozolins, J., Rungis, D., Aspi, J., Paule, L., et al. (2017). Wolf population genetics in Europe: A systematic review, meta-analysis and suggestions for conservation and management. *Biological Reviews of the Cambridge Philosophical Society, 92*(3), 1601–1629. https://doi.org/10.1111/brv.12298

Hofreiter, M., Capelli, C., Krings, M., Waits, L., Conard, N., Münzel, S., Rabeder, G., Nagel, D., Paunovic, M., Jambrěsić, G., Meyer, S., Weiss, G., & Pääbo, S. (2002). Ancient DNA analyses reveal high mitochondrial DNA sequence diversity and parallel morphological evolution of late pleistocene cave bears. *Molecular Biology and Evolution, 19*(8), 1244–1250. https://doi.org/10.1093/oxfordjournals.molbev.a004185

Jeong, D. H., Yang, J. J., & Yeon, S. C. (2021). Body mass and physical changes of reintroduced Asiatic black bears. *European Journal of Wildlife Research, 67*, 83. https://doi.org/10.1007/s10344-021-01514-1

Jiangzuo, Q., Werdelin, L., Sanisidro, O., Yang, R., Fu, J., Li, S., Wang, S., & Deng, T. (2023). Origin of adaptations to open environments and social behaviour in sabretoothed cats from the northeastern border of the Tibetan Plateau. *Proceedings. Biological sciences, 290*(1997), 20230019. https://doi.org/10.1098/rspb.2023.0019

Khaewphakdee, S., Simcharoen, A., Duangchantrasiri, S., Chimchome, V., Simcharoen, S., & Smith, J. L. D. (2020). Weights of gaur (Bos gaurus) and banteng (Bos javanicus) killed by tigers in Thailand. *Ecology and Evolution, 10*(11), 5152–5159. https://doi.org/10.1002/ece3.6268

Kilikar, D. (2024). *Does size matter? How leopards on forest fringes choose livestock kill*. Retrieved from https://india.mongabay.com/2024/07/does-size-matter-how-leopards-on-forest-fringes-choose-livestock-kill/

Kingdon, J. (1997). *The Kingdon field guide to African mammals*. Princeton University Press.

Kingdon, J. (2015). *The Kingdon field guide to African mammals* (2nd ed.). Bloomsbury Publishing.

Kitchener, A. C., & Dugmore, A. J. (2000). Biogeographical change in the tiger, Panthera tigris. *Animal Conservation, 3*(2), 113–124. https://doi.org/10.1111/j.1469-1795.2000.tb00236.x

Kluever, B. M., Breck, S. W., Howery, L. D., Krausman, P. R., & Bergman, D. L. (2008). Vigilance in cattle: The influence of predation, social interactions, and environmental factors. *Rangeland Ecology & Management, 61*(3), 321–328. https://doi.org/10.2111/07-087.1

Kruuk, H. (1972). *The spotted hyena: A study of predation and social behaviour*. University of California Press.

Kysely, R. (2008). Aurochs and potential crossbreeding with domestic cattle in Central Europe in the Eneolithic period. A metric analysis of bones from the archaeological site of Kutná Hora-Denemark (Czech Republic). *Anthropozoologica, 43*(2), 7–37.

Laundré, J. W., Hernandez, L., & Altendorf, K. B. (2001). Wolves, elk and bison: Reestablishing the "landscape of fear" in Yellowstone National Park, U.S.A. *Canadian Journal of Zoology, 79*, 1401–1409.

Linnell, J. D. C., Odden, J., Smith, M. E., Aanes, R., & Swenson, J. E. (1999). Large carnivores that kill livestock: Do "problem individuals" really exist? *Wildlife Society Bulletin, 27*, 698–705.

Lister, A. M., & Stuart, A. J. (2019). The extinction of the giant deer Megaloceros giganteus (Blumenbach): New radiocarbon evidence. *Quaternary International, 500*, 185–203. https://doi.org/10.1016/j.quaint.2019.03.025

Loog, L., Thalmann, O., Sinding, M. S., Schuenemann, V. J., Perri, A., Germonpré, M., Bocherens, H., Witt, K. E., Samaniego Castruita, J. A., Velasco, M. S., Lundstrøm, I. K. C., Wales, N., Sonet, G., Frantz, L., Schroeder, H., Budd, J., Jimenez, E. L., Fedorov, S., Gasparyan, B., Kandel, A. W., et al. (2020). Ancient DNA suggests modern wolves trace their origin to a Late Pleistocene expansion from Beringia. *Molecular Ecology, 29*(9), 1596–1610. https://doi.org/10.1111/mec.15329

Lu, D., Yang, Y., Li, Q., & Ni, X. (2020). A late Pleistocene fossil from Northeastern China is the first record of the dire wolf (Carnivora: Canis dirus) in Eurasia. *Quaternary International*. https://doi.org/10.1016/j.quaint.2020.09.054

Manning, K., Timpson, A., Shennan, S., & Crema, E. (2015). Size reduction in early European domestic cattle relates to intensification of neolithic herding strategies. *PLoS One, 10*(12), e0141873. https://doi.org/10.1371/journal.pone.0141873

Marciszak, A., & Gornig, W. (2024). From giant to dwarf: A trend of decreasing size in Panthera spelaea (Goldfuss, 1810) and its likely implications. *Earth History and Biodiversity*. https://doi.org/10.1016/j.hisbio.2024.100007

Marra, A. C., Villa, P., Beauval, C., Bonfiglio, L., & Goldberg, P. (2004). Same predator, variable prey: Taphonomy of two Upper Pleistocene hyena dens in Sicily and SW France. *Revue de Paléobiologie, Genève, 23*(2), 787–801.

McTavish, E. J., Decker, J. E., Schnabel, R. D., Taylor, J. F., & Hillis, D. M. (2013). New World cattle show ancestry from multiple independent domestication events. *Proceedings of the National Academy of Sciences of the United States of America, 110*(15), E1398–E1406. https://doi.org/10.1073/pnas.1303367110

Miller, C. S., Hebblewhite, M., Petrunenko, Y. K., Seryodkin, I. V., DeCesare, N. J., Goodrich, J. M., & Miquelle, D. (2013). Estimating Amur tiger (Panthera tigris altaica) kill rates and potential consumption rates using global positioning system collars. *Journal of Mammalogy, 94*(4), 845–855. https://doi.org/10.1644/12-mamm-a-209.1

Miquelle, D., Quigley, H., Smirnov, E. N., Merrill, T., Myslenko, A. E., & Hornocker, M. G. (1999). Hierarchical spatial analysis of Amur tiger relationships to habitat and prey. In J. Seidensticker, P. Jackson, & S. Christie (Eds.), *Riding the tiger* (pp. 71–99). Cambridge University Press.

Murtskhvaladze, M., Gavashelishvili, A., & Tarkhnishvili, D. (2010). Geographic and genetic boundaries of brown bear (Ursus arctos) population in the Caucasus. *Molecular Ecology, 19*(9), 1829–1841. https://doi.org/10.1111/j.1365-294x.2010.04610.x

Mulville, J. (2024). *Aurochs – 'A little below the elephant in size'*. Retrieved from https://www.cardiffsciscreen.co.uk/article/aurochs-little-below-elephant-size

Naha, D., Dash, S. K., Chettri, A., Chaudhary, P., Sonker, G., Heurich, M., Rawat, G. S., & Sathyakumar, S. (2020). Landscape predictors of human–leopard conflicts within multi-use areas of the Himalayan region. *Scientific Reports, 10*(1). https://doi.org/10.1038/s41598-020-67980-w

Németh, A., Bárány, A., Csorba, G., Magyari, E., Pazonyi, P., & Pálfy, J. (2017). Holocene mammal extinctions in the Carpathian Basin: A review. *Mammal Review, 47*(1), 38–52. https://doi.org/10.1111/mam.12075

Niedziałkowska, M., Hayward, M. W., Borowik, T., Jędrzejewski, W., & Jędrzejewska, B. (2019). A meta-analysis of ungulate predation and prey selection by the brown bear Ursus arctos in Eurasia. *Mammal Research, 64*, 1–9. https://doi.org/10.1007/s13364-018-0396-7

Paijmans, J. L. A., Barlow, A., Förster, D. W., Henneberger, K., Meyer, M., Nickel, B., Nagel, D., Worsøe Havmøller, R., Baryshnikov, G. F., Joger, U., Rosendahl, W., & Hofreiter, M. (2018). Historical biogeography of the leopard (Panthera pardus) and its extinct Eurasian populations. *BMC Evolutionary Biology, 18*(1), 156. https://doi.org/10.1186/s12862-018-1268-0

Palmeira, F. B. L., Crawshaw, P. G., Jr., Haddad, C. M., Ferraz, K. M. P. M. B., & Verdade, L. M. (2008). Cattle depredation by puma (Puma concolor) and jaguar (Panthera onca) in Central-Western Brazil. *Biological Conservation, 141*(1), 118–125. https://doi.org/10.1016/j.biocon.2007.09.015

Rafferty, J. P. (2023). *Cave bear*. Encyclopedia Britannica. https://www.britannica.com/animal/cave-bear

Reumer, J. W., Rook, L., van der Borg, K., Post, K., Mol, D., & de Vos, J. (2003). Late Pleistocene survival of the saber-toothed cat Homotherium in Northwestern Europe. *Journal of Vertebrate Paleontology, 23*(1), 260–262. https://doi.org/10.1671/0272-4634(2003)23[260:LPSOTS]2.0.CO;2

Rivals, F., Baryshnikov, G. F., Prilepskaya, N. E., & Belyaev, R. I. (2022). Diet and ecological niches of the Late Pleistocene hyenas Crocuta spelaea and C. ultima ussurica based on a study of tooth microwear. *Palaeogeography, Palaeoclimatology, Palaeoecology, 601*, 111125. https://doi.org/10.1016/j.palaeo.2022.111125

Saarma, U., Ho, S. Y., Pybus, O. G., Kaljuste, M., Tumanov, I. L., Kojola, I., Vorobiev, A. A., Markov, N. I., Saveljev, A. P., Valdmann, H., Lyapunova, E. A., Abramov, A. V., Männil, P., Korsten, M., Vulla, E., Pazetnov, S. V., Pazetnov, V. S., Putchkovskiy, S. V., & Rõkov, A. M. (2007). Mitogenetic structure of brown bears (Ursus arctos L.) in northeastern Europe and a new time frame for the formation of European brown bear lineages. *Molecular Ecology, 16*(2), 401–413. https://doi.org/10.1111/j.1365-294X.2006.03130.x

Schaller, G. B. (1976). *The Serengeti lion: A study of predator-prey relations*. University of Chicago Press.

Semenov, U. A. (2014). The wisents of Karachay-Cherkessia. *Proceedings of the Sochi National Park*, (8), 23–24.

Sinclair, A. R., Mduma, S., & Brashares, J. S. (2003). Patterns of predation in a diverse predator-prey system. *Nature, 425*(6955), 288–290. https://doi.org/10.1038/nature01934

Skinner, J. D., & Smithers, R. H. M. (1990). *The mammals of the southern African subregion*. University of Pretoria.

Stanton, D. W. G., Alberti, F., Plotnikov, V., Androsov, S., Grigoriev, S., Fedorov, S., Kosintsev, P., Nagel, D., Vartanyan, S., Barnes, I., Barnett, R., Ersmark, E., Döppes, D., Germonpré, M., Hofreiter, M., Rosendahl, W., Skoglund, P., & Dalén, L. (2020). Early Pleistocene origin and extensive intra-species diversity of the extinct cave lion. *Scientific Reports, 10*(1), 12621. https://doi.org/10.1038/s41598-020-69474-1

Stiner, M. C. (2004). Comparative ecology and taphonomy of spotted hyenas, humans, and wolves in Pleistocene Italy. *Revue de Paléobiologie, Genève, 23*(2), 771–785.

Stokstad, E. (2015). Bringing back the aurochs. *Science (New York, N.Y.), 350*(6265), 1144–1147. https://doi.org/10.1126/science.350.6265.1144

Underwood, R. (1982). Vigilance behavior in grazing African antelopes. *Behaviour, 79*, 81–107.

Upadhyay, M. R., Chen, W., Lenstra, J. A., Goderie, C. R., MacHugh, D. E., Park, S. D., Magee, D. A., Matassino, D., Ciani, F., Megens, H. J., van Arendonk, J. A., Groenen, M. A., Crooijmans, R. P. M. A., & European Cattle Genetic Diversity Consortium. (2017). Genetic origin, admixture and population history of aurochs (Bos primigenius) and primitive European cattle. *Heredity, 118*(2), 169–176. https://doi.org/10.1038/hdy.2016.79

Vuure, C. V. (2005). *Retracing the aurochs: History, morphology and ecology of an extinct wild ox*. Pensoft Publishers.

Wright, E., & Viner-Daniels, S. (2015). Geographical variation in the size and shape of the European aurochs (Bos primigenius). *Journal of Archaeological Science, 54*, 8–22. https://doi.org/10.1016/j.jas.2014.11.021

Modern Cattle and Their Predators

This part consists of five chapters that examine the ecology of domestic cattle and associated large carnivores in Asia, Europe, Africa, North America, and South America. The premise of these chapters is that cattle are the most important mammals domesticated for human consumption on all five continents, but similar but unique problems threaten the effective management of cattle populations. In the more advanced economies of Western Europe and North America, advanced management practices are being used, offset by increasingly sophisticated, successful, and publicly supported conservation measures and management plans, which may be at odds with livestock farming on favorable land. These measures have contributed to an increase in the populations of large carnivores, namely, brown bears and gray wolves on both continents, and additionally black bears, cougars, and jaguars in North America (including Mexico and Central America). Nevertheless, the progressive policies of Western Europe and Northern North America, supported by significant financial and technological resources, can make these regions role models for both agricultural development and large carnivore conservation and assessment technologies such as geomatics-based methods.

In tropical regions, common problems include the larger numbers of large carnivore species and poor, small-scale livestock farmers, and less effective conservation measures. Asia has the largest number of predator species capable of killing domestic cattle. These include highly charismatic tigers, and also leopards, snow leopards, brown and black bears, wolves and dholes, as well as several breeds of semi-domesticated cattle, the gayal and the domesticated water buffalo and yak, indicine and taurine breeds. In Africa, the charismatic lion and leopard, hyenas, wild dogs, and cheetahs face similar problems, including desertification, deforestation, and rapid population growth. Cattle, however, are indicine and taurine breeds, with the only species of wild cattle the African buffalo rarely introgressing. Predation of South American cattle is dominated by only two big cats, the jaguar, and the puma and to a lesser extent the spectacled bear. On this continent, the interface between large carnivore conservation and agricultural development is dominated by deforestation in the Amazon. The introduced taurine and indicine cattle and native water buffalo may be eaten primarily by jaguars and cougars..

The distinctive nature of conflicts between large carnivores and domestic cattle has encouraged the global adoption of more advanced research methods, including those based on geomatics, and comparative studies between the problems of the northern continents (where environmental changes may be less pronounced) and the problems of the Global South, where environmental changes and problems of social poverty may be prevalent. These five chapters explore these themes to build on the background of the first four chapters, which illustrated the beginnings of the bovine-carnivore-human complex. Geomatics-based research methods and applications are also increasingly being applied to these topics, contributing to greater involvement of people in this relationship by enabling the connection of larger groups of people who may be stakeholders in conservation and agricultural development.

Predators of Cattle in Asia

Introduction

Asia, the world's largest continent by area (44,579,000 km^2) and population (4.8 billion in 2024), also has the largest population of domestic cattle, with India alone having 193 million of Asia's 440 million and the world's 943 million cattle (National Geographic, 2006; UN Revision of World Population Prospects, 2022; Ganbold, 2024; Shahbandeh, 2024) (Table 1). Asia also has a larger number of predatory mammals capable of killing adult cattle (eight, namely the tiger, leopard, snow leopard, and perhaps the clouded leopard, and brown and black bears, wolves and dholes) than any other continent (for Africa the lion, leopard, and hyena, for Europe the brown bear and gray wolf, for North America the brown bear, black bear, gray wolf and cougar, for South America, the jaguar and cougar) (Campbell, 2017; Jha & Campbell, 2022). Asian cattle include the indigenous, domesticated zebu, the introduced domesticated taurine cattle, and the local, semi-domesticated water buffalo, gayal, and yak (Table 1). With the exception of the yak, which ranges in cold mountains and lowlands, the other species are across temperate and tropical modified landscapes.

In Asia, as elsewhere large carnivore conflicts emerge with human communities through livestock predation. Banerjee et al. (2013) note that only "rarely human communities coexist in harmony with large predators. Most often communities suffer due to predation on their stock while large carnivores suffer losses and at times extirpation due to retaliation." Major conflicts emerge from human and livestock presence in protected areas, in the margins of protected areas, and in some cases in unprotected areas with effective carnivore protection management, which may increase the possibilities of wildlife conflict with livestock managers (Campbell, 2017; Jha & Campbell, 2022). Farhadinia et al. (2022) note that Asia has 13.2% of terrestrial protected area (PA) coverage, and that "a higher agricultural land in 2015 was associated with lower PA coverage today." Pant et al. (2023), referring to protected areas in Asia, in this case Nepal, noted that "human-wildlife interactions occur where human and wildlife coexist and share common resources including

© The Author(s), under exclusive license to Springer Nature Switzerland AG 2025

M. O. Campbell, *Cattle, Their Predators and Geomatics Research*,
https://doi.org/10.1007/978-3-031-97363-5_5

Table 1 Common Asian cattle

Name	Scientific name	Range	Weight (kg)
Zebu (indicine cattle)	*Bos indicus* Linnaeus, 1758	Southern Asia, Tropics	<1600
Gayal	*Bos frontalis* Lambert, 1804	China, Northeast India, Bangladesh, Myanmar, Bhutan	400–700, <1000
Domestic water buffalo	Bubalus bubalis (Linnaeus, 1758)	South Asia, East Asia, tropical	300–550
Domestic yak	*Bos grunniens*, Linnaeus, 1766	North, Central, South Asia	225–585
Taurine cattle	*Bos taurus* Linnaeus, 1758	Global, especially Americas, Eurasia	<1500

Sources: Sanders (1980), Buchholtz (1990), Giasuddin et al. (2003), Bigi and Zanon (2008), Estes (2009), and Wang et al. (2017)

food or shelter. Increasing wildlife populations within protected areas also can increase interactions with humans living adjacent to these areas, resulting in conflicts including human casualties, livestock depredation, crop damage, and property loss." In addition, "understanding the effect of anthropogenic disturbance, and its interaction with carnivores and their prey, is crucial to support the conservation of threatened carnivores, particularly in rapidly changing landscapes" (Widodo et al., 2022).

This issue is especially prevalent in regions with high populations of humans and cattle, such as South Asia, where there is a combination of the largest populations of cattle, the world's most populous nations, India and China, and effective conservation of tiger and other predator populations (National Tiger Conservation Authority (NTCA), 2022; World Population Prospects, 2024). These cattle are of various types, taurine, indicine, and yak hybrids (see Figs. 1, 2, 3, 4, 5, 6, 7, and 8). In China and India, there are several effectively managed types of protected areas, including national parks, wildlife sanctuaries, biosphere reserves, reserved and protected forests, conservation reserves, community reserves, and private protected areas.

As of January 2023, the protected areas of India covered 173,629.52 sq.km, about 5.3% of the area of the country, and there were tiger reserves that were part of national parks and wildlife sanctuaries (National Tiger Conservation Authority, 2023; Wildlife Institute of India, 2023). In China, the world's second most populous country, the protected areas (mostly nature reserves) covered about 15% of the country and about 18% of the habitats for threatened mammals (Xu et al., 2017). Chen et al. (2022) also noted that from 2000 to 2020, the pressure of human activities in 75% of China's protected areas increased and accounted for 38% of the total area of the reserves. China's protected areas slightly relieved the pressure of human activities by a slight percentage, with regional differences in the effectiveness of the conservation policy.

In smaller nations such as Bhutan, "with rural people and their subsistence economy integrated into the conservation landscape, livestock predation is an inherent issue especially in northern regions where several predation hotspots are recognised. Tigers, snow leopards, leopards and dhole prey on livestock, largely due to

Fig. 1 Taurine bull, Hemu Village, Burqin County, Altay Prefecture, Xinjiang, China. (Permission Yang Song-unsplash)

untended free grazing in forests, inadequate penning at night, and lowered natural prey densities from competition with livestock" (Rajaratnam et al., 2016). In Nepal, even in protected areas, serious conflicts occur by predators such as snow leopards; in some areas, "local people held strongly negative attitudes towards snow leopards and most suggested that total extermination of leopards was the only acceptable solution to the predation problem" (Oli et al., 1994). Figel et al. (2023: 83) note that in Sumatra, "conflict with humans is a severe threat to critically endangered Sumatran tigers" and "livestock depredation (49%) and people reporting fear or anxiety upon tiger sightings (39%) comprised most incidents." Forest fragmentation in Sumatra is another factor for tiger decline (Luskin et al., 2017).

Wild cattle domestication is also an important aspect of the relations between humans, cattle, and large carnivores. For example, the domestication of the gayal and the yak has emerged as major development in South and Central Asia. In Central Asia, "with a current total population size of 14 million, the domestic yak constitutes one of the most important livestock genetic resources and plays an indispensable role in the life of pastoralists and agro-pastoralists in the region" (Qi et al., 2010: 242). The hybridization of cattle and yak may have begun about 3000 years

Fig. 2 Taurine cattle, Zuoyun County, Datong, Shanxi, China. (Permission Cajeo Zhang-unsplash)

BP (Zhang, 2000), but "whether taurine *B. taurus* or indicine *B. indicus* cattle were involved in the hybridization largely depends on the geographical area, e.g. taurine cattle were used in the Qinghai-Tibetan Plateau and Mongolian Plateau and indicine cattle used in the Himalayan areas and elsewhere" (Qi et al., 2010: 243). Chen et al. (2023) note that "scholars have also argued that hybridization between locally adapted Tibetan yak and introduced cattle facilitated a rapid adaptation of Tibetan cattle to hypoxia. This husbandry practice was used to extend the ranges of cattle and yak. Hybrids produce more milk and meat and are stronger than cattle at high altitudes. They are better adapted to low elevations than yak and expand possibilities for milk and traction at the mid-elevations" (2500–3500 m above sea level). In such areas, cattle predation is mostly attributed to the snow leopard and wolf (Maheshwari & Sathyakumar, 2020), while in the forests, grasslands, and agricultural areas at lower elevations in Asia, predation is attributed to tigers, leopards, brown and black bears, wolves, and dholes (Ramesh et al., 2020).

This chapter examines the relation between domestic cattle, their managers, and large predators in Asia. The chapter considers each of the predators that kill domestic cattle. More species of large predators may prey on domestic cattle than on wild

Fig. 3 Indicine cow, Hingoli, Maharashtra, India. (Permission Amol Sonar-unsplash)

cattle, as the domestic cattle such as the Zebu may be smaller and less aggressive than the wild cattle such as the guar and banteng. Smaller predators such as the clouded leopard and Eurasian lynx may kill young domestic cattle or calves. The current evidence suggests that conservation policies across Asia from Siberia to southwest and southeast Asia are effective, but less so than those of Europe, due to the larger land area and rural subsistence landusers, the greater number of potential predatory mammals, and the presence of large crocodiles in a few parts of the cattle range. There are also two species of canids capable of killing domestic cattle, gray wolves, and dholes. Financial and political issues also limit the implementation of conservation policies in Asia, compared with Europe, despite marked recent improvements in the former continent.

Tigers and Cattle Predation

The tiger is the main cattle killing mammal in Asia. The main areas of study are parts of India, where most tigers live, followed by nations with much smaller populations of tigers, such as Russia, Nepal, and Southeast Asia. As noted by Bargali and

Fig. 4 Taurine cow, West Bengal, India. (Permission Arpan Mondal-unsplash)

Ahmed (2018), India with an estimated tiger population of over 2000 tigers has over half of the current global wild tiger population, some living in 55 tiger reserves (see also National Tiger Conservation Authority, 2023; Wildlife Institute of India, 2023). The tiger populations in Asia are shown in Table 2 based on figures from Global Tiger Recovery Program 2.0 (2023–34). The tiger has been extensively studied in terms of its adaptation to human agricultural landcover, its choice between wild and domesticated prey, its requirements within a scenario of intensive human landuse, and the success of conservation policies in contributing to local increases in tiger populations (Campbell, 2017). For tiger attacks on people and livestock, mostly domestic cattle, Nyhus and Tilson (2010, 125) point out that the common hypotheses suggest that factors for tiger attacks on people and livestock include: "depletion of prey or inability to hunt prey effectively, hunger, old age, illness or injury, learned behavior from individual experience or from parents, older former residents that are no longer able defend a home range, and lack of aversion due to limited or ineffective retaliation."

Tigers are entirely capable of killing domesticated indicine and taurine cattle; this assessment is based on the tiger's ability to kill larger wild cattle, such as the gaur that may weigh several times more than the tiger; hence, they are the natural

Fig. 5 Domestic Yak-hybrid, Kathmandu, Nepal. (Permission Ashok Acharya-unsplash)

predators of the smaller, weaker domestic cattle. For example, Ahrestani (2018) notes that tigers take the largest wild prey, noting that for the gaur, "calves are preyed on by leopards, dhole, and tigers, although adults are preyed on primarily by tigers. Leopards and tigers typically kill calves and females by asphyxiation using bites to their throats; 88% of 33 B. gaurus that were recorded killed by tigers were found with bites to their throats in Nagarahole, India (see Karanth & Sunquist, 1995). Males, however, have a thick neck and often a dewlap, which makes killing them by asphyxiation difficult. Tigers have been known to overcome this by biting the hock, thereby hamstringing and bringing down male B. gaurus." In retaliation, "males are capable of killing tigers: corpses of a dead tigress and a male were both found after what was evidently a fight to death (Blackburn, 1934), and a radio-collared tiger that was found dead was apparently killed by a B. gaurus" (Ahrestani, 2018). Khaewphakdee et al. (2020: 5152), examining the weights of gaur and banteng killed by tigers, noted that "of all gaur killed, 45.2% were adults; of all banteng killed, 55.7% were adults. The average weight of banteng killed was 423.9 kg, which was similar to the 397.9 kg average weight for gaur." These weights would equate to large indicine or taurine cattle and indicate the possible predation by tigers on the latter breeds.

Fig. 6 Indicine cow, Bhutan. (Permission Ugyen Tenzin-unsplash)

Concerning the comparison of wild and livestock prey for tigers, some studies indicate that tigers kill livestock, especially larger animals such as cattle in areas where they are abundant. For example, Kolipaka et al. (2017) cite a study titled "wild versus domestic prey in the diet of reintroduced tigers in the livestock-dominated multiple-use forests of Panna Tiger Reserve, India" and note the possibility that "the presence of feral cattle, along with open-access grazing practices in multiple-use forests, would increase the incidence of predation on livestock by tigers, even when wild prey are available." The findings indicated that "sub-adult tigers and male tigers killed more livestock than wild prey, even when wild prey was available" and "with increasing distance from the core-zone boundary, all tigers killed more livestock, possibly because livestock were more easily accessible than wild prey." The authors state that these findings indicate "that open-access and unregulated livestock grazing is not currently compatible with large carnivore conservation in the same landscape" and suggested solutions including the corralling of valuable cattle, with less attention to feral and less valuable livestock for tigers (Kolipaka et al., 2017).

Kolipaka et al. (2017) used an analysis of 627 kills of 10 tigers from 2009 and 2014 and found that wild animals comprised 54% of the tigers' prey, while livestock

Fig. 7 Taurine bull in Phobjikha Valley, Bhutan. (Permission Bradford Zak-unsplash)

represented 43% of the prey specimens. The commonest wild animal (70%) killed was sambar, while domestic cows represented 87% of the domestic animals killed (Kolipaka et al., 2017). In addition, "male tigers killed a higher percentage of domestic animals in both the zones (66% and 79% in the core and buffer zone, respectively) than female tigers (29% and 39% in the core and buffer zone, respectively) and "a similar proportion of male and female domestic prey was killed in the core and buffer zones" (Kolipaka et al., 2017). Tigers mostly killed larger domestic and wild prey (cows and sambar), livestock were taken with increasing distance from the core protected area, male tigers killed livestock more than female tigers in core areas, although "the proportion of domestic prey killed by males and females in core and buffer zones was comparable," and that "open-access livestock grazing is not currently compatible with large carnivore conservation in the same landscape" (Kolipaka et al., 2017). These results align with other studies that found that tigers prefer large prey (Karanth & Sunquist, 1995; Chundawat et al., 1999; Sankar & Johnsingh, 2002; Bagchi et al., 2003; Kerley et al., 2015) and similar livestock

Fig. 8 Domestic Yak-hybrids in Shangri-La, Diqing Tibetan Autonomous Prefecture, Yunnan, China. (Permission Rogan Yeoh-unsplash)

predation patterns were also found for lions in similar environments in Western India (Banerjee et al., 2013).

Other studies included differences between tigers and leopards in cattle predation. For example, Bargali and Ahmed (2018) in a case study of the Corbett Tiger Reserve (CTR) noted that tigers move beyond the boundaries of the protected areas, into the surrounding multiple-use forest, which includes agricultural, urban, and other development landcover. The study "examined a total of 8365 incidents of livestock depredation between 2006 and 2015 with tigers killing more livestock in a year (573.3 ± 41.2) than leopards (263.2 ± 9.9)" (Bargali & Ahmed, 2018). The findings indicated that domestic cattle were the main tiger prey (75%), followed by buffaloes (24.5%), and smaller species (0.5%). Leopards took mostly cattle (90.5%). These studies aligned with other research findings on tiger and leopard predation on livestock (see Bagchi et al., 2003; Muhammed et al., 2007; Nyhus & Tilson, 2010; Ahmed et al., 2012). Tigers targeted larger prey than leopards as found in some

Table 2 Tiger populations

Country	Number of tigers
India	1411
Malaysia	500
Bangladesh	440
Russia	360
Indonesia	325
Thailand	200
Nepal	155
Myanmar	85
Bhutan	75
China	45
Lao PDR	17
Vietnam	10–30
Cambodia	<10<
Total	~3650

Note: Bargali and Ahmed (2018) put India's tiger population at 2226 (1945–2491), which is about 60% of wild tigers
Source: Global Tiger Recovery Program 2.0 (2023–34)

other studies (Biswas & Sankar, 2002; Bagchi et al., 2003; Hayward et al., 2012; Lamichhane & Jha, 2015). For example, Kumaraguru et al. (2010: 627) found that sambar, a large deer, comprised 35% of tiger diets, but only 17% of leopard diets in Anamalai Tiger Reserve, India. In this study area of Bargali and Ahmed (2018) livestock losses were economically balanced with local communities profits from wildlife tourism.

Another study along this line of tiger-leopard comparisons is that of Bhattarai and Fischer (2014: 522), which examined tiger-human conflicts in the buffer zone of Bardia National Park, Nepal, by interviewing 273 local householders and 27 key persons, where local contacts argued that about 2% of livestock was lost annually to tigers and "livestock predation rates were particularly high in areas with low abundance of natural prey." The authors note that "opportunistic killing of livestock by tigers is common as domestic animals are easy prey" and the data indicate that the amount of livestock predation by tigers is largely determined by the availability of natural prey; predation rates were higher in areas with low prey densities, but there is a possibility that leopards were partially responsible. Other theories, on the other hand, include variations in tiger densities and a greater presence of livestock in regions with low densities of prey (Bhattarai & Fischer, 2014: 525).

Leopards and Cattle Predation

Studies of leopard predation on cattle in Asia either focus on leopards or may compare such predation between leopards and tigers. Comparisons may also be made between taurine and/or indicine cattle, and domestic and/or wild water buffalo. Tigers may be described as taking larger cattle, such as adult males, including water buffalo and gayal, while leopards may be identified as preying on smaller prey such as females and juveniles, especially of taurine or indicine cattle (Harihar et al., 2011; Melletti & Burton, 2014). For example, Miller et al. (2016: 29) cite the case of cattle predation near the Kanha Tiger Reserve, Central India, which "has one of the highest rates of livestock depredation from tigers and leopards in India yet is also renowned as one of the most successful and stable sites of tiger conservation." In addition, the study "confirmed that tigers and leopards were the primary depredating carnivores, with tigers responsible for killing slightly more livestock than leopards," this difference contrasting "with reports from Corbett Tiger Reserve in northern India and Bhutan, where leopards kill substantially more livestock than tigers" (Miller et al., 2016: 30; citing Wang & Macdonald, 2006; Sangay & Vernes, 2008; Malviya & Ramesh 2015). Miller et al. (2016: 30) note that the difference "may be related to the lower availability of free grazing adult cattle in Corbett (where ~30% of households stall-feed livestock" based on a study by Malviya and Ramesh 2015) "and the lower density of tigers in Bhutan" (see Sanderson et al., 2006). "There was no evidence of other carnivores predation… and cattle were killed most frequently, followed by goats, buffalo, and pigs, respectively" (Miller et al., 2016: 30).

The study by Wang and Macdonald (2006) is based on villages in Jigme Singye Wangchuck National Park, central Bhutan, where tigers and leopards accounted for 82% of livestock losses in 2000. Sangay and Vernes (2008) describe a case study of predation activity over Bhutan by tigers, leopards, snow leopards, and Himalayan black bears on livestock during 2003–2005. Of 1375 livestock kills, leopards killed the largest number (70%) compared with tigers (19%), bears (8%), and snow leopards (2%). Cattle comprised about 50% of predated livestock and horses were 33%. For cattle, leopards killed more smaller animals, especially calves, and tigers killed more larger animals, especially adult males (Sangay & Vernes, 2008). The authors note that all the carnivores included in the study may take a range from small to large mammals, but larger prey are generally favored, with the upper size limit of prey condition partially by the size of the predator. Therefore, a large tiger (100–260 kg) may kill considerably larger (50–100 kg) mammals than a leopard, which weighs 30–90 kg and kills prey in the range of 25–50 kg (Sangay & Vernes, 2008: 1279).

Sangay and Vernes (2008: 1279) compare the results of their study on Bhutan with studies of India, which show that leopards in India mostly take prey weighing under 50 kg and tigers take prey about 82 kg in weight (see Johnsingh, 1992; Biswas & Sankar, 2002). According to Sangay and Vernes (2008: 1281), the data indicate that leopards are Bhutan's most common livestock predator, which is likely due to their greater numbers than other predators that are included in the study. Almost all of the calf and foal kills, most of the sheep and horse kills, and more than half of the cow kills in the study were likely caused by leopards, which presumably prefer to hunt the smaller or easier to tackle livestock. With leopards as the primary predator and cows, horses, sheep, and young livestock as the primary targets, policies intended to lessen livestock predation in Bhutan should concentrate on these animals. Arguably, it is possible that tigers would change to other livestock types if there were no larger livestock, but the data indicated that precautions should be taken to protect the larger and more valuable livestock, such as bull cattle, in areas where tigers are common. In a study on Central India, regarding the vulnerability of livestock, the authors also argue that "tigers and leopards were responsible for nearly all livestock losses and most often killed in the afternoon and early evening" and "cattle and buffalo were most at risk in dense forests away from villages and roads, whereas goats were most often killed in open vegetation near villages. A spatial predation risk model for cattle revealed high-risk hotspots around the core zone boundary, confirming the significant risks to livestock grazing illegally in the core" (Miller et al., 2016: 29).

Another form of comparative study between leopard and tiger cattle predation covers the reaction of leopards to tiger reintroductions (Harihar et al., 2011). This study covers the impact of a recovering tiger population on leopards in the Chilla Range of Rajaji National Park (RNP), India, justified on the grounds that many conservation efforts have been developed to support the declining populations of tigers. However, "possible cascading effects of intra-guild competition on other sympatric carnivores are rarely considered while planning such recovery programmes" (Harihar et al., 2011: 806). The hypothesis was that "there would be high dietary overlap between the two predators. Over time, optimal habitats would be dominated by tigers forcing leopards to the periphery of the protected area where they would subsist on small prey and domestic livestock. Consequently, leopards would occur at a lower density where sympatric with tigers" (Harihar et al., 2011: 806). The findings indicated an overlap in the diets of leopards and tigers, a shift in the diet of leopards towards domestic prey and small prey, a decline of leopard population density with increased tiger population densities, and therefore "the tiger population recovered but leopard densities declined sharply" (Harihar et al., 2011: 806).

Sijapati et al. (2021) take the case of rural Nepal where there are significant livestock depredations from large carnivores in and near Bardia National Park (BNP). Here, "drivers of carnivore depredations of livestock were influenced by carnivore species, animal husbandry practices, season, and deterrent technique" and "leopards killed more livestock than tigers, and the likelihood of livestock depredations was not affected by the number of livestock owned or preventative measures used to reduce depredations." The findings indicated that "1476 hoofed livestock were reportedly depredated by leopards, and 209 by tigers, during 2015–2019" and "the number of hoofed livestock killed by leopards each season was, at least, 86% higher than the number killed by tigers." The authors comment on the fact that "leopard depredations of livestock were substantially greater than depredations caused by tigers, irrespective of the season or proximity to BNP" in that they "suspect this marked difference in the frequency of depredations is a consequence of leopards being more common than tigers. The fact that more depredations from large carnivores overall occurred near BNP was likely due to the increased number of households moving nearer to BNP since the 1960s" and to the fact that "large carnivore populations including leopards and tigers have increased in Nepal, especially around PAs" (Sijapati et al., 2021).

Other studies focus principally on leopard attacks on cattle, with less focus on comparison with tigers. For example, Naha et al. (2020) examined attacks on livestock in eastern and western Himalaya between 2015 and 2018 and found that "the risk of a leopard killing livestock increased within a heterogeneous landscape matrix comprising of both closed and open habitats (very dense forests, moderate dense forests, open forests, scrubland and non-forests)" were "most of the attacks occurred when livestock were grazing freely within multi-use areas without supervision of a herder" and the "results suggest that awareness about high risk areas, supervised grazing, and removing vegetation cover around human settlements should be initiated to reduce predation by leopards" (Naha et al., 2020). Lamichhane et al. (2023) conducted a similar study, in the Annapurna Conservation Area, Nepal, between 2017 and 2020. The findings of this study indicated that the factors for leopard predation on livestock were "the area of forest, agricultural land, length of rivers, slope, proximity to settlements and protected areas, and elevation" which "significantly predicted the probability of leopard attacks on livestock" (Lamichhane et al., 2023). Other findings were an increase in the incidence of leopard predation on livestock with increased elevation and decreasing slopes, human settlements and protected areas proximity, and with variation based on livestock species, season, and time: "goats were the most highly predated livestock, followed by sheep, cow/ox, and buffalo" (Lamichhane et al., 2023).

Snow Leopards and Cattle Predation

Snow leopard predation on cattle (including taurine, indicine, and domesticated yak breeds) occurs through the snow leopard's range in Central Asia in India, China, Nepal, Russia, and Pakistan, with the main rivals being the wolf, and in the lower elevations, leopards, tigers, dhole, and black bear. Snow leopards in higher altitudes of the Himalayas are considered to be able to kill all livestock except adult male yak (Jackson et al., 1996). Dorji and Powrel (2022) cite a case of western Bhutan, where yak herders stated that a "total of 398 yaks were lost to snow leopards (78.86%) followed by dhole (18.3%), Himalayan black bear (Ursus thibetanus) (2.7%), and common leopard (Panthera pardus) (0.2%) in the past 5 years (2015–2019)" with "the majority (87.22.8%) of the kills by snow leopards" being "young yak and most (60.5%) kills were recorded during summer." Importantly, "snow leopards are considered harmful (73.2%), and herders (71.1%) are not in favour of snow leopard conservation" (Dorji & Powrel, 2022). The authors note that in their study area, there is no record of the presence of other predators such as wolves or tigers, and hence, "it was imperative to determine the status of livestock depredation by snow leopards in such conditions to implement site-specific mitigation measures" (Dorji & Powrel, 2022). The findings indicated that over 5 years prior to the study, snow leopards killed 313 yaks belonging to 56 yak herders. The yaks taken included young animals (87.22%), heifers (9.58%), and adults (3.19%). The killings occurred mostly during summer (60.5%) or spring (24.1%) and at night (76.4%). Half of the killings (49.8%) were identified by pugmark evidence and 28.9% were witnessed by herder (Dorji & Powrel, 2022).

Other studies have found snow leopard predation on cattle, usually less than on smaller livestock. For example, Anwar et al. (2011: 1077) found that prey mass consumed (70%) was from domestic livestock, including sheep (23%), goat (16%), cattle (10%), yak (7%), and cattle–yak hybrids (14%). The remaining 30% of the biomass consumed was from wild species, including the Siberian ibex (21%) and markhor (7%), and the attacks on domestic livestock influenced conflicts between the local people and predators. Augugliaro et al. (2020) found that snow leopard predation in Bayan Ölgii province, Mongolia, was significant, as "non-predatory causes of mortality were slightly higher than depredation cases, representing 4.5% and 4.3% of livestock holdings respectively," where the non-predatory factors concerned mostly winter severity (Augugliaro et al., 2020). Snow leopards were the second most prevalent predators in terms of livestock killings, after wolves, but ahead of wolverines. Based on these predation activities, "herders' attitudes towards the three large carnivores were negatively affected by the magnitude of the damages since they had a positive overall attitude towards both snow leopard and wolverine, whereas the attitude towards wolf was negative" (Augugliaro et al., 2020). This study found that "wolves and snow leopards showed a preference for large-sized livestock (while both selected yaks, wolves selected horses whereas snow leopards, cattle), even though, in terms of quantity, both killed more small-sized than large-sized livestock" and "Wolves were the most important livestock predators afflicting the highest losses in the goats and sheep category. They were reported to be responsible for more than 20 times more livestock losses than snow leopards. Similarly, more herders were affected by wolf predation than by snow leopard (92.3% for wolves versus 13.8% for snow leopards)," while wolverines did not contribute any significant predation (Augugliaro et al., 2020).

Din et al. (2017) found that "livestock mortality due to disease is the most serious threat to livestock (an average 3.5 animal heads per household per year) and ultimately to the rural economy (an average of US$352 per household per year) as compared to predation (1.78 animal heads per household per year, US$191) in the three study sites. Overall, 1419 (315 per year) heads of livestock were reportedly killed by snow leopards (47%) and wolves (53%) in the study sites… People with comparatively smaller landholdings and limited earning options, other than livestock rearing, expressed negative attitudes toward both wolves and snow leopards and vice versa. Education was an effective solution to dilute people's hatred for predators. Low public tolerance of the wolf and snow leopard in general explained the magnitude of the threat facing predators in the Pamirs" (Din et al., 2017; see also Bagchi & Mishra, 2006; Tiwari et al., 2020).

Regarding predation of yak, Dorji and Powrel (2022) found that "young yaks are more vulnerable to predation by snow leopards which constitute 94.5% of kills… adult yaks with an average body weight of 250 kg are too large and are less vulnerable to predation by snow leopards. The prey weight of up to 180 kg is considered the maximum size of a prey that a snow leopard can handle." However, local herders have argued snow leopards can kill adult yaks weighing more than 250 kg. Krofel et al. (2021) reported "a predation event of a male snow leopard killing a semi-domestic adult yak (about five times the body weight of the predator) from the Altai Mountain range in Mongolia" and "this study confirms that even the adults of large domestic ungulates are vulnerable to snow leopard predation" (Dorji & Powrel, 2022).

Lham et al. (2021: 275), in a study of Bhutan, argue that "retaliatory killing due to livestock depredation is a major threat to snow leopard conservation" and "identified five prey species" including of wild ungulates (85.8%), with yak being the livestock species (14.2%). In addition, the findings indicated that predation on livestock in Bhutan was lower than in the majority of studies conducted in some of the 12 snow leopard range countries possibly due to the higher proportion of livestock to wild prey in Bhutan (Lham et al., 2021). Yak were the only domestic species recorded in snow leopard scats in this study, although horses, sheep, and goats were reported in Nepal, India, Bhutan, and Pakistan (Lham et al., 2021: see also Bagchi & Mishra, 2006; Sangay & Vernes, 2008; Chetri et al., 2017; Khatoon et al., 2017).

Bano et al. (2021) present a study of livestock depredation by large carnivores between 2014 and 2019 based in the Karakoram-Pamir mountains of northern Pakistan, and "found a statistically significant increasing pattern of predation with influential factors such as age, gender, occupation, education of respondents, population of predators, threats index for predators and conservation efforts." The carnivores in the study area comprised snow leopards, Tibetan wolves, Himalayan lynx, brown bears, Tibetan red fox, and stone martin, but only the first are frequently livestock predators. More than half (54%) were attributed to wolves and 45.8% to the snow leopard. Of these killings, sheep constituted 35.16%, followed by goats at 27.74%, domestic yak at 20.64%, and cattle at 16.45%. As a result, threats to carnivores included retaliatory killing and habitat destruction. The majority of respondents thought that wolves and snow leopards were common, and the majority (53%) blamed the loss of natural prey, some (24%) believed that livestock is easier to kill than wild prey, and 23% thought that livestock was the preferred prey. About 26% of respondents perceived the threats of the snow leopard to be severe, while 30% thought the same for wolves.

Suryawanshi et al. (2013: 550) also present a similar study of snow leopard and wolf predation and the ecological factors that encourage such livestock depredation and human opinions of such predation, in the Upper Spiti Landscape of trans-Himalayan India. The findings indicated that "human perceptions can be at odds with actual patterns of livestock depredation, increases in wild prey populations will intensify livestock depredation by snow leopards, and prey recovery programmes must be accompanied by measures to protect livestock, compensation or insurance programmes should target large-bodied livestock in snow leopard habitats and sustained awareness programmes are much needed, especially for the wolf." In the study area, both snow leopards and wolves were perceived as threats in some settlements and not in others. The actual findings based on surveys recorded 809 livestock mortalities in 25 villages over 2 years. Mortalities were by feral dogs for 338 livestock, snow leopards for 194, and wolves for 173 livestock, most of the snow leopard killings occurred during the summer while the wolf killings covered the whole year. Most yak were killed by snow leopards, while cattle were equally killed by snow leopards, wolves, and feral dogs, and snow leopards showed a preference for yaks and horses over other livestock. The authors note that people disproportionately feared wolves, although the evidence was most of snow leopard killing of livestock (Suryawanshi et al., 2013: 550).

Khan et al. (2017: 519–525) also present a study of snow leopard and wolf predation in the Karakoram Mountains of Pakistan, which accounted for 375 kills during 2008–2012, or 4.3% annual losses of the livestock holdings in the village. Losses included domestic sheep (47%), goats (42%), cattle (8.3%), yak (1.9%), and equines (0.3%), with 58% females and snow leopards accounting for 60% of the losses and wolves for 37%. In addition, "the respondents considered snow leopards and wolves to be equally threatening to their livestock but were more afraid of snow leopards (53%) than wolves (42%)" (Khan et al., 2017: 521–522).

Tiwari et al. (2020) note that "conflict between snow leopards and humans across the trans-Himalaya is a pressing conservation concern," and in the study area of the Narphu valley of the Annapurna Conservation Area, Nepal, there were "poor herding and guarding practices with households owning larger numbers of total livestock (especially large bodied animals like yak (*Bos mutus* f. *grunniens*) and cattle (*B. primigenius* f. *taurus*)), which resulted in the higher loss" to snow leopards, as in the "2 years (2017/18 and 2018/19), 73.9% of the households interviewed lost livestock to snow leopards, with an annual average loss of two livestock per household" of which 55.4% were yak (mainly female: 79%), 31.7% goat, 6.8% sheep, 3.2% horse and 2.8% cattle (Tiwari et al., 2020). In addition, "41% of all herders considered snow leopard's preference for domestic livestock as the main factor in livestock predation, whereas only 5% perceived poor herding practice as the main reason for the loss" and "71% herders reported careful herding as a solution to snow leopard depredation, and 15% of herders considered the complete extermination of snow leopards as the best solution to the problem" (Tiwari et al., 2020). In addition, 92% of herders argued for "a decreasing trend of livestock numbers over the previous two years" due to predation by snow leopards, death of livestock due to natural disasters, starvation, accidents and poor pasturage, lack of human resources, and alternative income sources (Tiwari et al., 2020).

Lions and Cattle Predation

Lions in Asia are currently restricted to the Gir Forest of western India (Banerjee et al., 2013; Kolipaka et al., 2017). They are behaviorally like tigers, in attacking people and livestock, which influences human attitudes to lion conservation. Pocock (1939: 219–220) gave a historical reason for the near extinction of lions in India: "in all parts of the world occupied by Europeans where lions occur the disappearance of the lions is merely a question of time. They are a menace to human life, especially if they become man-eaters, and they ravage the flocks and herds of settlers, finding domesticated livestock easier to prey upon than wild game." In addition, concerning the lion's status in Asia, "there seems to be no reason to look beyond man as the prime cause of their extermination in India" (Pocock, 1939: 220).

Pocock (1939: 220) notes that "It is only necessary to cite the following instances attesting extensive slaughter in two localities." It is further noted that "during the Mutiny an English officer shot over 300 lions, 50 of them in the neighborhood of Delhi; and Capt. Mundy in his 'Pen and Pencil Sketches,' 1832, stated that before his time there were great numbers of lions in the jungle near Hansi in Patiala, but that owing to the rewards offered by Government for their destruction and the zest of English sportsmen they had entirely disappeared from the district. Incidentally he remarked that crack sportsmen considered the lion afforded better sport than the tiger because his attack is more open and his haunts less favorable for retreat. This opinion of the difference in character between lions and tigers agrees with that of Major Leveson, 'the old Skekarry,' who, after experiences in shooting on foot tigers in India and lions in Africa, came to the conclusion that the lion is the pluckier animal of the two. Perhaps he should have said less cautious and less cunning" (Pocock, 1939: 220). Currently, in the Gir Forest, lions "prey mainly upon the big game and domesticated livestock. It has been recorded that they lie up during the day in their lairs and issuing from them just before sunset make straight for the nearest hamlet of buffalo-herdsmen. If unsuccessful in their quest they move on to the next and wait for stragglers from the byres. Since the cattle are driven out to pasture long before daybreak, the lions have a good chance of a kill while it is still dark" (Pocock, 1939: 222).

More recently, Saberwal et al. (1994: 501) noted that in the environs of the Gir forest, "persistent attacks by lions on humans hinder support among local peoples for lion conservation… An average of 14.8 attacks by lions and 2.2 lion-caused deaths occurred annually between 1978 and 1991, and most attacks (82%) occurred on private lands outside the forest reserve. A drought in 1987–1988 precipitated an increase in rates of conflicts (from 7.3 to 40.0 attacks/year) and in the proportion of attacks that occurred outside the reserve (from 75% to 87%). The spatial pattern of lion attacks could not be distinguished from random before the drought, whereas attacks were clustered after the drought in village subdistricts with a higher ratio of revenue land to forest edge and those closer to sites where lions were formerly baited for tourist shows." The authors add that "subadult lions were involved in conflicts in disproportion to their relative abundance. A majority of villagers interviewed expressed hostile attitudes toward lions owing to the threat of personal injury and economic hardship (mainly livestock damage) posed by lions. The escalation in lion-human conflict following the drought probably resulted from a combination of increased aggressiveness in lions and a tendency for villagers to bring their surviving livestock into their dwellings." Government policies have been faulted, as "dissatisfaction with the government's compensation system for lion-depredated livestock was reported widely. The current strategy for coping with problem lions—that is, returning them to areas in the Gir forest already saturated with lions—is inadequate, as indicated by the sharp increase in lion-human conflict since 1988… Long-term alleviation may entail reducing the lion population by relocating or culling lions."

Banerjee et al. (2013) note that the coexistence of Asiatic lions and pastoral communities (Maldharis) in the Gir forests, India, created conflicts with high livestock density, with buffaloes being the most abundant cattle. In addition, lion density was higher in areas with Maldhari livestock than in areas without livestock. The Gir Protected Area at the time of writing included 50 Maldhari settlements (termed nesses), which consist of several buildings and families, with about 20–100 livestock, mostly buffalo (*Bubalus bubalis*) and cattle (*Bos indicus*). Regarding lion prey, a scat assessment indicated that "97.6% of lion scat contained a single prey type, while 2.4% of the scats had two prey items. Wild ungulates comprising chital, sambar, nilgai and wild pig together accounted for 76.4% of all prey occurrences, while domestic livestock (buffalo 13.7% and cattle 7.8%) contributed the rest" (Banerjee et al., 2013). In addition, "the percentage biomass contribution of different prey species to the lions' diet was most for livestock (33.7%) followed by chital (28.9%) and sambar (28.3%)." There was evidence of selective utilization of prey by lions. Chital, sambar, nilgai and wild pig were found to be utilized more than their availability while buffaloes were used less than their availability. Cattle were utilized in proportion to their availability. The order of prey preference by lions as estimated by Jacob's Index was sambar, wild pig, nilgai, chital and cattle (Banerjee et al., 2013).

The research recorded 308 livestock mortalities, with 58.4% due to lion predation, 3.2% by leopards and 38.4% by other natural causes, and lions killed mostly cattle (69.4%), buffaloes (29.4%), and camels (1.2%) (Banerjee et al., 2013). Where the sex of the lion could be determined, "female lions with dependent cubs were responsible for 54.4% of the attacks; single male or male coalitions were responsible for 19.1% of the attacks and mixed groups of lions made 26.4% of the kills. Lionesses in the study area were found to raid livestock in proportion to the prevailing adult sex ratio in the population. Thus, all lions were equally likely to predate livestock" (Banerjee et al., 2013). The authors conclude that "a crude estimate of total intake requirement of the lion population in the study area was about 124,733 kg for the study period of 28 months. Livestock were found to be contributing about 31,582 kg (25.3%) of biomass to the lion's diet… livestock composed 42% of lions' feeding events (16% from predation and 26% was from scavenging on livestock carcasses). Wild ungulates were found to compose the remainder 58% of lions' feeding events (47% predated and 11% appropriated from leopard kills or other lion kills)" (Banerjee et al., 2013).

Ram et al. (2023) present a more recent study of lion predation in the Gir Forest, comparing lions of the Gir PA and the surrounding multi-use land matrix. Based on scat analyses, we assessed the dietary composition of Asiatic lions. The findings suggested that "the Asiatic lion ate more wild prey (74%) than domestic livestock (26%) in the Gir PA. In contrast, lion diets comprised 51% wild prey and 42% domestic livestock in the multi-use land matrix. Sambar contributed most (38%) to the diet in PAs, while blue bull most (29%) in the multi-use land matrix. However, diet diversity and niche breadth were similar between these two areas. The dietary overlap in the lions' diet between these two areas was 0.52. The results suggest that large-size wild ungulates are the main prey species in Gir PA, while in the surrounding multi-use land matrix, both large wild ungulates and domestic livestock are the main prey. The present study may help future lion conservation management decisions."

The authors argue that the Asiatic lion is endangered, being restricted to the Saurashtra region of the Gujarat state, but recently due to successful conservation policies the lion populations have increased in the Gir National Park, Gir Wildlife Sanctuary, Paniya Wildlife Sanctuary, and Mitiyala Wildlife Sanctuary, which together comprise the Gir Protected Area (Ram et al., 2023). Lions have successfully moved into the surrounding multi-use land matrix, which includes forest, pasture, and suburban landcover, with variable prey species, including spotted deer or chital (*Axis axis*, Erxleben, 1777), sambar (*Rusa unicolor*, Kerr, 1792), and blue bull or nilgai (*Boselaphus tragocamelus*, Pallas, 1766), the latter especially in the surrounding landscape, which may affect lion diet composition (Ram et al., 2023). However, Ram et al. (2023) argue that there are few studies comparing lion diets between the protected area and the multi-use land matrix. The assumption of lion prey preferences was based on Carbone et al. (1999), who suggested that large carnivores over 21 kg usually take prey up to 45% of the carnivores' body weight, while the lions in the Gir Protected area averaged from 116 to 140 kg (Vasavada et al., 2022), therefore might take prey mostly sambar (130 kg) and blue bull (180 kg), over other, smaller species, and blue bull and domestic cattle (150 kg) outside the protected area.

The findings indicated that in the protected areas, lions killed sambar (35%), spotted deer (19%), domestic buffalo (13%), domestic cattle (13%), blue bull (8%), wild pig (6%), and other smaller species with domestic livestock (buffalo, cattle, goats) contributing 26%. However, in the multi-use land matrix, the main lion prey species were blue bulls (25%), domestic cattle (24%), domestic buffalo (17%), wild pigs (15%), sambar (4%), and some smaller species, with livestock (cattle, buffalo, goat, sheep) contributing 42%, while wild prey was 51% of the lions' diet. In terms of biomass in the diet in protected areas, the contribution was sambar (38%), spotted deer (18%), domestic buffalo (15%), and domestic cattle (14%) and domestic livestock comprised 29%. In terms of biomass in the diet in the multi-use land matrix, the contribution was blue bulls (29%), domestic cattle (28%), domestic buffalo (20%), and wild pigs (14%). Wild ungulates contributed 51%, while domestic livestock comprised 49%. The conclusion was that the hypothesis regarding the consumption of large prey was supported evidence by the predation on larger mammals including sambar, blue bull, and domestic livestock in both areas. Additional factors were that "domestic livestock generally lack antipredator responses in comparison with native wild prey species and represent easy targets as prey" and the variable effectiveness of the livestock management.

Bears and Cattle Predation

Asian bears, including brown and black bears, and less commonly sloth bears are capable of killing cattle (Chauhan, 2003; Dhamorikar et al., 2017; Kumar et al., 2022; Rawal et al., 2024). Studies either compare two species of bears or focus on one species. The brown bear, being the largest of these species may be considered the most capable of killing cattle or people engaged in cattle herding. For example, Kumar et al. (2022) note that "brown bear-mediated conflicts have caused immense economic loss to the local people living across the distribution range. In India, limited knowledge is available on the Himalayan brown bear, making human–brown bear conflict mitigation more challenging." The case study was of the Lahaul valley in Himachal Pradesh, India, where respondents complained about livestock depredation and/or crop damage, with many events occurring less than 500 m from the forested landcover. The livestock depredation targeted sheep, goats, and cattle and was reported by over a third of respondents. The authors noted that "the livestock depredation in high-altitude areas of the study landscape could be because of unsupervised livestock grazing practices in alpine pastures by the nomadic herders" (Kumar et al., 2022).

Chauhan (2003: 84) presents a study comparing brown and black bears in Great Himalayan National Park Conservation Area (GHNPCA), Himachal Pradesh, India, which involved "human casualties, livestock killings, and crop damage by the bears from 1989 to 1998." During this period there were 355 livestock killings by brown and black bears, with most killings occurring in alpine pastures and forests, and monthly variations in livestock killings associated with livestock seasonal movement from alpine pastures and forests to low altitude villages. The authors note that most cases of cattle depredation were in evening, night or early morning, and bears killed sheep (48%), goats (32%), cattle (18%) horses (<1%) and mules (<1%). In addition, "casualties were higher for unattended Herds… Two-hundred and thirty-six livestock were killed in the 38 incidents in which no grazers were present" and "63 cattle depredation cases occurred in the 26 incidents in which grazers were present" (Chauhan, 2003: 85–86). Singh et al. (2024: 26) also present a case of brown and black bear predation in the Chowkibal-Tangdhar sector of Kupwara district, Union Territory of Jammu and Kashmir, based on self-reported losses of livestock to large carnivores by local people. The authors note that "leopards reportedly killed livestock throughout the year at the relatively low altitude zones, but black bears and brown bears reportedly killed livestock in the monsoon months at the relatively higher altitude zone" and "leopards prefer to kill small to medium-sized livestock, but the bears had no such preferences" (Singh et al., 2024: 26). In addition, of the 389 livestock that were killed by bears and leopards, the breakdown was of 107 cattle, 139 goats, 6 horses, and 137 sheep. Brown bears killed 152 (39.07%), including 79 cattle, 37 goats, 1 horse, and 35 sheep. Black bears killed 120 (30.85%) including 28 cattle, 41 goats, 4 horses, and 47 sheep. Leopards killed 117 (30.08%) including 61 goats, 1 horse, and 55 sheep (Singh et al., 2024: 28).

Wolves and Cattle Predation

Wolves, the largest wild canids, are major predators of cattle in Asia, and their tendency to hunt in packs may increase their ability to kill larger prey (Newsome et al., 2016). Research indicates that wolf predation on livestock, sometimes in combination with other predators such as big cats or bears, affects human attitudes toward carnivore conservation. For example, Tiralla et al. (2021:83) present a case study of wolf predation on livestock in the Khentii Mountain range in Northern Mongolia between 2008 and 2012. The findings indicated that people were increasingly opposed to wolf presence, partly based on conflicts over livestock. It is also argued that the "information on the feeding ecology of wolves in Mongolia is rare, especially in the mountain taiga and mountain forest steppe regions of Northern Mongolia. Those regions are characterized by a relatively high wildlife diversity and are sparsely populated by humans" (Tiralla et al., 2021:83). The study in the Khentii Mountain range in Northern Mongolia between 2008 and 2012 found that "almost all wolf faeces contained remnants of wild ungulates, which made up 89% of the consumed biomass" (Tiralla et al., 2021:83). Prey species selected included Siberian roe deer (*Capreolus pygargus* Pallas, 1771), red deer (*Cervus elaphus* Linnaeus, 1758), wild boar (*Sus scrofa* Linnaeus, 1758), and other smaller mammal species. Importantly, "no evidence of domestic ungulates was found in the wolf diet. Thus, near-natural habitats with a diverse fauna of wild animals are important to limit livestock depredation."

The authors argue that in Mongolia, a nation with a population of 2.8 million, an ongoing decline of wildlife populations has been recorded during the last decades (Tiralla et al., 2021). It is caused by unsustainable hunting due to an increasing wildlife trade as a consequence of Mongolia's change to a free market economy with an increase in livestock from about 25 million to over 40 million, with a conversion of natural habitats into pasturelands, the increased utility of livestock as a food source and as source of income for over half of the Mongolian population. The authors note that little research has been undertaken in Mongolia, with the few studies showing wolf predation on livestock, but in the study areas, the populations and species diversity of wild ungulates were low. In the mountain taiga and mountain forest-steppe regions of northern Mongolia, where there is comparatively little human population and a high level of biodiversity, there is no information available regarding wolf feeding habits (Tiralla et al., 2021). Tiralla et al. (2021: 87) conclude that high natural prey biodiversity is a factor that reduces wolf predation on livestock: "the wide prey spectrum of the wolf in the Khentii Mountains indicates an almost untouched and healthy ecosystem with an adequate abundance of wild ungulates. In times of inaccessibility of common prey species, wolves fall back on buffer prey species such as lagomorphs or small mammals. Even though unprotected and free-ranging livestock herds occur within the research area, no evidence of livestock was found in the wolf diet."

This finding is supported by other studies that indicate that the low level of natural prey biodiversity influences wolf predation on livestock. For example, Nakazawa et al. (2008) studied wolves in the Bogdkhan Mountain Strictly Protected Area and found evidence of horse and goat predation. It is argued that "the size of the Bogdkhan Mountain Strictly Protected Area is not large enough to maintain the five wolf packs that roam there, nor is the number of wild prey in the area sufficient to support them. Consequently, the wolves also hunt outside the protected area and depend on free-ranging livestock within and around the protected area…the current study suggested that wolves caught fewyak and cattle in 2004 and none in 2005" (Nakazawa et al., 2008). Mahajan et al. (2022) note that "gray wolves are capable of adapting to human-dominated landscapes by utilizing domestic prey as a source of food. Livestock depredation by wolves incurs a heavy economic loss to the villagers, resulting in negative attitudes toward the species and leading to increased conservation conflict." In the case study of the environs of Kailadevi Wildlife Sanctuary, Rajasthan, India, in the year of study, 64% of respondents reported a loss of livestock, in which goats (63%) comprised the major share, followed by sheep (22%) and cattle calves (15%). Wolves tend to hunt medium-sized domestic prey (sheep and goats) that commonly graze in open agricultural areas. Respondents with prior experience of livestock loss held more negative attitudes. Shepherds held more negative attitudes than other occupations.

Dholes and Cattle Predation

The dhole (*Cuon alpinus* Pallas, 1811) is a canid smaller than the gray wolf, described by Heptner et al. (1998: 571) as "in general appearance the red wolf is unique and is not similar to the gray wolf—in its constitution, rather, mixes features of the wolf (or jackal) and fox." Kamler et al. (2020: 1396) note that "endangered dholes (*Cuon alpinus*) are restricted to small and declining populations in Southeast Asia, and little is known about how their ecology differs within the region." It is further noted that "India contains the largest remaining populations of dholes, and there have been numerous studies of dhole diets from this country. Researchers have attempted to determine the body mass of preferred prey of dholes based largely on previous studies of dhole diets from India." In addition, "taken together, it appears that dholes prefer medium (30–100 kg) and large-sized (>100 kg) ungulates, at least based on studies predominantly from the open and seasonally dry forest ecosystems of India" (Kamler et al., 2020:1398).

The extent of cattle predation by dholes varies according to different research areas, but generally sambar, approximately the size of a small cow, is the most predated species. For example, George et al. (2021) present a study of dhole predation in South Wayanad Forest Division, Kerala, India, from May 2017 to July 2017. The findings indicated 12 prey species, with "the most common and the largest ungulate species, sambar" comprising 69.68%, wild pig (15.48%), and domestic cattle (2.81%). A study by Selvan et al. (2013: 486) states that "sambar and chital were found to be significantly preferred by the Asiatic wild dog, and the preferred prey body mass range was between 40 and 60 kg. The prey preference of the Asiatic wild dog was mainly influenced by the availability of prey." Another study by Hayward et al. (2014: 255) states that a literature review of 24 studies from 16 sites from throughout the range of the dholes showed 19 prey species, with the commonest species being sambar and to a lesser extent, chital and wild boar, and the prey weight range being 30–235 kg, and the preferred weight range being 130–190 kg, which overlaps with the prey weight range of tigers and leopards and "although prey preferences are not affected by pack size, larger packs ultimately take larger prey" (Hayward et al., 2014: 255).

Kamler et al. (2020:1396) found that for prey selection of dholes in closed evergreen, hilly forests in northern Laos, and open deciduous, flat forests in eastern Cambodia, muntjac (20–28 kg) was the main prey species, with sambar in Laos, and wild pig and banteng in Cambodia being taken as well. In addition, "seasonal differences in overall diet occurred in Laos, but not Cambodia, primarily because of an increase in livestock consumption" (Kamler et al., 2020:1396) and "in India and Bhutan dholes often consume livestock in protected areas. In contrast, in previous studies conducted in Southeast Asia, dholes did not consume livestock, although livestock grazing was prohibited in the protected areas where those studies occurred" (Kamler et al., 2020:1397). Importantly, "persecution due to livestock predation is a major factor contributing to the decline of dholes" (Kamler et al., 2020:1397). Kamler et al. (2020: 1403) note that there were seasonal differences in the dholes' diet in Laos (for cool-dry and hot-dry seasons, not wet and dry) but not in Cambodia, and there was no seasonal difference in diets, and "among the prey categories, consumption of livestock (cattle and goats) contributed most to the seasonal difference in diets, with a significantly increased consumption during the hot-dry season" with a factor being the fact that in the Laos study site a majority of villagers removed their livestock from grazing areas inside the park during the dry season and placed the cattle in fields outside the park (Kamler et al., 2020: 1403).

Another study by Ghaskadbi et al. (2022) in Central India found that "wild ungulates like sambar and chital were the principal prey in terms of biomass (sambar 61.08%; chital 19.08%) and number of prey animals consumed (sambar 39.28%; chital 13.83%)." In addition, "the occasional scavenging of tiger kills, and improper disposal of cattle carcasses may justify cattle remains in the dhole scats. Dholes have been known to scavenge kills of larger co-predators in high tiger density PAs across the landscape." The authors "did not observe or find any forest department records of livestock depredation incidents by dholes across our landscape, suggesting that the 1.57% representation of cattle in our data could be from scavenging events, suggesting that the depredation of livestock by dholes is extremely low" (Ghaskadbi et al., 2022).

Thinley et al. (2011: 518–519) examined the seasonal diet of dholes in northwestern Bhutan in 2009, with the findings suggesting that large (>75 kg) ungulate species, especially sambar were main prey species (others including Serow (*Capricornis sumatraensis*), Muntjac (*Muntiacus muntjak*), Goral (*Nemorhaedus goral*), Wild pig (*Sus scrofa*), and cattle (Bos taurus)), and small ungulates (20–30 kg) were only 10% of the prey biomass, both sets during the wet and dry seasons. The findings indicated that "cattle were consumed only during the wet season, probably because herding practices reduced their availability in the dry season" and "dholes consumed twice as many individual wild pigs than individual cattle" (Thinley et al., 2011: 518). Thinley et al. (2011: 519) also note that "interestingly, livestock were not consumed by dholes during the dry season, although cattle comprised 22% of the biomass consumed in the wet season. Such a strong seasonal difference in consumption of cattle suggests herding behaviors and cattle availability might have strongly affected livestock predation by dholes." Factors for this may have been the relocation of cattle to lower elevations due to the cold temperatures and reduced forage, and in some cases to hay stocks near settlements.

During the wet season cattle generally foraged freely in the forests, while some other livestock such as horses, donkeys, mules, and domestic yaks in winter were more closely managed and not predated (Thinley et al., 2011: 519). According to Thinley et al. (2011: 520), during the dry and wet seasons, "cattle made up more than twice the biomass in the dhole diets than did wild pigs" although "when considering individual prey consumed over a year, twice as many wild pigs were consumed per dhole than were cattle" and "to reduce predation on cattle by dholes in our study area, we recommend farmers guard cattle more closely during the wet season."

Geomatics-Based Research on Carnivore Predation on Cattle in Asia

Geomatics (or geoinformatics)-based research has emerged as an important tool for the assessment of large carnivore predation on cattle (Campbell, 2018). Similar to other continents, the main contribution to geomatics-based research on this topic is the measurement of relevant landcover (e.g., suitable habitats for the carnivores, which may be near to cattle pasture, reduced suitable habitats, which may influence the carnivores to enter cattle pastures, increased cattle populations that may be allowed enter suitable carnivore habitats, or increased carnivore populations resulting from successful conservation policies and management, that may enter cattle foraging areas). Basic geomatics techniques involve measurement and analysis of landcover character (mostly vegetation types), while more advanced techniques may create predictive models for informing agricultural and/or conservation policies (Campbell, 2018).

Remote sensing and GIS applications have a wide application on environmental issues in Asia. What are cited here are a few critical selections of typical applications that contribute significantly to possible applications at the interface of landuse, agricultural development and cattle herding, large carnivore conservation, and related topics that can be used for such applications. An example is cited by Ranjan et al. (2016: 273), who note that "geo-informatics technology has dynamic role in mapping, monitoring and management of forest resources. The transformation of forest cover and its analysis on the earth's surface are essential for understanding the associations and interactions between natural phenomena and living organism, especially human. Also, deforestation is a major reason for global warming and one of the origin keys for the enhancement of greenhouse effect and climate change." Their case study is based on the 3S technology in the assessment of landcover changes within the forest cover area in Dalma Wildlife Sanctuary (DWLS) Jamshedpur-Jharkhand, India, using satellite coverage for forest cover dated 2009, 2011, 2013, 2015, and 2016. The authors note that "due to shifting cultivation, forest fire, and conversion of forest cover into crop land/bare land and settlement encroachments in forest region by villagers are rapidly increasing; as a result, deforestation is taking place. It is predicted that the study would demonstrate the effectiveness of 3S technology in forest renovation, planning and management" (Ranjan et al., 2016: 273).

Remote sensing and GIS-based decision support systems have also been developed and applied to predator-cattle interactions. For example, Ahmad et al. (2018: 66) argue that with wildlife habitat degradation, the "application of Remote Sensing and Geographic Information System has been widely accepted as a tool which has immense significance in wildlife habitat suitability modeling and mapping. Maps derived from analysis of remote sensing data and modeling in GIS are highly useful in making the strategies in wildlife management and conservation planning" (Ahmad et al., 2018: 66). This study aimed to identify wildlife habitat and "conservation hotspot grids," with the parameters being "forest cover, agriculture with settlement, forest fires, roads, streams and mines" (Ahmad et al., 2018: 66). The findings indicated that 42% of the studied area for wildlife habitat suitability in the Saranda Forest division had very high suitability for wildlife. The authors argue that such analyses will assist in policy-related decision-making regarding environment conservation (Ahmad et al., 2018: 66).

A study by Shivang and Dharaiya (2021:22) used "remote sensing and Geographical information system (GIS) techniques," including Resourcesat-2LISSIII (2008) and Sentinel 2A (2018) remote sensing data and supervised classification technique with Maximum Likelihood Algorithm (MLA) and field studies for change detection in the Taranga Hill Forest in the Aravali Mountain system in Gujarat, India. The authors justify the study on the recent "heavy biotic pressure" of this area, as the forest that "once had diverse vegetation and faunal diversity is now facing several issues like overgrazing, deforestation and human encroachment" (Shivang & Dharaiya, 2021:22). The findings indicated decreases in natural vegetation, including dense and open forest, and wetlands, and increased landcover characterized by agriculture, settlement and barren land over the previous decade. The use of the geomatics-based tools revealed that urbanization, transport development, and agricultural expansion could be significant threats to the biodiversity and wildlife in Taranga's remaining forested areas. A long-term study of wildlife habitat use, threat assessment, and seasonal biotic pressures on the forest are recommended by the work in order to develop long-term management and conservation strategies for this important wildlife habitat in northern Gujarat (Shivang & Dharaiya 2021:22).

A more advanced study is that of Naha et al. (2020), who use spatial risk modeling to examine carnivore (leopard) conflicts with human landuse stakeholders in the Indian Himalayan Region. The case study was based in western and eastern Himalayas, using information of 857 leopard attacks on livestock and multivariate analyses to detect the landscape features associated with increased probability of leopard attacks on livestock. The justification of this study is that "conflict with humans is a significant source of mortality for large carnivores globally. With rapid loss of forest cover and anthropogenic impacts on their habitats, large carnivores are forced to occupy multi-use landscapes outside protected areas" (Naha et al., 2020). It is argued that "Spatial risk modelling has been widely used to provide insights into habitats where carnivores kill livestock through both qualitative and graphical guides and help in prioritization of management interventions and mitigation measures. This is essentially a quantitative tool which uses location-specific data (kills) to predict the probability of a predator attack by relating landscape attributes at kill sites and comparing them to random sites representing landscape availability. Risk models help in quantifying the realized predation risk (where livestock or wild prey were killed) especially across heterogeneous landscapes" (Naha et al., 2020). The study examined livestock depredation by leopards in IHR, the husbandry practices which lead increased leopard attacks, the predation-risk zones associated with human-dominated landscapes, and mapping of "spatially map human–leopard conflict hotspots by modelling probability of livestock predation" (Naha et al., 2020).

The findings indicated that leopard predation "increased within a heterogeneous landscape matrix comprising of both closed and open habitats (very dense forests, moderate dense forests, open forests, scrubland and non-forests)" (Naha et al., 2020). The spatial risk maps identified "hotspots of human-leopard conflicts," usually where there was free grazing of livestock without herders. The findings also showed that "in 2015–2018, there was a total of 857 and 375 leopard livestock predation events recorded in North Bengal and Pauri Garhwal, respectively. 54% of the livestock killed were adult cows, 16% were goats, 14% calves, 6% adult pigs, 5% piglets, 3% ox, 2% goat yearlings in North Bengal. In Pauri Garhwal, 70% percent of the livestock predated were adult cows followed by 9% adult goats, 8% cattle calves, 7% ox, 1% horse, 2% donkey, 2% sheep and 1% mules" (Naha et al., 2020).

Another advanced study based on Bhutan in the eastern Himalayas examined human wildlife conflict using records from 2017 to 2020 and MaxEnt predictive modeling and ArcGIS (Letro et al., 2021). The main conflict types were livestock depredation and crop and property damage, and to a lesser extent attacks on humans. The predictive modeling suggested that out of the total area of Bhutan (38,394 sq.km), 2236.96 sq.km is under livestock predation risk and 1305.9 sq.km area is under high risk for crop and property damage (Letro et al., 2021). In terms of districts, Wangdue Phodrang, Trongsa, Trashigang, Lhuentse, and Trashi Yangtse account to 82% of the total predicted high risk areas for livestock depredation and Trashigang, Mongar, Wangdue Phodrang, Sarpang, and Samdrup Jongkhar districts accounting to 80.91% of the total predicted high risk areas for crop damage (Letro et al., 2021).

Regarding livestock predation, "the major conflicting carnivores were tiger, common leopard, Asiatic black bear, dhole, snow leopard, and wolves, which opportunistically preyed on domestic animals of varying size. Loss of cattle, buffalo, yaks, horses, sheep and goat were most common depending on the locality while predation of poultry and piggery by smaller felids were also reported. In the southern districts, major predators were tiger and common leopard, while in central belt, tiger, dhole, common leopard, and Asiatic black bear were the major predators of domestic animals. In the northern region, snow leopard, wolves and wild dog preyed on the sheep and yaks of the nomadic herders" (Letro et al., 2021:8). In addition, "loss of human lives to wildlife were occasionally reported from Bhutan. While the frequency of conflict is not very severe as compared to loss of crop and livestock to wildlife, such incidences risk human lives so it is of concern. Such cases were mostly accidental wherein wildlife especially Himalayan black bears maul the human in their self-defence when people are encountered out of sudden in the forests" (Letro et al., 2021:8). The authors recommend that "field verification of HWC prevalence in the hotspot areas for prioritizing HWC management interventions to reduce economic loss to the farmers and benefit conservation" (Letro et al., 2021: iii).

Conclusions

Asian domesticated and semi-domesticated hybrid cattle encounter more species of large carnivores than the cattle of any other continent, four species of big cats (tigers, lions, leopards, snow leopards, and possibly clouded leopards), three species of bears (brown, black and sloth bears), and two species of canines (wolves, dholes). The relationship between agricultural development manifested by cattle herding and landcover change from natural habitat to agricultural landscapes and conservation based on predatory species recovery and habitat regeneration are conflicted in Asia. This chapter has covered the main issues of cattle-predator relations at the ecological level. The main issues for the survival of cattle are the size of the cattle, as the larger the cattle, the greater their defense against predators, the larger the predators, the commoner the predation on cattle, and an indicator of predator capability for cattle predation was the size of their natural prey. The main issue for large carnivores concerned the human retaliation for cattle predation, which can result in local extinction of carnivore species, in some cases even in cases of conservation management. Habitat degradation is another serious issue. The complex issues in Asia are more complicated than those of Europe, but the relationships between the large carnivores and cattle are increasingly well managed. The simpler situation in Europe may yet serve as a model for the improvement of the complex dynamics in Asia.

References

Ahmad, F., Goparaju, L., & Qayum, A. (2018). Wildlife habitat suitability and conservation hotspot mapping: Remote Sensing and GIS based decision support system. *AIMS Geosciences, 4*(1), 66–87. https://doi.org/10.3934/geosci.2018.1.66

Ahmed, R. A., Prusty, K. P., Jena, J., Dave, C. V., Sahu, H. K., Rout, S., & Vihar, S. (2012). Prevailing human carnivore conflict in Kanha-Achanakmar corridor, Central India. *World Journal of Zoology, 7*(2), 158–164.

Ahrestani, F. S. (2018). Bos frontalis and Bos gaurus (Artiodactyla: Bovidae). *Mammalian Species, 50*(959), 34–50. https://doi.org/10.1093/mspecies/sey004

Anwar, M. B., Jackson, R., Nadeem, M. S., Janečka, J. E., Hussain, S., Beg, M. A., Muhammad, G., & Qayyum, M. (2011). Food habits of the snow leopard Panthera uncia (Schreber, 1775) in Baltistan, Northern Pakistan. *European Journal of Wildlife Research, 57*(5), 1077–1083. https://doi.org/10.1007/s10344-011-0521-2

Augugliaro, C., Christe, P., Janchivlamdan, C., Baymanday, H., & Zimmermann, F. (2020). Patterns of human interaction with snow leopard and co-predators in the Mongolian western Altai: Current issues and perspectives. *Global Ecology and Conservation, 24*, e01378.

Bagchi, S., & Mishra, C. (2006). Living with large carnivores: Predation on livestock by the snow leopard (Uncia uncia). *Journal of Zoology, 268*(3), 217–224. https://doi.org/10.1111/j.1469-7998.2005.00030.x

Bagchi, S., Goyal, S. P., & Sankar, K. (2003). Prey abundance and prey selection by tigers (Panthera tigris) in a semi-arid, dry deciduous forest in western India. *Journal of Zoology, 260*(3), 285–290. https://doi.org/10.1017/s0952836903003765

Banerjee, K., Jhala, Y. V., Chauhan, K. S., & Dave, C. V. (2013). Living with lions: The economics of coexistence in the Gir forests, India. *PLoS One, 8*(1), e49457. https://doi.org/10.1371/journal.pone.0049457

Bano, R., Khan, A., Mehmood, T., Abbas, S., Khan, M. Z., Shedayi, A. A., Zaman, S., & Muhammad Ali Nawaz, M. A. (2021). Patterns of livestock depredation and Human–wildlife conflict in Misgar valley of Hunza, Pakistan. *Scientific Reports, 11*, 23516. https://doi.org/10.1038/s41598-021-02205-2

Bargali, H. S., & Ahmed, T. (2018). Patterns of livestock depredation by tiger (Panthera tigris) and leopard (Panthera pardus) in and around Corbett Tiger Reserve, Uttarakhand, India. *PLoS One, 13*(5), e0195612. https://doi.org/10.1371/journal.pone.0195612

Bhattarai, B. R., & Fischer, K. (2014). Human–tiger Panthera tigris conflict and its perception in Bardia National Park, Nepal. *Oryx, 48*(4), 522–528. https://doi.org/10.1017/S0030605313000483

Bigi, D., & Zanon, A. (2008). *Atlas of native breeds: Cattle, horses, sheep and goats, pigs bred in Italy (in Italian)*. Edagricole.

Biswas, S., & Sankar, K. (2002). Prey abundance and food habit of tigers (Panthera tigris tigris) in Pench National Park, Madhya Pradesh, India. *Journal of Zoology, 256*(3), 411–420. https://doi.org/10.1017/s0952836902000456

Blackburn, H. V. (1934). A bull gaur (B. gaurus) and a tigress fight to death. *Journal of the Bombay Natural History Society, 37*, 950–951.

Buchholtz, C. (1990). True Cattle (Genus *Bos*). In S. Parker (Ed.), *Grzimek's encyclopedia of mammals* (Vol. 5, pp. 386–397). McGraw-Hill Publishing Company.

Campbell, M. (2017). *Biological conservation in the 21st century: A conservation biology of large wildlife*. Nova Science Publishers.

Campbell, M. (2018). *Geomatics and Conservation Biology*. New York, NY: Nova Science Publishers.

Carbone, C., Mace, G. M., Roberts, S. C., & Macdonald, D. W. (1999). Energetic constraints on the diet of terrestrial carnivores. *Nature, 402*(6759), 286–288. https://doi.org/10.1038/46266

Chauhan, N. P. S. (2003). Human Casualties and Livestock Depredation by Black and Brown Bears in the Indian Himalaya, 198998. *Ursus, 14*(1), 84–87. JSTOR. https://doi.org/10.2307/3872960

Chen, J., Shi, H., Wang, X., Zhang, Y., & Zhang, Z. (2022). Effectiveness of China's protected areas in mitigating human activity pressure. *International Journal of Environmental Research and Public Health, 19*(15), 9335. https://doi.org/10.3390/ijerph19159335

Chen, N., Zhang, Z., Hou, J., Chen, J., Gao, X., Tang, L., Wangdue, S., Zhang, X., Sinding, M. S., Liu, X., Han, J., Lü, H., Lei, C., Marshall, F., & Liu, X. (2023). Evidence for early domestic yak, taurine cattle, and their hybrids on the Tibetan Plateau. *Science Advances, 9*(50), eadi6857. https://doi.org/10.1126/sciadv.adi6857

Chetri, M., Odden, M., & Wegge, P. (2017). Snow Leopard and Himalayan Wolf: Food habits and prey selection in the Central Himalayas, Nepal. *PLoS One, 12*(2), e0170549. https://doi.org/10.1371/journal.pone.0170549

Chundawat, R. S., Gogate, N., & Johnsingh, A. J. T. (1999). Tigers in Panna: Preliminary results from an Indian tropical dry forest. In J. Seidensticker, S. Christie, & P. Jackson (Eds.), *Riding the tiger: Tiger conservation in human-dominated landscapes* (pp. 123–129). Cambridge University Press.

Dhamorikar, A. H., Mehta, P., Bargali, H., & Gore, K. (2017). Characteristics of human – Sloth bear (Melursus ursinus) encounters and the resulting human casualties in the Kanha-Pench corridor, Madhya Pradesh, India. *PLoS One, 12*(4), e0176612. https://doi.org/10.1371/journal.pone.0176612

Din, J. U., Ali, H., Ali, A., Younus, M., Mehmood, T., Norma-Rashid, Y., & Nawaz, M. A. (2017). Pastoralist-predator interaction at the roof of the world: Conflict dynamics and implications for conservation. *Ecology and Society, 22*(2). https://doi.org/10.5751/es-09348-220232

Dorji, P., & Powrel, R. B. (2022). A pattern of livestock depredation by snow leopard to the yak herding pastoralist in western Bhutan. *Pastoralism, 12*, 33. https://doi.org/10.1186/s13570-022-00247-3

Estes, R. (2009). *Gayal*. Encyclopedia Britannica. https://www.britannica.com/animal/gayal

Farhadinia, M. S., Waldron, A., Kaszta, Ż., Eid, E., Hughes, A., Ambarlı, H., Al-Hikmani, H., Buuveibaatar, B., Gritsina, M. A., Haidir, I., Islam, Z. U., Kabir, M., Khanal, G., Koshkin, M. A., Kulenbekov, R., Kubanychbekov, Z., Maheshwari, A., Penjor, U., Raza, H., Rosen, T., et al. (2022). Current trends suggest most Asian countries are unlikely to meet future biodiversity targets on protected areas. *Communications Biology, 5*(1), 1221. https://doi.org/10.1038/s42003-022-04061-w

Figel, J. J., Safriansyah, R., Baabud, S. F., & Herman, Z. (2023). Clustered conflicts in disturbed lowlands characterize human–tiger interactions in Aceh, Indonesia. *Wildlife Letters, 1*(2), 83–91. https://doi.org/10.1002/wll2.12016

Ganbold, S. (2024). *Cattle stock in the Asia-Pacific region in 2022, by country or territory*. Statistica. https://www.statista.com/statistics/658597/asia-pacific-cattle-production-by-country/

George, S., Mohanarangan, A., Thampy, D. R., & Chandy, G. (2021). Prey composition of post-denning dhole (Cuon alpinus) in South Wayanad Forest Division, Kerala, India. *Proceedings of the Zoological Society*. https://doi.org/10.1007/s12595-020-00357-z

Ghaskadbi, P., Bathla, N., Bhandari, A., Modi, S., Nigam, P., & Habib, B. (2022). Feeding ecology of the endangered Asiatic wild dogs (Cuon alpinus) across tropical forests of the Central Indian Landscape. *Scientific Reports, 12*(1). https://doi.org/10.1038/s41598-022-17906-5

Giasuddin, M., Huque, K. S., & Alam, J. (2003). Reproductive potentials of gayal (Bos frontalis) under semi-intensive management. *Asian-Australasian Journal of Animal Sciences, 16*(3), 331–334. https://doi.org/10.5713/ajas.2003.331

Harihar, A., Pandav, B., & Goyal, S. P. (2011). Responses of leopard Panthera pardus to the recovery of a tiger Panthera tigris population. *Journal of Applied Ecology, 48*(3), 806–814. https://doi.org/10.1111/j.1365-2664.2011.01981.x

Hayward, M. W., Jędrzejewski, W., & Jêdrzejewska, B. (2012). Prey preferences of the tiger Panthera tigris. *Journal of Zoology, 286*(3), 221–231. https://doi.org/10.1111/j.1469-7998.2011.00871.x

Hayward, M. W., Lyngdoh, S., & Habib, B. (2014). Diet and prey preferences of dholes (Cuon alpinus): Dietary competition within Asia's apex predator guild. *Journal of Zoology, 294*(4), 255–266. https://doi.org/10.1111/jzo.12171

Heptner, V. G., Naumov, N. P., Yurgenson, P. B., Sludskii, A. A., Chirkova, A. E., & Bannikov, A. G. (1998). *Mammals of the Soviet Union Volume II, Part 1a: Sirenia and Carnivora*. Amerind Publishing Co. Pvt. Ltd..

Jackson, R. M, Ahlborn, G. G, Gurung, M., & Ale, S. (1996). *Reducing livestock depredation losses in the Nepalese Himalaya*. Proceedings of the vertebrate pest conference, p. 17. Retrieved from https://escholarship.org/uc/item/111648sx

Jha, K. K., & Campbell, M. (2022). *Dynamics and interrelations between nature, science and society*. Nova Science Publishers.

Johnsingh, A. J. T. (1992). Prey selection in three large sympatric carnivores in Bandipur. *Mammalia, 56*(4). https://doi.org/10.1515/mamm.1992.56.4.517

Kamler, J. F., Thatdokkham, K., Rostro-García, S., Bousa, A., Caragiulo, A., Crouthers, R., In, V., Pay, C., Pin, C., Prum, S., Vongkhamheng, C., Johnson, A., & Macdonald, D. W. (2020). Diet and prey selection of dholes in evergreen and deciduous forests of Southeast Asia. *The Journal of Wildlife Management, 84*(7), 1396–1405. https://doi.org/10.1002/jwmg.21931

Karanth, K. U., & Sunquist, M. E. (1995). Prey selection by tiger, leopard and dhole in tropical forests. *Journal of Animal Ecology, 64*(4), 439–450. https://doi.org/10.2307/5647

Kerley, L. L., Mukhacheva, A. S., Matyukhina, D. S., Salmanova, E., Salkina, G. P., & Miquelle, D. G. (2015). A comparison of food habits and prey preference of Amur tiger (Panthera tigris altaica) at three sites in the Russian Far East. *Integrative Zoology, 10*(4), 354–364. https://doi.org/10.1111/1749-4877.12135

Khaewphakdee, S., Simcharoen, A., Duangchantrasiri, S., Chimchome, V., Simcharoen, S., & Smith, J. L. D. (2020). Weights of gaur (Bos gaurus) and banteng (Bos javanicus) killed by tigers in Thailand. *Ecology and Evolution, 10*(11), 5152–5159. https://doi.org/10.1002/ece3.6268

Khan, M. Z., Khan, B., Awan, M. S., & Begum, F. (2017). Livestock depredation by large predators and its implications for conservation and livelihoods in the Karakoram Mountains of Pakistan. *Oryx, 52*(3), 519–525. https://doi.org/10.1017/s0030605316001095

Khatoon, R., Hussain, I., Anwar, M., & Nawaz, M. A. (2017). Diet selection of snow leopard (Panthera uncia) in Chitral, Pakistan. *Turkish Journal of Zoology, 41*, 914–923. https://doi.org/10.3906/zoo-1604-58

Kolipaka, S. S., Tamis, W. L., van 't Zelfde, M., Persoon, G. A., & de Iongh, H. H. (2017). Wild versus domestic prey in the diet of reintroduced tigers (Panthera tigris) in the livestock-dominated multiple-use forests of Panna Tiger Reserve, India. *PLoS One, 12*(4), e0174844. https://doi.org/10.1371/journal.pone.0174844

Krofel, M., Groff, C., Oberosler, V., Augugliaro, C., & Rovero, F. (2021). Snow leopard (Panthera uncia) predation and consumption of an adult yak in the Mongolian Altai. *Ethology Ecology & Evolution*, 1–8. https://doi.org/10.1080/03949370.2021.1872709

Kumar, V., Sharief, A., Dutta, R., Mukherjee, T., Joshi, B. D., Thakur, M., Chandra, K., Adhikari, B. S., & Sharma, L. K. (2022). Living with a large predator: Assessing the root causes of Human–brown bear conflict and their spatial patterns in Lahaul valley, Himachal Pradesh. *Ecology and Evolution, 12*(7). https://doi.org/10.1002/ece3.9120

Kumaraguru, A., Saravanamuthu, R., Brinda, K., & Asokan, S. (2010). Prey preference of large carnivores in Anamalai Tiger Reserve, India. *European Journal of Wildlife Research, 57*(3), 627–637. https://doi.org/10.1007/s10344-010-0473-y

Lamichhane, S., & Jha, B. R. (2015). Prey selection by Bengal Tiger Panthera tigris tigris (Mammalia: Carnivora: Felidae) of Chitwan National Park, Nepal. *Journal of Threatened Taxa, 7*(14), 8081. https://doi.org/10.11609/jott.2424.7.14.8081-8088

Lamichhane, S., Bhattarai, D., Maraseni, T., Shaney, K. J., Karki, J. B., Adhikari, B., Pandeya, P., Shrestha, B., & Adhikari, H. (2023). Landscape predictors influencing livestock depredation by leopards in and around Annapurna Conservation Area, Nepal. *PeerJ, 11*, e16516. https://doi.org/10.7717/peerj.16516

Letro, L., Wangchuk, N., & Wangdi, S. (2021). *Identifying human-wildlife conflict hotspots in Bhutan: Implications for prioritizing sites for conflict management.* Nature Conservation Division, Department of Forests and Park Services.

Lham, D., Cozzi, G., Sommer, S., Wangchuk, S., Lham, K., & Ozgul, A. (2021). Ecological determinants of livestock depredation by the snow leopard Panthera uncia in Bhutan. *Journal of Zoology, 314*(4), 275–284. https://doi.org/10.1111/jzo.12883

Luskin, M. S., Albert, W. R., & Tobler, M. W. (2017). Sumatran tiger survival threatened by deforestation despite increasing densities in parks. *Nature Communications, 8*, 1783. https://doi.org/10.1038/s41467-017-01656-4

Mahajan, P., Chaudhary, R., Kazi, A. K., & Khandal, D. (2022). Spatial determinants of livestock depredation and human attitude toward wolves in Kailadevi Wildlife Sanctuary, Rajasthan, India. *Frontiers in Ecology and Evolution*. https://doi.org/10.3389/fevo.2022.855084

Maheshwari, A., & Sathyakumar, S. (2020). Patterns of livestock depredation and large carnivore conservation implications in the Indian Trans-Himalaya. *Journal of Arid Environments, 182*, 104241. https://doi.org/10.1016/j.jaridenv.2020.104241

Malviya, M. & Ramesh, K. (2015). Human–Felid Conflict in Corridor Habitats: Implications for Tiger and Leopard Conservation in Terai Arc Landscape, India. (DOAJ: Directory of Open Access Journals). https://doi.org/10.26077/x9tx-gb97

Melletti, M., & Burton, J. (Eds.). (2014). *Ecology, Evolution and Behaviour of Wild Cattle*. Cambridge University Press. https://doi.org/10.1017/cbo9781139568098

Miller, J. R. B., Jhala, Y. V., & Jena, J. (2016). Livestock losses and hotspots of attack from tigers and leopards in Kanha Tiger Reserve, Central India. *Regional Environmental Change, 16*(Suppl 1), 17–29. https://doi.org/10.1007/s10113-015-0871-5

Muhammed, N., Kamal, M. T., Haque, F., Chowdhury, M. S. H., & Koike, M. (2007). A study on the Royal Bengal Tiger (Panthera tigris tigris) of the Sundarbans in Bangladesh with special reference to tiger–human conflict. *Journal of Social Research and Development, 4*, 86–91.

Naha, D., Dash, S. K., Chettri, A., Chaudhary, P., Sonker, G., Heurich, M., Rawat, G. S., & Sathyakumar, S. (2020). Landscape predictors of human-leopard conflicts within multi-use areas of the Himalayan region. *Scientific Reports, 10*(1), 11129. https://doi.org/10.1038/s41598-020-67980-w

Nakazawa, C., Tungalagtuya, K., Maruyama, N., & Suda, K. (2008). Food habits of gray wolves in the Bogdkhan mountain strictly protected area, Mongolia. *Biosphere Conservation: For Nature, Wildlife, and Humans, 9*(1), 1–8. https://doi.org/10.20798/biospherecons.9.1_1

National Geographic. (2006). *Family reference atlas of the world*. National Geographic Society.

National Tiger Conservation Authority. (2023). *Tiger reserve list*. Government of India.

National Tiger Conservation Authority (NTCA). (2022). *Tiger status report-2018*. Government of India.

Newsome, T. M., Boitani, L., Chapron, G., Ciucci, P., Dickman, C. R., Dellinger, J. A., López-Bao, J. V., Peterson, R. O., Shores, C. R., Wirsing, A. J., & Ripple, W. J. (2016). Food habits of the world's grey wolves. *Mammal Review, 46*(4), 255–269. https://doi.org/10.1111/mam.12067

Nyhus, P. J., & Tilson, R. (2010). Panthera tigris vs Homo sapiens: Conflict, coexistence, or extinction. In P. J. Nyhus & R. Tilson (Eds.), *Tigers of the world: The biology, politics, and conservation of Panthera tigris* (2nd ed., pp. 125–141). Elsevier Press. https://doi.org/10.1016/B978-0-8155-1570-8.00008-6

Oli, M. K., Taylor, I., & Rogers, M. (1994). Snow leopard Panthera uncia predation of livestock: An assessment of local perceptions in the Annapurna Conservation Area, Nepal. *Biological Conservation, 68*(1), 63–68. https://doi.org/10.1016/0006-3207(94)90547-9

Pant, B., Sharma, H. P., Dahal, B. R., Regmi, S., & Belant, J. L. (2023). Spatio-temporal patterns of human-wildlife conflicts and effectiveness of mitigation in Shuklaphanta National Park, Nepal. *PLoS One, 18*(4), e0282654. https://doi.org/10.1371/journal.pone.0282654

Pocock, R. I. (1939). *Panthera leo. The Fauna of British India, including Ceylon and Burma. Vol. Mammalia 1* (pp. 212–222). Taylor and Francis Ltd..

Qi, X. B., Jianlin, H., Wang, G., Rege, J. E., & Hanotte, O. (2010). Assessment of cattle genetic introgression into domestic yak populations using mitochondrial and microsatellite DNA markers. *Animal Genetics, 41*(3), 242–252. https://doi.org/10.1111/j.1365-2052.2009.01989.x

Rajaratnam, R., Vernes, K., & Sangay, T. (2016). A review of livestock predation by large carnivores in the Himalayan Kingdom of Bhutan. In F. Angelici (Ed.), *Problematic wildlife*. Springer. https://doi.org/10.1007/978-3-319-22246-2_8

Ram, M., Sahu, A., Srivastava, N., Chaudhary, R., & Jhala, L. S. (2023). Diet composition of Asiatic lions in protected areas and multi-use land matrix. *Journal of Vertebrate Biology, 72*, 22065.1–22065.9. https://doi.org/10.25225/jvb.22065

Ramesh, T., Kalle, R., Milda, D., Gayathri, V., Thanikodi, M., Ashish, K., & Giordano, A. J. (2020). Patterns of livestock predation risk by large carnivores in India's Eastern and Western Ghats. *Global Ecology and Conservation, 24*, e01366. https://doi.org/10.1016/j.gecco.2020.e01366

Ranjan, A. K., Anand, A., Vallisree, S., & Singh, R. K. (2016). LU/LC change detection and forest degradation analysis in Dalma Wildlife Sanctuary using 3S technology: A case study in Jamshedpur-India[J]. *AIMS Geosciences, 2*(4), 273–285. https://doi.org/10.3934/geosci.2016.4.273

Rawal, A. K., Timilsina, S., Gautam, S., Lamichhane, S., & Adhikari, H. (2024). Asiatic black bear-human conflict: A case study from Guthichaur rural municipality, Jumla, Nepal. *Animals: An Open Access Journal from MDPI, 14*(8), 1206. https://doi.org/10.3390/ani14081206

Saberwal, V. K., Gibbs, J. P., Chellam, R., & Johnsingh, A. J. T. (1994). Lion-human conflict in the Gir Forest, India. *Conservation Biology, 8*(2), 501–507. https://doi.org/10.1046/j.1523-1739.1994.08020501.x

Sanders, J. O. (1980). History and development of zebu cattle in the United States. *Journal of Animal Science, 50*(6), 1188–1200. https://doi.org/10.2527/jas1980.5061188x

Sanderson, E., Forrest, J., Loucks, C., Ginsberg, J., Dinerstein, E., Seidensticker, J., Leimgruber, P., Songer, M., Heydlauff, A., O'Brien, T., Bryja, G., Klenzendorf, S and Wikramanayake, E. (2006). *Setting Priorities for the Conservation and Recovery of Wild Tigers: 2005-2015. The Technical Assessment. WCS, WWF, Smithsonian, and NFWF-STF*, New York and Washington, DC, USA.

Sangay, T., & Vernes, K. (2008). Human–wildlife conflict in the Kingdom of Bhutan: Patterns of livestock predation by large mammalian carnivores. *Biological Conservation, 141*(5), 1272–1282. https://doi.org/10.1016/j.biocon.2008.02.027

Sankar, K., & Johnsingh, A. J. T. (2002). Food habits of tiger (Panthera tigris) and leopard (Panthera pardus) in Sariska Tiger Reserve, Rajasthan, India, as shown by scat analysis. *Mammalia, 66*(2), 285–289.

Selvan, K. M., Veeraswami, G. G., & Hussain, S. A. (2013). Dietary preference of the Asiatic wild dog (Cuon alpinus). *Mammalian Biology, 78*(6), 486–489. https://doi.org/10.1016/j.mambio.2013.08.007

Shahbandeh, M. (2024). *Number of cattle worldwide from 2012 to 2024 (in million head)*. Statistica. https://www.statista.com/statistics/263979/global-cattle-population-since-1990/

Shivangi, V., & Dharaiya, N. (2021). Assessment of wildlife habitat in Taranga hill forest with reference to change in land use and land cover. *eJournal of Applied Forest Ecology (eJAFE), 9*(1), 22–28. www.ejafe.com

Sijapati, R. K., Sharma, H. P., Sharma, S., Subedi, J. R., & Belant, J. L. (2021). Livestock depredation by leopards and tigers near Bardia National Park, Nepal. *Animals: An Open Access Journal from MDPI, 11*(7), 1896. https://doi.org/10.3390/ani11071896

Singh, A. P., De, K., Uniyal, V. P., & Sathyakumar, S. (2024). Livestock depredation by large carnivores in Western Himalayan region of Jammu and Kashmir: Temporal adherence in predator's choice. *Tropical Ecology, 65*, 26–31. https://doi.org/10.1007/s42965-022-00290-6

Suryawanshi, K. R., Bhatnagar, Y. V., Redpath, S. M., & Mishra, C. (2013). People, predators and perceptions: Patterns of livestock depredation by snow leopards and wolves. *Journal of Applied Ecology, 50*(3), 550–560. https://doi.org/10.1111/1365-2664.12061

Thinley, P., Kamler, J. F., Wang, S. W., Lham, K., Stenkewitz, U., & Macdonald, D. W. (2011). Seasonal diet of dholes (Cuon alpinus) in northwestern Bhutan. *Mammalian Biology, 76*(4), 518–520. https://doi.org/10.1016/j.mambio.2011.02.003

Tiralla, N., Holzapfel, M., & Ansorge, H. (2021). Feeding ecology of the wolf (Canis lupus) in a near-natural ecosystem in Mongolia. *Mammalian Biology, 101*, 83–89. https://doi.org/10.1007/s42991-020-00093-z

Tiwari, M. P., Devkota, B. P., Jackson, R. M., Chhetri, B. B. K., & Bagale, S. (2020). What factors predispose households in Trans-Himalaya (Central Nepal) to livestock predation by snow leopards? *Animals, 10*(11), 1–14. https://doi.org/10.3390/ani10112187

UN Revision of World Population Prospects. (2022). United Nations Department of Economic and Social Affairs, Population Division.

UN Revision of World Population Prospects. (2024). United Nations Department of Economic and Social Affairs, Population Division.

Vasavada, D. T., Rana, V. J. & Ram, M. (2022). *Management plan for Gir Protected Areas*, vol. 1. Gujarat Forest Department, Gujarat, India.

Wang, S. W., & Macdonald, D. W. (2006). Livestock predation by carnivores in Jigme Singye Wangchuck National Park, Bhutan. *Biological Conservation, 129*(4), 558–565. https://doi.org/10.1016/j.biocon.2005.11.024

Wang, M. S., Zeng, Y., Wang, X., Nie, W. H., Wang, J. H., Su, W. T., Otecko, N. O., Xiong, Z. J., Wang, S., Qu, K. X., Yan, S. Q., Yang, M. M., Wang, W., Dong, Y., Wu, D. D., & Zhang, Y. P. (2017). Draft genome of the gayal, Bos frontalis. *GigaScience, 6*(11), 1–7. https://doi.org/10.1093/gigascience/gix094

Widodo, F. A., Imron, M. A., Sunarto, S., & Giordano, A. J. (2022). Carnivores and their prey in Sumatra: Occupancy and activity in human-dominated forests. *PLoS One, 17*(3), e0265440. https://doi.org/10.1371/journal.pone.0265440

Wildlife Institute of India. (2023). *List of tiger reserves in India*. https://wii.gov.in/wildlife_database_list_of_tiger_reserves_in_india

Xu, W., Xiao, Y., Zhang, J., Yang, W., Zhang, L., Hull, V., Wang, Z., Zheng, H., Liu, J., Polasky, S., Jiang, L., Xiao, Y., Shi, X., Rao, E., Lu, F., Wang, X., Daily, G. C., & Ouyang, Z. (2017). Strengthening protected areas for biodiversity and ecosystem services in China. *Proceedings of the National Academy of Sciences of the United States of America, 114*(7), 1601–1606. https://doi.org/10.1073/pnas.1620503114

Zhang, R. C. (2000). Interspecies hybridization between yak, Bos taurus and Bos indicus and reproduction of the hybrids. In X. X. Zhao & R. C. Zhang (Eds.), *Recent advances in Yak reproduction*. International Veterinary Information Service. http://www.ivis.org. Paper No. A1304.0900.

Predators of Cattle in Europe

Introduction

Europe is very effectively managed for both agricultural development and conservation policy (Pereira & Navarro, 2015; Gordon et al., 2021; Directorate-General for Agriculture and Rural Development, 2022; Hurduzeu et al., 2022; European Commission, 2024). Vinci (2022) writes that "in 2020, there were over 76 million cattle in the European Union (EU) and beef production reached 6.8 million tonnes—output that makes the EU the world's third largest producer, after the United States and Brazil." The agricultural sector in the European Union is also diverse for cattle herding, in terms of "herd size, farm structure and geographical distribution of farms between EU regions" with beef of high quality and exported globally (Vinci, 2022) (Figs. 1 and 2 show examples of European cattle). However, conflicts have emerged between efficient agricultural systems, including cattle herding, and wildlife conservation. For example, Gehring et al. (2010: 299) note that large carnivore and ungulate species have a similar history of extinction and subsequent recovery in North America and Europe. Managing human-animal conflicts at the intersection of livestock farming and wildlife conservation presents both opportunities and challenges for both continents. A recurring and growing worldwide issue is livestock predation and disease transmission between livestock and wildlife.

Regarding European wildlife, especially large carnivores, there have been recent population recoveries (Meriggi & Lovari, 1996; Pereira & Navarro, 2015; Stringham & Rogers, 2023). Gervasi et al. (2021) note that, referring to brown bears (*Ursus arctos* Linnaeus, 1758), lynx (*Lynx* Linnaeus, 1758), wolves (*Canis lupus* Linnaeus, 1758), and wolverines (*Gulo gulo* Linnaeus, 1758), "after centuries of decline due to multiple causes (extermination policies, habitat destruction, reduction in the prey base, etc.) all these four species have progressively regained space, expanded their numbers, and recovered much of their former distribution during the last 50 years." This occurred primarily as a result of several international conventions that elevated them from status as pest species to conservation priorities and established the framework for their national legal protection. It was also due to a number of larger social,

© The Author(s), under exclusive license to Springer Nature Switzerland AG 2025
M. O. Campbell, *Cattle, Their Predators and Geomatics Research*,
https://doi.org/10.1007/978-3-031-97363-5_6

Fig. 1 Cattle in Val d'Ayas, Italy. (Permission of Francesca Occhiuto-unsplash)

economic, and historical processes, such as reforestation and the progressive abandonment of agricultural lands, which reduced human impact and increased territory for large carnivores and their prey species (see also Chapron et al., 2014). This phenomenon of predator recovery may have positive or negative impacts, depending on the management. Some authors have cited effective management with positive influences, while others note negative effects, especially on livestock systems (Gervasi et al., 2021).

For example, Linnell and Cretois (2018) concur that large carnivore populations, including those of wolves, brown bears, Eurasian lynxes, and wolverines, have responded favorably to improved habitat conditions and supportive laws in recent decades. They have consolidated their presence in regions where they had declined and returned to numerous parts of Europe from which they had been absent for decades or even centuries. Large carnivores are becoming powerful symbols in the twenty-first century, but this recovery has also led to a number of conflicts with agricultural and rural stakeholders regarding the direct effects of large predators on livestock through predation, as well as a wider range of social conflicts that center on the difficulties facing rural communities. The populations of wildlife include 17,000 wolves, 15,000–16,000 brown bears, 8000–9000 Eurasian lynx, and

Fig. 2 Cattle in Sweden. (Permission Alex Kotomanov-unsplash)

1000–1250 wolverines in continental Europe outside Russia and Belarus (Linnell & Cretois, 2018).

These predatory species are mostly managed by the pan-European Bern Convention (CoE) and the Habitats Directive (EU), including the relationship between the carnivores and farming systems (Linnell & Cretois, 2018). An important consideration concerns the ability of large carnivores to adapt to variable land-cover, including human-dominated, agricultural landscapes (Linnell & Cretois, 2018). The authors point out that the main factors driving the recovery of predator populations include "greater legislative protection, reforestation, the recovery of wild herbivore populations, and reduced human impacts on the landscape associated with rural-urban migration, land abandonment, and urbanization in the twentieth century, which have facilitated a large-scale recovery of large carnivores across Europe" (Linnell & Cretois, 2018; see also Meriggi & Lovari, 1996; Pereira & Navarro, 2015; Stringham & Rogers, 2023). While this influences the success of conservation initiatives such as the Bern Convention and Habitats Directive, United Nations's Agendas, it is also evident that the existence of large carnivores is linked to a variety of effects on people's economic interests as well as pervasive social disputes between various interest groups with differing opinions about how to manage large carnivores and rural areas in Europe (Linnell & Cretois, 2018).

Linnel and Cretois (2018: 27) note that predation occurs wherever domestic animals and carnivores co-occur. However, the extent of the destruction and the species involved vary widely. The most fundamental factor leading to vulnerability is the size ratio between large carnivores and livestock. Small livestock species/breeds (e.g., sheep and goats) are more susceptible to death by more carnivore species than large livestock species/breeds (e.g., cattle, horses), and young animals of all species/breeds are more susceptible to death by more carnivores than adults. In general, in Europe, livestock predation is principally by wolves and affects principally sheep and, to a lesser extent, goats (Linnell & Cretois, 2018). A comparison of predation of livestock was compiled by Linnell and Cretois (2018) (Table 1). Linnell and Cretois (2018) note that "for the period 2012–2016 an annual average of 21,000 sheep were compensated as being killed by large carnivores within the EU countries, of which 92% were attributed to wolves." In addition, "the most basic factor leading to vulnerability is the size ratio between large carnivore and livestock. Small livestock species/breeds (e.g. sheep and goats) are vulnerable to being killed by more carnivore species than are large livestock species/breeds (e.g. cattle, horses), and juveniles of all species/breeds are vulnerable to more carnivores than are adults."

This means that in Europe, bears and wolves are typically the only animals that hunt cattle and horses, and they typically kill only calves and foals—rarely adults. Because of this, adult sheep and goats are especially vulnerable to wolves and bears; lambs are typically killed by lynxes and wolverines. Moreover, selecting large livestock species or breeds is the primary corrective measure that keeps predators from considering livestock as prey. Cattle greatly lessen vulnerability to wolves and bears and successfully eradicate lynx and wolverine predation, which is a problem when sheep or goats are used instead. When they are young, calves should be kept in stables or close to homes, as they may still be vulnerable. Cattle are less vulnerable, but when they die, their death costs are higher. Cattle, therefore, also require some level of protection. Additionally, there are significant advantages to changing to breeds or selectively breeding individuals that show robust anti-predator behavior, are good for herding, or are suitable for other mitigation strategies (Linnell & Cretois, 2018).

Table 1 Number of livestock compensated per year (average 2012–2016) in Europe

Predator	Cattle	Sheep	Goats	Horses	Reindeer	Dogs	Others	Total
Wolves	2329	22,407	4837	645	674	214	222	31,328
Bears	170	3920	31	20	826	140	80	5187
Lynx	14	5646	12	1	6885	3	13	12,574
Wolverines	0	7471	0	0	9000	0	2	16,473
Total	2513	39,444	4880	666	17,385	357	317	65,562

Source: Linnell and Cretois (2018: Tables 14, 15, 16, 17, pp. 45–47)
This table shows the number of livestock compensated per year (average 2012–2016) in Europe

Brown bears rank second after wolves in cattle predation (Fico et al., 1993). For example, Fico et al. (1993: 39) note that wolves "and brown bears (Ursus arctos) attack livestock in the Abruzzo region of Italy and using the compensation payments to owners for losses caused by predators as indicators, of the 4993 validated claims made between 1980 and 1988, the livestock losses ascribed to wolves, bears and wild boars were 92.1%, 7.8% and 0.1%, respectively." Cattle comprised 16.4% of the losses to wolves (the rest were sheep and goats 45.0%, equids 38.5% and rabbits, fowl and dogs 0.2%). Cattle also comprised 18.5% of the losses to bears (the rest were sheep and goats 67.6%, equids 9.5% and bee-hives 4.4%). Fico et al. (1993: 39) note that "a significant difference was observed between the number of attacks by wolves and bears; although both populations are of similar size… Among cattle and equids, victims are mostly new-born or young animals, while among sheep and goats only adult animals are attacked."

The rest of this chapter will examine large predator attacks on cattle in Europe, almost entirely by wolves and brown bears, with rare cases of lynx predation (Meriggi & Lovari, 1996; Pereira & Navarro, 2015; Linnell & Cretois, 2018; Khorozyan & Heurich, 2023; Stringham & Rogers, 2023). The main strands are the increased populations of bears and wolves, their rarer predation on cattle than on sheep, the high cost of cattle predation to the larger size and higher cost of cattle, and any ameliorative actions undertaken. Europe is cited as having perhaps the most effective predator conservation strategies; hence, conservation policies in this continent may be emulated in other continents.

Wolf Predation on Cattle

Wolves are recorded as the principal predators of livestock (Linnell & Cretois, 2018) (Figs. 3 and 4). Aligning with the general increase in predator populations, wolf attacks on cattle have increased in some localities. For example, Pimenta et al. (2017: 17) note that "wolf attacks on cattle greatly increased (>3 times) in Portugal in 1999–2013, there were far more wolf attacks on free-ranging than on semi-confined herds, but protecting free-ranging herds at night in winter would greatly decrease wolf predation risk, keeping calves <3 months old indoor would decrease wolf predation in semi-confined herds and under adequate management both free-ranging and semi-confined herds may be compatible with wolf presence." Pimenta et al. (2017: 17) add that "government records from 2012 to 2013 indicated that only <2% of cattle farms suffered wolf attacks, of which <4% had >10 attacks per year." Attacks by predators targeted the free-range system, which was defined by herds owned by multiple owners grazing mostly on public land far from cover and infrequently in enclosed areas. By altering the practices of about 25% of farmers in this system, it was possible to reduce wolf attacks by protecting these herds at night during the winter. Because the herds were frequently confined by fences or barns and grazed in pastures nearer shelters, attacks were much less frequent in the semi-closed system. About 50% of farmers in this system would be impacted if the

Fig. 3 Wolf in Abruzzo, Italy. (Permission Gabriele Brancati-pexels)

practice of releasing calves to pasture before 3 months of age were changed. This practice was linked to about 90% of attacks.

Rewilding Europe (2023) suggests that the European Commission should prioritize initiatives to enhance and facilitate improved livestock safeguards, enabling the coexistence of wolves in regions dominated by humans and livestock farming, instead of diminishing the wolf's protected status. It is crucial to note that livestock predation poses the most significant obstacle to harmonious living with wolves in Europe and worldwide. Rewilding Europe (2023) argues that this issue constitutes a significant conservation concern, as it may result in the unlawful or lawful hunting and trapping of wolves, as well as the use of illegal poison bait, posing threats to numerous endangered species and even human safety and domestic animals. Estimates suggest that wolves are responsible for the annual loss of between 30,000 and 40,000 European livestock, mainly sheep. Consequently, approximately eight million euros in compensation are paid to European livestock farmers annually. From an economic perspective, considering the billions of euros in agricultural subsidies distributed throughout Europe through the Common Agricultural Policy (CAP), this amount is relatively insignificant. The issue of livestock predation is further complicated by the scarcity of natural prey in numerous European

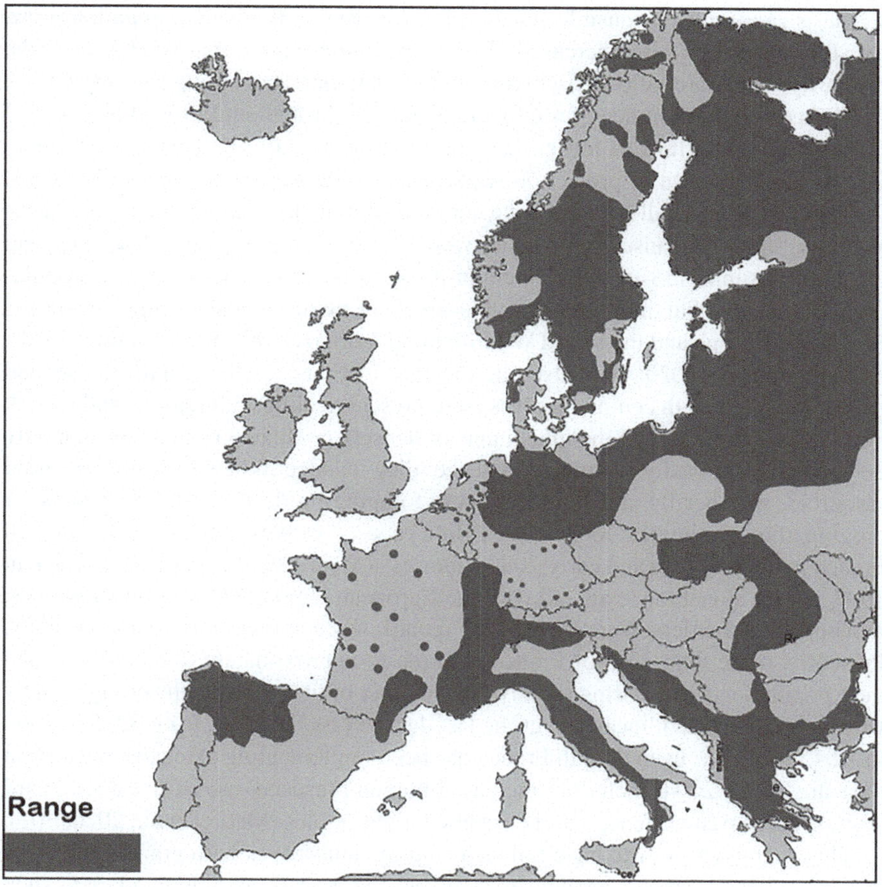

Fig. 4 Wolf Range CC0 1.0 Universal

landscapes, encouraging wolves to rely heavily on livestock as a food source, making it challenging to establish successful and widely accepted conservation strategies (Rewilding Europe, 2023).

Tomma (2024) comments that "the European Commission released a proposal to weaken protections for wolves living in the 27 nations of the European Union, drawing criticism from environmental groups. Just days later, environmentalists persuaded a court in Switzerland, which is not a member of the EU, to partially block a new government plan to kill up to 70% of the nation's wolf population." This would reduce the legal protection of the wolf from "strictly protected" to "protected." It is argued that in the 1970s, only a few isolated populations of wolves remained in Europe following centuries of hunting, but new research indicates that as many as 20,000 of these animals may now be found there. The Bern Convention on the Conservation of European Wildlife and Natural Habitats, a 40-year-old conservation agreement, provides protections for wolves and other large carnivores, which is primarily responsible for the increase. But as the wolf population grew, livestock predation also increased. The commission claims that 65,000 animals, primarily sheep, are killed by wolves annually. Despite accounting for only 0.07% of all sheep on the continent, farming organizations throughout Europe have exerted pressure on authorities to loosen laws prohibiting the killing of wolves (Tomma, 2024). The reduction in protection would permit EU nations to mass kill wolves, though they would still have to make sure the animals maintain a “favorable” conservation status. Because each country would set its own kill quotas, deadlines, and techniques, supporters of the plan claim it would be simpler to keep wolf populations at healthy but more manageable levels. Environmental groups, including Rewilding Europe and the World Wildlife Fund, criticized this plan (Tomma, 2024).

Meuret et al. (2020: 401) also cite the case of France, where “until the second half of the nineteenth century, wolves were present in almost all regions and mixed, not without conflict, with the inhabitants of densely populated rural areas. Not only do wolves adapt easily to human presence, they manage to benefit from their food resources, whether by attacking livestock, foraging on deceased farm animals, or digging through human food waste.” In France, “wolves were exterminated in France in the late 19th and early 20th centuries” (Meuret et al., 2020: 401). Thus, in 1993, a year after France agreed with the European Union (EU) to protect wolves, ranchers and herders were caught off guard when wolves arrived from Italy. Presently, more than 33% of France is home to approximately 5580 wolves, and their population is increasing rapidly. The number of livestock deaths brought on by wolves has increased linearly over the last decade (2009–2019), from 3215 in 2009 to 12,451 in 2019, even though France has been implementing extensive mitigation measures since 2004, including increased human presence, use of livestock guard dogs, secured pasture fences, and electrified night stables (Meuret et al., 2020: 401).

These efforts appear to have failed, as mosaic landscapes with grazing livestock provide suitable foraging grounds for wolves, and wolves are able to adapt to such methods. They no longer appear to link farm animals and people to danger. Meuret et al. (2020: 401) note that currently, despite the presence of dogs and people, half of successful wolf attacks on livestock take place during the day. France recently revised its wolf management policy, citing the high costs of insufficient protection. In addition to non-lethal precautions, breeders who have experienced repeated wolf attacks are now permitted to shoot wolves defensively, defying the law. Meuret et al. (2020: 411) note that herd protection in France, regulated by national regulations based on a European directive, is based on the premise that wolves are afraid of humans. In reality, this fear is not an intrinsic and permanent feature of the species. Rather, it is an at least partially acquired behavior that must be constantly strengthened by explicitly associating the close presence of a human with a real threat. Deterrence techniques or non-lethal defensive measures are only useful as a signal to remind people of the risk of death or serious injury due to non-compliance. Therefore, non-lethal and lethal techniques need to be better integrated or used in combination when necessary (Meuret et al., 2020: 411).

Brown Bear Predation on Cattle

Brown bears have been cited as important factors in predation on cattle, although to a lesser extent than wolves (Figs. 5 and 6). Rigg et al. (2011: 272) point out that gray wolves are blamed for four to six times more damage than brown bears, although in the spring and autumn, when sheep and cattle were most at risk, livestock were present in only 2 of 78 wolf feces. The main factor is that the recovery of large carnivores in Europe is accompanied by a resurgence of age-old conflicts with human interests. Carnivores prey on livestock, damage crops, compete with hunters for game, and occasionally injure or kill humans. The authors suggest nonlethal methods including separating livestock from carnivores (e.g., predator-proof fences, avoiding high-risk areas/seasons, confining livestock at night or in inclement weather) and deterrence from predators (e.g., electric fences, audio, visual and chemical repellents, conditioned taste aversion, proper carcass disposal, diversionary feeding, replacing at-risk stocks, adjusting calving/lambing times and location), livestock protection (e.g., guarding animals, harassing them, shooting predators with non-lethal projectiles, attaching protective collars, leaving horns on cattle) and

Fig. 5 Brown bear in Ukraine. (Permission Valeriy Pelts-pexels)

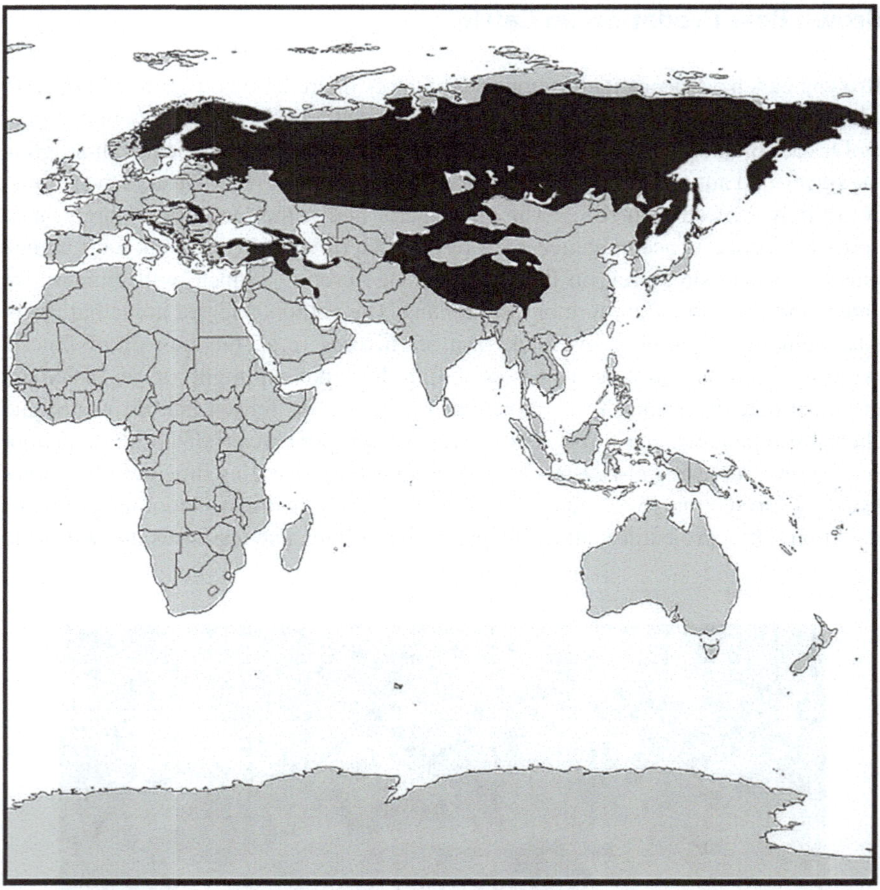

Fig. 6 Eurasian Brown bear range. (Permission Hannu Attribution-Share Alike 3.0 Unported)

compensation systems (Rigg et al., 2011: 272). The case is cited for bears in Slovakia, where a 30-year moratorium on bear hunting allowed the bear "to recover from 20 to 60 individuals in 1932 to 770–870" (Rigg et al., 2011: 272). The authors note that such large predators are largely restricted to the Carpathian Mountains, which cover 59% of Slovakia's 49,035 km^2 (Rigg et al., 2011: 273), and that the belief has even become more entrenched in Slovakia in recent years that there are too many bears even though the actual cost of economic damage has not increased. The results confirm that local conditions, rather than predator numbers, play a key role in determining the extent of losses (Rigg et al., 2011: 277).

Another study is cited of Norway, where it is noted that bears are less likely to prey on cattle than sheep (Zimmermann et al., 2003: 72). The theme was the replacement of sheep with cattle, on the premise that cattle would be less vulnerable to bear predation than sheep. The study was based in Hedmark County, south-eastern Norway, which "borders a reproducing brown bear population in Sweden and already contains several resident male brown bears"; in addition, "a brown bear conservation zone, within which bear density is planned to increase, covers 46% of the county's surface. There were about 7 times as many free-ranging sheep (128,600) as cattle (18,200) during summer 1998, with densities lowest inside the bear conservation zone. Estimated free-ranging cattle mortality was about 16 times lower than sheep mortality in summer 1998" (Zimmermann et al., 2003: 72). Importantly, for the 13 years prior to the publication date, "no cattle were confirmed as killed by brown bears" and "the perception of the brown bear as a threat to cattle is higher among farmers than among veterinarians or managers" (Zimmermann et al., 2003: 72).

Bear predation on cattle was considered unclear, but the authors point out that brown bear predation on cattle was relatively small, although there was evidence of bear predation on cattle in other countries. The authors pose the question: What might happen if sheep are removed and replaced by cattle? Which factors could expose those cattle to an increased brown bear predation risk? An optimal foraging strategy predicts the selection of easily accessible prey, which may involve prey that are locally abundant, less effective at avoiding predators, or less protected. In many areas with both sheep and cattle farming, brown bears appear to prefer sheep. If sheep are removed, cattle may be at greater risk (Zimmermann et al., 2003: 79). Cattle calves may be more vulnerable to bear predation due to their smaller size. In this case, large male brown bears may play a role due to their large size. Therefore, increased recovery of the Scandinavian Brown bear populations is hypothesized to increase the risk of predation for cattle and cattle ranges in forested areas, the dominant habitat of the brown bear sanctuary in Hedmark County, and therefore may be more exposed to predation (Zimmermann et al., 2003: 79).

The authors concluded that a larger proportion of farmers within the brown bear protection zone reported experiencing attacks by brown bears on cattle compared to those outside the brown bear protection zone (Zimmermann et al., 2003: 80). None of these farmers had electrically fenced their livestock and few used non-electric wire fencing, and free-roaming cattle could be a good alternative to sheep in Norway's brown bear sanctuaries. To date, brown bears have posed little threat to cattle within the bear protection zone. This could be partly explained by the lower frequency of uncontrolled calving within the brown bear protection zone than in other regions (Zimmermann et al., 2003: 80). Additionally, the risk of cattle predation may increase if sheep, a lighter and more accessible prey, are removed. The risk of cattle predation may also increase as the brown bear's reproductive range expands, resulting in an increase in older, large males (Zimmermann et al., 2003: 81).

Khorozyan and Waltert (2020) explore this thesis that sheep are more vulnerable to predation by bears than cattle and that replacing cattle with sheep can reduce bear predation on livestock. It is argued that livestock replacement aims to replace commonly killed livestock with livestock that are less likely to be killed by bears or other predators. One way this can be achieved is by replacing sheep with cattle or replacing heavy breeds of sheep with more agile ones, but only the former option has been tested with bears. While bears may kill fewer cattle than sheep, making replacing sheep with cattle numerically effective, this approach may be controversial in terms of financial benefits, as the market price of one head of cattle is equivalent to that of several sheep (Khorozyan & Waltert, 2020).

Taking the case of Eastern Europe, Pop et al. (2023) report that from 756 bear predation events (BPEs), cows were the most frequent livestock type reported with 410 incidents (54.2%), followed by sheep (n = 211, 27.9%) and other livestock (n = 135, 17.9%). Such bear predation events are influenced by complex factors, leading to a non-random distribution of BPEs. These factors were associated with brown bear life history, biology, and ecology, and resultant bear movements and food requirements, and associations with humans and livestock. It is argued that human actions and infrastructure, livestock management, and preventative measures can affect the occurrence of negative interactions and BPEs (Pop et al., 2023). In this study, the predictors for bear predation on cattle were those of the "global model" which included the predictive variables "relative bear abundance, proportion open habitat, proportion of agriculture, and Shannon Diversity Index (SDI), distance to forest, distance to settlement, and their interaction." In the case study, "all variables were significant for predicting BPEs with cows. Distance to forest had the strongest overall negative effect on the likelihood of cow predation, with the odds of damage decreasing in average by 75% for each additional Standard Deviation increase in distance to forest. However, this relationship changed based on distance to settlement; cow predation decreased with distance to forest for areas close to settlements but increases with distance to forest for areas far from settlements" (Pop et al., 2023). In addition to "distance from forest," which "was a strong negative predictor for cows and sheep," "distance to villages was a strong negative predictor for cows" and "landscape heterogeneity was positively associated with cow and other livestock" (Pop et al., 2023).

Geomatics-Based Assessment

Geomatics-based assessments may be used to determine predator populations, suitable habitats, habitat decline, and land-use conflicts between predators, livestock, and humans. Numerous studies have used geomatics (GIS, remote sensing) to record presence, track animals, and assess habitats. The geomatics-based tools include satellite images, aerial and field photographs, LiDAR, GPS, and drones. The European Environmental Agency (2019) states that "the EU Biodiversity Strategy to 2020 calls on Member States to carry out a mapping and assessment of ecosystems and their services. As such, an EU-wide ecosystem assessment was launched to provide harmonised information on the condition of ecosystems and biodiversity, and their capacity to provide ecosystem services." Regarding these types of assessments of landcover and habitats, Fedorca et al. (2019) present a study based in Romania, which has one of largest bear populations in Europe. It is argued that "landscape genetics is increasingly being used in landscape planning for biodiversity conservation by assessing habitat connectivity and identifying landscape barriers, using intraspecific genetic data and quantification of landscape heterogeneity to statistically test the link between genetic variation and landscape variability."

For example, Scharf and Fernández (2018: 1573) argue that "large carnivore populations in Europe are expanding into new areas. This generates opportunities to improve their conservation status, but also creates a need to address new conflicts with humans. Species management units are constrained by administrative boundaries, but effective conservation and conflict management require a continental-scale perspective on the opportunities and limitations for expanding populations." This study "evaluated the accuracy of local population models to predict European brown bear (Ursus arctos) distribution patterns in other populations and at the continental scale" and used GIS software (ArcGIS 10.1 (ESRI, Redlands, California, USA)) and statistical calculations using the r v3.3.3 statistical package (R Core Team, 2017; Scharf & Fernández, 2018: 1576). The results showed that determining the transferability of ecological models to predict habitat availability is crucial for guiding conservation management and policy. The transferability of the model to new areas can be influenced by a variety of factors, such as the model characteristics (including the discriminatory ability of the predictors), similarities between the area in which the model was adapted and the area to which the predictions are transferred, or species-specific adaptations to local environmental parameters (Scharf & Fernández, 2018: 1573). Importantly, brown bear populations expanded in recent years and recolonized their former habitat. The results suggested that bear populations may continue to expand into uninhabited suitable habitat (Scharf & Fernández, 2018: 1579). The results showed that integrating the results of habitat suitability models developed for multiple populations has clear advantages for providing more accurate, transferable, and informative predictions for conservation than models developed from local data were developed (Scharf & Fernández, 2018: 1580).

Regarding the resolution of satellite imagery used in GIS analysis and modeling, Cisneros-Araujo et al. (2021) note that computational approaches are recognized as one of the most useful and unbiased tools for guiding the design of nature management plans and policies, with habitat suitability and connectivity models particularly useful in producing spatially explicit information for conservation. Nevertheless, sufficient spatial resolution of the data and careful selection of environmental variables are essential to ensure reliable model performance. The use of remote sensing images is becoming increasingly important due to their potentially high spatial resolution as well as their broad temporal and spatial coverage and availability. The authors conclude that "brown bear habitat suitability and connectivity models" were developed from three datasets with different spatial coverage and accessibility: namely a Sentinel-1 and 2 land cover map (global coverage), pan-European vegetation and land cover layers (continental coverage) and LiDAR data, and the Forest Map of Spain (national coverage). The findings showed that the Sentinel imagery and pan-European datasets were suitable for habitat and connectivity modeling, but the Sentinel data was insufficient for detailed habitat–species associations were a wide range of vegetation types must be identified. Therefore, the thematic resolution of satellite-was important.

Steyaert et al. (2011) describe a GPS-based study on bear habitat and livestock management in Sweden. It is argued that in central Sweden, a traditional cattle herding system that permits free-range dairy cows to coexist with brown bears. Over the past 10 years, cattle predation remained consistent and was among the lowest rates recorded globally, despite an increasing brown bear population. However, important stakeholder support suggested a substantial reduction in brown bear populations to preserve the traditional husbandry system. During the free-range season, satellite tracking data was utilized to evaluate the resource selection of nine brown bears sympatric with six free-ranging cattle herds (Steyaert et al., 2011).

The methodology used a Global Positioning System-Global System for Mobile communication (GPS-GSM) collar on selected cattle in the herds and on bears (Televilt TellusTM GPS collars and Vectronic Aerospace GPS-plus collars), and the radio-locations were imported into a Geographical Information System (ArcGIS 9.3 and the ET Geowizards 9.9 extension), and the statistical modeling was based on generalized linear mixed models (GLMMs) using the statistical programming language and environment R 2.10.1 (R Development Core Team, 2009). The results suggested that there was a significant and negative relationship between the resource selection of brown bears and free-roaming cattle, mainly due to the inverse relationships between the species in vegetation density and human infrastructure, such as forest roads, buildings, and settlements. The probability of bear and free-ranging cattle encounters was mostly highest in areas with dense vegetation near the variables associated with human habitation. The authors conclude that with low reported predation rates and apparent habitat segregation between the bears and free-ranging cattle, the results do not support the argument to reduce brown bear numbers to protect cattle (Steyaert et al., 2011).

Conclusions

Wolves and brown bears are the main cattle predators in Eurasia. The principal factor for this predation is the increased numbers of both wolves and bears, and the effective management of agricultural development, which have contributed to the increased potential for cattle-predator contact. This contact has increased the potential of conflicts between the stakeholders of agricultural development and conservation management. This chapter has examined such situations in Eurasia before similar considerations for North America, which has a similar situation, and Asia, Africa, and South America, where conservation and agricultural management may be less efficient. Solutions concern more effective barriers between the predators and cattle, with less reliance of free ranging for livestock. These conflicts may point to the future of conservation and agricultural development.

References

Chapron, G., Kaczensky, P., Linnell, J. D. C., Arx, M. von, Huber, D., Andrén, H., López-Bao, J. V., Adamec, M., Álvares, F., Anders, O., Balčiauskas, L., Balys, V., Bedő, P., Bego, F., Blanco, J. C., Breitenmoser, U., Brøseth, H., Bufka, L., Bunikyte, R., & Ciucci, P. (2014). Recovery of large carnivores in Europe's modern human-dominated landscapes. *Science, 346*(6216), 1517–1519. https://doi.org/10.1126/science.1257553

Cisneros-Araujo, P., Goicolea, T., Mateo-Sánchez, M. C., García-Viñás, J. I., Marchamalo, M., Mercier, A., & Gastón, A. (2021). The role of remote sensing data in habitat suitability and connectivity modeling: Insights from the Cantabrian brown bear. *Remote Sensing, 13*(6), 1138. https://doi.org/10.3390/rs13061138

Directorate-General for Agriculture and Rural Development. (2022). *EU agricultural outlook 2022–32: Production growth of major agricultural sectors to slow down*. European Union. https://agriculture.ec.europa.eu/news/eu-agricultural-outlook-2022-32-production-growth-major-agricultural-sectors-slow-down-2022-12-08_en

European Environmental Agency (2019). *Mapping Europe's ecosystems*. https://www.eea.europa.eu/en/analysis/publications/mapping-europe-s-ecosystems

European Commission. (2024). *Strategic dialogue on the future of EU agriculture*. https://commission.europa.eu/strategy-and-policy/priorities-2019-2024/european-green-deal/agriculture-and-green-deal/strategic-dialogue-future-eu-agriculture_en

Fedorca, A., Russo, I. M., Ionescu, O., Ionescu, G., Popa, M., Fedorca, M., Curtu, A. L., Sofletea, N., Tabor, G. M., & Bruford, M. W. (2019). Inferring fine-scale spatial structure of the brown bear (Ursus arctos) population in the Carpathians prior to infrastructure development. *Scientific Reports, 9*(1), 9494. https://doi.org/10.1038/s41598-019-45999-y

Fico, R., Morosetti, G., & Giovannini, A. (1993). The impact of predators on livestock in the Abruzzo region of Italy. *Revue Scientifique et Technique (International Office of Epizootics), 12*(1), 39–50. https://doi.org/10.20506/rst.12.1.664

Gehring, T. M., VerCauteren, K. C., & Landry, J.-M. (2010). Livestock protection dogs in the 21st century: Is an ancient tool relevant to modern conservation challenges? *Bioscience, 60*(4), 299–308. https://doi.org/10.1525/bio.2010.60.4.8

Gervasi, V., Linnell, J. D., Berce, T., Boitani, L., Černe, R., Ciucci, P., Cretois, B., Derron-Hilfiker, D., Duchamp, C., Gastineau, A., Grente, O., Huber, D., Iliopoulos, Y., Karamanlidis, A. A., Kojola, I., Marucco, F., Mertzanis, Y., Männil, P., Norberg, H., Pagon, N., Pedrotti, L., Quenette, P. Y., Reljić, S., Salvatori, V., Talvi, T., von Arx, M., & Gimenez, O. (2021). Ecological correlates of large carnivore depredation on sheep in Europe. *Global Ecology and Conservation, 30*, e01798. https://doi.org/10.1016/j.gecco.2021.e01798

Gordon, I. J., Pérez-Barbería, F. J., & Manning, A. D. (2021). Rewilding lite: Using traditional domestic livestock to achieve rewilding outcomes. *Sustainability, 13*, 3347. https://doi.org/10.3390/su13063347

Hurduzeu, G., Pânzaru, R. L., Medelete, D. M., Ciobanu, A., & Enea, C. (2022). The development of sustainable agriculture in EU countries and the potential achievement of sustainable development goals specific targets (SDG 2). *Sustainability, 14*(23), 15798. https://doi.org/10.3390/su142315798

Khorozyan, I., & Heurich, M. (2023). Where, why and how carnivores kill domestic animals in different parts of their ranges: An example of the Eurasian lynx. *Global Ecology and Conservation, 46*, e02585. https://doi.org/10.1016/j.gecco.2023.e02585

Khorozyan, I., & Waltert, M. (2020). Variation and conservation implications of the effectiveness of anti-bear interventions. *Science Reports, 10*, 15341. https://doi.org/10.1038/s41598-020-72343-6

Linnell, J. D. C., & Cretois, B. (2018). *Research for AGRI Committee – The revival of wolves and other large predators and its impact on farmers and their livelihood in rural regions of Europe*. European Parliament, Policy Department for Structural and Cohesion Policies. Retrieved from https://www.europarl.europa.eu/RegData/etudes/STUD/2018/617488/IPOL_STU(2018)617488_EN.pdf

Meriggi, A., & Lovari, S. (1996). A review of wolf predation in southern Europe: Does the wolf prefer wild prey to livestock? *Journal of Applied Ecology, 1996*(33), 1561–1571.

Meuret, M., Moulin, C., Bonnet, O., Garde, L., Nozieres-Petit, M., & Lescureux, N. (2020). Missing shots: Has the possibility of shooting wolves been lacking for 20 years in France's livestock protection measures? *The Rangeland Journal, 42*(6), 401–413. https://doi.org/10.1071/RJ20046

Pereira, H. M., & Navarro, L. M. (Eds.). (2015). *Rewilding European landscapes*. Springer.

Pimenta, V., Barroso, I., Boitani, L., & Beja, P. (2017). Wolf predation on cattle in Portugal: Assessing the effects of husbandry systems. *Biological Conservation, 207*, 17–26. https://doi.org/10.1016/j.biocon.2017.01.008

Pop, M. I., Dyck, M. A., Chiriac, S., Lajos, B., Szabó, S., Iojă, C. I., & Popescu, V. D. (2023). Predictors of brown bear predation events on livestock in the Romanian Carpathians. *Conservation Science and Practice, 5*(3), e12884. https://doi.org/10.1111/csp2.12884

R Core Team. (2017). *R: A language and environment for statistical computing*. R Foundation for Statistical Computing. Retrieved from http://www.R-project.org/

R Development Core Team. (2009). *R: A language and environment for statistical computing*. R Foundation for Statistical Computing. Available at: http://www.R-project.org/

Rewilding Europe. (2023). *The key to living with wolves in Europe? Ramping up livestock protection measures*. https://rewildingeurope.com/blog/the-key-to-living-with-wolves-in-europe-ramping-up-livestock-protection-measures

Rigg, R., Find'o, S., Wechselberger, M., Gorman, M. L., Sillero-Zubiri, C., & Macdonald, D. W. (2011). Mitigating carnivore–livestock conflict in Europe: Lessons from Slovakia. *Oryx, 45*(2), 272–280. https://doi.org/10.1017/S0030605310000074

Scharf, A. K., & Fernández, N. (2018). Up-scaling local-habitat models for large-scale conservation: Assessing suitable areas for the brown bear comeback in Europe. *Diversity and Distributions, 24*(11), 1573–1582. https://doi.org/10.1111/ddi.12796

Steyaert, S. M. J. G., Støen, O.-G., Elfström, M., Karlsson, J., Lammeren, R. V., Bokdam, J., Zedrosser, A., Brunberg, S., & Swenson, J. E. (2011). Resource selection by sympatric free-ranging dairy cattle and brown bears Ursus arctos. *Wildlife Biology, 17*(4), 389. https://doi.org/10.2981/11-004

Stringham, S. F., & Rogers, L. L. (2023). *Are brown bears less aggressive in Europe than in North America?* Wild Research Institute. https://doi.org/10.13140/RG.2.2.32369.45923

Tomma, G. (2024). Plan to allow wolf hunting in Europe to spare livestock could backfire, some scientists say. *Science, 383*, 355. https://www.science.org/content/article/plan-allow-wolf-hunting-europe-protect-livestock-could-backfire-some-scientists-say

Vinci, C. (2022). *European Union beef sector: Main features, challenges and prospects*. European Parliamentary Research Service.

Zimmermann, B., Wabakken, P., & Dötterer, M. (2003). Brown bear-livestock conflicts in a bear conservation zone in Norway: Are cattle a good alternative to sheep? *Ursus, 14*, 72–83.

Predators of Cattle in Africa

Introduction

Cattle herding is a major livelihood activity across Africa, especially in the savanna regions where most large mammalian carnivores live (Campbell, 2017; Kim et al., 2020). Mwai et al. (2015: 911) note that "African cattle are now found all across the continent, with the exception of the Sahara and the river Congo basin." Dueppen (2012: 92) points out that "the West African Sahel and savanna are today home to diverse cattle breeds derived from complex social, political and environmental processes over at least the past four thousand years." Galal (2024) notes in a comparison of data on African cattle, that in 2023 "Africa had over 381 million heads of cattle, one of the major species raised for livestock farming on the continent" and "Ethiopia had the highest number of cattle in Africa as of 2022, roughly 68 million heads" and that Ethiopia had almost twice as many cattle as Chad, which had the second-highest cattle numbers in Africa. Erdaw (2023) notes that "the African livestock sector contributes about 30–80% to its Agricultural Gross Domestic Product From the world's total, around 85% of livestock keepers are found in subSaharan African countries. Although it produces only 2.8% of the world's meat and milk outputs, subSaharan Africa contributes more than 14% of the world's livestock resources."

A large proportion of African cattle are descended from the indicine zebu and taurine cattle, originating from southwest Asia (Campbell, 2013). Pérez-Pardal et al. (2018) note that "all tropically adapted humped cattle (Bos indicus or 'zebu'), descend from a domestication process that took place >8000 years ago in South Asia." Their study using Y-chromosome diversity and male-lineage zebu cattle history and diversity patterns showed the zebu Y-chromosome haplotypes in their dataset were within three lineages: "Y3A, the most predominant and cosmopolitan lineage; Y3B, only observed in West Africa; and Y3C, predominant in South and Northeast India" (Pérez-Pardal et al., 2018). Importantly, "the divergence times estimated for these three Zebu-specific lineages predate domestication" and therefore it is suggested that there was an "export of varied zebu lineages from domestication

© The Author(s), under exclusive license to Springer Nature Switzerland AG 2025

M. O. Campbell, *Cattle, Their Predators and Geomatics Research*,
https://doi.org/10.1007/978-3-031-97363-5_7

centres through time. The almost exclusive presence of Y3A haplotypes in East Africa is consistent with recent cattle restocking in this area" (Pérez-Pardal et al., 2018).

In addition, "the cryptic presence of Y3B haplotypes in West Africa, found nowhere else, suggests that these haplotypes might represent the oldest zebu lineage introduced to Africa ca. 3000 B.P. and subsequently replaced in most of the world" (Pérez-Pardal et al., 2018). The authors argue that Y-chromosome based studies are rarer than mtDNA4 and genome-wide information studies. Their study examined single nucleotide polymorphisms (SNPs) and identified a single zebu Y-chromosome lineage (Y3): "Admixture between B. taurus and B. indicus in Africa and differences in diversity parameters assessed using Y-chromosome markers have contributed to understand the complexity of domestication processes, the early spread of male zebu cattle in South Asia and the introduction of zebu to Africa through Indian Ocean trade" (Pérez-Pardal et al., 2018).

Mwai et al. (2015: 911) note that 150 or more indigenous African cattle breeds have been named, although it is suggested that there are more that have not been full studied, as while about 180 breeds of are recognized in sub-Saharan Africa and of these, there are 150 breeds of indigenous cattle and other more recently introduced "exotic and commercial composites." Mwai et al. (2015: 913) support the general classification note that indigenous African cattle may be classified into four categories: the humpless *Bos taurus*, humped *Bos indicus* sanga (African humpless *Bos taurus* × humped *Bos indicus* hybrid), and zenga, the last which may be defined as sanga × zebu backcross. Apart from the aforementioned primary cattle species, other breeds hybrid breeds, including sanga and zenga, which are hybrids of the taurine and zebu (sanga) and zebu and sanga (zenga), with the latter possessing a more advanced genetic heritage than Zebu (Mwai et al., 2015: 911). Mwai et al. (2015: 911) also note that "it should be noted that all African cattle do carry a taurine mitochondrial DNA, in other words they are no pure zebu Bos indicus on the African continent." Other African cattle breeds have been developed, mainly through crossbreeding with what are termed "exotic" breeds.

Kim et al. (2020: 1099) present a study that argues that despite the success of cattle breeding and pastoralism in Africa "the genetic history of its success remains unknown." The study is based on "whole-genome sequence analysis of 172 indigenous African cattle from 16 breeds representative of the main cattle groups" through the authors identified "a major taurine × indicine cattle admixture event dated to circa 750–1050 years ago, which has shaped the genome of today's cattle in the Horn of Africa" (Kim et al., 2020). They also identified "16 loci linked to African environmental adaptations across crossbred animals showing an excess of taurine or indicine ancestry" which included "include immune-, heat-tolerance- and reproduction-related genes" and identified "one highly divergent locus in African taurine cattle, which is putatively linked to trypanotolerance and present in crossbred cattle living in trypanosomosis-infested areas" (Kim et al., 2020). The authors conclude "that a combination of past taurine and recent indicine admixture-derived genetic resources is at the root of the present success of African pastoralism" (Kim et al., 2020).

Regarding the introduction of cattle breeds, the two main events would be of the taurine and indicine cattle. The African taurine cattle, which are also known as humpless cattle, are mostly found in West Africa, although some comparatively rarer commercial taurine breeds and crossbreeds are found across Africa (Mwai et al., 2015). Pérez-Pardal et al. (2018) note that "genetic data have demonstrated that African cattle mainly carry the maternal T1 taurine mtDNA lineage. However, the African zebu and sanga populations (crosses between African taurine and zebu cattle) mainly carry Y-chromosomes of zebu origin. The absence of zebu mitochondria in African cattle implies that introgression of the zebu lineage into Africa was mostly male mediated. A number of microsatellite-based studies have also shown that zebu introgression into Africa declines from East to West" (Pérez-Pardal et al., 2018). It is noted that most cattle breeds in Africa appear to be phenotypically the humped indicine zebu that have adapted to the local environmental variables, which may be unsuitable for European breeds, these variables including high temperatures, long droughts, and diseases transmitted by insect vectors (Mwai et al., 2015: 911).

Regarding the introduction of indicine cattle into Africa, Pérez-Pardal et al. (2018) describe the introduction of humped zebu-like cattle into Egypt western southwestern Asia (the Levant) in small numbers, with evidence from illustrations on tombs and temples 3400–3000 B.P. The argument is that despite these introductions, present African zebu cattle populations are the result of multiple introductions, possibly through eastern Africa rather than northern Africa. Notably, "pre-Aksumites, Aksumites and their trading partners in Yemen and the Red Sea brought zebu cattle to the Horn of Africa 2000–1600 B.P. following well known sea trade routes from South Asia to the Gulf of Oman. Humped cattle followed Sahelian routes, appearing in West Africa by 1000 B.P. The rinderpest panzootics of 1889–1896 are estimated to have annihilated up to 80% of herds in many regions or over 5.2 million African taurine cattle and resulted in zebu being massively reintroduced along the eastern coastline of Africa, largely replacing African taurine bulls" (Pérez-Pardal et al., 2018).

Regarding the resistance to diseases, trypanosomiasis remains a major issue. Different genetic strains have different resistance to trypanosomiasis (Campbell, 2013). Murray et al. (1984: 311) note that some livestock breeds have genetic resistance to African trypanosomiasis. The native West African taurine cattle breeds N'Dama and West African Shorthorn, which have been in Africa for 5000–7000 years, are the best examples of this trait, which is known as trypanotolerance. These breeds are increasingly being used in livestock projects in tsetse-infested areas as a result of confirmation that they are not only genetically resistant to trypanosomiasis but also productive enough to rival other native breeds.

The N'Dama is indigenous to the Fouta-Djallon highlands of Guinea and the dominant breed in western West Africa. This breed is believed to have been domesticated about 8000 years ago, and very importantly, acquired resistance to trypanosomiasis and other local diseases (Murray et al., 1981, 1982; Courtin et al., 2008; Campbell, 2013). The Zebu is found largely to the east of the N'Dama along the Sudano-Sahel belt and has been crossbred with some other local breeds. Unlike the

N'Dama, it is frequently seriously affected and sometimes killed by trypanosomiasis (Paling & Dwinger, 1993). There are Shorthorn cattle (Trail et al., 1979; Murray & Trail, 1984; Ganyo et al., 2018), which are also trypanotolerant, found mostly south of the Zebu in the Guinea savanna belt. Shorthorns can be divided into two types: the larger savanna type of the Guinean or Sudano-Guinean savannas from Côte d'Ivoire to Cameroon; and the smaller Dwarf (Forest) Shorthorns of the forested and coastal areas (Campbell, 2013).

Mwai et al. (2015: 911) argue for marked changes in African cattle biology over time due to local environmental variables, the details of which are debated, noting that "there is no doubt that African cattle population evolved dramatically over time" and "today, we find a mosaic of genetically diverse population from the purest Bos taurus to the nearly pure Bos indicus" (Mwai et al., 2015: 911). Kim et al. (2017) note that "indigenous cattle in Africa have adapted to various local environments to acquire superior phenotypes that enhance their survival under harsh conditions. While many studies investigated the adaptation of overall African cattle, genetic characteristics of each breed have been poorly studied" and that "N'Dama might have acquired distinct genotypes associated with growth and regulation of regional diseases including trypanosomiasis." Murray et al. (1981: 503) note that "the use of trypanotolerant livestock is considered to be an important strategy for the control of African animal trypanosomiasis" and their study "confirmed that N'Dama cattle are innately less susceptible to trypanosomiasis than Zebu cattle and can survive and be productive in endemic areas of trypanosomiasis where Zebu perish."

The African taurine and indicine cattle that have adapted to regional and local diseases and other environmental issues have also encountered predation by local predators (Campbell, 2017). In some areas, there are few predatory species, while in other numerous predatory species may have contact with cattle. For example, Sebsibe (2022) reports the case of the Central and Eastern Part of Bale Mountains National Park, Ethiopia, where there was pronounced conflicts between people and their livestock, and predators. Here, respondents reported that "within the past 10 years 1623 sheep, 741 cattle, 639 goats, 193 donkeys, and 124 horses were predated" and the responsible predators may have included lion, and leopard, spotted hyena, African wolf (Canis lupaster Hemprich and Ehrenberg, 1832), and Ethiopian wolf (Canis simensis Ruppell, 1840), with an "increasing trend of livestock predation" (Sebsibe, 2022).

Carnivore predation on livestock may be based on the carnivores' ability to take similar or larger, wild prey. As shown in Table 1, several African carnivores are able to kill large ungulates, with the lion being the dominant predator for all large ungulates below the size of the rhinoceros. Table 1 shows ungulates similar to and slightly larger than the domesticated taurine and indicine cattle. These wild ungulates are similar in size to domestic cattle from calves to adult males of indicine and taurine cattle. It must also be noted that wild ungulates may possess superior predator avoidance abilities to those exhibited by domestic cattle, which are mostly protected by human management. The ease of predation on domestic cattle is a factor for human-carnivore conflicts, which pervade much of the savanna vegetation, which is the primary pasture and large carnivore habitat. Table 2 shows some of the

Table 1 Common African prey species similar to cattle in size

Wild ungulate species	Scientific name	Weight range (kg)
Giant eland	*Taurotragus derbianus* (J.E. Gray, 1847)	300–1200
African buffalo	*Syncerus caffer* (Sparrman, 1779)	425–870
Common eland	*Taurotragus oryx* (Pallas, 1766)	500–600, up to 942
Bongo	*Tragelaphus eurycerus* (Ogilby, 1837)	150–405
Blue wildebeest	*Connochaetes taurinus* (Burchell, 1823)	140–410
Grévy's zebra	*Equus grevyi* (Oustalet, 1882)	350–450
Mountain zebra	*Equus zebra* (Linnaeus, 1758)	204–372
Plains zebra	*Equus quagga* (Boddaert, 1785)	175–322
Greater Kudu	*Tragelaphus strepsiceros* (Pallas, 1766)	120–270, up to 315
Waterbuck	*Kobus ellipsiprymnus* (Ogilby, 1833)	162–262
Gemsbok	*Oryx gazella* (Linnaeus, 1758)	100–240
Sable antelope	*Hippotragus niger* (Harris, 1838)	220–235
Scimitar oryx	*Oryx dammah* (Cretzschmar, 1827)	91–210
Hartebeest	*Alcelaphus buselaphus* (Pallas, 1766)	100–200

Sources: Ralls (1978); Kingdon (1988, 1989, 1997, 2015); Grzimek, 1990; Nowak (1999); Pappas (2002); Burnie and Wilson (2005); Mungall (2007); Estes (1999, 2024)
Lions may kill adults of all these species, while leopards and packs of hyenas and wild dogs may kill most, possibly excepting the larger species. Cheetahs may kill juveniles of the smaller species

Table 2 Examples of African domestic taurine and indicine cattle breeds

African cattle breed	Scientific name	Weight range (kg)
N'Dama	*Bos taurus*	250–60
Sanga	*Bos indicus* and *Bos taurus*	525
Zenga (Zebu and Sanga)	*Bos indicus* and *Bos taurus*	248+
Africander cattle (South African Sanga)	*Bos indicus* and *Bos taurus*	450–1090
Mashona cattle (Zimbabwe Sanga)	*Bos indicus* and *Bos taurus*	272–352

Sources: Mason (1996); Mwai et al. (2015); Farming Plan (2023); Animal World (2024); Division of Agricultural Sciences and Natural Resources (2024)
Variable weights are given for these African breeds, but the consensus appears to be in the range from 250 to just over 500 kg, with a few reaching around 1000 kg

domesticated breeds (mostly crosses between taurine and indicine ancestors), which weigh approximately the weight of medium large ungulates (Mwai et al., 2015). The wild ungulates and domesticated cattle are shown in Figs. 1, 2, 3, 4, 5, 6, 7, 8, 9, 10, 11, 12, 13, 14, 15, 16, 17, and 18.

Wild prey preferences of predators may influence the size of domestic livestock killed. As the tables above, the lion is capable of killing all the domestic cattle, which are well below the size of the largest buffalo and elands. For example, Clements et al. (2014) present "an objective approach to determining the weight ranges of prey preferred by and accessible to the five large African carnivores" arguing that "previous analyses used a subjective approach to identify prey weight preferences of the five large African carnivores, hence their accuracy is questionable." The study constructs a model that uses simulations of known predator prey preferences, which "when applied to the five large African carnivores, using carnivore diet information from across Africa, the model detected weight ranges of prey that are preferred, killed relative to their abundance, and avoided by each carnivore." The findings indicated that "prey in the weight ranges preferred and killed relative to their abundance are together termed 'accessible prey'. Accessible prey weight

Fig. 1 Giant eland. (Permission Charles J. Sharp. CC BY-SA 4.0. Clipped. https://creativecommons.org/licenses/by-sa/4.0/)

Fig. 2 Eland. Lewa Wildlife Conservancy, Isiolo, Kenya. (Permission David Clode-unsplash)

ranges were found to be 14–135 kg for cheetah, 1–45 kg for leopard, 32–632 kg for lion, 15–1600 kg for spotted hyaena and 10–289 kg for wild dog. An assessment of carnivore diets throughout Africa found these accessible prey weight ranges include 88 ± 2% (cheetah), 82 ± 3% (leopard), 81 ± 2% (lion), 97 ± 2% (spotted hyaena) and 96 ± 2% (wild dog) of kills" (Clements et al., 2014).

Owen-Smith and Mills (2008: 173) present a similar study assessing "the differing selection shown by carnivores for specific size ranges of prey" and "the extent to which the greater carcass mass of larger prey outweighs the greater numerical representation of smaller prey species in the predator diet." The study, based on the Kruger National Park, South Africa, examined five carnivore species, lions, leopards, cheetahs, African wild dog, and spotted hyena, and 22 herbivore prey species over 10 kg in adult body weight. The findings indicated that "these carnivores selectively favoured prey species approximately half to twice their mass, within a total prey size range from an order of magnitude below to an order of magnitude above the body mass of the predator. The three smallest carnivores, i.e. leopard, cheetah and wild dog, showed high similarity in prey species favoured. Despite overlap in prey size range, each carnivore showed a distinct dietary preference" (Owen-Smith & Mills, 2008: 173).

Fig. 3 African buffalo. Mahango, Namibia. (Permission Chris Stenger-unsplash)

In ungulate species up to the size of a giraffe (800–1200 kg), almost all mortality was caused by a predator (Owen-Smith & Mills, 2008: 173). Ungulates, which were more than twice the mass of the predator, contributed significantly to the lions' food intake, although the mortality rate of these species from predation was low. Predation became a negligible cause of death only in megaherbivores with adult body weights significantly greater than 1000 kg. Therefore, the relative size of predators and prey had a broad structuring influence on biomass fluxes within this large mammalian food web. Nevertheless, the large carnivore population was overwhelmingly dominated by the largest predator, which accounted for the majority of animals of various sizes killed (Owen-Smith & Mills, 2008: 173).

For small ungulates, there may be a similarity with small livestock such as sheep and goats, although small wild ungulates may exhibit superior evasive ability to predators than larger livestock such as calves. Calves may also be hunted on similar terms as small wild ungulates. For example, Sinclair et al. (2003: 288) present a study of the Serengeti savanna ecosystem in East Africa and explain that the main factors for mortality for adult ungulates are the species diversity of the predators and prey species and the body size of that prey species relative to other prey species and the predators. It is noted that "small ungulates in Serengeti are exposed to more

Fig. 4 Bongo. (Permission David Cashbaugh-unsplash)

predators, owing to opportunistic predation, than are larger ungulates; they also suffer greater predation rates, and experience strong predation pressure." A threshold is suggested at prey body sizes of about 150 kg, above which the ungulate species may have only a few natural predators and show food limitation. Thus, "biodiversity allows both predation (top-down) and resource limitation (bottom-up) to act simultaneously to affect herbivore populations" (Sinclair et al., 2003: 288).

Regarding conflicts between carnivores and people over livestock predation, the abilities of carnivores based on the features of the natural ecosystem are important. Human opinions and reactions to carnivores are based on physical factors and some cultural issues (Campbell, 2017; Mkonyi et al., 2017). For example, in a study on Namibia, Lindsey et al. (2013) found that land users noted significant conflicts with carnivores and that "changing land use patterns in southern Africa have potential to dramatically alter the prospects for carnivore conservation. Understanding these influences is essential for conservation planning." The predators included lions, cheetahs, leopards, brown hyenas, spotted hyanas, and wild dogs. Their findings indicated that farmers were most tolerant of leopards and least tolerant of lions, wild dogs, and spotted hyenas. In addition, "carnivores were more commonly present and/or tolerated where; wildlife diversity and biomass were higher; income from

Fig. 5 Blue wildebeest, South Africa. (Permission Piotr Usewicz-unsplash)

wildlife was higher; income from livestock was lower; livestock biomass was lower; in conservancies; game fencing was absent; and financial losses from livestock depredation were lower" (Lindsey et al., 2013). Lindsey et al. (2013) also found that local people's tolerance can vary by type of species, as in their study showed that leopards and brown hyenas were the most tolerated, because "leopards are perceived to have value through ecotourism and trophy hunting, whereas brown hyaenas are perceived to impose few costs," and the least tolerated where lions, wild dogs, and spotted hyenas. The reasons for low tolerance were livestock and wildlife killing, and financial consequences (Lindsey et al., 2013). These results for Namibia were similar to some findings of research on South Africa, Zimbabwe, and Kenya (Lindsey et al., 2013; Romañach et al., 2007).

This chapter examines the relationship between large carnivores and domestic cattle. Each wild carnivore capable of killing cattle is considered in turn. Because domestic cattle, like the N'Dama, shorthorn, and zebu, may be smaller and much less aggressive than African buffalo, they may be more easily killed by carnivores, including smaller carnivores such as leopards, cheetahs, spotted hyenas, and wild dogs. Based on available data, conservation efforts appear to be more successful in

Fig. 6 Grévy's zebra, Lewa Conservancy and Samburu National Park, Kenya. (Permission David Clode-unsplash)

Europe than Africa, due to the latter continent's larger land area, higher density of potential predatory mammals, lower financial resources, and poorer rural subsistence land users. Even though there have been some recent advancements in Africa, political and financial issues still restrict the application of conservation measures there in comparison to Europe.

Lions and Cattle Predation

Lions are discussed at length in this section because they are the main cattle predators in Africa. The lion is the dominant predator on the African continent outside the dense forest and the desert, partly due to its size, the fact that it usually hunts in cooperative groups (Figs. 19 and 20), and the absence of a similarly sized carnivore, such as the tiger and/or brown bears (Campbell, 2017). Lions in Africa are limited to the open savanna. Riggio et al. (2012: 17) define African savannas as variable vegetation belts with annual rainfall between 300 and 1500 mm, this vegetation type including most of the range of lions, and frequently including high human

Fig. 7 Mountain zebra, Etosha National Park, Namibia. (Permission Yathin S Krishnappa—CC BY-SA 4.0)

populations and human landuse that may challenge lion presence. Riggio et al. (2012: 17) further note that approximately 25% of the savannah area, or 3.4 million km^2, currently supports wild lion populations. There may be local extinctions and widespread declining populations. In addition, four of the ten lion-inhabited areas in East Africa and six in Southern Africa are regarded as lion strongholds, with roughly 24,000 lions ranging in strongholds, and possibly 4000 more. Nonetheless, the populations of more than 6000 lions may be threatened in terms of their long-term survival, and even in areas that are officially protected, lion populations in West and Central Africa are in grave danger of extinction (Riggio et al., 2012: 17).

Studies of African lion predatory behavior in relation to cattle either study lion predation or compare the predation of lions with those of the smaller predators. For the former theme, lions are argued to be natural cattle killers, due to similarities between adult and juvenile cattle and the natural prey of lions. As noted by (Weise et al., 2020), cattle "occupy the centre of the lion's preferred prey weight range," and therefore, "it is no surprise that their predation is a major source of human–wildlife conflict in Africa." In addition, as lions hunt mostly in groups, they can

Fig. 8 Plains Zebra Addo Elephant National Park, Addo, South Africa. (Permission Etienne Steenkamp-unsplash)

regularly kill the large and aggressive African buffalo and consequently can more easily kill the smaller taurine and indicine cattle. The lion may have a cattle-killing reputation even when little cattle killing is recorded. For example, Gebresenbet et al. (2018) write that "while livestock loss is a source of conflict, the degree to which livestock depredation is tolerated by people varies between regions and across cultures" and in a study of the environs of Gambella National Park and the Kafa Biosphere Reserve, both in western Ethiopia, the study found that "depredation was relatively low, and that disease and theft were the top factors of livestock loss." Notably, tolerance of lions was lower around Gambella National Park than in Kafa Biosphere Reserve. Education level, number of livestock per household, livestock loss due to depredation, and livestock loss due to theft were strong predictors of local agriculturists' attitude toward lion presence. Therefore, the level of livestock depredation alone was not the only factor for human attitudes to lions and the resultant human-lion conflicts, indicating the importance of cultural differences in lion-human relations and conservation (Gebresenbet et al., 2018).

An important point is that "lion-related benefits" for local people can affect local people's attitude to lions (Gebresenbet et al., 2018). For example, a study by

Fig. 9 Greater Kudu. Kruger National Park, South Africa. (Permission Dianne Gibson-unsplash)

Romañach et al. (2007: 185) notes that of "livestock farmers (community members) and commercial livestock ranchers in central Kenya" the "commercial ranchers had more positive attitudes and higher tolerance than community members. Community members said their tolerance would increase if they were to derive income from ecotourism or trophy hunting… community members with land title deeds were most tolerant of predators, stressing the importance of land security for conservation efforts." This study was based on "a novel index of tolerance by quantifying the number of livestock respondents would be willing to lose before killing the predator responsible" (Romañach et al., 2007: 185).

Patterson et al. (2004: 507) note that lions-human conflicts and the historical reduction of the lion's geographical range are largely caused by lion predation of livestock (2004: 507). For example, in Tsavo East National Park in Kenya, lions were responsible for 85.9% of attacks, with hyenas and cheetahs being the other predators. Lions and hyenas frequently attacked cattle at night, while cheetahs preyed on smaller sheep and goats. In addition, the cattle losses did not show temporal autocorrelation, indicating that the attacks were autonomous occurrences. The attacks were positively connected with rainfall and supported the trend of large

Fig. 10 Waterbuck, Masai Mara National Reserve, Kenya. (Permission Henrik Hansen-unsplash)

carnivores killing more livestock when native prey is hardest to locate and kill. Livestock accounted for 5–8% of ranch lion diets (Patterson et al., 2004: 507).

According to a study by Kissui (2008: 422), large carnivores in the Maasai steppe landscape of northern Tanzania have faced human retaliation as a consequence of their predatory activities. The consequences of this retaliation differ among species, being influenced by their unique predatory behaviors and cultural beliefs of pastoralists. The results indicated that lions primarily preyed on adult cattle and donkeys, whereas hyenas and leopards mainly focused on small livestock such as goats, sheep, calves, and dogs. Hyenas and leopards frequently attacked livestock during the night, while lions commonly targeted grazing livestock during the day. This behavioral pattern made lions more susceptible to direct retaliatory actions, although some villages resorted to using poison against hyenas and the cultural practices of herders amplified the incidence of retaliatory lion killings (as reported in Kissui, 2008: 422).

Lagendijk and Gusset (2008: 971) based on interviews examined the coexistence of humans and carnivores on communal land on the border of Greater Kruger, South Africa. The authors found that respondents expressed varying attitudes toward

Fig. 11 Gemsbok, Kimberley NC South Africa. (Permission Stephan Streuders-pexels)

predators, which were more positive among participants with higher education. The negative views were mainly due to fear of losses to people and livestock, especially lions. The results suggest that lions were considered the most common predator inside and outside the reserve. Lions were also the most well-known predators and were most often blamed for killing livestock. Despite these livestock losses and lack of education about conservation, most participants expressed positive opinions about large carnivore conservation, as carnivores were viewed as an integral part of the respondents' natural heritage. These positive views have been attributed to cultural tolerance and also to widely accepted management policies for predator control (Lagendijk & Gusset, 2008: 971).

Hazzah et al. (2009: 2428) conduct another study of lion conflicts and point out that African lion populations are declining in most of Africa, but the problem is particularly serious in southern Kenya, where the Maasai spear and poison lions on a scale that may ensure local extinction in the short term. Lion killing is shaped by Maasai perceptions of livestock predation, socioeconomic factors, and the complex relationship between Maasai and conservation. This affects tolerance toward lions and, consequently, Maasai behavior toward conservation initiatives and carnivores in general. This study was also based on interviews and questionnaires and focused

Fig. 12 Sable Antelope, Chobe National Park, Botswana. (Permission Chris Stenger-unsplash)

on the social and environmental predictors of lion killing and on studying the impact of a compensation system on individual tolerance (Hazzah et al., 2009: 2428). Individuals were likely to kill lions if they lost more livestock to predators as part of their overall livestock loss, as well as individuals who were close to an evangelical church, and individuals who were more likely to be involved in selling rather than accumulating livestock. The authors argue that the Maasai perceived high levels of conflict with lions; However, the actual rates (on a per capita basis for all MGRs) are quite low, at 5%. Only 25% of respondents said they would take revenge and kill a lion after their livestock was killed. This could be interpreted as this community being generally tolerant compared to others.

Gebresenbet et al. (2017: 751) present a comparative study of lion and leopard predation on livestock in the Adiyo and Gimbo districts in Kafa Biosphere Reserve, Ethiopia, using household and group interviews with local people. During the period of study (2009–2013), there were 350 attacks by lions and 62 attacks by leopards, respectively. Although lion attacks were more frequent than leopard attacks, the authors point out that their qualitative data indicate that locals are more tolerant of lion attacks because they are more revered in the community (Gebresenbet et al., 2017: 751). Between 2009 and 2013, lions were responsible for 85% of all

Fig. 13 Scimitar Horned Oryx. (Permission Geraldine Dukes-unsplash)

livestock losses (Gebresenbet et al., 2017: 755). Although lions posed a greater threat to the reserve than leopards, lions were still tolerated more than leopards. The assertion that there is a culture of tolerance toward lions was supported by the lack of a significant correlation between respondents' reported behavior after lion attacks and livestock loss. But attitudes toward leopards were less favorable, and the strong relationship between respondents' suspected behavior following leopard attacks and livestock loss indicates that the more households that were attacked by leopards, the stronger the behavioral intention of those households to retaliate (Gebresenbet et al., 2017: 757).

Tumenta et al. (2013) cite a study of Waza National Park, Cameroon, where the results "indicated high levels of livestock depredation around the Park, with attacks occurring most often at night. Lions were economically a substantial threat accounting for total losses of EUR 100,000 per annum. Per household, resident pastoralists lost one cow and nomadic pastoralists two cows per annum" (Tumenta et al., 2013: 237). Tumenta et al. (2013: 240) noted that lions were the most significant predator in terms of economic values, killing mostly cattle, which were the most valuable animals in terms of money. Raising livestock is a significant activity and a source of revenue for the herders in the Waza National Park area. In addition, because nomadic

Fig. 14 Hartebeest. (Permission Simon Brandintel-pexels)

herders usually owned twice as many cattle as local herders, they lost twice as many cattle to predators. Along with social reasons, they also seemed to visit the park more frequently for grazing and water. Significantly, most livestock losses happened at night. Additionally, it is possible that the disparity in predation between local and nomadic herders' herds resulted from different husbandry techniques. More resident herders used enclosures to keep their livestock safe at night. Because they were too mobile, nomadic herders found that constructing enclosures was costly and time-consuming. These results suggested that more enclosures to protect livestock at night, "although useful, might be difficult to adopt by nomadic pastoralists" (Tumenta et al., 2013: 240).

Weise et al. (2020) note that lions are increasingly endangered, in part due to the severe conflicts caused by cattle predation. Because cattle closely resemble the body size, shape, and herding behavior of prey, understanding lions' preferences for specific cattle is essential to improve appropriate predator control measures, reduce conflict, and ultimately improve lion conservation. The authors looked at the case of cattle predation patterns in the Okavango Delta in Botswana, and the results showed that lions killed cattle mostly at night and targeted the easiest prey, such as cattle without horns. However, prey preferences differed depending on the hunting

Fig. 15 Taurine cattle, South Africa. (Permission Khwezi Mvuyana-pexels)

circumstances. Where cattle were kept in enclosures, lions preferred young, inexperienced calves, often resulting in significant losses in a single incident. When cattle were left unprotected on pasture, lions preferentially killed cattle with mixed coat patterns and single males (Weise et al., 2020). In addition, losses in lions were caused by characteristics of cattle associated with ease of husbandry and resulting from domestication. The widespread availability of cattle and the attitude to conservation efforts led to losses as cattle no longer possess the key traits that allowed their ancestors to coexist with large predators (Weise et al., 2020).

The study examined whether lions showed specific preferences based on cattle demographic characteristics (sex and age) as well as morphological characteristics (body mass, horn length, and fur pattern) (Weise et al., 2020). The results showed that lions killed cattle predominantly at night (87.1%) and did not show a preference for either sex. Bulls and calves were most preferred, while heifers and cattle with uniform color patterns were significantly avoided. Cattle with mottled pelage patterns were most preferred, particularly in free-roaming herds. Preferences were context-specific, with lions favoring inexperienced calves (including several cases of excess killing) and free-roaming bulls and oxen when attacking the enclosure. About 13% of adult cattle did not have horns, and these were preferred by lions,

Fig. 16 Taurine watusi cattle (Sanga), Uganda. (Permission Michel Sautel—CC BY 3.0. https://creativecommons.org/licenses/by/3.0/)

while cattle with short horns were killed depending on availability and cattle with long horns were largely avoided (Weise et al., 2020).

In addition, cattle no longer exhibit the key traits that allowed their ancestors to coexist with large predators and now rely on humans to carry out important anti-predator activities (Weise et al., 2020). Therefore, the responsibility for mitigating human-animal conflict involving lions and cattle lies with humans, who either breed for traits that minimize predation or adequately protect their cattle (Weise et al., 2020). Cattle domestication has resulted in the loss of a number of behavioral and morphological traits that would have minimized predation—particularly increased aggressiveness, large body size, and large horns—in favor of traits that support husbandry practices and ease of management by pastoralists. Currently, domestic cattle rely almost exclusively on human herders and their various protective strategies to protect themselves from predators. This may be satisfactory where humans have eradicated large predators. However, the modern morphological characteristics exhibited by Tswana cattle in southern Africa did not provide effective protection against lions (Weise et al., 2020). The authors point out that most losses occurred while livestock were roaming unguarded from homesteads at night. This is

Fig. 17 Nguni cattle (Sanga breed), South Africa. (Permission holmespj-pixabay)

consistent with studies of other communal rangelands in southern Africa and Botswana. The widespread availability of cattle in mixed livestock-wildlife systems during peak lion activity periods allows for predation. Lions typically preferred wild prey over domestic prey when both are available, when migratory wild prey is scarce, stable livestock abundance on semiarid, seasonally flooded communal pastures in Botswana caused behavioral adaptation. The conflict is then exacerbated as the lions start favoring easily accessible livestock (Weise et al., 2020).

Crucially, domestication played a key role in the anti-predator capabilities of the studied cattle (Weise et al., 2020). Weise et al. (2020) found that about 31% of the study area's adult cattle, primarily females, lacked or had short horns, which made it more difficult for them to defend themselves against lions. The findings validated other research (Roberts, 1996) that the evolutionary function of long horns in female ruminants includes effective protection from predators. In some cases, they also avoided social groups, increasing their vulnerability to lions. Bulls because of their greater body mass may provide most energy resources to predators and therefore may be targets despite their more powerful horns. Calves might not have the experience necessary to avoid risky areas and were less able to injure attacking lions, making them possibly the easiest target and a favorite among lions, especially

Fig. 18 African Indicine cattle. (Permission Carlos Torres-unsplash)

during herd movements during lion attacks in stocking areas. The findings show that predator-prey interactions were altered during the domestication process and implied that all cattle were killed by lions, who generally displayed only minimal preferences, highlighting the danger of predation for any unguarded, free-roaming cattle (Weise et al., 2020).

Such studies may be evaluated in the context of lion predation on wild species, which may be equivalent or more difficult than domestic cattle. For example, Hayward et al. (2011) found that regarding lion predation on wild prey, "lions encountered preferred prey species far more frequently than expected based on their abundance, and they hunted these species more frequently than expected based on this higher encounter rate." Lions had different responses to the less preferred and avoided prey species but hunted avoided prey less frequently than expected in terms of the number of encounters. The authors concluded that there were preferred prey species, listed as blue wildebeest, Cape buffalo, gemsbok, giraffe, and plain's zebra, the hunting of these being termed a "preferential predation" but "there is conjecture over whether this involves conscious choice (by hunting in habitats most likely to contain preferred prey or hunting these prey more frequently than expected with random hunting) or is simply a reflection of those species that are easiest to kill

Fig. 19 Lion foraging savanna. (Permission Pixabay)

because of the morphological adaptations a predator has to hunting them and the responses of prey to these (optimal foraging)" (Hayward et al., 2011).

Hayward et al. (2011) note that lions hunted buffalo more frequently than expected and other species differed according to encounter and predation rates. For example, zebras were "encountered by lions far more frequently than expected, based on their relative abundance, and hunts were conducted as frequently as expected based on this high encounter rate"; lions hunted elands "far more than expected based on their encounter rate", but "only killed in proportion to the number of hunting attempts"; lions hunted hartebeest "far less than expected based on this high encounter rate and killed them as frequently as expected based on the number of hunts"; lions "encountered kudu in accordance with their abundance, they hunted kudu in accordance with their encounter rate, and they killed kudu in accordance with the number of hunts they conducted"; ostriches "were encountered as frequently as expected, however they were hunted far less frequently than expected based on the number of encounters, even though the number of kills were far more frequent than expected based on the hunting effort"; lions "encountered elephants far less frequently than expected based on their abundance and never hunted or killed them"; and "lions encountered warthogs more frequently than

Fig. 20 Lion group eating zebra prey. Naboisho, Narok, Kenya. (Permission Meg Von Haartman-unsplash)

expected based on their abundance, and they elected to hunt the encountered warthogs as expected based on the encounter rate" and "exhibited non-preferential predation on warthog in Addo, as they do throughout their range." The authors conclude that "both buffalo and zebra were preferentially preyed upon… This preferential predation arose through a high encounter rate (zebra and buffalo herds) and a high hunting rate (buffalo)" (Hayward et al., 2011).

These findings support the evidence that lions may favor larger, sometimes cattle-sized prey. Arguably, when larger prey species are present, lions prefer these species to the smaller species they take when the larger species are absent. For example, Fraticelli et al. (2024) note that as "lions and their prey are threatened across most of their range and especially in West and Central Africa" and "scarcity of large prey in many West and Central African ecosystems has been described as leading to a preference for hunting smaller prey in smaller groups." The study, based on Zakouma National Park in Chad, concluded that a "significant recovery in wildlife numbers" occurred and "compared to similar data collected prior to this significant recovery, lions preferred larger prey and fed in larger groups," and these "results show that diet shifts due to prey losses can be reversed with restoration of prey populations thanks to improved management, and we speculate that this may be true across large carnivores and across regions" (Fraticelli et al., 2024).

Wild prey adaptation to lion presence is another issue, which may be compared with the defensive capabilities of domestic cattle. For example, Tambling et al. (2012: 1297) examine the impact of the reintroduction of lions into the Addo Elephant National Park in South Africa, on Cape buffalo. The findings indicated that "following lion reintroduction, buffalo herds amalgamated into larger, more defendable units; this corresponded with an increase in the survival of juvenile buffalo. Current habitat preference of buffalo breeding herds is for open habitats, especially during the night and morning, when lion are active. The increase in group size and habitat preference countered initial high levels of predation on juvenile buffalo, resulting in a return in the proportion of juveniles in breeding herds to pre-lion levels." The authors predict that "the nature of the prey response to predator reintroduction is likely to reflect the trade-off between the predator selection and hunting strategy of predators against the life history and foraging strategies of each prey species" (Tambling et al., 2012: 1297).

Lion extinction is another issue in many parts of Africa, lions are extirpated and therefore may pose less of a threat to cattle and other livestock. Most lions currently live in East and South Africa. In West Africa, which has large numbers of N'Dama and sanga cattle, lions are currently very rare. The few lions in West Africa are still noted to be predators of livestock and even people. For example, Chardonnet et al. (2010: 12) wrote that "in Mali, for example, lion predation on cattle is one of the main reasons for the drastic reduction in the number of lions." In the savannas of Central African Republic, lion population decimation is mostly due to shooting by pastoralists during the dry seasons. The *Daily Telegraph* (Smith 2012), a leading British newspaper, noted that a lion "on a rampage" killed "two herders and 30 livestock in a settlement in northeastern Nigeria's Yobe State."

Howard (2014: 1) noted "lions may soon disappear entirely from West Africa unless conservation efforts improve, a new study predicts" and argues that studies have "found an estimated total of only 250 adult lions occupying less than one percent of that historic range. The lions form four isolated populations: one in Senegal; two in Nigeria; and a fourth on the borders of Benin, Niger, and Burkina Faso. Only that last population has more than 50 lions." Henschel et al. (2015: 4) estimate the West African lion population at just over 400 animals (<250 mature individuals) and in a declining situation. The lion was recorded as "native for Benin, Burkina Faso, Niger, Nigeria and Senegal; possibly extinct for Ghana and Guinea; and regionally extinct for Côte d'Ivoire, Gambia, Guinea-Bissau, Mali, Mauritania, Sierra Leone and Togo."

The factors in West Africa for lion population declines include declining populations of ungulate prey species, increasing human populations and intensity of land use, particularly agriculture, livestock grazing, hunting and urbanization, and human hunting of lions in response to perceived or actual threats to humans and livestock (Bauer et al., 2015). The United States Fish and Wildlife Service (2017), citing (Bauer et al., 2015), states that there are national territories in West Africa where lions have been recently extirpated, comprising Côte d'Ivoire, Gabon, Gambia,

Guinea-Bissau, Mali, Mauritania, Niger, Nigeria, Senegal, Sierra Leone, and other areas where lions may be considered possibly extirpated, comprising Ghana, Guinea, and Togo. There are also areas where lions still range, comprising Benin, Burkina Faso, Cameroon, Central African Republic, Chad, Congo, Democratic Republic of the Congo, Liberia, Niger, Nigeria, and Senegal. Table 3 shows that some protected areas in West Africa have no lions.

Importantly, lion prey species are declining in West Africa (Table 4). Bauer et al. (2015: 14,897) point out that "lion trends are consistent with time series data on their main prey species: whereas herbivore population sizes increased by 24% in southern Africa, herbivore numbers declined by 52% in East Africa and 85% in West Central Africa between 1970 and 2005" (see also Craigie et al., 2010) (Table 4). Henschel et al. (2014) add that the human killing of the wild prey species is even more serious than retaliatory killing of lions for livestock predation. Loomis (2014) suggests that in Senegal, the bush meat trade has decimated prey, especially buffalo and roan antelope in Niokolo Koba National Park by 95% over the past 20 years. Chardonnet et al. (2010: 42) note that in comparison with other African regions "in West Africa there is currently much less lion hunting, though pastoralists tend to shoot and poison lions to protect their stock. Little information is available on the effects of hunting in the region." There is also little information on retaliatory killing of lions, especially in Mali and the Central African Republic (Chardonnet et al., 2010).

Currently, lions in West Africa mostly range in isolated, unconnected isolated patches which are mostly not assessed, e.g., in Ghana and Guinea (Bauer & Van Der Merwe, 2004; Angelici et al., 2015; Henschel et al., 2015). Some lions are found in the WAP-Complex, comprising the W, Arli, and Pendjari National Parks in Burkina Faso, Benin, and Niger, and also in the Niokolo-Koba National Park in Senegal and

Table 3 Protected areas and lion presence in West Africa

Country	Area (km^2)	PA no.	PA area (km^2)	PA %	Lion presence
Niger	1,267,000	1	222,992	17.6	Yes
Mali	1,240,192	1	104,176	8.4	No
Mauritania	1,030,000	1	1600	1.7	Yes
Cameroon	475,442	12	51,823	10.9	Yes
Côte d'Ivoire	322,463	6	13,500	4	No
Ghana	238,534	7	36,019	15.1	No
Senegal	196,723	6	49,574	25.2	Yes
Liberia	111,369	1	2784	2.5	Yes
Sierra Leone	71,740	4	7820	10.9	No
Guinea-Bissau	36,125	5	5021	13.9	No
Gambia	10,380	2	436	4.2	No

Sources: Campbell (2018); Drewniak et al. (2012). Note the figure for Cameroon includes 9615 km^2 in the forest area outside normal lion habitat. International Union for the Conservation of Nature (1992), African Wildlife Foundation (2011). These numbers may vary, according to the definition of National Park, Conservation Area, and Nature Reserve
PA protected area

Table 4 Population status of lion prey species in West Africa

English name	Scientific name	Status
Roan antelope	*Hippotragus equinus* (É. Geoffroy Saint-Hilaire, 1803)	Least concern, declining in WA
African buffalo	*Syncerus caffer* (Sparrman, 1779)	Least concern, declining in WA
Common warthog	*Phacochoerus africanus* (Gmelin, 1788)	Least concern, declining in WA
WA giraffe	*Giraffa camelopardalis peralta* Thomas, 1898	Threatened, isolated populations
WA hartebeest	*Alcelaphus buselaphus major* (Blyth, 1869)	Least concern, declining
WA topi	*Damaliscus lunatus ssp. korrigum* (Ogilby, 1837)	Vulnerable to endangered
Dama gazelle	*Nanger dama* (Pallas, 1766)	Critically endangered
Dorcas gazelle	*Gazella Dorcas* (Linnaeus, 1758)	Vulnerable, declining
Red-fronted gazelle	*Eudorcas rufifrons* (Gray, 1846)	Least concern, declining
Oribi	*Ourebia ourebi Zimmermann*, 1782	Least concern, declining
Western giant eland	*Tragelaphus derbianus derbianus* (Gray, 1847)	Rare, declining
Kob	*Kobus kob* (Erxleben, 1777)	Least concern, vulnerable

Sources: de Jong et al. (2016); Campbell (2018); Scholte et al. (2025).

the Kainji Lake National Park. The numbers of lions in these areas are disputed. For example, DeMarco (2015) argues that currently the status of lions may be overestimated. The IUCN and Henschel et al. (2014) studied 17 protected areas and found lion presence in three of these protected areas, in Senegal (Niokolo-Koba National Park, with fewer than 60 lions), Benin, Burkina Faso and Niger (the tri-national W-Arly-Pendjari Complex, with 246–466 lions) and in Nigeria (the Kainji Lake National Park, 23–63 lions) (Henschel et al., 2014). No lions were found in some previous habitats, including the Mole National Park in Ghana and in Guinea's Haut Niger and Kankan Reserves, and other areas in Togo, Guinea-Bissau, Mali, and Côte d'Ivoire. The estimate was of a West African population of between 269 and 583 lions, with about 121–375 mature animals, and "continuing calamitous declines in prey populations are almost certainly causing concomitant declines in lions" (Henschel et al., 2014).

Regarding strategies to ameliorate lion attacks on cattle and consequent human retaliation, Frank (2023) cite a case study from commercial cattle ranches of Laikipia County, Kenya which have been "a unique exception, with low livestock stocking rates, careful range management, large populations of wild ungulates, and a full complement of large carnivores, including lions at a stable density of about 6/100 km^2. The Laikipia Predator Project, later Living with Lions, started in 1997 with two primary objectives: improving livestock husbandry to reduce lion predation losses and subsequent killing of lions, and studying behavioral adaptations of lions to human activities and persecution" (Frank, 2023).

Frank (2023) noted that "persecuted lions are secretive and nocturnal, so behavioral research was dependent upon radio collaring, requiring development an effective capture technique. Collars introduced ranchers to their lions as individuals, decreasing their propensity to shoot them after livestock predation. The most important breakthrough was the development of 'lion-proof' mobile bomas (corrals) which dramatically reduced nighttime losses and retaliatory killing. Global positioning system (GPS)-Iridium collars for research into lion movements, allowed development of a Lion Early Warning System to inform ranchers of morning lion locations, allowing them to avoid lions during daytime grazing. These measures reduced retaliatory lion killing by 90% between 1998 and 2017. Development of simple and inexpensive hyena-proof bomas for traditional pastoralists dramatically reduced their losses and motivation to poison predators. Studies of lion movements and ecological energetics in relation to human activities have revealed patterns of diel avoidance of humans/livestock by day and predation of wild prey near bomas at night, showing that lions partition their activities temporally in order to utilize high quality hunting habitat while minimizing risk of encountering humans" (Frank, 2023).

Leopards and Cattle Predation

The leopard, despite being smaller than the lion and much less social, is a highly efficient and tree-climbing predator (Campbell, 2017). Grey et al. (2017) note that the leopard is one of the big cats with the greatest geographic range. Leopards have survived for a long time in areas where there have been no other large predators because of their ability to withstand human activity, live in habitats that have been altered by humans, and remain elusive. The opportunistic hunting style and diverse diet of the leopard contribute to its persistence. Nevertheless, despite being relatively plentiful when compared to other large cat species, leopard populations have drastically decreased over the past century as a result of habitat loss and fragmentation, poaching, hunting for trade, retaliation for actual or perceived human-wildlife conflicts, and growing human population growth.

Leopards are generally regarded as the second predator of the African ecosystem, generally less capable of killing adult cattle, but capable of taking calves, adults of smaller cattle breeds, and small livestock (Campbell, 2017). Lamichhane et al. (2023) argue that leopard predation of livestock is a widespread problem in many Asian and African range countries, especially in and near protected areas. An analysis of the landscape characteristics influencing livestock depredation is necessary to develop effective conflict mitigation strategies. The Annapurna Conservation Area in Nepal is used as a case study. According to the findings, the likelihood of leopard attacks on livestock was significantly predicted by the area of forest, agricultural land, river length, slope, elevation, and proximity to populated and protected areas. Falling slopes and rising elevations were also found to be significant predictors (Lamichhane et al., 2023). Here, the frequency of predation by leopards on livestock varied significantly depending on livestock species, season, and time. Goats were the most predated livestock, followed by sheep, cattle, and buffalo. A total of 289.11 km^2 (or around 5% of the research area) was considered particularly at risk for leopard predation on livestock (Lamichhane et al., 2023).

Constant et al. (2015) cite the case of game farming in the Blouberg Mountain Range, South Africa, where leopards were responsible for 89% of reported wild game attacks, especially impala, and for 60% of reported livestock attacks. Regarding the economic costs of leopard predation, the highest figures were for nyala (*Tragelaphus angasii*, Angas, 1849) compared to other wild prey species. Significant costs were also incurred from predation on cattle and donkeys, especially for communal farmers compared to commercial farmers. The findings also indicated predation risk by leopards was influenced by increasing distance to water and human settlements (Constant et al., 2015). The livestock attacks were also associated with seasonal grazing changes, livestock management methods, seasonal breeding and calving, and seasonal water supply change (Constant et al., 2015). Subsistence agriculture, livestock farming, and commercial agriculture based on crop, cattle, and game farming, with over 50% of the commercial farmers engaged in cattle farming. In addition to leopards, predators cheetahs, caracals, brown hyena, spotted hyena, and black backed jackals.

The authors note that "of 37 reported game attacks, leopards were the greatest cause of game loss (89%), followed by cheetahs (8%) and caracals (3%) on commercial farms. Leopards predated on impala (28%), nyala (24%), warthog (21%), ostrich (9%), bushbuck (6%), kudu (6%), gemsbok (3%) and waterbuck (3%)" (Constant et al., 2015). In addition, leopards were held responsible for 67 of 112 of livestock attacks (60%), followed by brown hyenas (19%), jackals (12%), and caracals (9%). For cattle, leopards accounted for 87% of all cattle and 100% of all donkeys killed, with all of these predator attacks on young calves and foals (less than 3 months of age). No incidents of leopard attacks on goats or sheep were reported. Of the 69 livestock attacks by leopards, 97% were on free-ranging grazing herds and 3% in kraals; in all cases, shepherds and guardian dogs were absent during the attack and the average economic loss of cattle by leopards represented 2.6% and 58.3% of the estimated income on commercial farms and communal land, respectively (Constant et al., 2015). See Figs. 21 and 22 for images of leopards.

The risk of leopard predation on livestock was most significantly affected by distance to villages (30.9%), water proximity (23.3%), roadway proximity (21.2%), nature reserve proximity (15.4%), and elevation (9.2%) (Constant et al., 2015). Regarding the impacts of leopard predation on human livelihoods, the authors note that the predation had a greater economic cost for communal farmers who rely on cattle and donkeys for subsistence and functional uses than who derived an additional income from resources like pension grants, entrepreneurial activities, and other employment. Commercial farmers were less affected, as they owned larger livestock herds, and most households diversified their income with crop and game farming industries to ameliorate the effects of cattle predation (Constant et al., 2015).

In contrast, on the premise that livestock predation may be both real or imagined, Grey et al. (2017) present a similar study of leopard predation on wild prey and livestock in the Soutpansberg Mountains, South Africa, with the findings suggesting that Although landowners reported frequent incidents of leopards preying on livestock and introducing farmed game into the agricultural community of Soutpansberg, no evidence of these species was found in the leopards' diet. The most commonly

Fig. 21 Leopard foraging, Masai Mara National Reserve, Kenya. (Permission Bibake Uppal-unsplash)

eaten species by relative biomass were bushbuck, hyraxes, and vervet monkeys; in contrast, farmers reported that cattle and impala were frequently predated by leopards. Although the leopards in Soutpansberg share the landscape with domestic cattle and introduced game, they do not frequently use these species as prey and instead concentrate their diet on wild species. The premise of this study is that many large carnivore species have declined globally as a result of actual or perceived predation of economically valuable species, which is a serious conservation concern. This frequently results in unfavorable opinions and the killing of large predators, which may cause them to become extinct in the regions where it occurs. The actual or perceived predation of game and livestock by African leopards in South Africa is crucial because it influences attitudes against them negatively and makes farmers more inclined to take retaliatory action (Grey et al., 2017).

Grey et al. (2017) argue that, as a result of their study and other research findings, human-carnivore conflict can be ameliorated by reducing the discrepancy between actual and perceived levels of predation through landowner education, effective predator control measures, and improved and more informed government responses. There are a number of reasons why carnivores are often blamed for higher levels of

Fig. 22 Leopard with prey. Ruaha National Park, Iringa, Tanzania-unsplash

livestock and game predation than they actually cause. These include the mistaken identity of the responsible carnivore, misattribution of cause of death, and the socio-cultural and economic context in which the human-carnivore conflict occurs. For example, native predators such as feral domestic dogs can occur in the same area as large carnivores and kill calves, goats, and sheep. Deaths attributed to wild carnivores may, therefore, be caused by domestic predators rather than wild carnivores (Grey et al., 2017).

Some studies have also compared leopards with other predators in terms of livestock predation. For example, Kissui (2008: 422) presents a study based on the impact of prey preference, location, and timing of livestock attacks on the vulnerability of lions, leopards, and spotted hyenas to the occurrence of retaliatory killing in northern Tanzania. They note that all these predatory species are killed in the belief they kill livestock, but the persecution varies based on based on differences in the interspecies differences in predatory behavior and the cultural attitudes of pastoralists. The findings of this study indicated that lions took mostly adult cattle and donkeys, while hyenas and leopards mostly killed small stock such as goats, sheep and calves, and dogs. Lions often attacked grazing livestock in daylight, while hyenas and leopards mostly attacked livestock at night. These behaviors made lions the most vulnerable to direct retaliatory killing, although some hyenas were killed with poison. In addition, the local cultural traditions of pastoralists contributed to increased retaliatory killing of lions (Kissui, 2008).

The justification for the study (Kissui, 2008) was that lions, leopards, and hyenas coexist in the Maasai steppe landscape and kill livestock predation and "while several studies have documented retaliatory killing of African carnivores, no previous study has empirically compared the relative vulnerability of sympatric carnivore species to retaliatory killing" (Kissui, 2008). The author characterizes the three predators as follows: "Lions are social carnivores living in territorial groups, but individual lions can capture prey twice their size... Their most common prey are wildebeest, zebra and buffalo and warthog"; "Leopards ... occupy a broad variety of habitat from forest to desert and they seem to do better in human-dominated areas than lions and hyenas... Leopards exhibit remarkable behavioral plasticity in terms of habitat selection, activity patterns and prey selection"; "The spotted hyena is also a social carnivore living in territorial groups called clans ... The main prey for hyenas includes wildebeest, zebra and Thomson gazelle... They are highly adapted to human settlement and do not appear to be afraid of humans especially at night" (Kissui, 2008).

The findings of (Kissui, 2008) indicated that disease killed more cattle than predators: "Compared with predation, the impact of disease was 410 times greater for cattle and 45 times higher for goats and sheep." In addition, of the recorded livestock attacks, 58% were by hyenas, 25% by lions, and 17% by leopards. Lions attacked an average of 1.7 cattle, 1.8 calves, 4.5 goats and sheep, 1.3 donkeys, and 1 dog per event. Hyenas attacked an average of 1.2 cattle, 1.3 calves, 4.1 goats and sheep, 1.2 donkeys, and 4 dogs per attack event. Leopards attacked an average of 1.7 calves and 2.3 goat and sheep per event and 3 dogs (Kissui, 2008). Lions mostly preyed on cattle, whereas hyenas and leopards preyed more on smaller livestock (Kissui, 2008). Compared to lions, leopards (like hyenas) were more secretive, less likely to combat people over livestock carcasses, and less likely to kill cattle, which are more valuable to Maasai than smaller livestock (Kissui, 2008).

Cheetahs and Cattle Predation

The African cheetah is generally a predator of medium-sized ungulates, similar to those taken by leopards, despite the leopards' generally greater physical power (Campbell, 2017; Hayward et al., 2006a, b, c). For example, Hayward et al. (2006b: 615) conducted a review study and found that cheetahs target prey within a body mass range of 23–56 kg, "with a peak (mode) at 36 kg." Such medium-sized species include blesbok, impala, Thomson's and Grant's gazelles, and springbok. The authors argue that "the morphological adaptations of the cheetah appear to have evolved to capture medium-sized prey that can be subdued with minimal risk of injury" and "coincidentally, these species can be consumed rapidly before kleptoparasites arrive." Dreyer (2024: 4) compared cheetahs and lions and noted that "African lion and cheetah are ideal model species to contrast the impacts of predators on prey populations, as they differ in body size, hunting strategies, metabolic requirements, social behaviours and prey selection patterns." Regarding the lion, the main issues are large size, cooperative and ambush hunting methods, high metabolic requirements, and a wide possible prey weight range (32–632 kg). The result is a selection of prey species ranging from medium to very large, from antelopes to zebra and buffalo. Regarding cheetahs, the main issues are the specialized, solitary of social (male) hunting methods, smaller body size and fragile physique, and greater risk of injury. Consequently, the cheetah has a much narrower prey range, based on small to medium-sized prey such as Thomson gazelle, impala, and springbok, within the mass range of 14–135 kg. The result is that the cheetah more often targets smaller species and juveniles of small and medium prey species (Dreyer, 2024: 5) (Figs. 23 and 24).

However, cheetahs are capable of killing cattle, especially calves (Goldman, 2021). Goldman (2021) argues that some predation is due to overlaps between cheetah foraging patterns and cattle herding locations, taking a case study Central Namibia, a vital cheetah stronghold where cheetahs frequent private farmland more than national parks and have variable effects on livestock. In this area, "there were

Fig. 23 Cheetah, Maasai Mara National Reserve, Narok County, Kenya. (Permission Sutirta Budiman-unsplash)

Fig. 24 Cheetah feed on antelope in Lalibela Game Reserve, Grahamstown, South Africa. (Permission Matthias Wesselmann-unsplash)

farmers having huge problems with cheetahs, losing a lot of livestock, and there were other farmers who actually didn't have any problem at all." An important finding was that some cattle farmers had more problems with cheetahs than others, and some farms overlapped with cheetah "communication hubs" (Goldman, 2021).

This refers to an important study by Melzheimer et al. (2020), which was based on a long-term study with GPS-collared free-ranging cheetahs and found that such animals occupied communication hubs (CHs) in the core area of their territories. The CHs/territories were distributed in a pattern where they were separated by a surrounding matrix and were defended by the territory holders. The study assumed a role for such CHs pose in cattle predation, especially for young calves. In the study, farmers experimentally moved cattle herds further from the CHs during the calving season, which reduced calf losses, as cheetahs preyed on naturally occurring local wildlife prey in the CHs rather than following the domestic herds. The authors concluded that "this implies that in the cheetah system, there are 'problem areas,' the CHs, rather than 'problem individuals'" (Melzheimer et al., 2020).

The cheetah's preference for small ungulates is also studied in the context of livestock predation. For example, Winterbach et al. (2015) in a study in Botswana point out that cheetahs generally target wild prey, but they also take small livestock, which may contribute to conflicts with relevant people. The results showed that there is a broad prey base for cheetahs across much of Botswana's agricultural land, providing additional large areas of possible conservation management. Twenty percent of wild prey biomass appears to be the crucial point in distinguishing between high and low likely levels of human-cheetah conflict. Additionally, the study identified priority areas in agricultural zones where restoring wild prey populations, coupled with effective efforts to mitigate human-cheetah conflict, are the most effective and immediate conservation strategies required to sustain the large and widely ranging cheetah population in Botswana (Winterbach et al., 2015).

Mutoro et al. (2022) examined wild ungulate prey of cheetahs in a case study in Kenya and used these preferences to investigate possible predation on livestock. For example, the cheetah targeted small to medium-sized wild ungulates due to its morphological adaptations. However, the majority of the cheetah population occurred outside protected areas where habitat structures, species richness, and community composition are significantly influenced by human activities. The results showed that Grant's gazelle comprised the largest proportion of the cheetah's diet, although Thomson's gazelle was the most abundant medium-sized ungulate prey in the study area, in addition to other prey species, including two primate species, the yellow baboon and vervet monkey, and rock hyraxes. The study also documented livestock as a possible cheetah prey. Thuo et al. (2020) present a similar study and note that the livestock taxa have been identified in cheetah diets at a moderate frequency (12.8%), and these livestock killings were not affected by cheetah sex or season. The cheetahs targeted wild ungulates of various sizes, and occasionally domestic animals (Mutoro et al., 2022).

African Wild Dogs and Cattle Predation

The African wild dog is a social, cooperative hunter, usually preying on medium-sized antelopes, and may also take livestock and be persecuted by people as a result (Figs. 25 and 26). For example, Hayward et al. (2006c: 1122) note that African wild dogs generally target prey with "prey within a bimodal body mass range of 16–32 and 120–140 kg," and this indicates that this species prefers prey that is easier to kill with injuring to the dog and is relatively abundant. It is also noted that the current ecological status of this species is endangered, based on human actions and persecution, habitat change, competition with other predators, and disease. Tshimologo et al. (2021: 2) in a Zimbabwean study found that wild dogs in northern Botswana killed mostly medium-sized ungulates, including adult impala (more than any other species combined), kudu, and warthog (the average impala weighing 40–76 kg: Estes, 2004). Pole et al. (2004: 207) found in a study of the Savé Valley Conservancy (SVC) in the south-east lowveld of Zimbabwe that the diet of the dogs consisted mainly of impala (74%) and kudu (22%), which were the most abundant medium-sized ungulates in the area.

Fig. 25 African wild dog. (Permission Hal Cooks-unsplash)

Woodroffe et al. (2005: 226) point out that the wild dog is one of the world's most endangered large carnivore species, having been extirpated in 25 of the 39 countries where they were previously found, with a total population of less than 6000 animals. Conflict with humans is a current problem: shootings and poisonings were together responsible for the deaths of 26% of the 196 adult wild dogs that died in protected areas during the study. The deliberate killing of animals that stray beyond the boundaries of the protected area may be the main factor for the extirpation of wild dogs from all but the largest unfenced protected areas. Notably, wild dog populations require protected areas of more than 3500 km^2 for a 50% rate of survival, which is an area far larger than other African carnivores require (Woodroffe et al., 2005: 225).

Regarding livestock, in some cases, the wild dog is seen as a serious predator, while in some cases this assessment is overestimated (Rasmussen, 1999; Grey et al., 2017). Woodroffe et al. (2005: 225) suggested that although they are supposedly protected by reserves, endangered African wild dogs may be especially susceptible to deadly control by farmers due to their widespread distribution and propensity to stray outside the borders of protected areas. The study showed that "ranchers interviewed in South Africa and Zimbabwe ranked wild dogs as the least popular

Fig. 26 African wild dogs attacking an eland. (Permission AfricanConservation CC BY-SA 4.0)

predator species, disliked even more than spotted hyaenas, jackals, lions and leopards," but the "results show that, across most of the study area, wild dog predation was extremely uncommon, with approximately one attack reported per 1000 km^2 of occupied wild dog habitat per year" (Woodroffe et al., 2005: 231). The findings indicated that only six attacks on cattle were confirmed, with the majority of confirmed attacks on livestock (50 out of 56) involving sheep or goats. For every confirmed attack, wild dogs killed 3–2 sheep or goats on average. In confirmed attacks, 1 in 3 cattle were killed on average. Calves, not adults, were killed in each of the numerous cattle killings (Woodroffe et al., 2005: 228).

Gusset et al. (2009: 67) present a study of wild dogs, which are "persecuted throughout their shrinking range in sub-Saharan Africa for allegedly predating livestock" but were only minor predators in the study area. Here, black-backed jackals were blamed for 77% of the reported livestock losses, while wild dogs "accounted for only 2% of reported predator attacks and largely subsisted on wild prey. Nevertheless, most of these wild dogs were killed in the months following this survey." Gusset et al. (2009: 69) noted that "the five largest predator species (lion, leopard, cheetah, spotted hyaena and wild dog) combined from all three survey blocks constituted 22% of all predator attacks reported, with most attacks (77%)

being blamed on black-backed jackal." In addition, "the results indicate that wild dogs were not a major cause of livestock losses, despite the presence of two resident packs before and during the survey. Wild dogs accounted for only 2% of the reported predator attacks, although they were frequently located in the immediate vicinity of numerous cattle posts and farms…results indicate that respondents rarely kept records of their lost livestock. For example, only 37% of livestock reported to have been attacked by predators could be identified by age. Where age was specified, juveniles were the most commonly attacked age group (54%), followed by adults (43%). Predation on newborns was notably low (3%), although there was inconsistency in distinguishing newborns from juveniles in this survey" (Gusset et al., 2009: 67).

Lyamuya et al. (2014: 378–384) conducted a survey of livestock predation in the lands of the Maasai and Sonjo tribes in the eastern Serengeti ecosystem of northern Tanzania during 2007–2009. The findings indicated that "the annual rate of attack by African wild dogs Lycaon pictus was significantly higher (1.42 animals per 1000 livestock) in the Sonjo tribal area than in the Maasai tribal area (0.72 animals per 1000 livestock). In the Maasai tribal area there was a significant amount of predation by leopards Panthera pardus and spotted hyaenas Crocuta crocuta. In both tribal areas sheep Ovis aries and goats Capra aegagrus hircus were subject to predation more frequently than cattle" (Lyamuya et al., 2014: 378). The authors note that wild dogs were the most common livestock predator in both tribal areas, responsible for 47.8% of attacks on livestock in the Maasai tribal area and 99.1% of attacks in the Sonjo tribal area (Lyamuya et al., 2014: 380). In addition, "approximately 20% of the attacks in the Maasai tribal area were carried out by leopards and spotted hyaenas, compared to 0.9% (hyaenas) in the Sonjo tribal area. Lions were responsible for 5.8% of attacks in the Maasai tribal area, where cheetahs and striped hyaenas were together responsible for a further 5.8% of attacks… Most attacks on cattle, sheep and goats were by African wild dogs, although they killed more sheep and goats than cattle. Leopards and spotted hyaenas were also reported to kill more sheep and goats than cattle (18.3% and 16.7%), respectively whereas lions were the only carnivore species that killed more cattle than sheep and goats" (Lyamuya et al., 2014: 380).

This study has important conclusions, in that "wild dogs are probably the most serious predator in both the Maasai and Sonjo tribal areas following their reappearance in the eastern part of the Serengeti National Park. As a result of their large home range and their social lifestyle African wild dogs can consume large amounts of food relative to other carnivores" (Lyamuya et al., 2014: 381). In addition, "wild dogs and most of the other carnivores, except lions, preyed more frequently on sheep and goats than on cattle. This is because wild dogs, as well as leopards and both species of hyaenas, prefer medium-sized prey, within a bimodal body mass range of 16–32 kg to 120–140 kg" (Lyamuya et al., 2014: 382; see also Estes & Goddard, 1967). The conclusions suggest that the perception of the wild dog as a livestock predator is overstated: "losses of livestock to wild dogs and other carnivores are not critical in the study area. The approximate annual rate of loss is one animal per 1000 livestock. However, every animal has a cultural and sentimental value to the Maasai and Sonjo tribes. For this reason, any occurrence of predation causes a great outcry" (Lyamuya et al., 2014: 383).

Spotted Hyenas and Cattle Predation

Hyenas, despite their social hunting skills, generally take smaller wild prey and livestock than lions, with more overlap with leopards (Figs. 27 and 28). For example, Kissui (2008: 422) compared the killing by hyenas, lions, and leopards and found that hyenas, similar to leopards killed mostly small ungulates such as goat, sheep and calves, and dogs, while lions mostly killed adult cattle and donkeys. As hyenas, similar to leopards, mostly attacked livestock at night, they were less vulnerable to persecution than lions, which usually killed in daylight. The author notes that "these behaviors made lions the most vulnerable to direct retaliatory killing, although some villages specifically targeted hyenas with poison, and the cultural traditions of pastoralists also exacerbate the retaliatory killing of lions" (Kissui, 2008: 422). Koskey et al. (2021: 413) note that hyenas kill all livestock types

Fig. 27 Spotted hyena, Uetosha Park, Namibia. (Permission Alan J Hendry-unsplash)

Fig. 28 Spotted hyena with wildebeest remains, Karatu, Arusha, Tanzania. (Permission Marcel Oosterwijk: https://creativecommons.org/licenses/by-sa/2.0)

(horses, donkeys, mules, cattle, goats, and sheep), while leopards, common jackals, and servals (medium wildcats) killed mostly goats and sheep.

Kerley et al. (2018: 240) observed that spotted hyenas have a 25–35% success rate and are effective hunters, able to take down prey that is many times larger than themselves. They primarily consume large to medium-sized ungulates, up to 350 kg in weight. But they can also successfully hunt large animals like buffalo and giraffes. Furthermore, in KwaZulu-Natal, spotted hyenas consume livestock such as goats and cattle in the areas surrounding parks with hyena presence. Despite killing livestock, spotted hyenas are frequently falsely accused and persecuted for consuming the carcasses of animals that either died naturally or were killed by other carnivores. Because of this, spotted hyenas are frequently persecuted by poisoning the carcasses of livestock (Kerley et al., 2018: 240).

Kupika et al. (2016: 7) present a study from farming areas adjacent to Hurungwe and Charara Safari Areas, northern Zimbabwe, which investigated community perceptions toward livestock predation by lions and spotted hyenas from 2007 to 2011. The findings indicated that hyena were more likely to prey on smaller livestock such as goats and sheep, while lions killed a higher proportion of cattle. The local residents argued that the absence of game fences and the nature of the kill sites were the main factors for the hyena and lion livestock predation. The authors concluded that the members of the local communities had "positive attitudes towards lions and spotted hyenas despite the incidences of livestock loss through predation. Thus stakeholders in the area could incorporate the indigenous knowledge from elderly generations to educate the younger age groups on the significance of carnivores. Most people in Nyamakate and Vuti areas perceive broken game fence as the major cause of livestock lost regardless of distance to protected area. Seasonality and distance from protected area may not influence livestock predation but the environment (particularly kill site i.e. kraal, woodland, grassland) may influence the number of livestock kills in the study area" (Kupika et al., 2016: 7).

Koskey et al. (2021: 411) note that the spotted hyena is a livestock predator, which leads to conflicts with livestock owners. Their study showed that poor park fences are the main cause of conflict between humans and hyenas, according to 40% of respondents. The estimated total annual cost of hyena predation in this region is approximately $74,740, according to responses received in the surveys. The results suggest that the main livestock eaten were young pigs, cattle, and donkeys. Regarding hyena predation in the villages surrounding Lake Nakuru National Park and Soysambu Conservation Area, there were higher levels of predation on young pigs compared to cattle and donkeys. A total of 1020 livestock were wounded or killed. Of these, 777 were killed by hyenas, 75 were injured, and 7 were killed by unknown animals. People around Lake Nakuru National Park and Soysambu Conservation Area were affected by hyena attacks, including livestock and fear of walking at night (Koskey et al., 2021: 416). After young pigs (852 injured or killed), cattle were most often injured or killed (86), followed by dogs (68) and donkeys (14) (Koskey et al., 2021: 417).

Wolves and Cattle Predation

There are two main canine predators in Africa other than the wild dog: the African wolf (Canis lupaster, Hemprich and Ehrenberg, 1832) and the Ethiopian wolf (Canis simensis Ruppell, 1840) (Figs. 29 and 30). These two predators are principally killers of small and medium-sized ungulates. For example, Sebsibe (2022) present a study of the Central and Eastern Part of Bale Mountains National Park, Ethiopia, where cattle, sheep, goats, donkeys, and horses occur and note that the two wolf species predated sheep and goats, rather than cattle, horses, and donkeys, while cattle and donkeys were taken by hyenas and lions, and hyenas additionally killed horses. The African wolf has been classified as a golden jackal (and in some cases as the golden wolf, see Hunter & Barrett, 2020; Castelló & Sillero-Zubiri, 2018; Machado & Teta, 2020), but Rueness et al. (2011) note that "the Egyptian jackal (Canis aureus lupaster) has hitherto been considered a large, rare subspecies of the golden jackal (C. aureus). It has maintained its taxonomical status to date, despite studies demonstrating morphological similarities to the grey wolf (C. lupus)." In addition, "C. a. lupaster seems to represent an ancient wolf lineage which most

Fig. 29 African wolf in Ngorongoro Crater, Tanzania. (Permission Lee R. Berger—CC BY 2.5)

likely colonized Africa prior to the northern hemisphere radiation. We thus refer to C. a. lupaster as the African wolf."

The Ethiopian wolf is also referred to as the red jackal and "is the largest member of the genus in Africa." It is distinguishable from jackals (*C. aureus*, *C. mesomelas*, *C. adustus*) by its larger size, relatively longer legs, distinctive reddish coat, and white underparts, throat, chest, and tail markings (Sillero-Zubiri & Gottelli, 1994). Gottelli and Sillero-Zubiri (1992: 205) note that "the Ethiopian wolf is a social canid endemic to the highlands of Ethiopia but is the rarest and most likely endangered canid in the world, with only 500 individuals remaining today." Its range has already shrunk, and increased habitat loss from high-altitude subsistence farming and livestock overgrazing poses a threat. According to Sillero-Zubiri and Gottelli (1994), the most prevalent prey species in a case study were large rodents. According to Genbere (2017), wolves in Borena-Sayint National Park (BSNP) consumed rodents, medium-sized mammals, and other animals for 94.6% of their diet, compared to just 54.4% that consisted of livestock. Diurnal rodents accounted for 79.2% of all prey items. However, the presence of livestock indicated that the perception of wolves became more negative, due to livestock predation. Nonetheless, 71% acknowledged the need to preserve the ecosystem, and 66% said they had a

Fig. 30 African wolf with wildebeest, Ngorongoro, Tanzania. (Permission D. Gordon E. Robertson—CC BY-SA 3.0)

favorable opinion of wolves. Over the course of a year, 70 out of 140 herders lose 1–2 heads of small livestock to wolves and jackals, which equates to a 10% decrease in herd size or $92 per household. Nonetheless, 62.1% of herders felt positively about wolves (Figs. 31 and 32).

Geomatics and Large Carnivore Predation on Cattle in Africa

Geomatics techniques are increasingly used to evaluate agricultural landcover change in Africa, with important issues including habitat and farm losses to desertification and tracking of carnivores. Regarding the relationship between cattle and carnivores, habitat fragmentation, carnivore presence density, and carnivore intrusion into herding areas are among the more important, measurable issues. Regarding the habitat analysis, for example, Dolrenry et al. (2014) note that human pressures are causing African lion populations in Kenya and Tanzania to become more fragmented. In isolated habitats, lions inhabit a matrix of less suitable habitats. A metapopulation is described by a framework of habitat areas inside a less appropriate

Fig. 31 Ethiopian wolf bale, Ethiopia. (Permission Rod Waddington—CC BY-SA 2.0)

matrix. According to the findings, male lions' capacity for dispersal makes a significant contribution to population connectivity, whereas female lions' limited capacity for dispersal might be a limiting factor. Isolating areas is important, even for the largest lion populations. The ability of females to disperse and their proximity to the closest lion population are crucial factors for repopulating individual populations as well as for larger conservation initiatives, given the ongoing decline in lion populations and local extinctions (Dolrenry et al., 2014).

The study used a "incidence function model" (IFM, see Hanski, 1994a; Hanski et al., 1994), and the data analysis and patch measurement used ArcMap 10.1. The authors noted that "the IFM was created to use empirical data to examine the effects of patch area and isolation on patch occupancy and is one of the most commonly used metapopulation models. An IFM can be built with snapshot presence/absence data of a species at a particular site, the simplest form of data that can be collected during field studies" and "incidence function models assume that suitable habitat occurs in discrete patches surrounded by unsuitable matrix, and that occupancy of each patch is determined by local colonization and extinction events" (Dolrenry et al., 2014). Importantly, "extinction is negatively associated with patch area and colonization is negatively associated with patch isolation, so patch occupancy

Fig. 32 Ethiopian wolves and Zebu cattle. (Permission Laika ac—CC BY-SA 2.0)

should increase with area and decrease with isolation" and "Patch extinction and colonization are also assumed to depend on factors such as patch area (a proxy for local population size), spatial arrangement of patches, patch edge characteristics, dispersal ability of the species, and regional environmental stochasticity" (Dolrenry et al., 2014, see also Hanski, 1994b).

In a related study, the spatial distribution of six African herbivores that differ in body size and feeding strategy in response to environmental factors and direct predation risk from recently reintroduced lions was investigated using LiDAR, camera trap surveys, GPS telemetry, and kill site locations in the thicket biome of the Addo Elephant National Park, South Africa (Davies et al., 2016: 5728). The findings demonstrated that environmental factors, rather than direct risk of predation, largely determined the distribution of all species except buffalo. Rather, responses to predation risk were indirect, with diel patterns and environmental factors having a particularly strong influence on species distributions. Grazers were seen more often in open spaces and exhibited stronger responses to the variables that were measured. The distribution of browsers was more accurately predicted by terrain roughness than by vegetation density. The buffalo was the only animal to react to the possibility of encountering predators and stay away from areas where lions were more prevalent (Davies et al., 2016: 5728). In light of the larger context of impacts on predation risk, the authors conclude that direct effects on predation risk are relatively weak when predator densities are low and time since reintroduction is short. They also stress the necessity of thorough, long-term monitoring of predator reintroduction to prevent such events (Davies et al., 2016: 5728).

Conclusions

Cattle predation in Africa is largely due to lions and to a lesser extent to hyenas and leopards, and less to wild canids. Lions prey on all sizes of cattle, while leopards, hyenas, and other smaller predators take mostly calves, sheep, and goats. However, all types of cattle could be taken, as the canids and hyenas are social, cooperative hunters. Importantly, the predation on cattle was partly dependent of the similarity between the cattle and the wild prey taken by the particular predator. The evidence to increased possibilities for human-carnivore conflict, even with or near protected areas. Cattle herding is mostly practiced in savanna regions outside the forest, including Guinea, Sudan, and Sahel savanna belts, which coincide with preferred ranges of all the large African predators. Actions, mostly active persecution of carnivores, have been practiced reducing such predation. As in other continents, there are conflicts between conservation of large carnivores and expanding cattle herding, with added problems emerging from desertification, deforestation, and human population increase and generally higher levels of human poverty than in other continents. The extensive, complex environmental parameters allow the possible application of geomatics-based research methods. These methods are particularly important in Africa, as the continent with the largest number of carnivore species capable of killing cattle (lions, leopards, cheetahs, wild dogs, hyenas, two wolf species) compared with Europe (wolves, bears), South America (jaguars, cougars), North America (two bear species, cougars, wolves), and Asia (tigers, leopards, wolves, dholes, two bear species over a much larger land). The relative lack of conservation management increases the requirement for more effective assessment and monitoring methods. The chapter examined the trends of agricultural and conservation and the geomatics-based research methods that have enabled deeper understandings of social and environmental changes in contemporary Africa.

References

Animal World. (2024). *Africander cattle breed: Facts, uses, origins & characteristics (with pictures).* https://animal-world.com/africander-cattle-breed/

Angelici, F. M., Mahama, A., & Rossi, L. (2015). The lion in Ghana: its historical and current status. *Animal Biodiversity and Conservation, 38*(2), 151–162. https://doi.org/10.32800/abc.2015.38.0151

Bauer, H., & Van Der Merwe, S. (2004). Inventory of free-ranging lions Panthera leo in Africa. *Oryx, 38*(01). https://doi.org/10.1017/s0030605304000055

Bauer, H., Chapron, G., Nowell, K., Henschel, P., Funston, P., Hunter, L. T. B., Macdonald, D. W., & Packer, C. (2015). Lion (Panthera leo) populations are declining rapidly across Africa, except in intensively managed areas. *Proceedings of the National Academy of Sciences, 112*(48), 14894–14899. https://doi.org/10.1073/pnas.1500664112

Burnie, D., & Wilson, D. E. (Eds.). (2005). *Animal: The definitive visual guide to the world's wildlife*. DK Adult.

Campbell, M. (2013). *The political ecology of agricultural history in Ghana*. Nova Science Publishers.

Campbell, M. (2017). *Biological Conservation in the 21st Century: A Conservation Biology of Large Wildlife*. New York, NY: Nova Science Publishers.

Campbell, M. (2018). *Geomatics and Conservation Biology*. New York, NY: Nova Science Publishers.

Castelló, J. R., & Sillero-Zubiri, C. (2018). *Canids of the world: Wolves, wild dogs, foxes, jackals, coyotes, and their relatives* (Princeton Field Guides). Princeton University Press.

Chardonnet, P., Soto, B., Fritz, H., Crosmary, W., Drouet-Hoguet, N., Mésochina, P., Pellerin, M., Mallon, D., Bakker, L., Boulet, H & Lamarque, F, (2010). *Managing the conflicts between people and lion: review and insights from the literature and field experience*. Rome: Wildlife Management Working Paper 13.

Clements, H. S., Tambling, C. J., Hayward, M. W., & Kerley, G. I. (2014). An objective approach to determining the weight ranges of prey preferred by and accessible to the five large African carnivores. *PLoS One, 9*(7), e101054. https://doi.org/10.1371/journal.pone.0101054

Constant, N. L., Bell, S., & Hill, R. A. (2015). The impacts, characterisation and management of human-leopard conflict in a multi-use land system in South Africa. *Biodiversity and Conservation, 24*, 2967. https://doi.org/10.1007/s10531-015-0989-2

Courtin, D., Berthier, D., Thevenon, S., Dayo, G. K., Garcia, A., & Bucheton, B. (2008). Host genetics in African trypanosomiasis. *Infection, Genetics and Evolution, 8*(3), 229–238. https://doi.org/10.1016/j.meegid.2008.02.007

Craigie, I. D., Baillie, J. E. M., Balmford, A., Carbone, C., Collen, B., Green, R. E., & Hutton, J. M. (2010). Large mammal population declines in Africa's protected areas. *Biological Conservation, 143*(9), 2221–2228. https://doi.org/10.1016/j.biocon.2010.06.007

Davies, A. B., Tambling, C. J., Kerley, G. I., & Asner, G. P. (2016). Limited spatial response to direct predation risk by African herbivores following predator reintroduction. *Ecology and Evolution, 6*(16), 5728–5748. https://doi.org/10.1002/ece3.2312

DeMarco, E. (2015). *East, West Africa could lose 50% of their lions by 2035*. Science. https://doi.org/10.1126/science.aad4775

de Jong, Y.A., Cumming, D., d'Huart, J. & Butynski, T. (2016). *Phacochoerus africanus* (errata version published in 2017). The IUCN Red List of Threatened Species 2016. https://doi.org/10.2305/IUCN.UK.2016-2.RLTS.T41768A44140445.en

Dreyer, N. (2024). *The demographic-specific prey preferences of lions and cheetahs, and potential impacts on prey populations*. https://doi.org/10.13140/RG.2.2.33134.22080

Division of Agricultural Sciences and Natural Resources. (2024). *N'Dama cattle*. https://breeds.okstate.edu/cattle/ndama-cattle.html

Dolrenry, S., Stenglein, J., Hazzah, L., Lutz, R. S., & Frank, L. (2014). A metapopulation approach to African lion (Panthera leo) conservation. *PLoS One, 9*(2), e88081. https://doi.org/10.1371/journal.pone.0088081

Drewniak, Z., Finnegan, K., Miles, C., & Miles, M. (2012). National parks and protected areas in African countries. *SURG Journal, 6*(1), 23–30. https://doi.org/10.21083/surg.v6i1.1987

Dueppen, S. A. (2012). Cattle in the West African savanna: Evidence from 1st millennium CE Kirikongo, Burkina Faso. *Journal of Archaeological Science, 39*(1), 92–101. https://doi.org/10.1016/j.jas.2011.09.005

Erdaw, M. M. (2023). Contribution, prospects and trends of livestock production in sub-Saharan Africa: A review. *International Journal of Agricultural Sustainability, 21*(1), 2247776. https://doi.org/10.1080/14735903.2023.2247776

Estes, R. D. (1999). *The safari companion: A guide to watching African mammals, including hoofed mammals, carnivores, and primates* (pp. 98–100). Chelsea Green Pub. Co.

Estes, R. D. (2004). *The behavior guide to African mammals: Including hoofed mammals, carnivores, primates* (4th ed., pp. 158–166). University of California Press.

Estes, R. (2024). *Cape buffalo*. Encyclopedia Britannica. https://www.britannica.com/animal/Cape-buffalo

Estes, R. D., & Goddard, J. (1967). Prey selection and hunting behavior of the African wild dog. *The Journal of Wildlife Management, 31*, 52–70. https://doi.org/10.2307/3798360

Farming Plan. (2023). *Sanga cattle: Farming in Nigeria*. https://www.farmingplan.com/sanga-cattle/

Frank, L. G. (2023). Twenty years of lion conservation in a commercial rangeland. *Frontiers in Ecology and Evolution, 11*, 1141195. https://doi.org/10.3389/fevo.2023.1141195/

Fraticelli, C., Zayed, A. A., Leirs, H., & Bauer, H. (2024). Lions select larger prey in a Central African protected area with increasingly effective management. *Ecology and Evolution, 14*(7), e70062. https://doi.org/10.1002/ece3.70062

Galal, S. (2024). *Cattle population in Africa as of 2022, by country*. Retrieved from https://www.statista.com/statistics/1290046/cattle-population-in-africa-by-country/

Ganyo, E. Y., Boampong, J. N., Masiga, D. K., Villinger, J., & Turkson, P. K. (2018). Haematology of N'Dama and West African Shorthorn cattle herds under natural Trypanosoma vivax challenge in Ghana. *F1000 Research, 7*, 314. https://doi.org/10.12688/f1000research.14032.2

Gebresenbet, F., Baraki, B., Yirga, G., Sillero-Zubiri, C., & Bauer, H. (2017). A culture of tolerance: Coexisting with large carnivores in the Kafa Highlands, Ethiopia. *Oryx, 52*(4), 751–760. https://doi.org/10.1017/S0030605316001356

Gebresenbet, F., Bauer, H., Vadjunec, J. M., & Papeş, M. (2018). Beyond the numbers: Human attitudes and conflict with lions (Panthera leo) in and around Gambella National Park, Ethiopia. *PLoS One, 13*(9), e0204320. https://doi.org/10.1371/journal.pone.0204320

Genbere, G. E. (2017). *Ecology of the Ethiopian wolf (Canis simensis Rüppell 1835) in a changing landscape: Human carnivore interactions in Afroalpine ecosystems of Ethiopia*. University of Leiden.

Goldman, J. G. (2021). *How to Avoid Becoming a Meal for a Cheetah. Scientific American*. https://www.scientificamerican.com/podcast/episode/how-to-avoid-becoming-a-meal-for-a-cheetah/

Gottelli, D., & Sillero-Zubiri, C. (1992). The Ethiopian wolf – An endangered endemic canid. *Oryx, 26*(4), 205–214. https://doi.org/10.1017/S0030605300023735

Grey, J. N., Bell, S., Hill, R. A., & Hill, R. A. (2017). Leopard diets and landowner perceptions of human wildlife conflict in the Soutpansberg Mountains, South Africa. *Journal for Nature Conservation, 37*, 56–65. https://doi.org/10.1016/j.jnc.2017.03.002

Grzimek, B. (1990). *Grzimek's encyclopedia of mammals*. McGraw Hill Publishing Company.

Gusset, M., Swarner, M. J., Mponwane, L., Keletile, K., & McNutt, J. W. (2009). Human–wildlife conflict in northern Botswana: Livestock predation by endangered African wild dog Lycaon pictus and other carnivores. *Oryx, 43*(1), 67–72. https://doi.org/10.1017/S0030605308990475

Hanski, I. (1994a). A practical model of metapopulation dynamics. *Journal of Animal Ecology, 63*, 151–162. https://doi.org/10.2307/5591

Hanski, I. (1994b). Patch-occupancy dynamics in fragmented landscapes. *Trends in Ecology & Evolution, 9*(4), 131–135. https://doi.org/10.1016/0169-5347(94)90177-5

Hanski, I., Kuussaari, M., & Nieminen, M. (1994). Metapopulation structure and migration in the butterfly (Melitaea cinxia). *Ecology, 75*, 747–762. https://doi.org/10.2307/1941732

Hayward, M. W., Henschel, P., O'Brien, J., Hofmeyr, M., Balme, G., & Kerley, G. I. (2006a). Prey preferences of the leopard (Panthera pardus). *Journal of Zoology, 270*(2), 298–313. https://doi.org/10.1111/j.1469-7998.2006.00139.x

Hayward, M. W., Hofmeyr, M., O'Brien, J., & Kerley, G. I. H. (2006b). Prey preferences of the cheetah (Acinonyx jubatus) (Felidae: Carnivora): Morphological limitations or the need to capture rapidly consumable prey before kleptoparasites arrive? *Journal of Zoology, 270*(4), 615–627. https://doi.org/10.1111/j.1469-7998.2006.00184.x

Hayward, M. W., O'Brien, J., Hofmeyr, M., & Kerley, G. I. H. (2006c). Prey preferences of the African wild dog lycaon pictus (Canidae: Carnivora): Ecological requirements for conservation. *Journal of Mammalogy, 87*(6), 1122–1131. https://doi.org/10.1644/05-mamm-a-304r2.1

Hayward, M. W., Hayward, G. J., Tambling, C. J., & Kerley, G. I. (2011). Do lions Panthera leo actively select prey or do prey preferences simply reflect chance responses via evolutionary adaptations to optimal foraging? *PLoS One, 6*(9), e23607. https://doi.org/10.1371/journal.pone.0023607

Hazzah, L., Borgerhoff Mulder, M., & Frank, L. (2009). Lions and warriors: Social factors underlying declining African lion populations and the effect of incentive-based management in Kenya. *Biological Conservation, 142*(11), 2428–2437. https://doi.org/10.1016/j.biocon.2009.06.006

Henschel, P., Coad, L., Burton, C., Chataigner, B., Dunn, A., MacDonald, D., Saidu, Y., & Hunter, L. T. (2014). The lion in West Africa is critically endangered. *PloS one, 9*(1), e83500. https://doi.org/10.1371/journal.pone.0083500

Henschel, P., Bauer, H., Sogbohossou, E. and Nowell, K. (2015). *Panthera leo West Africa Subpopulation*. The IUCN Red List of Threatened Species. IUCN.

Howard, B. C. (2014). *Lions Approach Extinction in West Africa. Animals*. https://www.nationalgeographic.com/animals/article/140108-west-african-lions-endangered-conservation-science?loggedin=true&rnd=1752811949294

Hunter, L., & Barrett, P. (2020). *Field guide to carnivores of the world* (2nd ed.). Bloomsbury Publishing.

Kerley, G. I. H., Wilson, S. L., & Balfour, D. (Eds.). (2018). *Livestock predation and its management in South Africa: A scientific assessment*. Centre for African Conservation Ecology, Nelson Mandela University.

Kim, S. J., Ka, S., Ha, J. W., Kim, J., Yoo, D., Kim, K., Lee, H. K., Lim, D., Cho, S., Hanotte, O., Mwai, O. A., Dessie, T., Kemp, S., Oh, S. J., & Kim, H. (2017). Cattle genome-wide analysis reveals genetic signatures in trypanotolerant N'Dama. *BMC Genomics, 18*(1), 371. https://doi.org/10.1186/s12864-017-3742-2

Kim, K., Kwon, T., Dessie, T., Yoo, D., Mwai, O. A., Jang, J., Sung, S., Lee, S., Salim, B., Jung, J., Jeong, H., Tarekegn, G. M., Tijjani, A., Lim, D., Cho, S., Oh, S. J., Lee, H. K., Kim, J., Jeong, C., Kemp, S., et al. (2020). The mosaic genome of indigenous African cattle as a unique genetic resource for African pastoralism. *Nature Genetics, 52*(10), 1099–1110. https://doi.org/10.1038/s41588-020-0694-2

Kingdon, J. (1988). *East African mammals: An atlas of evolution in Africa, Vol. 3, Part B: Large mammals* (pp. 165–179). University of Chicago Press.

Kingdon, J. (1989). *East African mammals: An atlas of evolution in Africa, Vol. 3, Part D: Bovids*. University of Chicago Press.

Kingdon, J. (1997). *The Kingdon field guide to African mammals*. Academic Press.

Kingdon, J. (2015). *The Kingdon field guide to African mammals* (2nd ed., p. 601). Bloomsbury Publishing.

Kissui, B. M. (2008). Livestock predation by lions, leopards, spotted hyenas, and their vulnerability to retaliatory killing in the Maasai steppe, Tanzania. *Animal Conservation, 11*(5), 422–432. https://doi.org/10.1111/j.1469-1795.2008.00199.x

Koskey, C. C., Makanji, D. L., Wilkinson, C. E., & Eshiamwata, G. W. (2021). Socio-economic impacts of hyena predation on livestock around Lake Nakuru National Park and Soysambu Conservancy, Kenya. *Open Journal of Social Sciences, 9*, 411–425. https://doi.org/10.4236/jss.2021.99029

Kupika, O. L., Gwatsvaira, F., Kuvawoga, P. T., Muposhi, V. K., & Mbereko, A. (2016). Assessment of livestock depredation by lion and spotted hyena in farming areas adjacent to Hurungwe and Charara Safari Areas, northern Zimbabwe. *Annals of Social and Behavioural Sciences, 2*, 7–22.

Lagendijk, D. D., & Gusset, M. (2008). Human-carnivore coexistence on communal land bordering the greater Kruger area, South Africa. *Environmental Management, 42*(6), 971–976. https://doi.org/10.1007/s00267-008-9204-5

Lamichhane, S., Bhattarai, D., Maraseni, T., Shaney, K. J., Karki, J. B., Adhikari, B., Pandeya, P., Shrestha, B., & Adhikari, H. (2023). Landscape predictors influencing livestock depredation by leopards in and around Annapurna Conservation Area, Nepal. *PeerJ, 11*, e16516. https://doi.org/10.7717/peerj.16516

Lindsey, P. A., Havemann, C. P., Lines, R., Palazy, L., Price, A. E., Retief, T. A., Rhebergen, T., & Van der Waal, C. (2013). Determinants of persistence and tolerance of carnivores on Namibian ranches: Implications for conservation on southern African private lands. *PLoS One, 8*(1), e52458. https://doi.org/10.1371/journal.pone.0052458

Loomis, N. (2014). *West African Lions Under Threat in Senegal*. Voice of America; Voice of America (VOA News). https://www.voanews.com/a/west-african-lions-under-threat-in-senegal/1918938.html

Lyamuya, R. D., Masenga, E. H., Fyumagwa, R. D., & Røskaft, E. (2014). Human–carnivore conflict over livestock in the eastern part of the Serengeti ecosystem, with a particular focus on the African wild dog Lycaon pictus. *Oryx, 48*(3), 378–384. https://doi.org/10.1017/S0030605312001706

Machado, F. A., & Teta, P. (2020). Morphometric analysis of skull shape reveals unprecedented diversity of African Canidae. *Journal of Mammalogy, 101*(2), 349–360. https://doi.org/10.1093/jmammal/gyz214

Mason, I. L. (1996). *A world dictionary of livestock breeds, types and varieties* (4th ed.). C.A.B International.

Melzheimer, J., Heinrich, S. K., Wasiolka, B., Mueller, R., Thalwitzer, S., Palmegiani, I., Weigold, A., Portas, R., Roeder, R., Krofel, M., Hofer, H., & Wachter, B. (2020). Communication hubs of an asocial cat are the source of a human-carnivore conflict and key to its solution. *Proceedings of the National Academy of Sciences of the United States of America, 117*(52), 33325–33333. https://doi.org/10.1073/pnas.2002487117

Mkonyi, F. J., Estes, A. B., Msuha, M. J., Lichtenfeld, L. L., & Durant, S. M. (2017). Local attitudes and perceptions toward large carnivores in a human-dominated landscape of northern Tanzania. *Human Dimensions of Wildlife, 22*(4), 314–330. https://doi.org/10.1080/10871209.2017.1323356

Mungall, E. C. (2007). *Exotic animal field guide: Nonnative hoofed mammals in the United States* (1st ed.). Texas A&M University Press.

Murray, M., & Trail, J. C. M. (1984). Genetic resistance to animal trypanosomiasis in Africa. *Preventive Veterinary Medicine, 2*(1–4), 541–551. https://doi.org/10.1016/0167-5877(84)90100-4

Murray, M., Clifford, D. J., Gettinby, G., Snow, W. F., & McIntyre, W. I. (1981). Susceptibility to African trypanosomiasis of N'Dama and Zebu cattle in an area of Glossina morsitans submorsitans challenge. *The Veterinary Record, 109*(23), 503–510.

Murray, M., Morrison, W. I., & Whitelaw, D. D. (1982). Host susceptibility to African trypanosomiasis: Trypanotolerance. *Annals of Tropical Medicine and Parasitology, 21*, 1–68.

Murray, M., Trail, J. C., Davis, C. E., & Black, S. J. (1984). Genetic resistance to African trypanosomiasis. *The Journal of Infectious Diseases, 149*(3), 311–319. https://doi.org/10.1093/infdis/149.3.311

Mutoro, N. M., Chira, R., Gichuki, N., Kariuki, E., Eberle, J., Habel, J. C., & Wykstra, M. (2022). Dietary preference of cheetahs (Acinonyx jubatus) in south-eastern Kenya. *Ecology and Evolution, 12*(3), e8556. https://doi.org/10.1002/ece3.8556

Mwai, O., Hanotte, O., Kwon, Y. J., & Cho, S. (2015). African indigenous cattle: Unique genetic resources in a rapidly changing world. *Asian-Australasian Journal of Animal Sciences, 28*(7), 911–921. https://doi.org/10.5713/ajas.15.0002R

Nowak, R. M. (1999). *Walker's mammals of the world* (6th ed., pp. 1174–1175). Johns Hopkins University Press.

Owen-Smith, N., & Mills, M. G. (2008). Predator-prey size relationships in an African large-mammal food web. *The Journal of Animal Ecology, 77*(1), 173–183. https://doi.org/10.1111/j.1365-2656.2007.01314.x

Paling, R. W., & Dwinger, R. H. (1993). Potential of trypanotolerance as a contribution to sustainable livestock production in tsetse affected Africa. *The Veterinary Quarterly, 15*(2), 60–67. https://doi.org/10.1080/01652176.1993.9694374

Pappas, L. A. (2002). Taurotragus oryx. *Mammalian Species, 689*, 1–5. American Society of Mammalogists.

Patterson, B. D., Kasiki, S. M., Selempo, E., & Kays, R. W. (2004). Livestock predation by lions (Panthera leo) and other carnivores on ranches neighboring Tsavo National ParkS, Kenya. *Biological Conservation, 119*(4), 507–516. https://doi.org/10.1016/j.biocon.2004.01.013

Pérez-Pardal, L., Sánchez-Gracia, A., Álvarez, I., Traoré, A., Ferraz, J. B. S., Fernández, I., Costa, V., Chen, S., Tapio, M., Cantet, R. J. C., Patel, A., Meadow, R. H., Marshall, F. B., Beja-Pereira, A., & Goyache, F. (2018). Legacies of domestication, trade and herder mobility shape extant male zebu cattle diversity in South Asia and Africa. *Scientific Reports, 8*(1), 18027. https://doi.org/10.1038/s41598-018-36444-7

Pole, A., Gordon, I. J., Gorman, M. L., & MacAskill, M. (2004). Prey selection by African wild dogs (Lycaon pictus) in southern Zimbabwe. *Journal of Zoology, 262*(2), 207–215. https://doi.org/10.1017/s0952836903004576

Ralls, K. (1978). Tragelaphus eurycerus. *Mammalian Species, 111*, 1–4. American Society of Mammalogists.

Rasmussen, G. S. A. (1999). Livestock predation by the painted hunting dog Lycaon pictus in a cattle ranching region of Zimbabwe: A case study. *Biological Conservation, 88*(1), 133–139. https://doi.org/10.1016/s0006-3207(98)00006-8

Riggio, J., Jacobson, A., Dollar, L., Bauer, H., Becker, M., Dickman, A., Funston, P., Groom, R., Henschel, P., de Iongh, H., Lichtenfeld, L., & Pimm, S. (2012). The size of savannah Africa: A lion's (Panthera leo) view. *Biodiversity and Conservation, 22*(1), 17–35. https://doi.org/10.1007/s10531-012-0381-4

Roberts, S. C. (1996). The evolution of hornedness in female ruminants. *Behaviour, 133*(5–6), 399–442. https://doi.org/10.1163/156853996X00521

Romañach, S. S., Lindsey, P. A., & Woodroffe, R. (2007). Determinants of attitudes towards predators in central Kenya and suggestions for increasing tolerance in livestock dominated landscapes. *Oryx, 41*(2), 185–195. https://doi.org/10.1017/S0030605307001779

Rueness, E. K., Asmyhr, M. G., Sillero-Zubiri, C., Macdonald, D. W., Bekele, A., Atickem, A., & Stenseth, N. C. (2011). The cryptic African wolf: Canis aureus lupaster is not a golden jackal and is not endemic to Egypt. *PLoS One, 6*(1), e16385. https://doi.org/10.1371/journal.pone.0016385

Scholte, P., Pays, O., Chardonnet, B., Ouattara, A., & Tiomoko, D. (2025). Large mammal population trends in Comoé National Park (1958–2022): Towards understanding their asymmetric decline and recovery in West Africa's largest savanna park. *PLOS One, 20*(5), e0320455. https://doi.org/10.1371/journal.pone.0320455

Sebsibe, I. (2022). Humans-livestock predators conflict in the central and eastern part of Bale Mountains National Park, Ethiopia. *BMC Ecology and Evolution, 22*(1), 113. https://doi.org/10.1186/s12862-022-02065-y

Sillero-Zubiri, C., & Gottelli, D. (1994). Canis simensis. *Mammalian Species, 485*, 1. https://doi.org/10.2307/3504136

Sinclair, A. R., Mduma, S., & Brashares, J. S. (2003). Patterns of predation in a diverse predator-prey system. *Nature, 425*(6955), 288–290. https://doi.org/10.1038/nature01934

Smith, B. (2012). escenic. *The Telegraph*. https://www.telegraph.co.uk/news/worldnews/africaandindianocean/nigeria/9095288/Lion-on-a-rampage-kills-two-in-Nigeria.html?msockid=3355a8c55478673e1901bede5552667f

Tambling, C. J., Druce, D. J., Hayward, M. W., Castley, J. G., Adendorff, J., & Kerley, G. I. (2012). Spatial and temporal changes in group dynamics and range use enable anti-predator responses in African buffalo. *Ecology, 93*(6), 1297–1304. https://doi.org/10.1890/11-1770.1

Thuo, D., Broekhuis, F., Furlan, E., Bertola, L. D., Kamau, J., & Gleeson, D. M. (2020). An insight into the prey spectra and livestock predation by cheetahs in Kenya using faecal DNA metabarcoding. *Zoology (Jena, Germany), 143*, 125853. https://doi.org/10.1016/j.zool.2020.125853

Trail, J. C. M., Hoste, C. H., Wissocq, Y. J., Lhoste, P., & Mason, I. L. (1979). *Trypanotolerant livestock in west and central Africa. Vol. 1. General study*. International Livestock Research Institute.

Tshimologo, B. T., Reading, R. P., Mills, M. G. L., Bonyongo, M. C., Rutina, L., Collins, K., & Maude, G. (2021). Prey selection by African wild dogs (Lycaon pictus) in northern Botswana. *African Journal of Wildlife Research, 51*(1). https://doi.org/10.3957/056.051.0001

Tumenta, P. N., de Iongh, H. H., Funston, P. J., & Udo de Haes, H. A. (2013). Livestock depredation and mitigation methods practised by resident and nomadic pastoralists around Waza National Park, Cameroon. *Oryx, 47*(2), 237–242. https://doi.org/10.1017/S0030605311001621

Weise, F. J., Tomeletso, M., Stein, A. B., Somers, M. J., & Hayward, M. W. (2020). Lions Panthera leo prefer killing certain cattle Bos taurus types. *Animals, 10*(4), 692. https://doi.org/10.3390/ani10040692

Winterbach, H. E., Winterbach, C. W., Boast, L. K., Klein, R., & Somers, M. J. (2015). Relative availability of natural prey versus livestock predicts landscape suitability for cheetahs Acinonyx jubatus in Botswana. *PeerJ, 3*, e1033. https://doi.org/10.7717/peerj.1033

Woodroffe, R., Lindsey, P. A., Romañach, S. S., Stein, A. B., & Ranah, S. M. (2005). Livestock predation by endangered African wild dogs (Lycaon pictus) in northern Kenya. *Biological Conservation, 124*(2), 225–234. https://doi.org/10.1016/j.biocon.2005.01.028

Predators of Cattle in North America

Introduction

The history of domestic cattle in North America is currently described as beginning with the European colonization, during which period the indigenous bison declined sharply in numbers, due to extensive exploitation, and the introduction of European and African cattle (Bowling, 1942; Brand, 1961; Martínez et al., 2012; Alexander & Hernández Álvarez, 2018; Delsol et al., 2023). The introduction of mostly taurine cattle occurred through the dominant British colonization of northern North America and the Spanish colonization of southern North America. Bowling (1942: 151), for instance, reported that cattle were imported directly into Virginia, Massachusetts, New York, New Hampshire, Delaware, and possibly southern New Jersey from colonizing European nations. But many cattle were transported to the Southwest, the Gulf, Florida, and the Southeast from Mexico and Spanish possessions in the West Indies. Numerous cattle with at least some Spanish ancestry seem to have been transported to Massachusetts, Delaware, Virginia, and New Jersey. In addition, the initial mass introduction of cattle from Europe to the North American continent ceased around 1640. From this date until the American Revolution, the colonies' cattle needs were met through intercolonial trade or through trade with the Spanish colonies in the Western Hemisphere. A few cattle from the French colonies in the St. Lawrence River Valley found their way to the "Old Northwest." The era of livestock breeding in England did not begin until many years after the initial period of importation into America was completed, and in America, there was no fundamental work to improve livestock during this period (Bowling, 1942: 151). By 1800 the number had risen to several million head (Bowling, 1942: 152). The author notes that Columbus brought cattle with him from Spain in 1493 and that the cattle introduced later may have been for hides and, to a lesser extent, tongues, and by 1572 the cattle industry developed greatly and expanded in New Spain (Mexico), Central America, and the Caribbean later spread to the United States. Other cattle were imported from Holland and Sweden (Bowling, 1942).

© The Author(s), under exclusive license to Springer Nature Switzerland AG 2025

M. O. Campbell, *Cattle, Their Predators and Geomatics Research*,
https://doi.org/10.1007/978-3-031-97363-5_8

Currently, the domestic cattle industry is a massive undertaking. The United States is the world's largest producer of beef, mainly premium grain-fed beef for both domestic and international markets, and has the largest feedstock industry globally, according to the US Department of Agriculture (2024). However, the US imports a net amount of beef, buying grass-fed, lower-value beef for processing. North America currently has several dairy cattle breeds, including Jersey, Guernsey, American Brown Swiss, Ayrshire, and Holstein (Guinan et al., 2023) and beef cattle breeds including Simmental, Gelbvieh, Shorthorn, Black Angus, and Red Angus (Bhowmik et al., 2023) (Figs. 1, 2, 3, 4, and 5). In 2023, there were roughly 968,500 dairy cows in Canada, which is slightly more than twice as many dairy heifers, according to Shahbandeh (2024). Nonetheless, on January 1st of that year, there were more than 3.5 million beef cows in Canada. Since 2001, there have been roughly three times as many beef cows as dairy cows in the USA annually. There were roughly 9.4 million dairy cows and approximately 29 million beef cows in 2023, according to Shahbandeh (2024). In 2022, there were approximately 8.5 million cattle in Mexico (Navarro Villa, 2024).

After the introduction of cattle and the massive expansion of cattle range lands, large carnivore predation on cattle became an issue, as cattle were generally more

Fig. 1 Ayrshire cow. (Permission Matthias Zomer-pexels)

Fig. 2 American Brown Swiss. (Permission Malcolm Morley-Brown. Public Domain)

vulnerable to predation than the indigenous, wild bison. According to Wuerthner (2020), bison no longer need to practice predator avoidance because most of the larger predators that threatened them have gone extinct since the end of the last Ice Age. The only predators that threaten bison besides humans are wolves. The bison's size serves as a strong deterrent against other predatory animals. According to recent research on the predator-prey relationship between wolves and bison, healthy adults are comparatively safe from wolf attacks, but calves and older adults make up the majority of the prey of wolves. For domestic cattle, some reports indicate that predation may be increasing.

For example, according to the United States Department of Agriculture (2022), livestock losses due to predators occur nationally. Historically, this has primarily been an issue in the West, but as populations of certain predators spread their range, the occurrence is growing in the East (e.g., coyotes). According to a recent study, there is little evidence that killing predators benefits ranchers (Hall, 2016). A contentious plan to kill a pack of wolves implicated in a string of attacks on cows in Washington State was approved by wildlife officials. According to the state, eliminating the carnivores could help stop further losses in livestock. As a result, thousands of coyotes, wolves, bears, and other predators were killed by the USA using

Fig. 3 Guernsey cow. Man vyi—Public Domain

this rationale. For this reason, other countries have also permitted predator hunts, such as Canada and Finland (Hall, 2016).

There are also compensation programs for livestock farmers' losses. For example, the Government of Alberta administers the Wildlife Predator Compensation Program, which stipulates that cattle, bison, sheep, pigs, and goats are protected from attacks by wolves, grizzly bears, black bears, cougars, and eagles and covers the costs of veterinary care and medications in connection with this with the incident or loss of an animal, up to the value of the animal, based on the average of the animal species and class. Evidence of live livestock robberies includes evidence that the livestock bled from the attack. The blood can be on the body and around the area of attack. Below the tears and puncture wounds in the skin, the tissue is damaged and bruises or bleeds. The body may be stretched in an unnatural position. There may be evidence of a struggle, such as broken vegetation or torn earth (Government of Alberta, 2024).

Different types of injuries can be used to identify the predator. Wolf attacks can cause injuries to the lower parts of the body, traces of blood on the carcass, multiple wolf tracks, as well as cuts, punctures, and damage more serious than those caused by coyotes (Government of Alberta, 2024). Bear hunting may involve injuries to the

Fig. 4 Jersey. (Permission Man vyi—Public Domain)

skull, neck, withers, and spine; dragging the carcass; and covering the carcass with debris; in addition, bears can attack all ages of cattle, pigs, and sheep. Grizzlies are more likely to hunt larger animals, while black bears are more likely to hunt calves (Government of Alberta, 2024). In cougars, the injuries usually occur to the throat, throat, and skull, and an entry wound into the chest cavity can occur by dragging the carcass for cover and becoming covered by debris. Most attacks occur on smaller livestock such as sheep and goats, horses, and exotic animals (Government of Alberta, 2024). Attacks by coyotes are described as being similar to those of wolves, except that the smaller coyote can be a single attacker, causes less damage, and attacks smaller livestock such as sheep, goats, or young calves (Government of Alberta, 2024).

In the same province, Alberta, Lee et al. (2017: 10) note that the majority of beef producers (64%) in Alberta, Canada, reported losses as a result of carnivore predation, and the region shares a landscape with large carnivores, where interactions can have detrimental effects. Although the rate varied greatly among producers (ranging from 0% to 25% annually), the average calf loss rate was reported to be 2%. For survey participants, the direct yearly economic loss as a result of looting was $2

Fig. 5 Holstein. (Permission Matthias Zomer-pexels)

million. Several assumptions can be used to extrapolate this to $22 million on a provincial level. Given how important Alberta's beef industry is, these problems are significant. In addition to supporting more than 62,000 jobs and contributing $5.87 billion to the national gross domestic product, the province produces almost 40% of all cattle in Canada. In Alberta, there are roughly 20,000 cattle producers with 5.2 million cattle. Numerous carnivore species, such as wolves, coyotes, cougars, black bears, and grizzly bears, can be found in Alberta. Wildlife offers a multitude of benefits and is a significant component of Alberta's natural heritage. Ranchers, cattle, and/or the carnivore may suffer as a result of interactions between large carnivores and cattle (Lee et al., 2017: 10).

Lee et al. (2017: 12) further note that 74% of respondents experienced impacts from carnivores, with the coyote being the species of most concern. In the predation of cattle, 64% of producers reported losing at least one animal between 2011 and 2013. The predation was predominantly on calves, with 61% of producers reporting losses. The average reported predation rates were consistent with other studies (Lee et al., 2017: 13: see also Muhly & Musiani, 2009; Hoag et al., 2011). Furthermore, the authors note that prevention (e.g., removal of cattle carcasses from the landscape, moving calving areas closer to housing, and electric fencing of calving areas) has been identified as the best way to reduce negative interactions between producers and carnivores (Lee et al., 2017: 15).

An interesting study of South Saskatchewan Region (SSR), Canada, compares the livestock predation of the main carnivores in North America (grizzly bear, wolf, black, cougar, eagle), through the Wildlife Predator Compensation Claim program, which covers cattle, bison, sheep, swine and goats, but not horses, donkeys, or "exotic animals such as llamas, alpacas or wild boar" (Government of Alberta, 2020: 5). The findings of this study were that the eight districts that fall within SSR accounted for 508 primary claims approved between 2015 and 2019. Grizzly bears were involved in the largest number of all depredation incidents, followed by wolves, cougars, unknown bears, black bears, eagles, and unknown species. There were a total of 751 livestock animals involved. Sixty-two percent of depredation involved cattle and 35% involved sheep, and 3% involved goats (Government of Alberta, 2020: 16). Of the eight districts covered, the breakdown is shown in Table 1 (Table 2).

These examples illustrate the important role of predation on cattle in North America, and possible escalation of conflicts between conservation management, habitat and agricultural landcover changes, and agricultural development, with serious financial implications. Similar to Europe, compensation may play an important role, but the larger number of mammal carnivore species capable of killing cattle complicates the North American scenario compared with Europe (5 and 2, respectively). The rest of this chapter examines the impact of predation on cattle in North America, including Mexico and Central America. The predation by wolves is selected first because most sources cite the wolf as the principal predator of cattle. The brown and black bears are then examined, followed by the cougar and jaguar, the latter occurring only in Latin America and the southwestern edger of the United States. Coyotes, which generally take only calves, are considered last. The predation ecology of North America shows similarity with Europe, but the predators are much more numerous, especially the bears, and the big cats introduce a new dimension, with a similar ecology to the Old World leopard, and even the extinct tigers of southwest and western Asia. The competition between sophisticated and advanced

Table 1 Primary claims for livestock predation. South Saskatchewan Region, Canada

Districts	Wolf %	Brown Bear %	Cougar %	Black Bear %	Unknown Bear %
Rocky Mountain House District	**45**	21	28	4	2
Sundre-Olds District	25	4	**50**	4	17
Cochrane/ Canmore District	**37**	27	34	0	0
High River District	**49**	23	23	3	3
Claresholm District	16	**54**	9	9	12
Blairmore District	38	**55**	0	2	5
Pincher Creek District	17	**72**	4	1	7
Cardston District	20	**70**	3	3	2
Average %	**31**	**41**	**19**	**3**	**6**

Unknown bear refers to a situation where the bear could not be identified but appears to be added to brown bears in the text. For Cardston District, eagles are included with Unknown Bear. The primary predator is in bold and gray. Source: Derived from Government of Alberta (2020), claims 2015–2019

Table 2 Livestock killed by predators in South Saskatchewan Region, Canada

Districts	Cattle %	Sheep	Goats
Rocky Mountain House District	54	46	0
Sundre-Olds District	22	76	2
Cochrane/ Canmore District	39	53	7
High River District	74	23	3
Claresholm District	68	27	4
Blairmore District	100	0	0
Pincher Creek District	94	6	0
Cardston District	48	52	0
Average %	**62**	**35**	**2**

For Claresholm, brown bears were mostly responsible for the cattle incidents (52%) while cougars were responsible for most (90%) of the sheep-related incidents. For Pincher Creek District, brown bears were responsible for more than half of the incidents, cattle more than sheep, with wolves and cougars next, respectively. For Cardston District, brown bears were involved in nearly all the sheep incidents (unusually, as cougars are generally responsible), and 70% of the cattle predations, with wolves responsible for 24% of cattle predations
Source: Derived from Government of Alberta (2020)

conservation policies and management for large carnivores, and the increasingly advanced cattle industry, is also comparable to that of Europe.

Wolf Predation and Cattle

Gray wolves range through most of Canada and parts of the northwest United States, in forested, mixed, and open vegetation cover (Boitani et al., 2023). In the twenty-first century, wolves in Alaska and northern Canada are widely distributed and abundant, unlike those of southern Canada and the contiguous United States, which have been reduced or extirpated (Musiani & Paquet, 2004: 52). However, there is substantial conflict between the issues of wolf conservation and livestock management, especially with the success of conservation management. For example, Musiani and Paquet (2004: 50) note that "wolf management can be controversial, reflecting a wide range of public attitudes" and there is a "spectrum of approaches in Canada and the United States." Although "during the early 20th century, wolves were considered undesirable" and "were subject to persecution and were extirpated from large areas of their original range" "with increased environmental awareness in the 1970s, attitudes toward wolves began to change. Wolf conservation became a focus of public interest, providing conditions that favored regional wolf recovery. However, in regions where livestock production or big-game hunting is valued, wolves have continued to be controlled by management authorities or through the actions of individual citizens. With US wolf populations recovering in the conterminous states, a rule was approved to delist the species from endangered to threatened status under the Endangered Species Act" (Musiani & Paquet, 2004: 50). The authors acknowledge that in Canada and the United States, attitudes toward gray wolves have changed drastically in recent years. Now, preservationist groups and large sectors of the public regard the wolf positively, viewing its preservation as a high priority for wildlife conservation efforts. At the same time, however, wolf reintroduction continues to receive energetic opposition from individuals and organizations (Musiani & Paquet, 2004).

Concerning livestock, it is noted that in North America, wolves have been killed to protect livestock and even wild ungulates, and for fur, disease control and out of fear (Musiani & Paquet, 2004). Importantly, "In some US states, private landowners are authorized to kill wolves in cases of demonstrable threats by wolves toward humans and human property, including livestock." The example is cited of the state of Montana, where ranchers may kill wolves during livestock predation, and some wolves may be killed illegally. Similar events may occur in Canada. Musiani and Paquet (2004: 54) noted that "wolves thrive in areas with high ungulate densities but tend toward local extinction in areas with high densities of livestock, owing to conflicts with ranchers. Positive relationships have also been found between wolf presence and areas with forest cover, few roads, and low human density." In addition, "wolves typically prey on all ungulates present within their range. Often included are domestic ungulates… Now, as the result of wolf recovery, wolves are moving into areas of intense livestock production. In these areas, conflicts and associated costs of livestock protection are on the increase" (Musiani & Paquet, 2004: 56). It is also noted that there are compensation programs in the USA and Canada for livestock losses to wolves, which may or may not be effectively managed (Musiani & Paquet, 2004. 57).

There are two main issues: the laws protecting wolves and human attitudes to the protection of wolves. Musiani and Paquet (2004, 58) note that "in the conterminous US states, the federal Endangered Species Act (ESA) has protected all wolves since 1974. The ESA is the strongest species protection legislation in the United States, usually superseding state laws. In addition, there are currently large differences in attitudes towards wolves in Canada and the United States. Such attitudes tend to be positive and related to attitudes toward other carnivores and environmental issues. However, local community resistance to wolf presence (particularly in agricultural areas) may indicate social and economic pressures that are incompatible with successful wolf and wilderness conservation" (Musiani & Paquet, 2004. 58). The authors conclude that in North America, rural agricultural areas where wolves occur are frequently frequented by wild and domestic ungulates, both of which are preyed upon by the wolves. Managing high densities of wild ungulates could result in reduced predation of livestock by wolves (Musiani & Paquet, 2004. 59).

Blackley (1990) present a study of wolves in Alberta, Canada, where "wolf predation of livestock is common and normally occurs on forested grazing leases and farmland at the forest fringe. Wolves prey on a variety of domestic animals including cattle, sheep, swine, horses, poultry and dogs. Cattle and other livestock in grazing leases and private pastures are readily available to wolves and other predators. Cattle, primarily calves, are the most common livestock prey. Alberta data suggest that wolves select young, inexperienced or disabled cattle as prey much more often than healthy adult cattle. Calves comprised about 70 per cent of 1581 cattle killed by wolves in Alberta from 1980 to 1987… Most calves of domestic cattle killed by wolves are from 6 to 9 months old" (Blackley, 1990). Wolf attacks are contrasted with those of coyotes, which occur throughout Alberta, and may prey on "sheep, goats, cattle, pigs, poultry, dogs and cats" and is the "major predator of sheep in Alberta" and "kill lambs in preference to ewes or rams." In addition, "coyotes rarely attack adult cattle unless they are predisposed to attack due to birthing problems, illness or injury. Most coyote predation of cattle happens during the spring through fall pasture season. Few losses occur in winter when livestock are generally kept and fed at close quarters" (Blackley, 1990). Wolves are also compared with domestic dogs in terms of livestock predation, as dogs may attack livestock "throughout Alberta's farming community. Cattle, sheep, goats, swine, horses and poultry may be attacked. Several stray or neighborhood dogs, occasionally accompanied by the livestock owner's dog, are usually involved" (Blackley, 1990).

Some studies have investigated the link between wolf predator control (including culling) and the killing of livestock. For example, Poudyal et al. (2016) found that "the culling of one wolf the previous year would decrease the expected number of cattle killed this year by 1.9%, and the expected number of sheep killed by 3.4%. However, for every wolf killed there is a corresponding 2.2% increase in the expected number of sheep killed in the same year. The increase in sheep depredation appears to be a short term phenomenon." Kompaniyets and Evans (2017) noted that "wolf control to reduce cattle depredation is an important issue to ecology and agriculture in the United States." And their study found "a positive link between wolf control and cattle depredation. However, it would be incorrect to infer that wolf control has a positive effect on the number of cattle depredated. We maintain that this link comes from a growing wolf population, which increases cattle depredation, and in turn, causes an increase in the number of wolves killed. While the wolf population is growing, we see both wolf removal and cattle depredation simultaneously grow. It is not until the wolf population growth nears the steady state, that removal of wolves has a sufficient negative effect to reduce or stabilize the number of cattle depredated."

Several other studies examine the link between wolf control and livestock predation. For example, Wielgus and Peebles (2014) present a study of the effects of wolf mortality on reducing livestock depredations in Idaho, Montana, and Wyoming from 1987 to 2012 using a 25 year time series, on the premiss that such studies on lethal control of wolves are rare and unclear: "The number of livestock depredated, livestock populations, wolf population estimates, number of breeding pairs, and wolves killed were calculated for the wolf-occupied area of each state for each year." The findings indicated that "we found that the number of livestock depredated was positively associated with the number of livestock and the number of breeding pairs. However, we also found that the number of livestock depredated the following year was positively, not negatively, associated with the number of wolves killed the previous year. The odds of livestock depredations increased 4% for sheep and 5–6% for cattle with increased wolf control—up until wolf mortality exceeded the mean intrinsic growth rate of wolves at 25%." It is argued that possibilities for these outcomes, the increased livestock depredations at ≤25% mortality include "compensatory increased breeding pairs and numbers of wolves following increased mortality" and "after mortality exceeded 25%, the total number of breeding pairs, wolves, and livestock depredations declined" but the authors caution that "mortality rates exceeding 25% are unsustainable over the long term. Lethal control of individual depredating wolves may sometimes be necessary to stop depredations in the near term, but we recommend that non-lethal alternatives also be considered" (Wielgus & Peebles, 2014).

A possibility is the removal of only those wolves judged as killing cattle. For example, Furber (2020) describes an example in British Columbia: here selective removal of the identified wolves, which resulted in a predation-free pasture. The author notes that "this was of one of the earliest successes of a new approach to predator control introduced to B.C. in 2003. The WPLCCP evolved from public pressure against traditional wolf control methods and evidence that none of the historic methods were truly effective in resolving predator-livestock conflicts... The aim was to reduce cattle losses due to predation through the use of preventative measures, education, targeting offending predators, and compensating ranchers for kills verified by a wildlife control specialist" (Furber, 2020). This system was later replaced with the Wild Predator Loss Prevention-Mitigation Pilot Program, under the Agricultural Research and Development Corporation under the Ministry of Agriculture (Furber, 2020).

A similar study by Bedayn (2023) reported that cattle farmers in the US state of Colorado, namely he Gunnison County Stockgrowers' Association and Colorado Cattlemen's Association were opposed to the reintroduction of wolves, despite the voters approval in a 2020 ballot: "similarly to EU initiatives on wolf protection—it has become a divisive issue, sharpening divides between rural and urban residents. City dwellers largely voted to reintroduce the apex predators into the rural areas where prey can include livestock that help drive local economies. Now, representatives of the cattle industry association are suing state and federal agencies in the hopes of delaying the predators' release" (Bedayn, 2023). Bedayn (2023) poses the question "Are wolves likely to attack livestock?" The answer reported was that "an analysis of state and federal data by news agency The Associated Press found that, in 2022, grey wolves attacked domesticated animals hundreds of times across 10 states in the contiguous US, including Colorado. Data showed that attacks killed or injured at least 425 cattle and calves, 313 sheep and lambs, 40 dogs, 10 chickens, five horses and four goats."

Brown Bear Predation and Cattle

Brown bears are also cattle predators in North America (Fowler, 2016; Kienlen, 2022; Thomas, 2022; Spearchief-Morris, 2023). For example, Blackley (1990) notes that conflicts with farm animals occasionally occur. Problems usually occur in peripheral areas where human settlements border wilderness. When food is scarce, predation can increase. Cattle are the most common domestic prey, likely because cattle are more common in grizzly bear territory than other livestock. Grizzlies appear to prefer younger animals but also kill adult cattle. Additionally, of the 17 recorded cattle kills in Wyoming, 4 were yearlings and 13 were calves. Similarly, 15 calves, 10 yearlings, and 1 cow were killed by grizzly bears in southern Alberta between 1986 and 1987. Sheep, goats, pigs, and poultry are taken occasionally. Grizzly bears "do not select for any age or size of sheep, probably because even the largest sheep are easy prey for bears" (Blackley, 1990). It is also mentioned that grizzlies are more capable of predation because they are larger and stronger than black bears. As a result, grizzly bears typically kill more cleanly than black bears and typically bite the top of the neck or the back, close to the withers, to seize their prey. The spine and surrounding tissues sustain significant harm as a result. Less frequently, tooth wounds occur on the head or lumbar area of the back. The face, neck, shoulders, and other body parts of cattle may bear claw marks. Cattle are usually bitten by grizzly bears on the top of the back, neck, and frequently on the withers. The remains of a carcass are typically covered with debris by grizzly bears. Grizzlies are known to grab cattle by the forelegs and bite them on the back or top of the neck. Bears rarely show signs of a lengthy chase because they are reluctant to pursue prey. Grizzly bear kills frequently take place in or near cover. Before feeding, bears typically drag their prey into cover (Blackley, 1990).

Lewis (2018) cites the conflicts between conservation and bear predation issues and describes the killings of cattle by bears in the state of Montana, USA, and that the effects of an increasing grizzly population are also being felt in Wyoming. According to reports, the Fish and Wildlife Service decided to give three states—Wyoming, Montana, and Idaho—control over grizzly bears. Political action organizations decided to fight against Wyoming Game and Fish's grizzly population management plan. The action promised to lessen what they view as a gory predator for cattle producers in Wyoming, particularly those in the northwest, where the majority of grizzlies live. Later on, though, a judge in Missoula, Montana, decided in favor of a group of environmental organizations that had filed a lawsuit, arguing that grizzly hunting would once more put the species in danger of going extinct. More recently, the hunt was entirely halted, and the grizzly bear population in the Yellowstone region was granted Endangered Species Act status of grizzly bears was restored.

Northern Ag Network (2016) compared bears and wolves as cattle predators, noting that "despite the record number of livestock killed already this year by the bears and the attention they've received for ranging farther out on the Rocky Mountain plains this year and into towns, they are still not Montana's number one predator of livestock. Not even close. Wildlife Services has had their hands full this year. According to John Steuber, Montana Wildlife Services State Director, the number of livestock killed by Grizzly bears has already topped numbers from 2015 which was a record high year. In 2015, there were 90 head of livestock confirmed killed and probable killed by Grizzly bears. As of May 31 of this year, 94 head of livestock have been confirmed killed and probable; a number that doesn't include the 40 sheep killed near Valier over Memorial Weekend, the 2 alpacas and a couple calves." Nevertheless, the authors cite Steuber: "Coyotes are still the primary predator in the state. And a lot of that can be attributed to their range and numbers are certainly higher. Wolf range is still limited, primarily in the western half of the state, as is grizzly bear range, although both are spreading." A problem concerns the legal protections of grizzly bears: "grizzly bears, unlike coyotes or even wolves in some cases, come with few management options for the rancher, and that increases the frustration. Unlike coyotes, ranchers can't shoot grizzly bears when they are harassing livestock. Producers can do little besides hope the bears move on before killing livestock. When the bears are relocated… they often return" (Northern Ag Network, 2016).

Anderson et al. (2002: 247) note that grizzly bear depredation on livestock on public land grazing allotment increased within the Greater Yellowstone Ecosystem (GYE) in the western United States as the grizzly bear population has grown in number and occupied range. The study used generalized linear mixed models to examine spatio-temporal relationships between grizzly bear depredation of livestock and United States Department of Agriculture Forest Service and National Park Service grazing allotments in the GYE during 1992–2014. The findings indicated that more grazing allotments were occupied by grizzly bears during the study period, and most livestock depredations were associated with these areas of population expansion, and the number of livestock and grizzly bear were the most significant factors for the number of livestock depredation events relative to other allotment attributes. The estimated number of depredation events increased by approximately 20% when cow-calf pairs increased by 100 pairs and grizzly bear density index increased by 1 bear/196 km^2 (the average annual home-range size of a female grizzly bear in the GYE). In addition, grazing allotment size was positively related to the number of depredation events, and the presence of bull cattle or horses was associated with a 50% reduction in depredations. Livestock depredation events were more numerous for allotments with lower road density, less rugged terrain, higher vegetative primary productivity, and more whitebark pine coverage (Anderson et al., 2002: 247). The authors suggest that the increase in "the grizzly bear population, the environmental managers, and livestock producers may focus efforts on allotments with a higher density of grizzly bears, fewer roads, and quality grizzly bear habitat, and higher vegetative productivity, when attempting the management of depredation" (Anderson et al., 2002: 247).

One method of reducing grizzly bear predation on livestock is termed intercept feeding, where the bears are fed alternative food sources. For example, Morehouse and Boyce (2017: 66) argue that "supplemental and diversionary feeding can reduce conflicts between wildlife and people. However, feeding also can increase species abundance, survival, and reproductive success, which might increase human–wildlife conflicts." In the case study of southwestern Alberta, Canada, it is noted that the "provincial government fed road-killed ungulates to grizzly bears (Ursus arctos) each spring during 1998–2013 attempting to reduce spring depredation of livestock by grizzly bears" (Morehouse & Boyce, 2017: 66). The study used "non-invasive genetic sampling, remote trail cameras, and complaint records to evaluate the efficacy of Alberta's intercept-feeding program," and incidents were examined before, during, and after the intercept-feeding program, where an "incident" was defined as "where the grizzly bear caused property damage, obtained anthropogenic food, or killed or attempted to kill livestock or pets" (Morehouse & Boyce, 2017: 66). The findings indicated that "grizzly bear–livestock incidents did not decrease during the intercept-feeding program" and when the intercept-feeding program was suspended in 2014 and 2015, there was no increase in spring livestock depredation (Morehouse & Boyce, 2017: 66). The authors note that "in total, approximately $720,600 CAD has been spent on the intercept-feeding program between 1998 and 2013. Intercept feeding did not decrease spring livestock depredation; therefore, other mitigation efforts, including electric fencing and deadstock removal, might be a more cost-effective long-term solution" (Morehouse & Boyce, 2017: 66).

Black Bear Predation and Cattle

Black bears may also kill cattle, although this species is smaller and less powerful than the brown bear (Campbell, 2017). Blackley (1990), in a Canadian study, notes that "black bears inhabit the western and northern forest areas of Alberta. They are better adapted to living near people than grizzly bears. Predation on livestock commonly occurs on farms that are close to forests. Black bears prey on cattle, swine, sheep and poultry. Cattle are taken more often than other livestock. Black bears tend to prey on calves less than 6 months of age (Fig. 6). Calves, yearlings, and adults comprised 71, 11 and 18 per cent, respectively, of the 466 cattle killed by black bears in Alberta from 1974 to 1978." In addition, "black bear/cattle problems peak during May–July in Alberta. Predation in late summer may be dependent on the berry crop. Fewer livestock problems develop if berries are readily available. Black bears bite and claw the top of the neck and back of cattle, pigs and sheep. Smaller prey are usually bitten or sometimes killed by a blow to the head or neck. Inexperienced bears may rip open the underparts of prey and expose the viscera. Black bear kills usually have more claw marks on them than on grizzly bear kills. Black bears are generally not as large as grizzly bears, and probably have more difficulty killing larger animals" (Figs. 7, 8, 9, and 10)

In another Canadian report, CBC News (2015) noted that "about half of attacks on calves are from black bears this year, as opposed to wolves, coyotes and cougars. Black bears are killing more calves than usual this year in the Williams Lake area of B.C., and conservation officers suspect the unusual weather may have played a role." In this report, a conservation officer was cited as point to black bears as "responsible for 47 to 50 per cent of all predator attacks on cattle in the Cariboo region… Usually, attacks by coyotes, cougars and wolves are more prevalent… Although black bears can often be spotted grazing on grass on the side of the road, they are omnivores and will attack other mammals for food" (CBC News, 2015). In an old report from Alberta, this position of strong attacks by black bears was reinforced. Horstman and Gunson (1982: 34) noted that "compensation was paid on 541 approved black bear-livestock predation claims in Alberta during 1974–79. Cattle accounted for 81% of all cases; sheep and swine 9% each. Multiple kills were more common than single kills in sheep and swine cases and infrequent in cattle cases. Most of the cattle killed (71%) were calves. Sex and ages of known predators (18) indicated that most predation is by mature and old males. Twenty-five percent of all compensated losses occurred on grazing leases on public lands in the forested

Fig. 6 Red Angus Cow and Calf, Oregon. (Permission Darcy Vial, USDA Farm Service Agency Images, Public Domain)

(unsettled) part of the province. Although bear predation on livestock is probably underestimated by compensation statistics, such predation is relatively uncommon considering the numbers of livestock and bears on shared pastures."

In a study based on northern Mexico, Juárez-Casillas and Varas (2013: 970) note that black bears are "considered endangered in Mexico. In northern Mexico, bears are considered responsible for preying on livestock, damaging private property, and crops; therefore, they are killed to protect locals' livelihoods." The study examined "the diet of black bears in the Sierra de Picachos located in the Northeast portion of the State of Nuevo León, Mexico" to "determine the diet of black bears during 3 consecutive seasons (i.e., dry, rainy, and post-rainy), and to determine the presence and frequency of livestock and crops remains in bear scats," but examination of the bear diets using scats found that "plants were the most common item (73.4%), animals (16.8%), and undetermined wood fragments (9.3%), no "statistically significant difference between seasons" and no "cattle residues in any of the scats; therefore, black bears in the study site either do not or rarely feed upon livestock" (Juárez-Casillas & Varas, 2013: 970). The authors suggested that "management plans include informing local people on bear behavior. Also, specific actions to

Fig. 7 Black Angus Bull. (Permission Roger Starnes-unsplash)

avoid bear proximity should be implemented along with government actions aimed at improving bear habitat that minimize bear-human conflict in the area" (Juárez-Casillas & Varas, 2013: 970).

Cougar Predation and Cattle

The cougar may fulfil a similar ecological role as the leopard in Africa and Asia, and in prehistoric Europe, as an apex predator one step down from the huge bears and lion-sized cats, and the social canids (Campbell, 2017, 2022). They are known cattle killers, perhaps less than wolves. Blackley (1990) present a case of livestock in Alberta, Canada, where it is noted that "cougars primarily prey on deer; however, they will occasionally kill cattle, colts, sheep, goats, swine, dogs and cats" as in Arizona, where "80 per cent of the cattle killed by cougars were under six months of age." In Alberta, "cougar predation of livestock usually occurs along the agricultural-forest fringe… Confined livestock are often attacked. Corrals and fences offer little resistance or discouragement to cougars, which easily jump or climb" (Blackley, 1990). In terms of prey size, "cougars typically kill small prey

Fig. 8 Red Shorthorn Bull. (Permission Robert Scarth -CC BY-SA2.0)

such as sheep, goats or calves with a bite to the top of the neck or head, usually severing the spinal column. Cougars leap on the back or shoulders of larger animals such as cattle, horses, elk and deer and kill with a bite to the neck" (Blackley, 1990).

Sport hunting is a major component of cougar predator control. For example, Peebles et al. (2013) present a report on this topic and note that sport hunting of predators often serves to reduce predator populations and the associated discomfort and predation on livestock. The study examined the impact of remedial sport hunting on reducing cougar complaints and livestock thefts in Washington from 2005 to 2010 (Peebles et al., 2013). This was based on the number of complaints, livestock thefts, cougars harvested, and estimated cougar populations. Human population and livestock for all 39 counties and the 136 wildlife management units in Washington (Peebles et al., 2013). Results showed that complaints and depredation were positively associated with human population, livestock, and cougar populations, as well as with "opportunities." Complaints and livestock theft increased dramatically as cougar harvest increased (36–240%) (Peebles et al., 2013).

Peebles et al. (2013) "suggest that increased young male immigration, social disruption of cougar populations, and associated changes in space use by cougars—caused by increased hunting resulted in the increased complaints and livestock depredations. Widespread indiscriminate hunting does not appear to be an effective

Fig. 9 Black Simmental Bull. (Permission Mark Sneed - CC BY-SA 3.0)

preventative and remedial method for reducing predator complaints and livestock depredations" (Peebles et al., 2013). The findings also show that "there were several different factors that influence the number of cougar complaints and depredations across the state of Washington. In increasing order of importance these include: the human population, the number of livestock, number of cougars, the number of cougars killed, and proportion of cougars killed" (Peebles et al., 2013). Importantly, "consistent with expectations, each additional cougar on the landscape increased the odds of a complaint or livestock depredation by about 5%. However, contrary to expectations, each additional cougar killed on the landscape increased the odds by about 50%, or an order of magnitude higher. By far, hunting of cougars had the greatest effects, but not as expected. Very heavy hunting (100% removal of resident adults in 1 year) increased the odds of complaints and depredations in year 2 by 150% to 340%. It appears that remedial sport hunting to reduce complaints and depredations is actually associated with increased, not decreased, complaints and depredations the following year" (Peebles et al., 2013). Therefore, "remedial hunting of cougars, in Washington, was associated with increased, not decreased, complaints and depredations" (Peebles et al., 2013).

Fig. 10 American Longhorn Bull, Texas. (Permission Ray_Shrewsberry-Pixabay)

Laundré and Papouchis (2020) present a similar study of California, which reports that cougar "management in the western United States is highly contentious, particularly with regard to the use of sport hunting as a management tool. Since the 1970s, puma in ten western states have been managed by state fish and game agencies through the use of a sport hunt." The authors note that "the rationale presented by wildlife managers is that sport hunting, in addition to providing recreational hunting opportunities, also reduces threats to human safety and livestock safety, and increases populations of the puma's ungulate prey, namely deer (*Odocoileus* sp.) and elk (*Cervus elepus*)" (Laundré & Papouchis, 2020). Legal instruments in California prohibited the sport hunting in 1972. The study tested the hypotheses that sport hunting reduces cougar populations, cougar-human interactions, livestock predation, and wild ungulate predation. The findings indicated that compared with the other 10 states where cougar hunting was practiced, in California there was "no evidence that sport hunting of pumas has produced the management outcomes sought by wildlife managers aside from providing a sport hunting opportunity. Consequently, and particularly because other research suggests that sport hunting actually exacerbate conflicts between pumas and humans, we recommend that state agencies re-assess the use of sport hunting as a management tool for pumas" (Laundré & Papouchis, 2020).

Laundré and Papouchis (2020) argue that an increasing amount of scientific research is casting doubt on the idea that sport hunting helps increase ungulate populations or decrease cougar populations. Numerous studies demonstrate that sport hunting exacerbates the very issues it is meant to mitigate by increasing the frequency of interactions between cougars and livestock and humans (Beier, 1991; Lambert et al., 2006; Laundré et al., 2007). The increasing volume of information casts doubt on the efficacy of sport hunting as a cougar management strategy. It is argued that cougars account for less than 0.2% of all cattle losses, which is a very small percentage. Comparing California to eight other states, it had a higher percentage of cattle losses than those of two other states. The patterns in the calves were comparable. There are no appreciable variations in the average losses of either cattle or calves between the five-year study period in California and the other ten states. It was concluded that the hypothesis that the percentage of calves lost would be inversely connected to the total number of cougars killed showed no discernible correlation. Contrary to expectations, there was a correlation between higher calf losses and cougar deaths in one state, Wyoming. The idea that cougars killed for sport resulted in lower per capita losses of sheep or cattle was not supported by the results of comparisons of livestock losses by cougars. The exact opposite of what was expected was discovered in some cases: higher livestock losses were linked to higher cougar mortality rates.

Jaguar Predation and Cattle

Jaguar predation on cattle occurs in jaguar territory in Mexico and Central America, which jaguars share with the cougar (Campbell, 2017, 2022). For example, Ceballos et al. (2024) point out that the jaguar originally ranged from the southwestern United States to central Argentina but is currently considered near-endangered worldwide, with approximately 55% of its historical range having been extirpated and many of its populations now either endangered or endangered. Habitat destruction and fragmentation, rancher retaliation, illegal hunting, prey declines, and introduced diseases are the greatest threats to the long-term viability of jaguar populations. The authors argue that the development and expansion of agricultural land and livestock pastures, human settlements, roads, and other infrastructure are the main causes of the loss and fragmentation of the jaguar's habitat. In many communities, jaguars are seen as a threat and are killed by ranchers in retaliation, as the loss of livestock to predators is usually viewed only from the perspective of economic loss. Furthermore, jaguars are still hunted (albeit illegally) as trophies throughout their range, and recently, the illegal trade of their bones and teeth for the Chinese black market has become an increasing threat (Ceballos et al., 2024).

Rosas-Rosas et al. (2008: 554–560) point out that in northeastern Sonora, Mexico, livestock production frequently faces conflict due to puma and jaguar predation, and there are biological, social, and economic repercussions from these conflicts because jaguars are endangered in Mexico. In northeastern Sonora, home to the northernmost jaguar breeding population in North America, the study examined the rate of cattle predation by pumas and jaguars between 1999 and 2004 (Rosas-Rosas et al., 2008: 554). According to the findings, only calves younger than 12 months were killed by jaguars and pumas, and calves made up 58% of the prey biomass consumed by jaguars and 9% by pumas each year. The findings indicated that "jaguars and pumas killed only calves <12 mo old, and calves constituted 58% of prey biomass consumed by jaguars and 9% by pumas. Annual cause-specific mortality rates of confirmed jaguar predation (≤0.018), confirmed and suspected jaguar predation (≤0.018), and all confirmed and suspected large felid predation (≤0.018) were low and cattle calf survival was high (0.89–0.98 annually). If calves reported as missing but for which no evidence of mortality could be found were classed as large felid predation, annual cause-specific rates increased to 0.006–0.038. Collectively, confirmed jaguar and puma predation accounted for <14% (57/408) of total cattle losses, with jaguars responsible for 14% of all calf losses; this could increase to a maximum of 36% (146/408) if missing calves were included in the totals." Because most of the confirmed calf kills in the study were associated with three individual jaguars, the authors suggest that targeting problem cats rather than large-scale predator control may be a viable alternative to managing chronic predator problems. In addition, as most cases of jaguar predation occurred along dense riparian habitats during the dry season, modified livestock management measures, such as establishing permanent water sources in uplands and away from dense vegetation cover, could alleviate many cases of jaguar predation on cattle.

The authors conclude that jaguars were the primary predators of cattle in northeastern Sonora although the annual confirmed kill rate (0–1.8%) was much lower than estimated by ranchers who attributed all calf mortality to jaguars and cougars. This was true even when all confirmed and suspected large carnivores were included (cause-specific mortality rates ≤1.8%) and even when unknown missing calves were included as prey of the big cats. It is argued that jaguar predation on livestock in northeast Sonora should be closely monitored because reported losses from predation vary greatly from actual ones, and confirmed losses have generally been associated with a few specific individual jaguars or locations. Identification of both predator species and individuals is critical to managing predator conflict, as illegal killings due to livestock predation were the leading cause of death for jaguars, an endangered species, in the study area (Rosas-Rosas et al., 2008: 554–560).

Soto-Shoender and Giuliano (2011: 561) present a study of ranchers in the lowlands of Guatemala to determine whether ranchers used recommended management practices to prevent such predation by large carnivores, comparing ranches with and without attacks on livestock, to determine whether ranch characteristics and landscape structure near ranches explain differences in the occurrence of carnivore predation. Importantly, livestock losses to the predators accounted for only 0.7% of the livestock on the ranches studied, and jaguars were most frequently charged with livestock attacks (suggesting that this big cat has a negative perception in the area), followed by cougars and coyotes. Males and smaller cattle were most commonly attacked, and general attack patterns on livestock were similar to previously studied

Neotropical sites. The landscape structure around ranches (e.g., forest cover, distance to forests, waterways, and human settlements) best explains the likelihood of livestock predation (Soto-Shoender & Giuliano, 2011: 561). The authors note that outreach programs and conflict mitigation measures need to be implemented for those ranches that are far from human settlements but close to forest areas and water sources. The presence of predation by jaguars, pumas, and coyotes is locally specific in Mesoamerica, and conflict mitigation strategies suggested in other studies may need to be modified and evaluated to be effective in this region (Soto-Shoender & Giuliano, 2011: 561).

The United Nations Environment Programme (UNEP) notes that in Panama, and in Central America in general, "jaguars have been known to prey on livestock. Over 40 per cent of the big cats' habitat in Panama has been lost due to increased urbanisation, infrastructure projects, agriculture and cattle ranching. This loss has put an increased strain on the species and those that live and work alongside their habitat. Desperate to protect their livestock, cattle ranchers have been killing jaguars in retaliation. 96% of the 339 jaguar deaths reported by Yaguara Panama Foundation from 1989 to 2019 in the country were due to killing by humans after livestock death, with an average of 20–44 jaguar deaths per year." It is argued that some cattle ranchers have adopted some actions and "anti-predation measures help protect cattle from jaguars; they can include electric fences, enhanced jaguar monitoring with camera traps and GPS technology, strategic pasture placement and modifying livestock rotation patterns" (UNEP).

Steinberg (2016) cites a study of jaguar conservation in southern Belize focused on Mayan hunters. Interestingly, "Belize has emerged as an international leader in jaguar conservation through the creation of numerous protected areas that contain prime cat habitat and by strengthening conservation laws" citing the Cockscomb Basin Jaguar Preserve (1984), which was the first special jaguar protection area in the Americas, the adjacent Chiquibul National Park (1995) and the Labouring Creek Jaguar Corridor Wildlife Sanctuary in central Belize (2010). However, it is noted that "outside national parks and communities that directly benefit from ecotourism, it is less clear how supportive rural residents are of cat conservation. It is also not clear if jaguars persist outside protected areas in locations such as southern Belize, where the environment has been significantly altered by human activities" (Steinberg, 2016). The study contrasts lowland Guatemala, with "high deforestation rates in and out of protected areas has led to greater forest fragmentation. As more people have resettled in the lowlands, more conflicts with jaguars have also occurred, including poaching and retribution killings in response to cattle predation." However, in the case study of Belize "most Mayan villagers in southern Belize keep few or no cattle, so, unlike in other parts of its range such as Mexico, South America, or even central Belize, conflicts with jaguars have little to do with cattle losses" and "therefore, among the Mayans in southern Belize, dog predation, not cattle, drives conflict with jaguars. These numbers should be treated with some caution though, because when a dog disappears for any reason, jaguars are often blamed without any real evidence" (Steinberg, 2016).

Regarding the role of habitat in cattle predation, Rosas-Rosas et al. (2017) investigated habitat characteristics of kill sites of cattle in Sonora, Mexico, from 1999 to 2004 to see whether habitat management or cattle distribution could be used as effective nonlethal methods to limit predation. The findings indicated that "kill-sites were positively associated with oak, semitropical thorn scrub, and xeric thorn scrub vegetation types, and permanent water sources and roads, and negatively associated with upland mesquite" (Rosas-Rosas et al., 2017). The study found that their model was valid for 93% of the kill-site locations. The authors conclude that "because kill-sites were associated with specific habitat attributes, management practices that alter cattle distribution, such as placement of permanent water sources in uplands, herding, and fencing riparian areas characterized by frequent depredations, can be used to minimize co-occurrence of jaguars and cattle and, thus, potentially limit predation without illegal killing of jaguars" (Rosas-Rosas et al., 2017). In addition, "these practices could also lead to more uniform use of pastures and, consequently, higher stocking rates, resulting in increased profitability to landowners. Managing habitat attributes that predispose cattle to predation may provide a viable alternative for maintaining both livestock enterprises and a large, endangered carnivore in areas of conflict" (Rosas-Rosas et al., 2017).

Also, regarding jaguar ecology and livestock prey opportunities, Anile et al. (2020: 667) used camera traps and modeled factors affecting jaguar detection and occupancy in Sonora and found that jaguar occupancy increased with increasing prey abundance, the frequency of domestic cattle calves, and the proportion of subtropical vegetation. Jaguar numbers were lower with increased abundance of adult cattle and peccaries and higher fragmentation. Therefore, jaguar presence in Sonora was influenced by several factors, and it is suggested that conservation measures and reduction of predation on livestock could include effective fencing, as fences increase wildlife prey for jaguars, thereby reducing calf predation by jaguars while preserving habitat against overgrazing. Additionally, jaguar conservation management in Sonora could consider increased financial support for ecotourism and predator compensation to offset potential jaguar impacts (Anile et al., 2020: 667).

Gutiérrez-González & Lopez-Gonzalez (2017) present a study of interactions between jaguars and pumas in the northernmost part of the jaguar range, in a region where both species are persecuted for livestock predation. The study examined whether jaguars are the dominant species in this unique ecosystem where: (1) pumas are more numerous than jaguars, (2) pumas are better adapted to dry environments, and (3) jaguars and pumas are similar in size. The study found that where both species were detected, the presence of peccaries was positively correlated with the presence of jaguars and cougars, whereas in areas where jaguars but no cougars were detected, the presence of deer increased the likelihood of presence of jaguars and the likelihood of both predators being present was higher—are more likely to occur together than to be found independently, and so we rejected the hypothesis that jaguars were the dominant species in our study area. The activity patterns of the predators overlapped by 60%, and they were primarily nocturnal. While cougar activities overlapped with calves more than with other prey, jaguar activities overlapped more with deer and calves than did cougar activities, which may suggest a preference.

In support of their study, the authors note that both jaguars and cougars can hunt cattle, as due to the domestication of cattle, jaguars can kill larger individuals compared to their natural prey and cougars tend to select young adults or young people. Calves are at higher risk of predation, especially in places where cattle are allowed to roam freely and are easier prey than deer. Although cougar activity patterns overlapped with calves than jaguars with calves, the findings indicated an average of 0.40 overlap between jaguars and calves, similar to the overlap between jaguars and deer. It is argued that jaguars could select calves and deer in equal proportions if cattle were available (Gutiérrez-González & Lopez-Gonzalez, 2017).

Regarding human tolerance of jaguars as a preventive measure in conflicts over livestock predation, Lecuyer et al. (2023) note that their study "developed an experimental tool with a gamified survey-based field experiment to test multiple intended behaviours along a gradient of livestock loss to predation by jaguars (Panthera onca) in Calakmul, Mexico. Using a within-subject design, participants played a card-choice game against livestock loss scenarios." These choices were three: "No action [taken]," "Protect [domestic animals]," and "Kill [the jaguar]" (Lecuyer et al., 2023). The authors add that the "our experimental design was validated by increased frequencies of 'Kill' responses in reaction to hypothetical jaguar incursion with increasing damage. The greatest difference occurred between zero and minimal damage (one chicken), showing that whether damage occurs at all is at least as important as its extent. Secondly, average frequency of intentional retaliation to jaguar incursion was higher for participants who had already killed a jaguar. Finally, age, education, and ethnicity all correlated with intentional behaviours to implement preventive measure or kill jaguars. Past experience of attack had no effect" (Lecuyer et al., 2023). The results warranted a new approach to conservation in the face of possible predator conflict, as the results demonstrated that surveys/game-based experiments can assess tolerance to potentially dangerous interactions with wildlife while providing a more nuanced understanding of people's tolerance on a loss gradient. This enables a movement from the dualistic model of "conflict" versus "coexistence" to a more complex understanding that incorporates individual attitudes and experiences, cultural meaning, and socioeconomic context (Lecuyer et al., 2023).

Geomatics Methods and Cattle Predation

Geomatics is an important scientific development in North America. Brown (2018) notes that "more common in North America than in many parts of Europe, geomatics can be said to refer to the measurement and analysis of the properties of the Earth." Geomatics-based methods have been used in studies of the factors for cattle predation in North America, including habitat change, agricultural landcover expansion, and impacts of conservation policies. In one example, Laporte et al. (2010) "compared elk and cattle anti-predator responses to wolves in southwest Alberta,

Canada within home ranges and livestock pastures, respectively." The methods included the use of satellite- and GPS-telemetry collars on wolves, elk, and cattle and the measurement of prey response variables during periods of wolf presence and absence, which included "speed, path sinuosity, time spent head-up, distance to neighboring animals, terrain ruggedness, slope and distance to forest." The findings regarding cattle indicated that "individual as well as group behavioral analyses were feasible, and these indicated increased path sinuosity and decreased distance to neighbors. In addition, cattle groups showed a number of behavioral changes concomitant to wolf visits, with variable direction in changes" (Laporte et al., 2010). The authors concluded that the "results suggest both elk and cattle modify their behavior in relation to wolf presence, with potential energetic costs. Our study does not allow evaluating the efficacy of anti-predator behaviors but indicates that artificial selection did not result in their absence in cattle. The costs of wolf predation on livestock are often compensated considering just the market value of the animal killed. However, society might consider refunding some additional costs (e.g., weight loss and reduced reproduction) that might be associated with the changes in cattle behaviors that we documented" (Laporte et al., 2010).

In another study, Clark et al. (2020: 30) present GIS/GPS-based study of wolf cattle predation in the western United States, noting that "spatial variability in terrain, vegetation, and other features affect cattle and wildlife distribution on mountainous grazing lands of the western United States... Research addressing this problem was conducted in 2009–2011 at four study areas in western Idaho to evaluate models and mapping tools for predicting spatial risk of wolf-cattle encounters. The methods included GPS collars on lactating beef cows, resource selection function (RSF) models, based on negative binomial regression from the GPS data, and a wolf based RSF model for mapping the habitat suitability for wolves. The authors also used "spatial relationships between cattle and wolf selectivity patterns were used to classify and map wolf-cattle encounter risk to 5 classes (very high to very low) across each study area during the wolf rendezvous period. The analyses using the GPS-based, wolf-cattle encounter observations revealed 84% of observed encounters occurred in areas of high- or very high−encounter risk (class 4 or 5). About 75% of confirmed wolf depredations recorded among three of four study areas were located in areas of high or very high risk" (Clark et al., 2020: 30). The authors note that their case study "revealed the spatial risk of wolf-cattle encounters can be accurately predicted, and although this knowledge is quite important for management, a critical question is still unanswered. Does increased spatial knowledge of encounter risk lead to greater predictability of where depredation events will occur within rugged and remote cattle grazing lands?" (Clark et al., 2020: 48). It is concluded that "this new predictive understanding of wolf-cattle encounter risk will greatly aid livestock producers, resource managers, and policy makers in more effectively applying husbandry practices, allocating mitigation resources, and developing conflict mitigation plans and policies applicable throughout the mountainous western United States and potentially other regions of the world where wolves and cattle come into conflict" (Clark et al., 2020: 30).

Teichman et al. (2013) present a study of "cougar habitat, human landscape characteristics and the combination of habitat and human features on the temporal and spatial patterns of cougar-human conflicts in British Columbia." In this study, "cougar-human 'conflict' as any incident of cougar road mortality, depredation on livestock or attack on humans" (Teichman et al., 2013). Spatial GIS techniques were used to analyze spatial and temporal characteristics of cougar–human conflicts. In addition, these methods were used to "estimate which variables adequately predict conflict and map the relative probability of cougar-human conflict" (Teichman et al., 2013). Human factors for conflict included human population density, density and distance to roads, and cattle density. In addition, "habitat variables hypothesized to affect cougar and prey density and thus conflict levels included elevation and terrain ruggedness, land cover reclassified to represent natural habitat occupied by cougars (conifer, deciduous, mixed forest and shrubland), forest edge density, distance to water combined as the minimum of the distances to nearest river and lake, protected area and distance to northern edge of the cougar range" (Teichman et al., 2013).

The findings of this study indicated that "conflicts involved similar numbers of male and female cougars with conflict rate decreasing over the past decade. Conflicts were concentrated within the southern part of the province with the most conflicts per unit area occurring on Vancouver Island. For both sexes, the most supported spatial models for the most recent (1998–2007) conflicts contained both human and habitat variables. Conflicts were more likely to occur close to roads, at intermediate elevations and far from the northern edge of the cougar distribution range in British Columbia. Male cougar conflicts were more likely to occur in areas of intermediate human density. Unlike cougar conflicts in other regions, cattle density was not a significant predictor of conflict location. With human populations expanding, conflicts are expected to increase. Conservation tools, such as the maps predicting conflict hotspots from this study, can help focus management efforts to decrease carnivore-human conflict" (Teichman et al., 2013). Regarding the utility of the GIS-based methods, the authors note that "the predictive capacity for carnivore-human conflict may be increased by including reproductive status and age structure in conflict modeling. On the other hand, habitat data are increasingly available and we have shown that broad habitat patterns can be adequate predictors of cougar-human conflict. Predictions of potential areas that are likely to have cougar-human conflicts can help target management strategies that reduce cougar mortalities. With reduced occurrences of cougar-human conflicts, people are more likely to accept and thus coexist with cougars in a shared landscape such as southern British Columbia."

Christoff and Devenish-Nelson (2024) present a study of habitat suitability and protected area coverage for an expanding cougar Puma concolor population in Canada. It is noted that "in Canada, cougars are expanding their range eastwards and little research is available for use in decision making by land managers and conservation planners. To inform proactive management regarding expanding populations of cougars in Canada, we utilized open-source cougar presence and land-cover data in a maximum entropy habitat suitability model to determine potentially suitable habitat for cougars across the country. We then used a gap analysis to determine the effectiveness of existing formal protected areas to protect potential cougar habitat." The findings indicated that "while the habitat is highly fragmented, the highest suitability occurs in areas of medium road density, indicating that the potential for new human–cougar conflict will likely involve residents of exurban and rural areas. Protected areas offered 16% coverage of suitable habitat, although most protected areas that overlap predicted cougar habitat are not large enough to effectively conserve the large home range requirements of cougars."

Ceballos et al. (2021: 3287–3309) present a similar assessment of jaguars in Mexico, which involved "a database of jaguar occurrence records, geospatial dataset of all Natural Protected Areas and using the expert criterion of the Mexican jaguar specialists to define Jaguar Geographic Regions; i.e. well-defined large units with similar ecological characteristics across the geographic range of jaguars in Mexico." Jaguar Conservation Units, i.e., core units with jaguar populations of 15 or more individuals, were then identified. Biological Corridors were measured with Least Cost Path and Circuit Theory analyses "to define the i.e. regions with enough habitat to allow for jaguar dispersal along their geographic range but putting emphasis in connecting Jaguar Conservation Units." It is argued that "the results showed that JGRs cover 25% of the country but only 21% of those regions are currently protected in nature reserves" illustrating "the importance of the creation of new protected areas or the implementation of other schemes of conservation to protect larger extensions of the jaguars' habitat in Mexico" (Ceballos et al., 2021: 3287–3309).

Treves et al. (2011: 451) also present "a risk map for gray wolf (Canis lupus) attacks on livestock in Wisconsin between 1999 and 2006 that correctly identified risk in 88% of subsequent attack sites from 2007 to 2009." The findings indicated that "more-open habitats farther from any forest and closer to wolf pack ranges were the riskiest for livestock" and the authors concluded that "prediction promotes prevention. We recommend that the next generation of risk mappers employ several criteria for model selection, validate model predictions against data not used in model construction before publication, and integrate predictors from organismal biology alongside human and environmental predictors." The authors constructed a model "for gray wolf attacks on livestock that took place from 1999 to 2006 in Wisconsin, and its predictions were verified by the data from subsequent affected sites from 2007 to 2009" (Treves et al., 2011: 451). This involved information from the Wisconsin Department of Natural Resources (WDNR), which had mapped and verified wolf attacks on domestic animals (including radio-collared wolves, aerial telemetry and GPS location error, and the estimation of the ranges of wolf packs without radio-collared individuals through track surveys during snow-cover periods) here livestock comprising cattle, mostly calves, and sheep and horses.

Parameters considered in this study included "(1) the distance to the nearest wolf pack range in kilometers (DW), (2) the number of pack members, (3) the area of the pack range, and (4) the number of wolves per square kilometer" and landcover using ArcGIS Version 9.1 (ESRI) (grass/pasture, forest distance, wolfpack distance, open water, wooded wetlands, length of forest edge, coniferous forest cover, deciduous forest cover), and estimations of the density of people, houses, farms, roads, deer, cattle, and livestock premises" (Treves et al., 2011: 453). The authors note that "the probability of wolf attack on livestock was higher at open habitat sites (which may correlate with livestock availability on pastures), closer to a known wolf pack range, and farther from any type of forest, with an interaction between the last two predictors such that sites far from forest and far from wolf packs were less risky. The mechanism underlying this interaction remains obscure" (Treves et al., 2011: 454). The best predictors that distinguished the affected and unaffected sites included grass/pasture/hayfield (percentage of area), distance to forest (in kilometers), distance to nearest wolf pack (in kilometers), open water (percentage of area), wooded wetlands (percentage of area), length of the forest edge (in kilometers), the number of wolves per 10 square kilometers, coniferous forest (percentage of area), emergent wetlands (percentage of area) and deciduous forest (percentage of area) (Treves et al., 2011: 453). The authors cite an important limitation and caution, in that "interpreting risk maps requires care. These maps are correlational, not causal; therefore, any landscape predictor is best interpreted as a complex association of environmental variables. In our case, we cannot disentangle a wolf pack's history and individual membership from its landscape correlates, because the wolf pack ranges did not change much from year to year" (Treves et al., 2011: 455). It is however argued that "the predictive power of the associations that we demonstrated leads us to recommend that wildlife managers and livestock owners act to mitigate the risk posed by high concentrations of grassland, pasture, or hayfield far from forest and near wolf packs" (Treves et al., 2011: 455).

Spatial distribution and presence models may also predict predator habitats, which may be useful for general prediction useful for studying possible interactions with livestock. For example, Mladenoff et al. (1999: 37) describe a test of a predictive spatial model of gray wolf habitat suitability. The model "is based on logistic regression analysis of regional landscape variables in the upper Midwest, United States, using radiotelemetry data collected on recolonizing wolves in northern Wisconsin since 1979. The model was originally derived from wolf packs radio-collared from 1979 to 1992 and a small test data set of seven packs. The model provided a 0.5 probability cut level that best classified the landscape into favorable (road density < 0.45 km/km^2) and unfavorable habitat (road density > 0.45 km/km^2) and was used to map favorable habitat with the northern Great Lake states of Wisconsin, Minnesota, and Michigan." The authors note that their purpose was "to provide a better validation test of the model predictions based on data from new packs colonizing northern Wisconsin from 1993 to 1997," and in the "test, the model correctly classified 18 of 23 newly established packs into favorable areas." The authors conclude that "Wolves may continue to occupy areas of slightly lower habitat probability if adequate population source areas are present to offset the greater mortality in these lower quality areas. The model remains quite robust at predicting areas most likely to be occupied by wolves colonizing new areas based on generally available road network data" (Mladenoff et al., 1999: 37).

Sells et al. (2022) argue that some models may be overly rigid and neglect the individualistic tendencies of some species. The study cites predictive modeling of grizzly bear foraging, in a case study of the Northern Continental Divide Ecosystem of northwest Montana. It is argued that mechanistic modeling of grizzly bear movements has shown that grizzly bears have highly individualistic spatial behavior. Some individuals avoided, while others preferred areas with green vegetation, rugged terrain, forest edges, riparian areas, dense development, and safe habitats. This individualism underpinned the need for an individual modeling approach to understand and predict grizzly bear behavior (Sells et al., 2022). Parameters included robustness, distance to forest edge, density of forest edge, density of banks, density of buildings, and distance to safe habitat. The results suggested individualism, as women's and men's responses varied greatly for each habitat variable measured (Sells et al., 2022). The authors conclude that their method for understanding and predicting habitat use for species of conservation importance provided a better understanding of grizzly bear spatial behavior and strong predictive ability and showed grizzly bears in the Northern Continental Divide ecosystem extremely exhibited individualistic spatial behavior. The authors argue that their method provides a powerful way to understand and predict habitat use for conservation-important species by following a multi-stage approach to testing hypotheses about animal behavior, simulating movements, and translating results into validated prediction maps (Sells et al., 2022).

Conclusions

The relationships between domestic cattle in North America and large carnivores are similar to those in South America and Europe, but differ in that there are more species of large carnivores capable of killing cattle in North America. Wolves and brown bears may kill domestic cattle as in Europe, jaguars and pumas as in South America, and the black bear found in North America may also kill cattle. Similar to Europe, success in protecting large carnivores has created conflicts between humans and carnivores, which may conflict with expanding livestock populations and ranges. In North America, there are large open areas of land for cattle production that require very strict monitoring and evaluation. Extensive conservation efforts have successfully increased the number of carnivores, while the cattle population is also increasing. Current evidence suggests that gray wolves and brown bears are the main cattle predators, while black bears are the least likely to kill cattle. The vast areas and associated complicated ecologies require detailed application of geomatics-based surveys and analysis tools. Geomatics as a science is perhaps most advanced in North America because the United States has greater financial and technological resources compared to Europe. Geomatics-based methods are also embedded in conservation management, which includes large-scale conservation management that can include large carnivore reintroductions, expanded protected areas, and legal conflicts between livestock management and powerful conservation actors. The extremely complex and large-scale problems of North American agricultural development and conservation management could support the claim that North America could be a model for such developments on other continents, apart from the smaller scales of Europe and the deforestation problems of Asia and South America and Africa, as well as desertification in Africa.

References

Alexander, R. T., & Hernández Álvarez, H. (2018). Agropastoralism and household ecology in Yucatán after the Spanish invasion. *Environmental Archaeology, 23*(1), 69–79. https://doi.org/10.1080/14614103.2017.1342396

Anderson, C. R., Ternent, M. A., & Moody, D. S. (2002). Grizzly Bear-Cattle interactions on two grazing allotments in Northwest Wyoming. *Ursus, 13*, 247–256. http://www.jstor.org/stable/3873205

Anile, S., Greenspan, E., & Nielsen, C. K. (2020). Determinants of jaguar occupancy at the northern range edge. *Mammal Research, 65*, 667. https://doi.org/10.1007/s13364-020-00511-0

Bedayn. (2023). Do wolves attack livestock? US cattle ranchers sue over planned release of apex predators. Retrieved from https://www.euronews.com/green/2023/12/14/do-wolves-attack-livestock-us-cattle-ranchers-sue-over-planned-release-of-apex-predators

Beier, P. (1991). Cougar attacks on humans in the United States and Canada. *Wildlife Society Bulletin (1973–2006), 19*(4), 403–412. http://www.jstor.org/stable/3782149

Bhowmik, N., Seaborn, T., Ringwall, K. A., Dahlen, C. R., Swanson, K. C., & Hulsman Hanna, L. L. (2023). Genetic distinctness and diversity of American Aberdeen cattle compared to common beef breeds in the United States. *Genes, 14*(10), 1842. https://doi.org/10.3390/genes14101842

Blackley, K. (1990). *Methods of investigating predation of livestock*. Alberta Agriculture and Rural Development Information Management Division.

Boitani, L., Phillips, M., & Jhala, Y. (2023). Canis lupus (amended version of 2018 assessment). The IUCN Red List of Threatened Species 2023: https://doi.org/10.2305/IUCN.UK.2023-1.RLTS.T3746A247624660.en.

Bowling, G. A. (1942). The introduction of cattle into Colonial North America. *Journal of Dairy Science, 25*(2), 129–154. https://doi.org/10.3168/jds.S0022-0302(42)95275-5

Brand, D. D. (1961). The early history of the range cattle industry in northern Mexico. *Agricultural History, 35*(3), 132–139. http://www.jstor.org/stable/3740624

Brown, I. (2018). How can we attract students to geomatics? https://www.gim-international.com/content/article/how-can-we-attract-students-to-geomatics

Campbell, M. (2017). *Biological conservation in the 21st century: A conservation biology of large wildlife*. Nova Science Publishers.

Campbell, M. (2022). The two tigers: The jaguar ('El Tigre') of the Americas and the Tiger of Asia. In K. K. Jha & M. Campbell (Eds.), *Dynamics and interrelations between nature, science, and society*. Nova Science Publishers.

CBC News. (2015). Black bear attacks on calves increasing in B.C.'s Cariboo region. Retrieved from https://www.cbc.ca/news/canada/british-columbia/black-bear-attacks-on-calves-increasing-in-b-c-s-cariboo-region-1.3133185.

Ceballos, G., de la Torre, J. A., Zarza, H., Huerta, M., Lazcano-Barrero, M. A., Barcenas, H., Cassaigne, I., Chávez, C., Carreón, G., Caso, A., Carvajal, S., García, A., Morales, J. J., Moctezuma, O., Monroy-Vilchis, O., Ruiz, F., & Torres-Romero, E. J. (2021). Jaguar distribution, biological corridors and protected areas in Mexico: From science to public policies. *Landscape Ecology, 36*(11), 3287–3309. https://doi.org/10.1007/s10980-021-01264-0

Ceballos, G., Zarza, H., González-Maya, J. F., Torre, J. A. de la, Arias-Alzate, A., Alcerreca, C., Barcenas, H. V., Carreón-Arroyo, G., Chávez, C., Cruz, C., Medellín, D., García, A., Huerta-García, M. A., Lazcano-Barrero, M. A., Medellín, R. A., Moctezuma-Orozco, O., Ruiz, F., Rubio, Y., Luja, V. H., & Torres-Romero, E. J. (2024). Correction: Beyond words: From jaguar population trends to conservation and public policy in Mexico. *PLoS One, 19*(6), e0305642. https://doi.org/10.1371/journal.pone.0305642

Christoff, J. A., & Devenish-Nelson, E. S. (2024). Habitat suitability and protected area coverage for an expanding cougar Puma concolor population in Canada. *Ecology and Evolution, 14*(9), e70228. https://doi.org/10.1002/ece3.70228

Clark, P. E., Chigbrow, J., Johnson, D. E., Larson, L. L., Nielson, R. M., Louhaichi, M., Roland, T., & Williams, J. (2020). Predicting spatial risk of wolf-cattle encounters and depredation. *Rangeland Ecology & Management, 73*(1), 30–52. https://doi.org/10.1016/j.rama.2019.08.012

Delsol, N., Stucky, B. J., Oswald, J. A., Cobb, C. R., Emery, K. F., & Guralnick, R. (2023). Ancient DNA confirms diverse origins of early post-Columbian cattle in the Americas. *Scientific Reports, 13*, 12444. https://doi.org/10.1038/s41598-023-39518-3

Fowler, S. (2016). *Bear attacks kill and wound livestock in Florenceville-Bristol*. Retrieved from https://www.cbc.ca/news/canada/new-brunswick/bear-attack-dead-cows-1.3754265.

Furber, D. (2020). *Focus on removing the real predators, the ones that are killing cattle, and ignore the rest*. The Canadian Cattlemen. Retrieved from https://www.canadiancattlemen.ca/features/focus-on-removing-the-real-predators-the-ones-that-are-killing-cattle-and-ignore-the-rest/

Government of Alberta. (2020). *SSR Predator compensation 2015 to 2019: South Saskatchewan Region*. Environment and Parks, Government of Alberta.

Government of Alberta. (2024). *Wildlife Predator Compensation Program*. Retrieved https://www.alberta.ca/wildlife-predator-compensation-program

Guinan, F. L., Wiggans, G. R., Norman, H. D., Dürr, J. W., Cole, J. B., Van Tassell, C. P., Misztal, I., & Lourenco, D. (2023). Changes in genetic trends in US dairy cattle since the implementation of genomic selection. *Journal of Dairy Science, 106*(2), 1110–1129. https://doi.org/10.3168/jds.2022-22205

Gutiérrez-González, C. E., & López-González, C. A. (2017). Jaguar interactions with pumas and prey at the northern edge of jaguars' range. *PeerJ, 5*, e2886. https://doi.org/10.7717/peerj.2886

Hall, J. (2016). The Case for Mass Slaughter of Predators Just Got Weaker. *National Geographic*. Retrieved from https://www.nationalgeographic.com/animals/article/wildlife-lethal-nonlethal-predator-control-hunting-evidence?loggedin=true&rnd=1708207495694.

Hoag, D. L. K., Boone, R. B., & Keske, C. M. H. (2011). The cost for agriculture to coexist with wildlife in Colorado. *Human Dimensions of Wildlife, 16*, 318–329.

Horstman, L. P., & Gunson, J. R. (1982). Black Bear predation on livestock in Alberta. *Wildlife Society Bulletin (1973–2006), 10*(1), 34–39. http://www.jstor.org/stable/3781799

Juárez-Casillas, L. A., & Varas, C. (2013). Evaluation of black bear (Ursus americanus) diet and consequences in its conservation in Sierra de Picachos, Nuevo León, Mexico. *Revista Mexicana de Biodiversidad, 84*(3), 970–976. https://doi.org/10.7550/rmb.33376

Kienlen, A. (2022). Total ban on killing grizzly bears isn't working, say cattle producers. Alberta Farmer. https://www.albertafarmexpress.ca/livestock/total-ban-on-killing-grizzly-bears-isnt-working-say-cattle-producers/.

Kompaniyets, L., & Evans, M. A. (2017). Modeling the relationship between wolf control and cattle depredation. *PLoS One, 12*(10), e0187264. https://doi.org/10.1371/journal.pone.0187264

Lambert, C. M., Wielgus, R. B., Robinson, H. S., Katnik, D. D., Cruickshank, H. S., Clarke, R., & Almack, J. O. N. (2006). Cougar population dynamics and viability in the Pacific Northwest. *The Journal of Wildlife Management, 70*(1), 246–254. https://doi.org/10.2193/0022-541X(2006)70[246:CPDAVI]2.0.CO;2

Laporte, I., Muhly, T. B., Pitt, J. A., Alexander, M., & Musiani, M. (2010). Effects of wolves on elk and cattle behaviors: Implications for livestock production and wolf conservation. *PLoS One, 5*(8), e11954. https://doi.org/10.1371/journal.pone.0011954

Laundré, J. W., & Papouchis, C. (2020). The elephant in the room: What can we learn from California regarding the use of sport hunting of pumas (Puma concolor) as a management tool? *PLoS One, 15*(2), e0224638. https://doi.org/10.1371/journal.pone.0224638

Laundré, J. W., Hernández, L., & Clark, S. G. (2007). Numerical and demographic responses of pumas to changes in prey abundance: Testing current predictions. *The Journal of Wildlife Management, 71*(2), 345–355. https://doi.org/10.2193/2005-627

Lecuyer, L., Börger, T., Calmé, S., Schmook, B., & White, R. M. (2023). Cattle ranchers' tolerance to jaguar revealed through a spectrum of intended behaviours across a gradient of losses in a gamified survey-based field experiment. *Biological Conservation, 285*, 110253. https://doi.org/10.1016/j.biocon.2023.110253

Lee, T., Good, K., Jamieson, W., Quinn, M., & Krishnamurthy, A. (2017). Cattle and carnivore coexistence in Alberta: The role of compensation programs. *Rangelands, 39*(1), 10–16. https://doi.org/10.1016/j.rala.2016.11.002

Lewis, T. (2018). *Grisly outlook: Bears kill more and more livestock as their population grows.* Retrieved from https://www.tsln.com/news/wyoming-grizzly-bear-update/

Martínez, A. M., Gama, L. T., Cañón, J., Ginja, C., Delgado, J. V., Dunner, S., Landi, V., Martín-Burriel, I., Penedo, M. C., Rodellar, C., Vega-Pla, J. L., Acosta, A., Alvarez, L. A., Camacho, E., Cortés, O., Marques, J. R., Martínez, R., Martínez, R. D., Melucci, L., Martínez-Velázquez, G., et al. (2012). Genetic footprints of Iberian cattle in America 500 years after the arrival of Columbus. *PLoS One, 7*(11), e49066. https://doi.org/10.1371/journal.pone.0049066

Mladenoff, D. J., Sickley, T. A., & Wydeven, A. P. (1999). Predicting gray wolf landscape recolonization: Logistic regression models vs New field data. *Ecological Applications, 9*(1), 37–44. https://doi.org/10.1890/1051-0761(1999)009[0037:pgwlrl]2.0.co;2

Morehouse, A. T., & Boyce, M. S. (2017). Evaluation of intercept feeding to reduce livestock depredation by grizzly bears. *Ursus, 28*(1), 66–80. https://doi.org/10.2192/URSU-D-16-00026.1

Muhly, T. B., & Musiani, M. (2009). Livestock depredation by wolves and the ranching economy in the Northwestern U.S. *Ecological Economics, 68*, 2439–2450.

Musiani, M., & Paquet, P. C. (2004). The practices of wolf persecution, protection, and restoration in Canada and the United States. *Bioscience, 54*(1), 50–60. https://doi.org/10.1641/0006-3568(2004)054[0050:TPOWPP]2.0.CO;2

Navarro Villa, P. (2024). *Production volume of cattle in selected Latin American countries in 2023.* Retrieved from https://www.statista.com/statistics/1005380/latin-america-cattle-heads-number-country/

Northern Ag Network. (2016). Grizzly Bears Kill Record Number of Livestock, Still Not Number 1 Predator. https://northernag.net/grizzly-bears-kill-record-number-of-livestock-still-not-number-1-predator/.

Peebles, K. A., Wielgus, R. B., Maletzke, B. T., & Swanson, M. E. (2013). Effects of remedial sport hunting on cougar complaints and livestock depredations. *PLoS One, 8*(11), e79713. https://doi.org/10.1371/journal.pone.0079713

Poudyal, N., Baral, N., & Asah, S. T. (2016). Wolf lethal control and livestock depredations: Counter-evidence from respecified models. *PLoS One, 11*(2), e0148743. https://doi.org/10.1371/journal.pone.0148743

Rosas-Rosas, O. C., Bender, L. C., & Valdez, R. (2008). Jaguar and puma predation on cattle calves in Northeastern Sonora, Mexico. *Rangeland Ecology & Management, 61*(5), 554–560. https://doi.org/10.2111/08-038.1

Rosas-Rosas, O. C., Bender, L. C., & Valdez, R. (2017). Habitat correlates of jaguar kill-sites of cattle in Northeastern Sonora, Mexico. *Directory of Open Access Journals.* https://doi.org/10.26077/7cx9-rn90

Sells, S. N., Costello, C. M., Lukacs, P. M., Roberts, L. L., & Vinks, M. A. (2022). Grizzly bear habitat selection across the Northern Continental Divide Ecosystem. *Biological Conservation, 276*, 109813. https://doi.org/10.1016/j.biocon.2022.109813

Shahbandeh, M. (2024). Cattle inventory in Canada 2023. Retrieved from https://www.statista.com/statistics/442302/cattle-inventory-by-type-canada/#statisticContainer

Soto-Shoender, J. R., & Giuliano, W. M. (2011). Predation on livestock by large carnivores in the tropical lowlands of Guatemala. *Oryx, 45*(4), 561–568. https://doi.org/10.1017/s0030605310001845

Spearchief-Morris, J. (2023). *Alberta conservation program looks to change how ranchers deal with grizzly bear conflicts*. Retrieved from https://www.theglobeandmail.com/canada/article-alberta-conservation-program-looks-to-change-how-ranchers-deal-with/.

Steinberg, M. (2016). Jaguar conservation in southern Belize: Conflicts, perceptions, and prospects among mayan hunters. *Conservation and Society, 14*(1), 13. https://doi.org/10.4103/0972-4923.182801

Teichman, K. J., Cristescu, B., & Nielsen, S. E. (2013). Does sex matter? Temporal and spatial patterns of cougar-human conflict in British Columbia. *PLoS One, 8*(9), e74663. https://doi.org/10.1371/journal.pone.0074663

Thomas, H. S. (2022). *Spring calving heralds beginning of predator season*. Canadian Cattlemen. Retrieved from https://www.canadiancattlemen.ca/livestock/spring-calving-heralds-beginning-of-predator-season/

Treves, A., Martin, K. A., Wydeven, A. P., & Wiedenhoeft, J. E. (2011). Forecasting environmental hazards and the application of risk maps to predator attacks on livestock. *Bioscience, 61*(6), 451–458. https://doi.org/10.1525/bio.2011.61.6.7

United States Department of Agriculture. (2022). *Protecting livestock from predators*. Retrieved from https://www.aphis.usda.gov/wcm/connect/APHIS_Content_Library/SA_Our_Focus/SA_Wildlife_Damage/SA_Operational_Activities/SA_Livestock/CT_Protecting_Livestock_Predators?presentationtemplate=APHIS_Design_Library%2FPT_Print_Friendly

Wielgus, R. B., & Peebles, K. A. (2014). Effects of wolf mortality on livestock depredations. *PLoS One, 9*(12), e113505. https://doi.org/10.1371/journal.pone.0113505

Wuerthner, G. (2020). *Bison ecology, ecological influence, behavior, and decline*. The Wildlife News. Retrieved from https://www.thewildlifenews.com/2020/09/02/bison-ecology-ecological-influence-behavior-and-decline/

Predators of Cattle in South America

Introduction

South America is a key cattle production region, containing 23% of the global cattle stock, especially in Brazil and Argentina (FAO, 2019; Table 1 and Figs. 1 and 2). Cattle production plays a key role in the economies of South American countries, and cattle sectors have increased by 69% since 1961 (FAO, 2019). For example, Zia et al. (2019) point out that "Brazil has the world's second-largest cattle herd—232 million head" and that "in 2018, Brazil was the world's largest exporter of beef, providing close to 20 percent of total global beef exports, outpacing India, the second-largest exporter, by 527,000 metric tons carcass weight equivalent (CWE)." In addition, the USDA has projected that the economy of Brazil will continue its export growth direction for at least the next decade, and may reach 2.9 million metric tons, or 23% of global total beef exports, by 2028. Brown (2022) also mentions the direction of cattle production and ecology in Brazil, noting that free grazing traditional cattle farming is the main ranching style across Brazil. The example is cited of Acre, a state in the Brazilian Amazon, where government data indicates that the number of cattle increased by 8.3% to 3.8 million in 2020, or four times its human population. The cattle industry has become a key component of Acre's economy and contributes to the state's aims for agricultural development (Brown, 2022) (Figs. 3, 4, 5, and 6).

While acknowledging the growth of this type of cattle industry, the National Wildlife Federation (2023) points out that the extensive growth of cattle farming in South America has had a negative impact on the environment due to the logging, burning, and pasture conversion of large tracts of tropical forests. Burning land releases massive amounts of carbon emissions into the atmosphere, destroys wildlife habitat, and lowers biodiversity. The demand for beef, leather, and other products is predicted to keep rising, placing a great deal of strain on the ecologically significant areas of Colombia, Argentina, Brazil, and Paraguay. Furthermore, one of the main causes of tropical deforestation is cattle ranching. To accommodate the 60 million cattle that currently range in the Brazilian Amazon, more than 50 million

© The Author(s), under exclusive license to Springer Nature Switzerland AG 2025

M. O. Campbell, *Cattle, Their Predators and Geomatics Research*,
https://doi.org/10.1007/978-3-031-97363-5_9

Table 1 Cattle and large predator populations in South America

Country	Cattle (million)	Jaguar	Cougar	Spectacled bear
Brazil	234.35	66,865–106,105	289,536.078	ND
Argentina	54.24	107–550	13,902–22,521	ND
Chile	2.96	0	ND	ND
Peru	5.86	17,843–26,788	30,845	5000
Bolivia	10.74	10,260–15,449	54,930–87,886	3000
Paraguay	13.51	708–2497	ND	ND
Uruguay	11.57	0	ND	ND
Ecuador	3.86	1586–2359	ND	1200–2000
Colombia	29.64	11,724–21,311	ND	3000–6000
Venezuela	16.64	8761–14,334	ND	1100–1600
Guyana	0.98	(3233–5462	ND	ND
Suriname	0.34	2275–4081	ND	ND
Fr. Guiana	ND	1097–2105	ND	ND

Sources: Janson and Emmons (1990), Peyton (1999), Cuesta and Suárez (2001), Ruiz Garcia (2003), Ríos-Uzeda et al. (2007), Viteri (2007), Kelly et al. (2008), Negrões et al. (2010), Velez-Liendo (2010), Garshelis (2004), Jędrzejewski et al. (2018), Food and Agriculture Organization of the United Nations (2023)—with major processing by Our World in Data. Regarding the spectacled bear figures, Velez-Liendo and García-Rangel (2017) caution that "all abundance and density estimates for this species have known biases, so consistency among the estimates is not verification of their accuracy" and these are "national assessments applying different approaches"

Fig. 1 Cattle in Santa Rosa de Cabal, Risaralda, Colombia. (Permission Daniel Quiceno-Unsplash)

Fig. 2 Bull in Ipiales, Nariño, Colombia. (Permission Sebastian Tovar-Unsplash)

hectares of forest have been cut down and burned. Conflicts with the larger predators that live in these forests may unavoidably arise from such land use (National Wildlife Federation, 2023).

Ventura Vieira et al. (2020: 45) note that South America is a major global producer of food and that the region's diverse animal production systems—such as pasture, mountainous areas, and feeding grounds—make it more challenging to gather particular phenotypes and necessitate that genetic assessment models take interactions between genotype and environment into consideration. The dairy and beef industries are the two primary forces behind farming and livestock production. The beef industry is based on a variety of systems, ranging from small farmers with low productivity and simpler management issues and technologies to very large producers with intensive and complex animal production technologies. South America is known as a major producer of beef, with Brazil, Argentina, and Chile being the largest producers. Zebu cattle, the most significant breed in South America and primarily found in tropical areas, is one type of beef cattle, mainly in forested Venezuela, Brazil, Paraguay, Colombia, Peru, and Bolivia, and taurine crosses or breeds also occur in subtropical areas (e.g., Chile, Argentina, and Uruguay). The most significant breed of cattle raised in South America is zebu cattle. The majority

Fig. 3 Cattle in Columbia. (Permission Julian-Unsplash)

of dairy cattle, including domestic water buffalo, are housed on pasture (Debaky et al., 2019; Vieira Ventura et al., 2020: 46; da Silva et al., 2021).

South America is also a heavily forested continent, with 20% of the global forests, and the Amazon and associated forests covering about 8.42 million square kilometers or 47% of the area of the continent (FAO, 2020; Salas, 2024). As a consequence, ranching expansion has been a major cause of deforestation, with areas of pronounced deforestation including Amazonia (Nepstad et al., 2009), the Cerrado (Brannstrom et al., 2008), the Chiquitanía (Muller et al., 2012), and the Gran Chaco (Baumann et al., 2017; Gasparri et al., 2013). There are many different types of cattle production systems, with different impacts on the natural environment. For example, some systems have relatively low stocking rates and use natural vegetation, which is abundant in many areas (Grau & Aide, 2008; Grau et al., 2015; Mazzini et al., 2018).

The Brazilian Amazon forest and linked forest ecosystems is possibly the most researched forest area in South America, with most such research pointing to severe deforestation, mostly under intensive agricultural development (Heilmayr et al., 2020; Silva Junior et al., 2021). Figures 7, 8, 9, 10 and 11 show forests and forest savanna/open landcover in South America. Cattle herding has been mooted as an

Fig. 4 Cow in La Rioja, Argentina. (Permission Nahir Confesor-Unsplash)

important factor for such landcover change, with variable impacts on the natural vegetation and ecosystems (World Wildlife Fund, 2024). For example, according to the World Economic Forum (2024), cattle ranching is one of the main causes of deforestation and CO_2 emissions in Brazil. It has historically contributed to 24% of annual tropical deforestation worldwide and significantly to greenhouse gas emissions. Since 2001, the state of Pará has been responsible for 40% of all deforestation in the Brazilian Amazon, with cattle ranching being by far the largest cause of forest clearance.

According to the World Wildlife Fund (2024), large-scale cattle ranching is the primary cause of deforestation in almost every Amazonian nation and is responsible for 80% of the deforestation that occurs currently. Three hundred and forty million tons of carbon dioxide are released into the atmosphere annually as a result of cattle ranching deforestation alone; this amounts to 3% of the world's current emissions. Beyond the conversion of forests, cattle pastures raise the danger of wildfires and seriously harm aquatic and riparian ecosystems by causing soil erosion, river siltation, and organic matter contamination (see also Nepstad et al., 2009). With 88% of the cattle herds in the Amazon, Brazil is said to be the primary driver of this change, followed by Bolivia and Peru. In addition, Brazil may make global

Fig. 5 Cattle in San Martín de los Andes, Neuquén Province, Argentina. (Permission Mateo Krossle –Unsplash)

headlines during the dry season (May to September) due to wildfires, which are caused by slash-and-burn agriculture, a method of opening cattle pastures and basic subsistence plantations. The primary cause of the direct conversion of rainforests is cattle ranching, which is made possible by these fires. Then, taking advantage of the cattle ranchers, soy developers appropriate their land and drive cattle ranching (as well as deforestation) to newly established areas (World Wildlife Fund, 2024) (Figs. 12 and 13).

Faddul (2024) notes that the Brazilian rainforests have been under deforestation-related practices for years, with satellite images showing that the boundaries of the Amazon jungle are slowly but steadily receding and in 2022 deforestation in the country was the highest in the world, according to Faddul (2024). In addition, livestock farming contributes to approximately 65% of deforestation in Brazil. Forests are converted into fields for various reasons, such as mining or the construction of hydroelectric power plants. However, the most important purpose is to provide pastures for livestock, as Faddul (2024) states. Brazil is rated as the world's largest beef exporter, with exports to over 150 countries. During 2023, around 2.5 million tons of processed and fresh beef were exported. Although it is common knowledge,

Fig. 6 Cattle in Colombia. (Permission Sebastian Laverde-Unsplash)

especially among Brazilians, that deforestation and beef sales are linked, it is not always easy to find evidence in records and trace the supply chain from point of origin to final consumer. In some cases, cattle are slaughtered after passing through several farms a process termed "cattle laundering" that makes tracking livestock difficult and impossible. To track these cattle, researchers have connected the supply chain from the supermarket to the grazing fields using data on environmental fines and satellite images of agricultural sites. This has allowed researchers to identify the connections between Brazilian pastures and slaughterhouses as well as supermarket chains in the United States and Europe. According to Fadul (2024), these investigations produced proof of illicit deforestation in forested areas like the Amazon, Cerrado, and Pantanal (Faddul, 2024).

Additionally, Kimbrough (2022) offers a study on livestock raising in protected areas. According to available data, 1.1 million cattle were bought straight from protected areas, and an additional 2.2 million cattle grazed in protected areas and indigenous territories for at least some portion of their lives (Kimbrough, 2022). Furthermore, it mentions that millions of cattle grazed in protected areas (PAs) in the Brazilian Mato states of Grosso, Pará, and Rondônia between 2013 and 2018, and that some of the biggest beef exporters in the world continue to purchase cattle

Fig. 7 Map and satellite image of South America. Note the deforestation in south-central Brazil and the Brazilian-Bolivian border, compared with the dark green of the remaining Amazon forest

grazing in protected areas of the Amazon rainforest despite pledges to cease the practice. The primary cornerstone of Brazil's conservation efforts, protected areas, are arguably the best ways to preserve forests and the biodiversity they contain, according to sources cited by the authors. The meat processors purchase cattle from owners whose land is strictly protected raises concerns (Kimbrough, 2022: see also Cerri et al., 2018; West et al., 2022).

Regarding the impact of cattle grazing on forest ecosystems, Mazzini et al. (2018: 913) conducted a broad, literature-based study of southern South America and found that impacts of cattle grazing may have variable effects on forest ecosystems, sometimes negative and even positive, with case studies mostly from the forests of Patagonia, Chaco, and Monte. The authors noted that about 66% were negative vegetation effects mostly in the Patagonian Forest and Chaco, and positive effects of about 16% were mostly in Monte and neutral effects of about 18% were mostly fauna-related variables in the Uruguayan savanna. The authors argue that the effects of cattle grazing on forests in southern South America are not homogeneous and depend on the particular forest ecosystem considered and the forest property measured. The different impacts on livestock may be partly explained by differences in grazing history and varying productivity of ecosystems (Mazzini et al., 2018: 913).

Nepstad et al. (2014: 1118) present the case that deforestation in the Brazilian Amazon may be reversed. The study reports that the Brazilian Amazon has recently seen a 70% decrease in deforestation, indicating that a large agricultural frontier can be successfully managed for a reduction in deforestation. This decrease in

Fig. 8 Amazon rainforest canopy, San Juan Bautista, Peru. (Permission David-Geere-unsplash)

deforestation appears to have been influenced by a decrease in the demand for new deforestation as well as legal enforcement, intervention in the supply chains for soy and beef, limitations on credit availability, and the extension of protected areas. Public policy relied on punitive measures, and supply chain interventions that have contributed to this slowdown largely depend on companies' risk management. Although they have been developed, systems to provide farmers with positive incentives to refrain from deforestation have not yet been fully implemented. Territorial strategies for combating deforestation have proven successful and have the potential to build on recent gains in slowing the process while offering a framework for tackling other crucial aspects of sustainable development (Nepstad et al., 2014: 1118).

Regarding the position of predators, South America has only three species of mammalian predator capable of killing cattle; the jaguar, cougar, and spectacled bear. The jaguar appears to be the main predatory mammal judged as a cattle killer. For example, Quigley et al. (2017) write that "jaguars are opportunistic hunters. Over 85 prey species (wild and livestock/domestic), including mammals, reptiles and birds, have been recorded in their diet across their geographic range" and that "in some areas cattle (Bos sp.) comprise an important item in their diet… The decline of wild prey below a minimum threshold increases livestock depredation by

Fig. 9 Amazon rainforest, Madre de Dios, Peru. (Permission Renting-unsplash)

big cats, which in turn may lead to retaliatory killings of felids by livestock owners." Cattle killing may also be due to the preference for larger prey, as "the intake of large prey likely helps Jaguars save energy, since it is estimated that 50% of kills are larger prey on which they will feed for up to 4 days" (Quigley et al., 2017; see also Nunez et al., 2000; Scognamillo et al., 2003a; Polisar et al., 2003a; Palmeira et al., 2008a; Cavalcanti & Gese, 2010a; Khorozyan et al., 2015).

For the cougar or puma, Nielsen et al. (2015) point out that the population is of Least Concern but declining, and have wide variable habitats, and then to occupy more open landcover than the sympatric jaguar (see also Sunquist & Sunquist, 2002a). Similar to the slightly larger jaguar, cougars may take large prey but generally favor medium-sized or smaller prey. Nevertheless, they may take cattle and may be persecuted for it (Nielsen et al., 2015). Laundré and Hernández (2010: 85) note that several studies have assessed the rate of cattle predation by cougars in Latin America (see for example, Hoogesteijn et al., 1993; Mazzolli et al., 2002; Conforti & Azevedo, 2003a; Polisar et al., 2003a; Bueno-Cabrera, 2004; Michalski et al., 2006). Several studies may not clearly distinguish between cougars and jaguars in cattle killing, but for cougars predation on cattle is less common and predation on sheep and goats is commoner and generally localized (Nielsen et al., 2015). Polisar

Fig. 10 Deforestation in Para State, Brazil. (Permission Ibama from Brasil - Operação Onda Verde, 2014, CC BY 2.0. httpscommons.wikimedia.orgwindex.phpcurid=51685892)

et al. (2003a) reported a 13% loss of cattle to cougars and jaguars over ten years in north-central Venezuela. Other cattle losses to big cats are reported by Michalski et al. (2006) in central Brazil, and Mazzolli et al. (2002) in southern Brazil. Conforti and Azevedo (2003a) reported that jaguars killed 39 times more livestock than cougars on a ranch in southern Brazil.

For the spectacled or Andean bear, according to Velez-Liendo and García-Rangel (2017), the species numbers are 2500 to 10,000 individuals, and it is considered Vulnerable. This bear has a broad diet, mostly omnivorous. The authors also note that "Andean bears also opportunistically prey on mammals, including rabbits and mountain tapirs, but most notably free-ranging domestic cattle" (Velez-Liendo & García-Rangel, 2017). However, Andean bears may be killed due to the belief, real or imagined that they kill cattle and other livestock. The authors point out that, on average, 180 bears are killed annually throughout the range; however, the real number is believed to be much higher and is likely to rise. They contend that, as more and more people turn to dairy farming for a living, recent surveys in northern Ecuador have revealed historically high numbers of cattle killed by bears (at least

Fig. 11 Deforestation in Brazil. By Ibama from Brasil - Operação Hymenaea, Julho2016, CC BY 2.0, httpscommons.wikimedia.orgwindex.phpcurid=51685896Operação_Hymenaea,_Julho-2016_(29399454651)

320 in 2009–2014) (Velez-Liendo & García-Rangel, 2017; see also Jampel, 2016; Zukowski & Ormsby, 2016).

The rest of this chapter will examine in detail the relationship between domestic cattle and their predators in South America. The jaguar, as the main cattle predator, is examined in detail, followed by the cougar and Andean bear. The predation rates, trends, and impacts are compared with Asia, Africa Europe, and North America.

Jaguars and Cattle

Jaguar Predation on Cattle

The jaguar is the main cattle hunter and killer among the predatory mammals of South America, largely because of its size, strength, and the proximity of cattle as agricultural development intrudes into the jaguars' forested habitat (Cavalcanti & Gese, 2010a; Quigley et al., 2017). Typical jaguar habitat is shown in Figs. 14, 15,

Fig. 12 Deforestation in Brazil. Planet Labs, Inc. –httpswww.planet.comgallerysugarcane-deforestation (direct link) with NC license. (With CC-BY-SA)

16, and 17. The main issues of jaguar predation concern predation on cattle, abundance of the jaguar prey, the jaguar's preference for larger prey, and human attacks on real or imagined cattle predators. Cavalcanti et al. (2010: 383) mention that the jaguar, possibly the most charismatic conservation species in Central and South America, is the least tolerated carnivore in Brazil. The reason for this basic incompatibility is that jaguars require both a large and abundant amount of unspoiled prey as well as a large and open habitat. Conflicts with ranchers are identified by the World Wildlife Fund (2023) as one of the primary issues endangering the jaguar's survival. Contrary to popular belief, cattle can be attacked by jaguars because of their limited resources and the frequent human invasion of their habitat. Ranchers frequently murder jaguars as a form of retaliation because of this, upsetting the delicate balance that keeps ecosystems intact.

Several factors may combine to influence jaguar predation on cattle. Cavalcanti et al. (2010) take the example of Brazil, where the largest populations of jaguars in South America are found, especially in the Pantanal and the rainforests of Amazonia, where outside protected areas in logging and agricultural areas, and mixed land-cover, with conflicts most common in areas with large cattle ranches, low human

Fig. 13 Forest settlement, Puyo, Pastaza, Ecuador. (Permission Ben-Traveling-pexels)

densities, moderate cattle densities, and there are small patches of wilderness with natural prey. Brazil also has "the world's largest commercial cattle herd (>200 million head)" and an "overlap between areas where beef production flourishes and jaguars survive, namely, the Pantanal and the agricultural frontier of southern Amazonia" (Cavalcanti et al., 2010). In the study area, about 32% of the prey taken by jaguars were domestic cattle (Cavalcanti, 2008). In another area in the Pantanal, cattle comprised 46% of the prey taken (Crawshaw & Quigley, 2002). In another study in north-east Argentina, young cattle and calves were the majority of jaguar kills (Perovic, 2002), as was also found in central-western Brazil (Palmeira et al., 2008a). Cavalcanti et al. (2010) note that jaguars are capable of killing adult male cattle, but younger cattle appear to be taken more often. In the northern Pantanal jaguar predation issues were reported by 82% of the landowners and jaguars were considered dangerous human livelihoods by most stakeholders, with up to 11% of their livestock taken (Cavalcanti et al., 2010).

Cavalcanti and Gese (2010a: 722) monitored jaguar prey in an area of the southern Brazilian Pantanal with access to both natural and domestic prey and found that of "438 prey items. Kills were composed of 31.7% cattle (9.8% adults and 21.9% calves), 24.4% caiman (*Caiman crocodilus yacare*), 21.0% peccaries (mostly

Fig. 14 Pantanal Poconé MT Brazil. (Permission Abraao-Paes-unsplash)

Tayassu pecari), 4.1% feral hogs (*Sus scrofa*), 3.9% marsh deer (*Blastocerus dichotomus*), 3.2% giant anteaters (*Myrmecophaga tridactyla*), 2.0% capybaras (*Hydrochoeris hydrochaeris*), 1.6% brocket deer (*Mazama americana* and *M. gouazoubira*), and other avian, mammalian, and reptilian species." In addition, "although all 10 cats killed cattle, 5 killed a high proportion of cattle (>35% of kills), and 3 killed few cattle (<15%). Males (27%) and females (35%) killed cattle in similar proportions" (Cavalcanti & Gese, 2010a: 722).

Zimmermann et al. (2005a: 406) also note that most domesticated animals including cattle, horses, donkeys, dogs, and pigs are important prey items for jaguars. Many cattle ranches in South America occupy formerly prime jaguar habitat, and therefore jaguars may invade such livestock ranges, and the livestock has become easy prey for jaguars, especially in Brazil and Venezuela (see also Crawshaw & Quigley, 1991; Hoogesteijn et al., 1993). As a result, human attitudes to jaguars are generally negative, due to cattle predation. For example, Zimmermann et al. (2005a: 409) note that in their case study of cattle owners in the Pantanal of Brazil, "most (82%) respondents have suffered cattle losses to jaguars and most (66%) believed that jaguar attacks were becoming more common. Most respondents (62%) reported that jaguar attacks did not show any clear seasonal pattern. There was a

Fig. 15 Forest understory habitat Amazon, Peru. Madeline Hogan-unsplash

positive relationship between cattle lost and both ranch size and number of cattle. However, the proportion of cattle lost declined with increasing ranch size and with cattle numbers."

Comparisons have been made of jaguar predation on domestic cattle and domestic water buffalo. For example, Hoogesteijn and Hoogesteijn (2008: 132) in a case study of Venezuela found that domestic cattle were far more predated by jaguars and domestic water buffalo, largely due to the buffaloes' superior anti-predator behavior, size, and aggression. In this study, more cattle were killed than buffalo in all the studied ranches. Cattle were more vulnerable than buffalo, and importantly, buffalo showed much more defensive behavior toward predators, to the extent that "livestock mortality associated with jaguar and puma may be reduced by keeping buffaloes and cattle in the same paddock, or by keeping only buffalo" (Hoogesteijn & Hoogesteijn, 2008: 132). There were significant behavioral differences between the cattle and buffalo: "regardless of habitat and/or husbandry system, predation was greater among cattle and their crosses than among buffalo" and "the odds of cattle being predated by jaguar or puma were 25 times higher than for buffalo" (Hoogesteijn & Hoogesteijn, 2008: 135). In addition, "when in danger buffalo cows form a circle around their calves, and the bulls go around this circle ready to confront any

Fig. 16 Capybara, common jaguar prey. (Guarapiranga, São Paulo - SP, Brasil- Permission Jaime Dantas-unsplash)

predator or menace in an active, aggressive way. Cattle, on the other hand, tend to scatter and flee in the presence of a predator, leaving young animals confused and vulnerable to predation."

The difference between cattle and buffaloes was that the cattle were "raised by humans for at least 7000 years with emphasis on domestication in absence of predators. In contrast, there probably exists a long evolutionary relationship between buffalo and the Asian tiger in which buffalo may have evolved social and defensive behaviour" (Hoogesteijn & Hoogesteijn, 2008: 135). Hoogesteijn and Hoogesteijn (2008: 135) note that in part of their study area, predation was significantly reduced in a group of calving cows when 40 female buffalo and one male buffalo were placed in the maternity paddock. Although cattle depredations were common in this ranch, no depredations were observed on the buffalo herd, which grew steadily after 2001. In another study cited by Farrell (1999), predation also immediately declined in Venezuela when cows in a pasture were replaced by buffalo and our key observation is that the presence of buffalo tends to be enough to deter large predators from preying on cattle to attack. Therefore, further studies are needed to determine how

Fig. 17 Giant Anteater, common jaguar prey. (South Pantanal-Brazil- Permission Nareeta Martin-unsplash)

many buffalo per hectare or what ratio of cattle to buffalo would be needed to prevent jaguar and puma predation.

Human Attitudes to Jaguar Killing of Cattle

Jaguar killing of cattle and other livestock is arguably a major determinant of human attitudes to jaguars (Campbell, 2022). However, some authors argue that killing cattle is only one of the reasons why people kill jaguars. For example, Marchini and Macdonald (2012: 213–221) note that "the killing of jaguars by ranchers in Amazonia and the Pantanal is a major threat to the species." The findings of the study were that "the impact of jaguars on livestock is not the only predictor of a rancher's intention to kill jaguars. Fear, personal and social motivations, and internal and external barriers (e.g. lack of skills and force of law, respectively) to killing jaguars can also influence jaguar killing" and "the relative importance of these factors in determining intention to kill varies with region and affluence." It is also noted that because jaguars are considered cattle killers, it may be assumed that people kill jaguars in an effort to reduce the economic losses that result from livestock predation. The authors argue that "the main focus of conservationists was the economic and ecological issues of human–wildlife conflicts, but little has been done to understand the link between jaguars killing livestock and people killing jaguars. The killing of jaguars may not be strictly retaliatory and might have motivations besides the economic. If we are to curb the killing of jaguars, first we have to understand the underlying causes of this behavior" (Marchini & Macdonald (2012: 213–221). The authors conclude that ranchers may kill jaguars for other reasons besides the damage they cause to livestock. Fear, social and personal incentives, and both internal and external obstacles to killing jaguars all influence the intention to kill them. Region and wealth have an impact on how important these factors are in determining intent (Marchini & Macdonald, 2012: 213–221).

Cavalcanti et al. (2010: 385) also point out that efforts in Brazil to stop human persecution of jaguars have been grounded in what could be called a "bio-rational" understanding of the issue, which recognizes that livestock killing is the primary cause of the problem and may involve the use of electric fences, aversive conditioning, and translocation. It's possible, though, that livestock depredation has less to do with human persecution of jaguars than previously thought. The authors contend that cultural and societal perceptions of the potential threat that jaguars pose when attacking people and killing livestock may be more to blame for such human actions than the actual effects of jaguars on livestock or human safety. The perceived effects of conflicts between humans and carnivores frequently outweigh the available data, and other elements, such as the excitement of the chase and the perceived social standing of jaguar hunters in the community, might also be important (Cavalcanti et al., 2010: 385). One could refer to this as a "perception blight" since it fails to see the causes of jaguar persecution from a broad perspective. The significance of cattle to the ecology of jaguars, ranchers' perceptions of the exploitation of cattle by jaguars, and other perceptions about jaguars and jaguar hunting that are relevant to managing conflicts between jaguars and humans that may be relevant to ranchers' perceptions and persecution of jaguars were all examined in the case study conducted in the Pantanal, Brazil (Cavalcanti et al., 2010: 385).

Comparing jaguars with cougars, Conforti and Azevedo (2003a: 215) present what is "apparently the first study on local perceptions towards large carnivores in Brazil" and note that the inhabitants within and outside of Brazil's Iguaçu National Park (INP) have killed jaguars. Although cougar predation has not been frequently documented, both jaguars and cougars can be found in the area. Two primary factors were found to distinguish people's perceptions of the two species: most people thought that local authorities had released jaguars into the INP, but not cougars. Additionally, fewer people expressed fear of cougars than they did of jaguars. Attitudes toward the two species tended to be similar despite these stark differences in perceptions. The lack of correlation between the predation history of the properties and perceptions of jaguars suggested that the influence of predators was not great enough to change the way that locals felt about carnivores.

Jaguar Predation on Wild Animals by Size

Jaguars have been recorded as preying on variably sized prey, and it has been argued that their natural preferences for wild prey may influence a preference for cattle as prey. Some research suggests that jaguars prefer small prey. For example, Miranda et al. (2018: 724–731) studied jaguars in the Caatinga dry forest, in northeastern Brazil, and found "that over 90% of the diet of jaguars was comprised of prey under 5 kg, mainly armadillos." It was nevertheless argued that such small prey was not the norm: the "mean prey size (5.23 kg) in our study area are among the lowest ever described for jaguars in the literature ..." showing "that jaguars are able to shift their diet to small prey when larger quarry is scarce. However, subsisting in such a stressful trophic position may lead to decreased levels of recruitment and low emigration rate" and "breeding females would have difficulty raising cubs without abundant large prey" (Miranda et al., 2018: 724).

Hayward et al. (2016a) reviewed 3214 jaguar kills recorded in the literature over the species range and found that "jaguars significantly preferred capybara *Hydrochaeris hydrochaeris* and giant anteater *Myrmecophaga tridactyla*, and avoided agoutis, carnivorans, primates, black-eared opossum *Didelphis marsupialis* and tapirs. Generalized linear models showed that jaguars select prey primarily based on socio-ecological and behavioral traits (abundance and herd size), rather than morphological characteristics (body size)." However, the authors found that "their accessible prey weight range was 6–60 kg, preferred prey weight range was 45–85 kg, and mean mass of significantly preferred prey was 32 ± 13 kg leading to a predator to prey body mass ratio of 1:0.53, which is much less than that of other solitary felids (although 1:0.84 may be the relationship with the smallest jaguars)" (Hayward et al., 2016a). The findings suggested therefore that "compared with other large, solitary felids, jaguars have an unusual predator to prey body mass ratio, show limited effect of prey morphology as a driver of prey selection, lack evidence of optimal foraging beyond their preferred prey, and a lack of preferential hunting on Cetartiodactyla herbivores" (Hayward et al., 2016a).

Weckel et al. (2006: 25–30) found that despite a high correlation between the prey used and available, which was suggestive of opportunistic foraging, "jaguars exhibited selective tendencies in discriminating between larger prey. Jaguars used collared peccary Tayassu tajacu greater than its availability, while preying upon the equally abundant and similarly distributed white-lipped peccary Tayassu pecari and tapir less than predicted based upon availability. Armadillo Dasypus novemcinctus and paca Agouti paca, 56.6% of total consumption, were consumed relative to availability but exhibited low spatial overlap with jaguar." These findings point to selective predation by the jaguar on the collared peccary, which is a medium-sized mammal weighing up to 35 kg (Sutherland-Smith, 2015).

Cougars and Cattle

Cougar Predation on Cattle

The cougar has a similar status as the jaguar as a cattle killer but is generally perceived as less powerful and may be less feared for potential attacks on people, as it may be in North America in the presence of other apex predators (Campbell, 2022). In terms of ecological status, Elbroch and Kusler (2018) note that throughout its range, the cougars share habitat with six apex predators, namely, the gray wolf, grizzly bear, American black bear, jaguar, coyote, and maned wolf) and three felid mesocarnivores, namely, the ocelot, bobcat, and Canada lynx, and "evidence suggested that wolves, grizzly bears, black bears, and jaguars are dominant" over cougars, which are dominant over coyotes and maned wolves and the three felid mesocarnivores with which they share range. Cougars "are subordinate to at least one other apex carnivore in 10,799,252 (47.5%) of their 22,735,268 km^2 range across North and South America" (Elbroch & Kusler, 2018). Cougars are more

Fig. 18 Cougar at Torres del Paine National Park, Patagonia, Chile. (Permission Wolves201. https://creativecommons.org/licenses/by-sa/4.0/)

Fig. 19 Cougar habitat, San Carlos de Bariloche, Río Negro, Argentina. Damaris Azocar-unsplash

generalist in habitat than jaguars and occur in a wide range of habitats. Typical cougar habitat is shown in Figs. 18, 19, 20, 21, 22, 23, 24, and 25.

Studies of cougar predation on livestock generally either study only the cougars or compare the cougar and jaguar predation with variable conclusions (Palmeira et al., 2008a). In an example based on a cross-continental, cougar-based study, Palmeira et al. (2015: 903) used a social method (semi-structured interviews with 42 livestock breeders in 36 rural properties) to determine cougar predation on livestock in a protected area of the highlands of a southeastern Brazilian Atlantic Forest. In this area, cougar predation was primarily focused on horses (51%), followed by cattle (28%), sheep (17%), and goats (4%); the cougar predation being 0.4% of the total livestock population, and the main predictors for predation probability were property elevation and distance from the urban center.

In another similar study by Schulz et al. (2014), the authors note that in Southern Brazil, where in traditional pasture systems, livestock may forage without close management, and the animals are vulnerable to cougar attacks. Using social interviews, the study found that cougars favored attacks during low visibility such as thick fog, in fields near forested areas, and also attacked easier prey such as sheep rather than cattle. Some farmers even ceased sheep farming due to cougar predation.

Fig. 20 Cougar habitat open forest. -Mendoza, Argentina. (Permission Bibiana Davila-unsplash)

Consequences included human killings of cougars, and factors for predation included forest fragmentation and agricultural landcover in proximity to forested cougar habitats (Schulz et al., 2014). Cougar predation would be more serious, as "when pumas start to attack cattle's calves, the conflict with local population could become more serious. A six-month-old calf is worth approximately double of an adult sheep and the cattle's breeding management is extensive and without a controlled breeding season in the region" (Schulz et al., 2014).

In a comparative study, Polisar et al. (2003a: 297) note that the following factors may affect the depredation of livestock by jaguars and cougars: (1) innate and learned behavior; (2) individual cat health and status; (3) allocation of space and resources between jaguars and pumas; (4) cattle husbandry practices; and (5) abundance and distribution of natural prey. Michalski et al. (2006: 179) also conducted a comparative study of cougar and jaguar predation on livestock in Amazonia, taking a case study of 236 cattle ranches. The results showed that the distance to the closest riparian forest corridor and the percentage of forest area left were important positive and negative determinants of predation events, respectively (Michalski et al., 2006: 179). There were significant predation loss peaks during the main calving period at the end of the dry season. The size of the cattle herd and the proportion of forest area

Fig. 21 Cougar habitat, grass and shrubland, guanaco grazing. (Permission Hector Perez-pexels)

had positive effects on predation rates in 60 cattle farms. The average proportion of cattle lost to big cats within 24 months varied across the region depending on herd class size. The highest annual monetary costs were found in large cattle ranches, and the authors noted that predation patterns could be related to landscape and livestock management variables such as the proportion of forest area, distance to the nearest riparian corridor, annual calving peak, and cattle herd size (Michalski et al., 2006: 179).

Palmeira et al. (2008a: 118–125) compared cattle depredation by cougars and jaguars over a six-year period (1998–2003) in central-western Brazil. The findings indicated that about 19% of cattle (especially calf) mortality was due to predation, especially near forests. It was noted that "the ranch-hands were not able to routinely distinguish between jaguar and puma kills" but to do so "external or internal lesions caused by bites or claw marks, especially in the head, neck and legs" were used to identify the predator (Palmeira et al., 2008a: 118–125). In particular, jaguar attacks result in dislocation or fracture of the cervical vertebrae, either from the animal's impact with the ground or from a bite to the base of the skull or neck." Typically, the jaguar begins to eat the head, neck, and chest area, with the rear parts often remaining intact. The pumas, on the other hand, kill their prey by suffocation by biting

Fig. 22 Guanaco, common prey of cougar. Puerto Madryn, Chubut, Argentina. (Permission Gonzalo Esteguy –pexels)

their throats and also leaving marks from their claws on the shoulders and back of their prey. They begin to eat their prey from the abdominal side, reaching the ribs and the muscles of the hind legs (Palmeira et al., 2008a: 118–125; Pitman et al., 2002).

Palmeira et al. (2008a: 124) found that "calves were the most vulnerable age category to all causes of mortality" with "calves younger than 9 months were the most frequently killed age class" and "the majority of the kills checked were confirmed to have been by puma tended to kill smaller calves than jaguars. The average weight of calves lost to pumas and to jaguars was 74 kg (0–6-months old) and 144 kg (3–9-months old), respectively" and cougars caused "most of the cattle depredation" (Palmeira et al., 2008a: 124). In addition, cougars killed calves aged 0 to 6 months, while jaguars killed calves aged 3–9 months. This was consistent with cougar and jaguar predation on wildlife, with the latter cat taking larger prey (Palmeira et al., 2008a: 124; see also Scognamillo et al., 2003a).

Kissling et al. (2009: 807) studied the spatial risk of livestock to cougar depredation in Patagonian ranches in Argentina. The presence of cougars on ranches was found to be positively correlated with the topography, forested vegetation cover, and

Fig. 23 Guanaco, common prey of cougar. Guanaco en Esquel Chubut Argentina (Permission Emareynares- pexels)

proximity to the predicted regional cougar habitat. The authors note that the possibility of livestock exposure to cougars can be decreased by varying livestock management throughout the season, according to a comparison of the risk of predation by cougars in simulated livestock management scenarios (Kissling et al., 2009). The attitude of ranchers toward cougar was generally negative, and some ranching associations in some provinces of Patagonia even established financial incentives for cougar hunting because they believed that cougars posed threats to livestock, especially sheep. The paucity of knowledge regarding cougar ecology in Patagonia was cited as a significant obstacle to the creation of conservation policies and initiatives meant to lessen human-cougar conflict and support healthy cougar populations in Patagonia (Kissling et al., 2009).

Human Attitudes to Cougar Killing of Cattle

Cattle killing by cougars may create similar attitudes from people as jaguar kills of cattle (Campbell, 2017a). Nanni et al. (2020: 429), in a case study of cougar

Fig. 24 Guanaco, common prey of cougars. Arequipa, Perú. (Permission Maria Camila-pexels)

predation and human tolerance in the Argentine Dry Chaco, noted that predation by cougars on livestock was thought to be a significant factor in livestock losses. Although there was a negative attitude toward pumas, there was no significant correlation with predation. Retaliatory hunting, however, was more prevalent among those who reported livestock losses to cougars. Low cougar presence reduced the likelihood of livestock predation, which was increased with the breeding of goats and sheep and by the absence of corrals. Furthermore, cougar predation may have a particularly negative impact on the livelihoods of the local peasants in the Argentine Dry Chaco, who frequently rear goats and cattle in open forests. This could exacerbate preexisting negative attitudes toward predators (Nanni et al., 2020: 431). In this study, "55% of respondents raised cattle, almost 40% had goats, and 65% had two or more types of livestock" and "puma predation was reported as the main cause of livestock loss by 40% of respondents" (Nanni et al., 2020: 434). In addition, there was widespread intolerance toward cougars among respondents that was not strongly associated with predation, although neutral and positive attitudes toward the cougar were more common among respondents who did not report cougar predation on livestock. The widespread negative attitudes toward cougars, even among

Fig. 25 White llama, common prey of cougars. Tupiza, Bolivia. (Permission Paz Arando-unsplash)

respondents who had not suffered livestock losses, should be confirmed with a larger sample. However, the lack of a strong link between tolerance and predation in livestock could be due to deeply held beliefs about the negative impacts of this species (Nanni et al., 2020: 434).

In another study of the High Andes of Argentina, Lucherini and Merino (2008: 81) in a multi-species comparative study found that cougars and foxes are considered pests for hunting livestock and are actively hunted as a result. Although the perception of the Andean cat and the Pampas cat was more positive, especially among schoolchildren, they are also often killed by locals. It is also noted that only a minority of respondents passively accepted the loss of livestock, and the slight difference between the proportions of adults who viewed cougars as pests and those who reported killing cougars was primarily due to the fact that some herders reported having difficulty finding/killing cougars (Lucherini & Merino, 2008: 82–83).

Cougar Predation on Wild Animals by Size

Similar to studies of big cat predation on livestock in South America, studies of cougar predation on wild prey may either study only the cougars or compare the cougar and jaguar predation. Karandikar et al. (2022) give a study of cougars and note that the cougar has a highly flexible diet, preying on 232 different animal species. Cougars killed more larger-bodied prey species with increasing distances from the equator, but consumption of medium-sized species showed an opposite trend. Importantly, the killing of domestic species was negatively correlated with that of medium-sized wild species, thus indicating the importance of maintaining native prey presence and ecology (Karandikar et al., 2022). It is argued that as the predation of wild species is significantly linked to the predation of livestock, future studies should examine the relationships between the availability and selection of primary and other wild prey and the resulting impacts on the predation of native species to guide conservation efforts and reduce conflict between cougars and humans (Karandikar et al., 2022).

Llanos and Travaini (2020) described cougars in the arid environments of central Patagonia, Argentina, and noted that "the main prey consumed by pumas in livestock ranches was the guanaco" but "sheep contribution to diet was four times less than that of the guanaco." It is argued that sheep predation by cougars "is a socio-environmental problem that has intensified in recent decades in Patagonia" and "a common recommendation to limit livestock depredation by large carnivores is to maintain suitable levels of wild prey populations." In the northeastern regions around Chubut, Argentina, guanaco constituted about 75% of the total biomass consumed; the next most significant prey was sheep, which was consumed four times less by biomass than guanaco. According to the authors, these findings clarify the role of this large native ungulate as a primary prey and its potential importance as a buffer against predation in sheep (Llanos & Travaini, 2020). See Figs. 19 and 20 for the guanaco.

A comparative study of cougars and jaguars is that of Schulz et al. (2014), which compared cougar and jaguars in terms of predation and noted that in Brazil, Jaguars usually kill animals larger than pumas such as 8- to 10-month-old calves. Pumas mostly kill sheep, goats and younger calves (see also Pitman et al., 2002). Scognamillo et al. (2003a: 269) compared the prey between cougars and jaguars in Venezuelan llanos and noted that for their presence, "a high degree of spatial overlap was observed between jaguars and pumas, which may be related to the abundance and distribution of prey species" and "major segregation was found in food habits" as "jaguars selected for large prey and pumas for medium-sized prey. Jaguars selected for capybara Hydrochaeris hydrochaeris and collared peccary Tayassu tajacu and consumed caiman Caiman crocodilus and white-tailed deer Odocoileus virginianus less than expected. Pumas selected just for collared peccary and also killed caiman less than expected" (Scognamillo et al., 2003a: 269).

Silveira et al. (2008) present a study that aims to determine the parameters for identification of predators at kills. It is noted that "determining which type of predator is responsible for a kill can be difficult, although size can be an indicator. Prey smaller than 250 kg can be taken by any of the sympatric predators (e.g., jaguars, pumas, domestic dogs), while kills greater than 250 kg can be definitively attributed to jaguars, since above this biomass the jaguar is virtually the only predator capable of preying on such large animals." Silveira et al. (2008) note that certain details should be considered when analyzing a jaguar kill, mentioning four such markers: (a) Jaguars typically leave tooth marks in dense, heavy bones like the femur, the base of the skull, or the upper and lower part of the neck, which can cause fractures and vertebral fractures; (b) many large prey animals have their necks broken by jaguars, which frequently results in the animal's head being turned backward. In contrast to pumas, jaguars seldom bite their prey's throat in order to suffocate it. (c) Like pumas, jaguars typically start feeding on the front of the prey and work their way up to the ribs and central part of the carcass; they prefer to eat the neck, lower part of the neck, hump, and chest; small animals, like calves, are typically eaten whole; (d) Unlike pumas, jaguars can drag a carcass up to 1.5 km and conceal it in the canopy of vegetation (Silveira et al., 2008).

Guerisoli et al. (2021: 33–43) describe the diet of cougars in the South American temperature region, comparing the occurrence of native prey and livestock. The study was based on 18 publications dated between 1991 and 2020 and fecal-based sample from rangeland of central Argentina. The primary livestock were sheep and to a lesser extent cattle, while wild prey were guanaco and armadillos, introduced wild boar and European hare. The authors note that "livestock frequency of occurrence was greater inside protected areas than outside" and that "although the puma is considered in this large area as conflictive, wild prey (exotic and native) are its main food sources."

Other studies have documented cougar prey and compared these with those of jaguars. For example, Elbroch and Kusler (2018) compared the ecological roles of cougars and jaguars and found that in South America, jaguars dominated pumas in 60% ($n = 15$) of 25 sources, and 10 sources where neither species dominated where jaguars and pumas were sympatric. They also found that "the size difference between pumas and other apex predators does correlate with dominance. For example, evidence that jaguars are dominant is strongest in areas where jaguars are large and weigh considerably more than pumas, but more ambiguous where the two species are more similar in size" (Elbroch & Kusler, 2018; see also López-González & González-Romero, 1998a; Gutiérrez-González & López-González, 2017a). The authors also note that "jaguar-puma interactions were the most frequently studied competitive interaction among apex predators in the Americas. In contrast to wolf-puma research which has been conducted with marked individuals, telemetry, and GPS technology, most jaguar-puma research has been conducted with non-invasive camera trapping assessing spatial and temporal time-sharing of habitats, and to a lesser extent, dietary overlap through scat analysis … Whereas some authors were certain they found evidence that jaguars were dominant over pumas, other researchers, from areas where jaguars are smaller and sometimes numerically fewer than pumas, suggested the two species may be at least equal in competitive interactions, or even that pumas may be dominant over jaguars. We did discover five references that report incidents of jaguars killing pumas, whereas we did not find a single documentation of the reverse" (Elbroch & Kusler, 2018; see also Crawsha Jr & Quigley, 1984; Harmsen et al., 2009a; Gutiérrez-González & López-González, 2017a).

Another comparative study of cougars and jaguars is that of Polisar et al. (2003a: 300–301), who argue that "despite considerable overlap in prey species, jaguar and puma diets were significantly different whether examined with livestock included or excluded. Jaguar showed a preference for large-bodied prey, with puma taking more medium-sized and smaller prey wild prey than jaguar. Puma dietary niche breadth exceeded that of jaguar." In addition, "the most important wild prey in the jaguar diet in terms of frequency of occurrence and biomass consumed were collared peccary, capybara, and white-lipped peccary. Puma took most large prey in proportion to availability, but selected for collared peccaries, and took caiman less than expected according to availability" and "the four most important wild prey items in puma diet on bases of both frequency of occurrence and biomass were collared peccary juveniles, deer, capybara, and caiman" (Polisar et al., 2003a: 300–301).

Another study by Foster et al. (2013: 373) examined the predation patterns of pumas and jaguars and found that they had similar nocturnal and crepuscular activities, with common diurnal activity only in the Pantanal and little temporal separation, as well as a significant overlap between the cat's activity patterns and that of their prey, indicating that they adjust their activity to reduce their energy expenditure when foraging. The authors note that both cats use the same environmental resources and have similar life strategies and resource use, but the jaguar is generally considered to be more specialized than the cougar; It prefers dense habitats (forests and shrublands as opposed to open grasslands and swamps) near water and avoids anthropogenic habitats, whereas the cougar is more adapted to a wider range of habitats and is less dependent on water bodies (Foster et al., 2013: 373; see also Iriarte et al., 1990a; Cullen, 2006; Sollmann et al., 2012a).

Foster et al. (2013: 373) argue for differences between the two cats: As generalist predators, jaguars and pumas consume a variety of prey species, but the majority of their diets consist of peccaries, deer, large caviomorph rodents, and armadillos. Due to their larger body mass, jaguars are believed to rely more on larger prey than pumas, which tend to choose smaller prey species when they are present with jaguars, presumably to reduce competition, although some studies show high dietary overlap (see Iriarte et al., 1990a, Taber et al., 1997, Scognamillo et al., 2003a, Núñez et al., 2000; Haines, 2006). Jaguars and cougars may also avoid each other and also adapt to prey behavior, including that of "nonvigilant" prey animals, which may react slowly (Foster et al., 2013: 374). Foster et al. (2013: 378) note that these results support the opportunistic hunting theory, "which states that predators adjust their activity to reduce foraging energy expenditure" and note examples where the main prey for both jaguars and cougars is capybara, and jaguars tend to hunt larger capybaras (>13 kg) than those hunted by cougars (1–15 kg) (Iriarte et al., 1990a; Scognamillo et al., 2003a). Also cited are cases where peccaries are the main prey, the jaguar's generalist prey base shows predation on lesser anteaters and cattle (Foster et al., 2013: 378).

Fig. 26 Spectacled bear, in natural habitat, Huila, Colombia. (Permission Robert Acevedo-Pexels)

Spectacled or Andean Bears and Cattle

Spectacled Bear Predation on Cattle

Spectacled bears have been recorded preying on cattle, largely due to increased human populations in Andean countries, leading to agricultural and settlement expansion into the Andean cloud forest habitat of bears (Goldstein et al., 2002). Such conflicts may occur when livestock rearing intrudes into bear habitats, which may be dense forests or mountain forests such as the Peruvian Andes (Peyton, 1980). Typical bear habitat is shown in Figs. 26, 27, 28, and 29. The National Wildlife Federation (2023) notes that "the large-scale expansion of cattle ranching in South America has come at great environmental cost, as vast expanses of tropical forests have been cut, burned, and converted to pastures. This process of slash and burn land conversion destroys wildlife habitat, reduces biodiversity, and releases tremendous amounts of carbon emissions into the atmosphere. Growing demand for beef, leather, and other products is expected to continue to exert significant pressure on ecologically important regions in Brazil, Argentina, Paraguay and Colombia."

Fig. 27 Chingaza, La Calera, Colombia, known spectacled bear range. (Permission Hector Orjuela-unsplash)

For example, Peyton (1981: 48) points out that "land nationalisation, starting in 1963, brought settlers and cultivation into mountain areas of Peru where hitherto spectacled bears had been disturbed only by occasional hunting. This greatly reduced the numbers of bears, and also habituated those that survived to eating the farm crops and farm animals that replaced their natural foods, thus incurring the wrath of the farmers who replied with guns." Wordley (2021) argues that "as deforestation and climate change make life harder for them, bears increasingly search for food around human settlements, where they devour fields of maize and sometimes kill livestock… While it's not clear if bears really kill as many sheep and cows as people think they do, there have been several well-documented cases of people killing bears."

Goldstein et al. (2002) note that spectacled bear-cattle conflicts are recorded throughout the species distribution: and since 1980, field evidence suggests that bears eat cattle carcasses and cattle meat can be important for the spectacled bear diet with evidence from Bolivia, Colombia, and Venezuela. The authors nevertheless take a critical look at the evidence, noting that "many questions are still unanswered. Are spectacled bear predators of cattle or feed upon occasional cattle

Fig. 28 Machu Picchu, Cusco, Perú, known spectacled bear range. (Permission Eddie Liszka-unsplash)

carcasses encountered? How many bears are involved in cattle predation or carcass feeding? Is the predatory or carcass feeding behavior of bears a common behavior over the whole bear distribution?"

Studies on the predation of cattle by bears indicate that bears do predate cattle, but only rarely (Goldstein et al., 2009: 11). For example, Goldstein et al. (2009: 11) note that in studies of bear predation of cattle in 140 Venezuelan villages during 1986–89 and 1996–2000, "only 12 localities reported cattle depredation by bears during the first survey, and of these only 7 reported new cattle depredations during the follow-up survey" (see Goldstein, 1991, 2002). Goldstein et al. (2009: 11) argue that these numbers are low: "if cattle were the preferred prey for Andean bears in general, as is the case with the western Cantabrian brown bear population in Spain, we would have expected cattle losses in the Andes to be greater than those reported." Therefore, "cattle are clearly not a preferred prey item; livestock appears in the Andean bear diet only at specific localities and times" (Goldstein et al., 2009: 11; Goldstein, 1991, 2002).

Regarding cattle predation in Venezuela, Goldstein (1991: 233) reported that only 16 of the 34 claims of predation that were filled in the case study area were

Fig. 29 Baños, Ecuador, known spectacled bear range. (Permission Kiyoshi- unsplash)

recent enough to warrant field verification. Using the traditional Páramo herding methods, which involve leaving small herds (5–30 animals) of young and unproductive animals to graze in páramo areas 2800 m above sea level, distant from human settlements, carcasses or skeletal remains of cattle were found at all 16 recent claim sites, and predation was reported in all localities. Goldstein (1991: 233) therefore concluded that the data acquired from the spectacled bear survey indicates that spectacled bears actively prey on cattle. The conclusion was reached that the overall impact of spectacled bear predation on livestock in the Venezuelan Andes is negligible, based on the number of predatory episodes recorded, in particular, the finding of signs of struggle at the feeding sequences inspected and the cessation of predatory events after the killing of the bear involved in the episodes (Goldstein, 1991: 234). Goldstein (2002: 371) adds that such bear-cattle conflicts in Bolivia and Venezuela occur in remote cattle pastures in open grassland areas near large tracts of cloud forest. In Venezuela, all but one location where bear-livestock conflicts occurred was within national parks. The lack of information on bear-cattle conflicts is due to a lack of monitoring by national park authorities. Livestock farmers deal with any incident of bear predation without notifying the authorities, and usually, the "problem bears" and any other bears that happen to be in the area are killed.

There is a long history of interactions between bears and cattle, which appear to occur at intervals of up to 10 years.

Jampel (2016: 84) uses the Ecuadorian Andes as an example of bear predation on cattle, pointing out that since 2009, locals have reported numerous Andean bear attacks on cattle. This is because of market conditions, rural outmigration, and climate change, which have all contributed to non-migrants in rural areas raising cattle as a means of subsistence. Given that cattle are delicate capital assets, the region's shift to cattle-based livelihoods has increased conflict between people and bears. It should be mentioned that more than 250 accounts of Andean bear attacks on cattle on the eastern flank of the northern Ecuadorian Andes were made between early 2009 and spring 2012. This is a sharp rise over earlier figures both here and across the nation (Jampel, 2016: 84). Jampel (2016: 90) adds that the most straightforward explanation for bear attacks on cattle is that the bears are starving as a result of habitat loss. In one attack area, it is also observed that the secondary forest encircled the pasture vegetation, and tree vegetation recolonized the abandoned housing. This created a comfortable environment for bears to attack cattle and then drag the carcasses. Moreover, bull calves are frequently housed in these marginal pastures with infrequent check-ups (Jampel, 2016: 90–91). Furthermore, because of outmigration and the low labor needs of cattle, there is less human activity in these pastures and the nearby forest, as well as less noise (Jampel, 2016: 91).

Predation on Cattle and Human Attitudes

Bears are possibly overly blamed for cattle predation. Goldstein et al. (2002) note that "local accounts or folktales, report that bears pursue cattle on steep slopes or near cliffs to make them fall, or to pull the cattle by the tail to make them fall, or even to handle them by the horns, and any cattle disappearance in bear country is blamed on predatory bears." Goldstein et al. (2009: 10) also examine the public perception of the spectacled bear as a predator of livestock, and note that "people who live far from conflict areas often view large carnivores such as Andean bears as charismatic symbols of wilderness" and that "to these ex situ residents, the Andean bear is generally seen as a nonaggressive vegetarian animal, and complaints about conflicts or cattle depredation are often viewed as a ruse by complainants to gain benefits or justify the killing of bears." By contrast, "people who live in areas where cattle are unattended in the mountain pastures for weeks at a time often perceive Andean bears as cattle predators" (Goldstein et al., 2009: 10; see also Goldstein, 1991; Paisley, 2001). In addition, "to these in situ residents cattle losses are often automatically attributed to bear depredation. Bears in those areas can be perceived as pests that should be killed as a preventive measure. Even in localities with no recent history of bear–cattle conflicts, as soon as a problem develops and bears are associated with the disappearance of cattle, the formerly positive perception is reversed and bears become subject to persecution and poaching," although "the perceived conflict is greater than the real conflict" (Goldstein et al. 2009: 11; see also Torres et al., 1995; Castellanos, 2001; Galasso, 2002).

Goldstein et al. (2009: 8) found in case studies from Bolivia, Colombia, Ecuador, Peru, and Venezuela that Andean bears are perceived as livestock predators where herding is common. Bears are often blamed for the disappearance or death of livestock. This negative perception often leads to poaching of Andean bears by local residents and disregard for protected areas and conservation efforts. An important topic is the concept of the problem bear. It is argued that anecdotal information about predatory bears in Venezuela, Colombia, Ecuador, Peru, and Bolivia always assumes that a so-called problem bear—usually an adult male—was solely responsible for the predation of livestock. After the removal or disappearance of the problem bear, predation usually ceased, confirming the problem bear hypothesis, until a new problem bear emerged. Cases of elimination of problem bears in Venezuela and Ecuador are cited.

Goldstein (2002: 371) notes that information collected in bear-cattle conflict areas in Bolivia and Venezuela shows that most livestock deaths are due to predation by bears, an assumption that is not always true, as the experiences in Páramo La Fiera, Venezuela, and Pasto Grande in Bolivia have shown. All of the locations where there were spectacled bear-livestock predation claims, which were visited in Bolivia and Venezuela for this study only had very old, decomposing cattle carcasses linked to bear activity indicators, and there was no proof that they were the result of scavenging or predation. Additionally, bear-cattle conflict persists throughout the range of spectacled bears in Venezuela and Bolivia, where highland farming practices are prevalent (Goldstein, 2002: 371).

Opinions on bear predation on cattle also derive from the value of cattle as capital (Jampel, 2016). For example, Jampel (2016: 91) notes that due to the function of livestock as a store of wealth, bear attacks are more costly per incident than bears eating corn. The phenomenon of human-bear conflict is a product of changing livelihoods and land use, spatial arrangements related to deforested and recovered areas, and the endangered capital available to bears. In retaliation for bear attacks on cattle, some pastoralists promised retaliation, contributing to greater concerns about the peaceful coexistence of humans, cattle, and Andean bears on the agricultural frontier (Jampel, 2016: 91). Furthermore, the human-bear "conflict" represents a longer-term political-economic conflict over territory, as it is not just about profits and material well-being, but also about placemaking (Jampel, 2016: 91).

Rojas-VeraPinto et al. (2020. 25) give the example of Peru, where "in 2015, the Ministry of Agriculture of Peru received complaints about Andean bears' livestock attacks in central Peru." Notably, there is currently a widespread belief in some parts of its range that Andean bears hunt cattle in order to survive. Conflict and the preservation of this threatened species may result from this. There is typically a lack of conclusive evidence, though, as some reported events may be related to bears scavenging on already-dead cattle and are not necessarily related to bear attacks. In Peru, it is commonly believed by rural settlers that bears attack cattle so they can feed on them (Rojas-VeraPinto et al., 2020: 25). Using a carcass as an example, these authors also look at the evidence of bear attacks on cattle. They observe that, in contrast to predatory attacks by dogs or cougars, where the bite sites are typically on the neck or legs, bear attack signs include shattered bones, clotted blood on the back and withers, extremities scattered in the forest and in nests, signs of battle, and the absence of internal organs (Rojas-VeraPinto et al., 2020: 28).

Spectacled Bear Predation on Wild Animals by Size

Spectacled bears have a variable carnivorous, omnivorous, and herbivorous diet, and are mainly frugivorous/folivorous (Rodriguez et al., 2014: 25; see also García-Rangel, 2012). Peyton (1980: 639) describes the diet of the spectacled bear, based on scats, finding that "83 foods were confirmed in the diet of spectacled bears. These include insects, rodents, livestock, corn, berries, tree wood, bamboo hearts, palm frond petioles, 22 species of Bromeliaceae, 11 species of Cactaceae, and the fruits of 31 species of trees." Regarding the carnivorous diet, Pisso-Florez et al. (2021: 19–25) record a rare incident of an attack on a mountain tapir (*Tapirus pinchaque*, Roulin, 1829), noting that "despite their overlap in distribution, little is known about interactions between both species, with few reported cases of Andean bear attacks on the Mountain tapir" and that this is a "possible attack by an Andean bear on a Mountain tapir in the northern part of Puracé National Natural Park, Colombia based on strong wounds and marks on a tapir's back and rump. The wounds match typical attack patterns generated by Andean bears and corroborates previous camera traps records of bears attacking tapirs in this locality." Mountain tapirs are very large animals, weighing up to 250 kg and 1.8 m in length (Lizcano et al., 2017). Evidence of bear predation included claw marks, reports of attacks, and Mountain tapir hair in bear feces (Castellanos, 2011; Rodriguez et al., 2014; Castellanos, 2019; López-Ordóñez et al., 2020).

Rodriguez et al. (2014) present a photographic study of a bear attack on a tapir, clearly showing an Andean bear on the back of an adult mountain tapir. The Andean bear hangs on its paws, its head is visible, and does not bite the tapir. The attacked animal appears to be walking or running quickly and its head is blurred, yet a mountain tapir in the image is easily identified by the size, color, and shape of its legs. There is insufficient evidence that the mountain tapir was killed as a result of the attack, so we are not reporting it as a predation event. However, the event involves a predation attempt by an Andean bear on a mountain tapir, an event that has never been recorded before (Rodriguez et al., 2014:25). The authors conclude that bears killed a cow or horse The image taken in Purace NP, along with previous records of tapir hair in bear scat samples, confirms that Andean bears do indeed attack mountain tapirs, thus supporting the hypothesis that Andean bears carry prey on mountain tapirs (Rodriguez et al., 2014: 26).

Geomatics Assessments

Geomatics-based contributions to assessments of livestock landuse and predator ecology are largely based on satellite imagery, aerial photography, drones, electronic tracking and tagging, camera traps, radar, and LiDAR (Campbell, 2018). For satellite imagery-based research, Finer et al. (2018: 1303) note that "recent years have seen major strides in documenting historical and annual tropical forest loss with satellites. Now, a convergence of satellite technologies and analytical capabilities makes it increasingly possible to monitor deforestation in near real time, on the scale of days, weeks, or months, rather than years." In addition, "it is critical to capitalize on continually improving satellite technology to better detect, understand, and prioritize deforestation events."

Such geomatics-based methods have been applied to the South American environment, with positive results. For example, Voiland (2020) notes that "scientists have used satellites to track the deforestation of the Amazon rainforest for several decades — enough time to see some remarkable shifts in the pace and location of clearing." These studies showed massive deforestation for several decades, after which the evidence pushed the Brazilian government to adopt the Action Plan for the Prevention and Control of Deforestation in the Legal Amazon (PPCDAm), which involved the creation of national and state parks, indigenous protected territories, stronger environmental enforcement agencies, and satellite monitoring systems, and restricted the export of resources derived from illegally deforested land (Voiland, 2020). The result was positive, with large-scale declines in deforestation. In addition, in 1998, the government established PRODES, a data-collection system satellite Landsat 5 and 7 records. Data was collected from Landsat images every few weeks, with annual updates on deforestation. The Brazilian National Space Agency (INPE) could use the information to calculate the decline in rainforest cover (Voiland, 2020).

PRODES data was initially limited to government organizations, but because of public pressure concerning deforestation, in 2002 INPE published the dataset online, with explanatory deforestation maps of the total Brazilian rainforest, enabling informed engagement between the government and scientists, nongovernmental organizations (NGOs), and the general public (Voiland, 2020). DETER, a second more advanced satellite-based system, began operation in 2004, capable of daily recordings of deforestation, vegetation health parameters, and fire events from lower-resolution sensors (MODIS) on NASA's Terra and Aqua satellites. The daily measurements compensated for the lower resolution. Voiland (2020) notes that "the daily coverage allowed forest monitors to identify some newly cleared areas in near-real time. When DETER was first set up, it sent deforestation alerts to enforcement officers within two weeks. By 2011, the system was sending alerts within a day." Therefore, "rather than doing random patrols and checking for permits, rangers would send squads of enforcement officers — with satellite images in hand — to investigate areas of recent, suspicious activity. That made it much easier to remove or apprehend equipment" (Voiland, 2020). Limitations included the difficulty in determining smaller clearance patches (mostly below 25 hectares), but the ability of DEMETER to detect deforestation still influenced a decline in clearance patches, which was a strategy to avoid detection from satellites. Between 2001–2007 and 2008–2014, the number of large clearings (>50 hectares) declined by 46 percent and the number of small clearings (<1 hectare) increased by 34 percent. Rajão et al. (2017) give a description of the main satellite systems used in Brazil (Table 2).

Table 2 Satellite-based systems used in Brazil

Name	Origin	Satellites	Frequency	Resolution (m)
PRODES	1988	Landsat, ResourceSat, CBERS	Annually	20–30
DETER	2004	Terra	Daily/Monthly	250
DEGRAD	2007	Landsat, CBERS, ResourceSat	Annually to 2014	20–30
TerraClass Amazônia	2008	Landsat, Terra, ResourceSat, and CBERS	Biennial	20–250
DETEX	2009	Landsat, CBERS, ResourceSat	NA	20–30
TerraClass Cerrado	2013	Terra, CBERS, ResourceSat Landsat	Biennial	20–250
DETER B	2015	CBERS, ResourceSat	Daily/Monthly	56–64
UNFCCC	2016	Landsat, ResourceSat	Five years	24–30

Source: Rajão et al. (2017). The land-use detected were simple for PRODES and DETER (mostly clear-cut), but the later systems were more sophisticated, with detection of more sophisticated detection, such as effects of fire and more selective logging. UNFCCC refers to the Third National Communication of Brazil to the United Nations Framework Convention on Climate Change

Radar technology is also used, an advantage because radar can penetrate clouds, especially the use of synthetic aperture radar (SAR) data which can also detect forest canopy details beyond the ability of Landsat nor MODIS technology, for the detection of less evident deforestation. Voiland (2020) notes that "in a few years, the launch of the NISAR Mission should sharply increase the amount of freely available radar data needed to monitor deforestation in the Amazon. In the meantime, an updated version of DETER is now ingesting higher-resolution data from the ResourceSat-2 and CBERS satellites, bringing the deforestation detection threshold down to 3 hectares (7 acres)."

However, the impact of such technologies on the studied environments, especially in terms of the relevant social relations, has been questioned (Campbell, 2018). For example, a study by Slough et al. (2021) notes that "despite substantial investments in high-frequency, remote-sensed forest monitoring in the Amazon, early deforestation alerts generated by these systems rarely reach the most directly affected populations in time to deter deforestation." The case study is of a "community monitoring program that facilitated transfer of early deforestation alerts from the Global Forest Watch network to indigenous communities in the Peruvian Amazon and trained and incentivized community members to patrol forests in response to those alerts" with the findings suggesting that "the program reduced tree cover loss, but the estimated effects from the experiment are imprecise" (Slough et al., 2021). The reason for the limitations were that "community members perceived that the program's monitors were new authorities with influence over forest management and that the monitors' incentivized patrols were substitutes for traditional, unincentivized citizen patrols that suffer from free riding and inhibit timely community detection of and responses to deforestation" (Slough et al., 2021).

For satellite image-based assessments, Kattan et al. (2004: 155) present a study of landcover assessments relevant to the conservation status of the spectacled bear in the northern Andes, a region including Colombia, Ecuador, and Venezuela, using satellite image-based analysis of vegetation (Andean forests, high elevation shrublands, and grassland) fragmentation and estimations of habitat availability. Satellite images were used to construct maps of remnant habitat patches, which were then overlain with a road map and a human accessibility model for human presence impacts. The findings indicated that the range is fragmented through landcover change and roads into 113 blocks greater than 100 km^2 mostly less than 5000 km^2 covering 42% of the original bear habitat. Some of the fragmented blocks were also internally changed with human presence of change, which was especially fragmented in smaller fragments. Bear population sizes were estimated using possible densities, at high (0.25), medium (0.11), and low (0.04), the results suggesting that 29 populations would have over 500 individuals with the high estimate (the largest possible population would be 9048), nine with the medium and one with the low estimate, with lower estimates if the core area was measured. The authors conclude that bear conservation must be based on the protection or even regeneration of suitable habitat, increased size of fragmented forest blocks, and reduced human actions such as hunting.

LiDAR is also used to monitor the South American forest (Campbell, 2018). For example, Ometto et al. (2023) cite a "new forest above-ground biomass map for the Brazilian Amazon," which uses the "largest airborne LiDAR database ever collected in the Amazon, mapping 360,000 km^2 through transects distributed in all vegetation categories in the region. The map uses airborne laser scanning (ALS) data calibrated by field forest inventories that are extrapolated to the region using a machine learning approach with inputs from Synthetic Aperture Radar (PALSAR), vegetation indices obtained from the Moderate-Resolution Imaging Spectroradiometer (MODIS) satellite, and precipitation information from the Tropical Rainfall Measuring Mission (TRMM)." In addition, "this unique dataset enabled a better representation of the regional distribution of the forest biomass and structure, providing further studies and critical information for decision-making concerning forest conservation, planning, carbon emissions estimate, and mechanisms for supporting carbon emissions reductions." It is argued that "the current map enables more detailed characterization of the forest structure, allowing for estimation of biomass within a wider range of values than the maps previously used in National Inventories," this is largely because "Airborne Laser Scanning (ALS) has evolved to provide an important contribution to the scientific debate and improve forest biomass representation" (Ometto et al., 2023).

For studies including wildlife and satellite images, environmental parameters must be identified and measured using satellite images and field assessments (Campbell, 2018). For example, Meza Mori et al. (2020) present a study of the distribution of the spectacled bear, which is relatively unknown, using georeferenced locations, MaxEnt entropy modeling, and environmental variables including precipitation and temperature change in Amazonas, northeastern Peru. The findings of the study indicated that "most of the important habitats of the spectacled bear are not part of the protected natural areas of Amazonas, under current as well as under future scenarios" (Meza Mori et al., 2020). The classified areas of "high," "moderate," and "low" potential habitat currently conditions cover 1.99% (836.22 km^2),

14.46% (6081.88 km^2), and 20.73% (8718.98 km^2), respectively, of the Amazon forest, where the area classified as "High" potential habitat will increase under climate change possibilities, and "moderate" and "low" potential habitat, as well as total habitat, decreases over the time. Therefore, bear distribution decreases in areas with higher precipitation, either increased annual precipitation or during the driest month. Bears also avoided drier and colder areas. It is argued that "from 66.7% to 75.2% of the potential distribution of the spectacled bear, in all scenarios, was driven by three environmental variables: mean temperature of driest quarter, precipitation of driest month and forest cover. The environmental variables that were among the three of least contribution, in three or more models, were mean diurnal range, temperature seasonality, elevation and distance to water." These findings indicate that "spectacled bear habitats have suffered severe by anthropogenic disturbance," including clearing for farming and pasture, fallow, and secondary vegetation, and other studies have shown human presence and infrastructure affect the spectacled bear presence and create human-bear conflicts (Figueroa et al., 2016; Rojas-VeraPinto et al., 2019). Meza Mori et al.'s (2020) comparison of their predictive model with the IUCN expert-based map produced differences, and a comparison of the predicted model with the map of Natural Protected Areas suggested that the current protected areas are too small to cover most of the important bear habitats.

Another GIS-based model was developed by Kissling et al. (2009: 807–817) in a Patagonian, Argentinian case study, which was "a rule-based habitat model in a Geographic Information System (GIS) to predict the distribution of puma habitat at a regional scale in Patagonia" (Kissling et al., 2009: 807). The study "used empirically derived puma occurrence records from Patagonian ranches 1) to test our regional habitat predictions, and 2) to evaluate if paddock characteristics (vegetation cover, topography, and distance to roads) contribute to explain puma occurrences within ranches" and "simulated three livestock management scenarios differing in their spatial and seasonal allocation of livestock to paddocks and compared the likelihood of livestock exposure to pumas among scenarios" (Kissling et al., 2009: 807). The prediction was that 22% of the study region was suitable for puma home ranges, noting that within ranches "puma occurrences were positively associated with paddock topography, woody vegetation cover on paddocks, and proximity to predicted regional puma habitat."

The rule-based model also predicted that approximately 25% of the study region in western Patagonia would be suitable habitat for pumas, according to the authors. Because most of the large areas of forest and shrubland are found along the eastern side of the Andes, the puma's predicted habitat was concentrated in the southwest of this region (Kissling et al., 2009). Typically, cougar occurrences were more prevalent in enclosures that were closer to predicted habitats, featured a higher proportion of woody vegetation, and exhibited more intricate terrain than one might anticipate given the resources available. On the other hand, there were no discernible differences in distance to roads or the proportion of shrub steppes, grass steppes, and prairies. Examining various potential models for puma presence in enclosures, we discovered that the most effective models consistently took into account the distance to rule-based predicted habitats and the elevation range of the enclosure (Kissling et al., 2009).

Regarding satellite telemetry in global and possibly South American applications, Hofman et al. (2019) note that "satellite telemetry is an increasingly utilized technology in wildlife research, and current devices can track individual animal movements at unprecedented spatial and temporal resolutions. However, as we enter the golden age of satellite telemetry, we need an in-depth understanding of the main technological, species-specific and environmental factors that determine the success and failure of satellite tracking devices across species and habitats." This study was conducted at a global scale, using over 3000 devices applied to the study of 62 terrestrial species in 167 projects, testing GPS fixes and failure rates. The findings suggested that species and unit characteristic applications were more successful than applications to environmental characteristics, and "that the performance of satellite telemetry applications has shown improvements over time, and based on our findings, we provide further recommendations for both users and manufacturers" (Hofman et al., 2019).

Regarding drones, Wirsing et al. (2022) note that "drones have emerged as a popular wildlife research tool, but their use for many species and environments remains untested and research is needed on validation of sampling approaches that are optimised for unpiloted aircraft." Arana (2020) cite the example of developments in the nations containing parts of the Amazon forest. It is argued that "Brazil and Peru are the most relevant actors in terms of the use of drones to safeguard natural and water resources. Since 2016, countries that share the Amazon basin, such as Brazil, Colombia, and Peru, have been developing joint projects for purely ecological purposes to safeguard the Amazon region from threats such as illegal mining, species trafficking, fires, and deforestation." These systems may be used for sophisticated landcover and ecological assessments, although the applications listed are mainly military and law enforcement: "both Brazil and Peru have focused on carrying out joint surveillance projects essentially for the Amazon region through the use of a complex satellite system backed by the use of drones which serve to identify and analyze movements, illicit operations and activities" and "Peru uses its drones to monitor the border areas that it shares with Ecuador and Colombia. In addition to that, it deploys from Lima, specialized drone management and control teams to tackle illegal mining that seriously compromises Peruvian biodiversity."

Regarding camera trapping to detect carnivore presence, Negrões et al. (2010: 1195) used remotely operated cameras to gather data on the puma population and occupancy in a tropical forest area of Brazil, where the cougar population and presence status is relatively unclear. To analyze factors influencing cougar presence, data from five monitoring projects spanning three consecutive years (2005–2007) and two seasons (wet and dry) in a state park and a private forest reserve were utilized. Cougar population and density estimates for the 2007 data were derived from developing a standardized method for individual identification. This identification process relied on (1) persistent features (SP; unchanging physical characteristics) and (2) alterable features (VP; traits that may alter over time, such as scars and botfly markings). After individual recognition, a capture-recapture history was compiled and analyzed using closed population capture-mark-recapture models. The likelihood of capturing a cougar was influenced by camera placement (along roads versus trails), the year of data collection, and prey availability. The occurrence of cougars was linked to increased species diversity, and a relationship was identified between the proportion of cougars and jaguars. A density of approximately 3.4 cougars per 100 km^2 was estimated. Information obtained using camera traps can be employed to approximate the size of cougar populations in tropical forests (Negrões et al., 2010: 1195). The authors conclude that as habitat is being lost and South American cats face heightened vulnerability, these findings underscore the vital significance of private forest preserves for conservation purposes (Negrões et al., 2010: 1195).

Regarding GPS tracking, Palomares et al. (2023) present "a new system for capturing large Neotropical felids. For the first time, a minimally invasive capture system was used to capture a jaguar and facilitate its posterior GPS tracking in a fragment of Atlantic Forest in southeastern Brazil. The jaguar was a 16-year-old adult male, who mainly moved over an area of 175 km^2 in protected areas. The GPS collar worked optimally, and it located the jaguar on 86% of expected occasions" and "using a minimally invasive capture system is more efficient, selective, portable, and less risky for both animals and trappers, and causes less stress to animals compared to other capture methods used for the species." This was termed a "minimally invasive capture system (MICS)" and the jaguar was monitored "using a GPS Vertex Plus Iridium collar with an optimal performance of 86% in expected locations. The jaguar's home range was within the observed range for the species and the animal was primarily maintained in protected areas" mostly within native grassland, marsh, and dense lowland forest. It is concluded that "the use of a MICS for trapping jaguars is a promising technique that shows advantages in terms of efficiency, selectivity, portability, reduced potential risk of injury to animals or trappers, and animal stress compared to other capture methods used for the species" (Palomares et al., 2023).

Using geographical coordinates and human participatory record taking, De Angelo et al. (2011: 534) note that "participatory monitoring… can be a useful tool for obtaining long-term data across large areas," although "their secretive behaviour and low densities make research on, and monitoring of, large carnivores difficult." The methodology was based on the Jaguar Project Monitoring Network (which involved 40 institutions, including governmental, non-governmental, and private organizations and individual farmers, students, and others) and was developed to survey the presence of jaguars and cougars a participatory network of volunteers (320 people trained for the exercise, including park rangers, timber and forestry employees, NGO members, students, researchers, local government staff, army or border security officers, tourism guides and farmers), and a partnership with researchers in the three countries that share the Upper Paraná Atlantic Forest (Argentina, Brazil, and Paraguay) was established (De Angelo et al., 2011).

Researchers collected cat droppings and tracked the footprints of large cats. The study utilized the geographic coordinates of each dataset provided by collectors, local maps, and satellite imagery (De Angelo et al., 2011). With the aid of the kernel density estimation tool, they estimated the area covered and created distribution maps for each species using occurrence points from July 2002 to July 2008. These points identified the presence of cougars or jaguars, accurately marked on the map. The native forest cover was determined using supervised classification of Landsat satellite images from 2004, and protected areas were identified using maps for the Upper Paraná Atlantic Forest region (as mentioned in De Angelo, 2009). Out of the collected data, 61.2% was deemed reliable evidence for the presence of cougars or jaguars (De Angelo et al., 2011).

The findings indicated that jaguars are present in Argentina, Brazil, and Paraguay, while pumas had a more extensive territory and were documented in parts of these countries where jaguars were not recorded (De Angelo et al., 2011). Both species, specifically jaguars, are mainly found in the remaining medium and large forest fragments within this forest. It is argued that despite frequently causing conflicts with local populations, their charm and cultural importance make them symbolic species that inspire volunteer participation and institutional dedication (De Angelo et al., 2011). The results further suggested that cougar sightings were more frequent than jaguar sightings. The cougar was prevalent across the vast majority of the study area, excluding regions with substantial human encroachment. However, jaguars were primarily spotted in the Argentina-Brazil Green Corridor and the largest forest fragments in Brazil and Paraguay. In Brazil, jaguars occur near or within the largest protected areas. In Paraguay, most jaguars were located within these protected areas. The southernmost jaguars in the Upper Paraná Atlantic Forest were recorded in Brazil and Argentina. The cougar's records extended further south in both Argentina and Brazil (De Angelo et al., 2011).

Most cougar and jaguar sightings (about 83% each) occurred in regions dominated by native forests or wetlands, and cougars inhabited more fragmented forests than jaguars (De Angelo et al., 2011). Both species were found in all substantial fragments (over 100 square kilometers), but less than 2% of forest fragments with an area of less than 10 square kilometers demonstrated signs of their presence. Cougars were detected in a higher percentage of smaller and medium-sized fragments than jaguars, and the total area encompassed by fragments with confirmed cougar presence (19,266 square kilometers) was larger than the total area with confirmed jaguar presence (16,585 square kilometers). Jaguars were spotted in seven large forest fragments (over 100 square kilometers) located outside the designated jaguar conservation zones for this region (De Angelo et al., 2011).

Using "spatial predation risk modeling," Miller (2015: 2887) note that "spatial predation risk modeling recently emerged as a tool for predicting and mapping hotspots of livestock depredation using locations where carnivores attacked livestock in the past." Carvalho et al. (2015: 413) present a study using such ecological niche modeling of the jaguar in Brazil livestock depredation and consequent human hunting retaliation. This research utilized field information on livestock predation, alongside environmental and human-induced factors, to develop a model illustrating jaguar predation risk in the Trans-Amazon Highway region, an area significantly impacted by livestock farming-driven deforestation. Five algorithms and a concluding ensemble model were employed to construct a map depicting depredation risk and pinpoint related factors (Carvalho et al. (2015: 413). The variables that displayed the most significant correlation with predation risk were distance to deforestation (with a negative relationship) and cattle density (with a positive relationship). The primary factors driving livestock predation were linked to human-induced disturbance, suggesting that the spatial distribution of risk is not a fixed attribute but rather changes with shifts in deforestation boundaries and livestock herd sizes. The authors propose that an increase in livestock predation in this area is probable, and the application of ecological niche models to identify potential hotspots for livestock damage is an innovative strategy, offering the potential to regulate the distribution of conservation and management efforts across extensive territories at a reasonably low expense (Carvalho et al., 2015: 413).

Conclusions

Jaguars, cougars, and Spectacled bears are the only predators capable of killing cattle in South America. The evidence recorded suggests that jaguars are the most serious predators, followed by cougars and then bears. Factors for cattle predation include the number, size, and strength of predators; the shared ranges of predators and cattle herds; human reactions; and herding patterns. More accurate assessments may be made by distinguishing between actual kills of livestock and scavenging on carcasses. Current evidence suggests that despite South America's status as a major, expanding cattle-producing continent, the predation of cattle may be within manageable limits. Nevertheless, serious conflicts may emerge, due to the increased actions for the preservation of large carnivores, increased numbers of cattle, use of free-ranging methods for cattle herding, and negative human attitudes to large carnivores that develop with real or imagined cattle predation. Effective management of both agricultural and conservation systems and of the relations between local and governmental stakeholders remain necessary for the management of these issues.

GPS, Animal Tracking and Conservation Biology

Possible Applications to the Activities of Dangerous Animals

There are myriad possible applications of GPS, accelerometers, and acoustic recording devices. One method for examining future applications is to look at cases where such sophisticated techniques were not used and identify the parameters that could be measured using such techniques in the future. Three examples are cited: the complexities of human-big cat relations in Central and South America; the urban behavior and spatial ecology of urban cougars in North America; and the spatial conflicts of vultures and airplanes at airports worldwide. In each of these cases, all the types of technological applications discussed above are relevant. Parameters to be measured include the spatial location of the animals, their population clusters or densities, their feeding habits and habitat preferences, and the locational character of their conflicts with people. These parameters, measured with advanced geomatics techniques, may be documented and then compared with human perceptions, the better to find solutions to problems and correct misconceptions based on human error and biases.

Jaguars and Cougars and Human Perceptions

The relationship between jaguars (*Panthera onca*, Linnaeus, 1758) and cougars (*Puma concolor*, Linnaeus, 1771) and people in the Americas is a complex issue that centers on their size and perceived power, and their ability to intrude into human spaces and access human assets, such as safety, livestock, and even human lives. Although North American big cats generally have less violent relations with people than those in Africa and Asia, substantial conflict emerges (Campbell & Torres Alvarado, 2011; Campbell, 2017b). The jaguar, the world's third largest cat and mostly larger than the cougar, has a wide range from the southern United States, southwards to Central and South America and on to northern Argentina (Sanderson et al., 2002; Morrison et al., 2007; Zeller, 2007; McCain & Childs, 2008). Jaguars are generally less studied than lions, tigers, and leopards (Brodie, 2009) with most studies focused on their non-modified habitats (Sanderson et al., 2002; Sollmann et al., 2008; Tôrres et al., 2008; Cavalcanti et al., 2010; Rabinowitz & Zeller, 2010; Campbell & Torres Alvarado, 2011). Most of the problematic episodes between people and jaguars concern the reduction of the latter's habitat through deforestation and the conversion of mixed forest/grass landcover into cultivated and urban landcover. This results in the hunting of the big cats and the extirpation of their prey species through over-hunting and habitat landcover conversion (Conforti & Azevedo, 2003b; Zimmermann et al., 2005b; Caso et al., 2008; Campbell, 2017b). Concerning the justification of human fear for the jaguar, the big cat was feared and worshiped in prehistoric and historic times by Native American tribes; through these times and up to the current era, jaguars were more feared than cougars (Benson, 1998; Saunders, 1998; Santos et al., 2008). The recorded attacks on people are however rare, and less frequent than those by tigers, lions, and leopards in Asia and Africa (Guggisberg, 1975; McDougal, 1987; Bailey, 1993).

The smaller, more widely distributed cougar is similar in size to the leopard (Lopez-Gonzalez & Gonzalez-Romero, 1998b; Campbell & Torres Alvarado, 2011). Having a wider range than the jaguar, the cougar occurs through the whole of the jaguar's range (Scognamillo et al., 2003b). The cougar is considered more elusive, adaptable, and less dangerous to people and possibly livestock (Campbell & Torres Alvarado, 2011; Campbell, 2017b). The perception of the jaguar as more dangerous is mostly due to its larger size and stronger build, and preference for larger-sized prey, including larger livestock predation (Farrell et al., 2000; Polisar et al., 2003b; Scognamillo et al., 2003b; Palmeira et al., 2008b; Rosas-Rosas et al., 2008; Rosas-Rosas & Bender, 2012).

One important application of geomatics, in this case GPS tracking satellite mapping of habitats, concerns the disputed range of this species in the southwestern United States. The jaguar's past range included much of the southwestern United States, but now most records are in northwestern Mexico (Brown &

López-González, 2001; Gutiérrez-González & López-González, 2017b). This Mexican population has the lowest density across the entire range of this species in Central and South America (Gutiérrez-González et al., 2012). The survival rates are mostly unknown, but generally considered to be low, possibly due to the dry climate and open, human-modified, non-forested ecological system (Brown & López-González, 2001; Campbell, 2017b). In this area, there are more records of cougars than jaguars, possibly due to their greater adaptability to both human proximity and open, human-modified landcover (Logan & Sweanor, 2001; Sunquist & Sunquist, 2002b; Campbell, 2017b; Gutiérrez-González & López-González, 2017b).

Another important application for geomatics techniques is the documentation of the relations between jaguars and cougars, especially concerning conflicts, competitive exclusion, differences in prey selection, and reactions to human presence. This would require very detailed studies of location, stress, food conflicts, and avoidance strategies, all of which would be applications of the sophisticated GPS techniques mentioned above. The relationship between the jaguar and cougar creates several ecological intersections such as the possible dominance of the smaller predator (the cougar) by the larger predator, the more adaptable species (the cougar) contrasted with the less adaptable, and possible differences and similarities in prey species between the predatory species. Parameters for the assessment of such relations include avoidance behavior by the smaller predator, through different foraging areas or times, numbers of outright conflicts and the outcomes of such conflicts, or the permanent exclusion of one species from the conflict zones (Donadio & Buskirk, 2006; Ramesh et al., 2012; Vanak et al., 2013). Another parameter concerns how such relations may change with human proximity or interference, which may involve habitat degradation and landcover change, or the killing of one or both predators, or prey species, or the introduction of vulnerable livestock. In the case of large competing predators, the presence of large herbivores is important for the coexistence of large carnivores (Odden et al., 2010; Mitchell & Hebblewhite, 2012; Carter et al., 2015; Gutiérrez-González & López-González, 2017b).

Detailed studies of the parameters of such large carnivore conflict are largely absent (Campbell, 2017b). Competition between jaguars is even disputed (Yáñez et al., 1986; Nowell & Jackson, 1996; Farrell et al., 2000). Competition avoidance has been argued by some, as the jaguar may take larger prey, and the cougar more wide-ranging prey (Jaksic et al., 1981; Rabinowitz & Nottingham, 1986; Iriarte et al., 1990b; Emmons, 1991). Scognamillo et al. (2003b; 269) argued that where jaguars and cougars coexist, "although several studies have focused on the interactions between these two predators, the ecological and behavioral factors that promote their coexistence remain unclear." This study examined the factors for the coexistence of jaguars and cougars in the Venezuelan llanos, finding both presence overlaps and some interspecies avoidance. Jaguars took larger- and medium-sized prey (capybara *Hydrochoerus hydrochaeris,* Linnaeus, 1766, collared peccary *Pecari tajacu,* Linnaeus, 1758) and to a lesser extent spectacled caiman (*Caiman crocodilus,* Linnaeus, 1758) and white-tailed deer (*Odocoileus virginianus,* Zimmermann, 1780). Cougars took medium-sized prey such as the collared peccary and occasionally, caiman. Abundant medium-sized prey and habitat heterogeneity, favored by both species, possibly encouraged the cats' coexistence (Scognamillo et al., 2003b). Nevertheless, Harmsen et al. (2009b: 612) on the Cockscomb Basin of Belize noted that "apart from their overall spatial similarities, jaguars and pumas avoided using the same location at the same time. This interspecific segregation was detectable over and above the spatial and temporal segregation of individual jaguars" (Harmsen et al., 2009b: 612). Evidently, the "physically smaller pumas avoid jaguars"; evidence from Mexico indicated that may even be eaten by jaguars.

Another important parameter concerns the predation by both species on livestock (Campbell, 2017b). This is a key area that would benefit from geomatics and GPS-based studies, as no definitive conclusions have been reached using other means. Therefore, the extent of predation and prey selection is disputed in the literature. As noted by Azevedo (2008, 494) who studied "the last significant population of jaguars in southern Brazil… located inside a protected area, the Iguacu National Park" and pointed out that the "importance of livestock predation on jaguar and puma coexistence is poorly known." Poachers and other types of hunters killed off most of the natural prey of the big cats, including the white-lipped peccary (*Tayassu pecari*, Link, 1795). The dominant prey for jaguars were collared-peccary, coati, and deer (66.7% of prey biomass). Large wild species (50%) and medium-sized wild species (49.5%) comprised the main prey biomass. The dominant prey for cougars were large species (44.0%) and medium-sized species (55.4%) (Azevedo, 2008). Oliveira (1994, 2002) and Rosas-Rosas and Valdez (2010) also found that jaguars preferred peccary species, such as the collared peccary and white-lipped peccary, and cougars preferred deer species (e.g., *Odocoileus* spp.). Livestock predation was one consequence of the decline of these species.

Cattle (*Bos taurus*, Linnaeus, 1758), mostly calves, are predated by both jaguars and cougars. Azevedo's (2008) study found that jaguars were more likely than cougars to prey on livestock, but this was not related to the numerical presence of livestock, hence the conclusion was that the return of large wild ungulates might reduce wildlife predation (Azevedo, 2008: 494). Gutiérrez-González and López-González (2017b) also argued that big cat presence was positively correlated with peccary presence, and jaguar presence was positively correlated with deer presence, with cougar absence. Both jaguars and cougars may prey on livestock, but he apportionment of blame for livestock killings is unclear (Oliveira et al., 2002; Polisar et al., 2003b; Palmeira et al., 2008b; Rosas-Rosas & Bender, 2012).

There is insufficient evidence of interspecies differences in livestock predation and of inter-species differences in adaptation to human reactions to livestock predation (Saenz & Carillo, 2002; Sanderson et al., 2002). In cases, the big cats may be blamed for livestock kills by other species, such as feral dogs (*Canis lupus familiaris*, Linnaeus, 1758), coyotes (*Canis latrans*, Say 1823), and even ocelots (*Leopardus pardalis*, Linnaeus, 1758) (Campbell & Torres Alvarado, 2011). Some studies have used identification techniques that are arguably less effective than the current geomatics techniques, including scat and genetic (deoxyribonucleic acid (DNA)) studies (these may use bile powder for DNA analyses) and animal remains (Höss et al., 1992; Paxinos et al., 1997; Wasser et al., 1997; Farrell et al., 2000). Cameras may also be used to identify individuals, using fur patterns (Sarmiento, 2004; Silver et al., 2004; Salom-Perez et al., 2007; Maffei et al., 2011; RosaRosas & Bender, 2012). These techniques may have limitations, such as the failure to record individual behavioral differences, such as evasiveness and territorial movements, which may differ between males and females (Wallace et al., 2003; Salom-Perez et al., 2007). Surveys of human opinions may also be unreliable (Campbell & Torres Alvarado, 2011).

Another important issue, possibly resolvable by detailed geomatics surveys, was that jaguars and cougars are more likely to share habitat than to avoid each other. There is no conclusive answer using current survey methods. This may be disputed by some as the jaguar's superior strength is perceived as a general factor for competitive exclusion (Donadio & Buskirk, 2006). Some authors point to the competitive exclusion of the cougar, especially in the forests where jaguars attain their largest size (Emmons, 1987; Scognamillo et al., 2003b; Sollmann et al., 2012b). Other authors argue against the competitive exclusion of cougars. Romero-Muñoz et al. (2010) found jaguars were not strongly dominant over cougars in a Bolivian dry forest, possibly due to the latter's greater adaptation to the dry landcover. Gutiérrez-González and López-González (2017b: 7) also determined that "jaguars were not dominant over pumas in our study area" of arid Northern Mexico, unlike the findings of other studies (Novack et al., 2005; Harmsen et al., 2009b; Romero-Muñoz et al., 2010). Gutiérrez-González and López-González's (2017b) hypothesis was that the cougar's superiority in prey selection, and adaptation to environmental variability and human presence, compensates for the jaguar's superiority in strength and thick forest specialization (see also Iriarte et al., 1990b; Polisar et al., 2003b; Scognamillo et al., 2003b; Ruth & Murphy, 2010; Tortato et al., 2015; Hayward et al., 2016b).

Other studies argue that jaguars are less able to cope with human-induced landcover change, due to their generally larger size, preference for denser vegetation, and larger prey preferences (deer *Capreolinae* spp., Brookes, 1828; tapirs (*Tapirus terrestris*, Linnaeus, 1758), peccaries, giant anteaters (*Myrmecophaga tridactyla,* Linnaeus, 1758), capybaras, caimans and turtles (*Podocnemis*, Wagler, 1830 and *Trachemys,* Agassiz, 1857 spp.), and livestock, and very importantly, the more negative public image of the jaguar (Campbell & Torres Alvarado, 2011). However, in the relative absence of detailed, advanced geomatics-based studies, it is difficult to determine the impact of human-induced landcover change on jaguars (Campbell, 2017b). Boron et al. (2016, 1) note that "information on jaguar densities across unprotected landscapes is still scarce and crucially needed to assist management and range-wide conservation strategies." These authors further note that the commonest habitat for jaguars is "wetter and prey-rich habitats such as lowland tropical forests" (Boron et al., 2016: 7). Jaguars are usually rare where there is human hunting activity (Quiroga et al., 2013) and forest degradation (Paviolo et al., 2008), but commoner in areas of cattle rearing (Arispe et al., 2007; Soisalo et al., 2006; Tobler et al., 2013; Boron et al., 2016).

Cougars in Urban Land

Geomatics applications are particularly relevant to cougars in urban substrates, because of the elusive behavior of the urban. The elusive behavior of the urban cougar is extreme to the point of invisibility, and there is insufficient understanding of the cougar's foraging and even reproductive behavior in urban contexts (Campbell, 2012a). Urban cougars are much more invisible than urban bears in North America, and unlike the latter animals, do not target human waste and gardens. Important general issues concerning the urban cougar are: (1) the interactions between cougars and people; (2) the urban substrates that attract or repel cougars; and (3) cougar adaptation to urban landcover (Campbell, 2012b). Recent literature records increasing interactions between people and cougars in urban North America (Beier, 1993; Torres et al., 1996; Weaver et al., 1996; Campbell, 2012a,b). Some cougars use urban substrates for transit rather than habitation (Dickson & Beier, 2002, Dickson et al., 2005). Beier et al. (2010) observe that "probably every adult mountain lion in the coterminous United States has seen humans, crossed paved roads, and encountered human settlements." Cougar penetration into urban substrates is especially common in the cities of Denver/Boulder, Missoula, Los Angeles, and San Diego, which have seen conflicts with people since the 1970s than other cities with similar records of cougar presence (Shuey, 2008).

The topic of cougar presence in urban areas, and the need for more advanced geomatics-based studies on urban behavior and location, is justified largely by social attitudes to this species. Wolch et al. (1997, 106) pointed to a dominant narrative in the American media, which described the cougar as "a disruption to urban life, a nuisance to society, or a threat to humans." Cougars were linked to the themes of "serial killers," premeditated criminal behavior," and "popular worries about rising crime and lawlessness" (ibid., 109). Notably, as argued by Perry (1994), "in any civilization, killers aren't allowed to run loose." Cougars might also be developing a taste for companion animals because they were easier prey to wild animals (Wolch et al., 1997, 110). Large carnivore conservation for social benefit has nevertheless been defended if public support can be garnered (Linnell et al., 2001; Karanth & Chellam, 2009). This links to urban planning and design, which may require detailed ecological information at species and population levels (Fonturbel & Simonetti, 2011; Campbell, 2012a, b). Human attitudes to the big cats must be compared with the facts of aggressive behavior.

As cougars share urban North America with other large mammals, especially black bears, an important research area that would complement geomatics-based tracking of urban predators concerns comparisons of social attitudes to the different species (Campbell, 2012a, b, 2017b). Several studies have found that cougars, compared with black bears, are perceived as more predatory, unpredictable, secretive, and more a threat to humans especially children (Davis et al., 2001; Campbell & Lancaster, 2010; Campbell, 2012a). Compared to brown or grizzly bears, the cougar is, however, less feared (Kellert et al., 1996: 5). This is because of the brown bear's larger size, reputed ferocity, and greater visibility, although the threat of an attack from an elusive cougar is often rated at similar levels (Pelton et al., 1976; Kellert,

1994; Campbell, 2012a). Brown bears rarely penetrate urban substrates, so the comparison is usually confined to rural or conservation areas (Campbell, 2017b).

An important parameter for geomatics studies would be to investigate if such conflicts emerge from human penetration of cougar habitats through urbanization and suburban developments, or if it is the cougars that adapt to human presence and invade previously unsuitable land. Traditionally, the latter view was more popular; urban expansion was to blame (Spencer et al., 2001, Dickson & Beier, 2002; Thornton & Quinn, 2009). Studies have argued that all types of cougars, not only sub-adults and transients occur in exurban, suburban, and residential areas (Shuey, 2008). Some studies veer from the traditionalist view and argue for cougar adaptation to human presence. For example, Keston et al. (2013) argue that in some cases where sub-adult cougars penetrate urban landscapes more frequently than adult cougars, it may be due to their attempt to establish home ranges near their previous territory (see also Beier et al., 2010; Logan & Sweanor, 2001; Maehr et al., 2002; Ruth, 1991; Van Dyke et al., 1986). Kertson et al. (2013) further attribute cougar presence in urban areas to individualist rather than population-level behavior, which makes prediction difficult. This unpredictable, elusive behavior of the cougar contrasts with that of the urban black bear, which predictably invades urban substrates for food; seasonal fruits, urban green spaces, and organic garbage (Merkle et al., 2013). Black bears are much more frequently linked to property damage and general nuisance behavior (Decker et al., 1981; Herrero & Higgins, 1999; Freedman et al., 2003; Morzillo et al., 2007).

Another important, potentially measurable parameter concerns the correlations between cougar presence and variations in urban landcover (Campbell, 2012a, b). Put simply, do cougars prefer some parts of cities more than others? This is relevant to urban planning if the urban areas preferred or avoided by cougars are planned for future development. Important externalities would be killings of prey species and companion animals and changing public attitudes to cougar presence in these different landcover types. Generalizable hypotheses and findings from the current literature cluster around three main points. Firstly, cougars appear to prefer urban green spaces to dense urban substrates, especially when connected along corridors with favorable cover, and conflicts may emerge as these green spaces are also valued for human recreation health. Secondly, as cougars usually do not scavenge (unlike bears), their motives are more like predatory, justifying concerns for companion animals and children in such recreational, residential, and transit areas (Campbell & Lancaster, 2010; Campbell, 2012a). The third hypothesis or finding emerges from the first two, in that these issues may affect human quality of life, through human fear and location avoidance and fuel negative public attitudes to cougars. Geomatics-based surveys in this regard would provide the definitive evidence to either support or deny the veracity of these perspectives, but most studies have yet to use the advanced methods described in this chapter (Campbell, 2017b).

Taking the case of urban green spaces, as likely the preferred cougar urban habitat, there are several relevant parameters. Urban green spaces are areas with variable proportions of trees, shrubs, and grasses, surrounding urban areas (greenbelts), inside urban areas (greenhearts), and in scattered areas within an urban area (in the case of smaller, isolated green spaces) (Kuhn, 2003; Campbell, 2009). Urban green areas are mostly recorded as expanding, due to the increased rates of urbanization, the recreational and health interests of urban dwellers, and wildlife conservation (Bird et al., 1996; Belant, 1997; Jokimaki, 1999, 2003; Pirnat, 2000; Fernandez-Juricic, 2000; Savard et al., 2000; Chiari & Seeland, 2004; Cornelis & Hermy, 2004; Erickson, 2004).

An important parameter for tracked cougars would be the duration of the time they spend in the green spaces under study, after locating them and recording their movements. This could be covered by the latest GPS technology, as outlined by Augliere (2015), Wilmers et al. (2015), and Lynch et al. (2014). This is important because cougars in urban North America may be classified as transients, occasional inhabitants, or even permanent residents (Campbell, 2012b). Cougars may use urban green spaces for hunting, transit, territorial marking, relaxing, family support, and/or breeding, largely due to the presence of cover, and prey species (Burdett et al., 2010; White et al., 2011; Keston et al., 2011a, b, 2013). If time and location are important parameters, these must also be linked to green space size and vegetation configuration, two issues largely neglected in the literature (Campbell, 2017). Studies have noted that cougars prefer green spaces and tend to frequent areas with higher heterogeneity, including both green and built landcover (Campbell, 2017, 2012a)

These studies complicate the green space hypothesis. Shuey (2008) found no linear relation between cougar presence and green areas, as there were more cougar-human encounters in grassland and suburban landcover than there were in the preferred cougar habitat of evergreen tree growth, even when such tree growth covered more land than the suburban land-cover. In heterogeneous landcover, more cougar-human encounters occurred near human-modified land cover distant from natural cougar habitat. Therefore, urban centers with highly heterogeneous landcover might have more cougar-human conflicts in developed areas. Cougars may use residential landcover when green spaces and corridors are absent, rapidly moving between green spaces through urban substrates by night to reduce (Kertson et al., 2011b). This allows a "high potential for coexistence" between people and cougars (Kertson et al., 2013: 1). Kertson et al. (2013) further argue that "cougars and people appear to coexist better than previously perceived and use of areas close to people is likely to continue if these environments provide ample stalking and security cover, adequate prey resources, and limited human interference." This adaptive behavior is important because it disproves the common assumption correlations between green spaces and urban cougar presence and may impact human quality of life and human-cougar conflicts in non-green space areas (Campbell, 2012a).

References

Arana, F. P. (2020). The Geostrategic use of Drones in South America. https://www.wgi.world/the-geostrategic-use-of-drones-in-south-america/

Baumann, M., Israel, C., Piquer-Rodriguez, M., Gavier-Pizarro, G., Volante, J. N., & Kuemmerle, T. (2017). Deforestation and cattle expansion in the Paraguayan Chaco. *Regional Environmental Change, 17*, 1179–1191.

Brannstrom, C., Jepson, W., Filippi, A. M., Redo, D., Xu, Z., & Ganesh, S. (2008). Land change in the Brazilian Savanna (Cerrado), 1986 – 2002: Comparative analysis and implications for land-use policy. *Land Use Policy, 25*, 1986–2002.

Brown, S. (2022). Cattle boom in Brazil's Acre spells doom for Amazon rainforest, activists warn. https://news.mongabay.com/2022/01/cattle-boom-in-brazils-acre-spells-doom-for-amazon-rainforest-activists-warn/

Bueno-Cabrera, A. (2004). *Impacto del puma (Puma concolor) en ranchos ganaderos del área natural protegida "Cañón de Santa Elena"*. Tesis de Maestría. Instituto de Ecología, A.C. Xalapa, Veracruz, México.

Campbell, M. (2017a). *Biological conservation in the 21st century: A conservation biology of large wildlife*. Nova Science Publishers.

Campbell, M. (2022). The two tigers: The Jaguar ('El Tigre') of the Americas and the Tiger of Asia. In K. K. Jha & M. Campbell (Eds.), *Dynamics and interrelations between nature, science, and society* (pp. 83–123). Nova Science Publishers.

Carvalho, E. A., Zarco-González, M. M., Monroy-Vilchis, O., & Morato, R. G. (2015). Modeling the risk of livestock depredation by jaguar along the Transamazon highway, Brazil. *Basic and Applied Ecology, 16*(5), 413–419. https://doi.org/10.1016/j.baae.2015.03.005

Castellanos, A. (2011). Do Andean bears attack mountain tapirs. *International Bear News, 20*(4), 41–42.

Castellanos, A. (2019). The threat of road-kill to Andean tapirs: The case of 'Jorgito', the Andean tapir that lived beside the Quito-Amazon highway, Ecuador. Boletín técnico 14. *Serie Zoológica, 14–15*, 1–5.

Castellanos, A., Altamirana, M., & Tapia, G. (2001). *Ecología y Comportamiento de Osos Andinos Reintroducidos en la Reserva Biológica Maquipucuna*. Implicaciones en Conservación.

Cavalcanti, S. M. C., & Gese, E. M. (2010a). Kill rates and predation patterns of jaguars (Panthera onca) in the southern Pantanal, Brazil. *Journal of Mammalogy, 91*(3), 722–736. https://doi.org/10.1644/09-MAMM-A-171.1

Cavalcanti, S. M. C., Marchini, S., Zirnmerrnann, A., Gese, E. M., & Macdonald, D. W. (2010). Jaguars, livestock and people in Brazil: Realities and perceptions behind the conflict. In D. Macdonald & A. Loveridge (Eds.), *The biology and conservation of wild felids* (pp. 383–402). Oxford University Press.

Cerri, C. E., Cerri, C. C., Maia, S. M., Cherubin, M. R., Feigl, B. J., & Lal, R. (2018). Reducing Amazon deforestation through agricultural intensification in the Cerrado for advancing food security and mitigating climate change. *Sustainability, 10*(4), 989. https://doi.org/10.3390/su10040989

Conforti, V. A., & Azevedo, F. C. (2003a). Local perceptions of jaguars (Panthera onca) and pumas (Puma concolor) in the Iguaçu National Park area, South Brazil. *Biological Conservation, 111*(2), 215–221. https://doi.org/10.1016/S0006-3207(02)00277-X

Crawshaw, P. G., & Quigley, H. B. (2002). Hábitos alimentarios del jaguar y el puma en el Pantanal, Brasil, con implicaciones para su manejo y conservación. In R. Medellín, C. Equihua, C.-L. Chetkiewicz, P. Crawshaw, A. Rabinowtz, K. Redford, J. Robinson, E. Sanderson, & A. Taber (Eds.), *El Jaguar en el Nuevo Milenio, 223–235*. Universidad Nacional Autonoma de Mexico, Wildlife Conservation Society, and Fondo de Cultura Economica.

Crawshaw, P. G., & Quigley, H. B. (1984). A ecologia do jaguar ou onça pintada no Pantanal. In *Relatório Final* (p. 112). Instituto Brasileiro De Desenvolvimento Florestal.

Crawshaw, P. G., & Quigley, H. B. (1991). Jaguar spacing, activity and habitat use in a seasonally flooded environment in Brazil. *Journal of Zoology, 223*(3), 357–370. https://doi.org/10.1111/j.1469-7998.1991.tb04770.x

Cuesta, F., & Suárez, L. (2001). *Oso de anteojos (Tremarctos ornatus)*. SIMBIO/EcoCiencia/Ministerio del Ambiente/UICN.

Cullen Jr., L. (2006). *Jaguar as landscape detectives for the conservation inthe Atlantic Forest of Brazil*, Ph.D. thesis University of Kent, Cantebury, UK.

da Silva, J. A. R., Garcia, A. R., de Almeida, A. M., Bezerra, A. S., & de Brito Lourenço Junior, J. (2021). Water buffalo production in the Brazilian Amazon Basin: A review. *Tropical Animal Health and Production, 53*(3), 343. https://doi.org/10.1007/s11250-021-02744-w

De Angelo, C., Paviolo, A., Rode, D., Cullen, L., Sana, D., Abreu, K. C., et al. (2011). Participatory networks for large-scale monitoring of large carnivores: Pumas and jaguars of the Upper Paraná Atlantic Forest. *Oryx, 45*(4), 534–545. https://doi.org/10.1017/S0030605310000840

Debaky, H. A., Kutchy, N. A., Ul-Husna, A., Indriastuti, R., Akhter, S., Purwantara, B., & Memili, E. (2019). Review: Potential of water buffalo in world agriculture: Challenges and opportunities. *Applied Animal Science, 35*(2), 255–268. https://doi.org/10.15232/aas.2018-01810

Elbroch, L. M., & Kusler, A. (2018). Are pumas subordinate carnivores, and does it matter? *PeerJ, 6*, e4293. https://doi.org/10.7717/peerj.4293

Faddul, J. (2024). *Investigating 'Cattle Laundering' and deforestation in the Amazon: Interview with the 2024 Goldman Prize Winner*. Global Investigative Journalism Network. https://gijn.org/stories/investigating-cattle-laundering-deforestation-amazon-goldman-prize-2024/

FAO. (2019). *Food and Agriculture Organization of the United Nations*. FAOSTAT statistics database. Retrieved from http://www.fao.org/faostat/en/.

FAO. (2020). *Forest Resources Assessment*. Food and Agriculture Organization of the United Nations.

Farrell, L. E. (1999). *The ecology of the puma and the jaguar in the Venezuelan Llanos*. MSc thesis, University of Florida, Gainesville, USA.

Figueroa, J., Stucchi, M., & Rojas-VeraPinto, R. (2016). Modelación de la distribución del oso andino Tremarctos ornatus en el bosque seco del Marañón (Perú). *Revista Mexicana de Biodiversidad, 87*(1), 230–238.

Finer, M., Novoa, S., Weisse, M. J., Petersen, R., Mascaro, J., Souto, T., Stearns, F., & Martinez, R. G. (2018). Combating deforestation: From satellite to intervention. *Science (New York, N.Y.), 360*(6395), 1303–1305. https://doi.org/10.1126/science.aat1203

Food and Agriculture Organization of the United Nations (2023) – with major processing by Our World in Data. "Number of cattle – FAO" [dataset]. Food and Agriculture Organization of the United Nations, "Production: Crops and livestock products" [original data]. Retrieved August 16, 2024 from https://ourworldindata.org/grapher/cattle-livestock-count-heads

Foster, V. C., Sarmento, P., Sollmann, R., Torres, N., Jacomo, A. T. A., Negrões, N., Fonseca, C., & Silveira, L. (2013). Jaguar and Puma activity patterns and predator-prey interactions in four Brazilian biomes. *Biotropica, 45*(3), 373–379. https://doi.org/10.1111/btp.12021

Galasso, L. (2002). *The spectacled bear's impact on livestock and crops and use of remnant forest fruit trees in a human altered landscape in Ecuador*. Thesis, University of Wisconsin, Madison, Wisconsin, USA.

García-Rangel, S. (2012). Andean bear Tremarctos ornatus natural history and conservation. *Mammal Review, 42*, 85–119.

Garshelis, D. L. (2004). Variation in ursid life histories: Is there an outlier? In D. Lindburg & K. Baragona (Eds.), *Giant pandas. Biology and conservation* (pp. 53–73). University of California Press.

Gasparri, N. I., Grau, H. R., & Gutierrez Angonese, J. (2013). Linkages between soybean and neotropical deforestation: Coupling and transient decoupling dynamics in a multi-decadal analysis. *Global Environmental Change, 23*, 1605–1614.

Goldstein, I. (1991). Spectacled bear predation and feeding behavior on livestock in Venezuela. *Studies on Neotropical Fauna and Environment, 26*(4), 231–235. https://doi.org/10.1080/01650529109360857

Goldstein, I. (2002). Spectacled bear–cattle interactions and tree nest use in Bolivia and Venezuela. *Ursus, 13*, 369–372.

Goldstein, I., Cuesta, F., Poveda, J., Torres, D., Jorgenson, J., & Paisley, S. (2002). *Spectacled bear-cattle conflicts: A regional perspective*. Conference: 14th International Conference on Bear Research and Management, July 28 – August 3. Steinkjer, Norway.

Goldstein, I., Paisley, S., Wallace, R., Jorgenson, J., Cuesta, F., & Castellanos, A. (2009). Andean bear-livestock conflicts: A review. *Ursus, 17*, 8–15. https://doi.org/10.2192/1537-6176(2006)17[8:ABCAR]2.0.CO;2

Grau, H. R., Torres, R., Gasparri, N. I., Blendinger, P. G., Marinaro, S., & Macchi, L. (2015). Natural grasslands in the Chaco. A neglected ecosystem under threat by agriculture expansion and forest-oriented conservation policies. *Journal of Arid Environments, 123*, 40–46.

Grau, R., & Aide, T. M. (2008). Globalization and land-use transitions in Latin America. *Ecology and Society, 13*(2), 16.

Guerisoli, M. M., Gallo, O., Martinez, S., Luengos Vidal, E. M., & Lucherini, M. (2021). Native, exotic, and livestock prey: Assessment of puma Puma concolor diet in South American temperate region. *Mammal Research, 66*, 33–43. https://doi.org/10.1007/s13364-020-00549-0

Gutiérrez-González, C. E., & López-González, C. A. (2017a). Jaguar interactions with pumas and prey at the northern edge of jaguars' range. *PeerJ, 5*, e2886.

Haines, A. M. (2006). Is there competition between sympatric jaguar Panthera onca and puma Puma concolor? *Acta Zoologica Sinica, 52*(6), 1142–1147.

Harmsen, B. J., Foster, R. J., Silver, S. C., Ostro, L. E., & Doncaster, C. P. (2009a). Spatial and temporal interactions of sympatric jaguars (Panthera onca) and pumas (Puma concolor) in a neotropical forest. *Journal of Mammalogy, 90*, 612–620.

Hayward, M. W., Kamler, J. F., Montgomery, R. A., Newlove, A., Rostro-García, S., Sales, L. P., & Van Valkenburgh, B. (2016a). Prey preferences of the Jaguar Panthera onca reflect the post-Pleistocene demise of large prey. *Frontiers in Ecology and Evolution, 3*, 148. https://doi.org/10.3389/fevo.2015.00148

Heilmayr, R., Rausch, L. L., Munger, J., & Gibbs, H. K. (2020). Brazil's Amazon Soy Moratorium reduced deforestation. *Nature Food, 1*, 801–810. https://doi.org/10.1038/s43016-020-00194-5

Hofman, M. P. G., Hayward, M. W., Heim, M., Marchand, P., Rolandsen, C. M., Mattisson, J., Urbano, F., Heurich, M., Mysterud, A., Melzheimer, J., Morellet, N., Voigt, U., Allen, B. L., Gehr, B., Rouco, C., Ullmann, W., Holand, Ø., Jørgensen, N. H., Steinheim, G., Cagnacci, F., et al. (2019). Right on track? Performance of satellite telemetry in terrestrial wildlife research. *PLoS One, 14*(5), e0216223. https://doi.org/10.1371/journal.pone.0216223

Hoogesteijn, R., & Hoogesteijn, A. (2008). Conflicts between cattle ranching and large predators in Venezuela: Could use of water buffalo facilitate felid conservation? *Oryx, 42*(01), 132. https://doi.org/10.1017/s0030605308001105

Hoogesteijn, R., Hoogesteijn, A., & Mondolfi, E. (1993). Jaguar predation and conservation: Cattle mortality caused by felines on three ranches in the Venezuelan Llanos. In N. Dunstone & M. L. Gorman (Eds.), *Mammals as predators* (pp. 391–407). Symposia of the Zoological Society of London.

Iriarte, J. A., Franklin, W. L., Johnson, W. E., & Redford, K. H. (1990a). Biogeographic variation of food habits and body size of the America puma. *Oecologia, 85*(2), 185–190. https://doi.org/10.1007/bf00319400

Jampel, C. (2016). Cattle-based livelihoods, changes in the taskscape, and human–bear conflict in the Ecuadorian Andes. *Geoforum, 69*, 84–93. https://doi.org/10.1016/j.geoforum.2016.01.001

Janson, C. H., & Emmons, L. H. (1990). Ecological structure of the nonflying mammal community at Cocha Cashu Biological Station, Manu National Park, Peru. In A. H. Gentry (Ed.), *Four neotropical forests* (pp. 314–338). Yale University Press.

Jędrzejewski, W., Robinson, H. S., Abarca, M., Zeller, K. A., Velasquez, G., Paemelaere, E. A. D., et al. (2018). Estimating large carnivore populations at global scale based on spatial predictions of density and distribution – Application to the jaguar (Panthera onca). *PLoS One, 13*(3), e0194719. https://doi.org/10.1371/journal.pone.0194719

Karandikar, H., Serota, M. W., Sherman, W. C., Green, J. R., Verta, G., Kremen, C., & Middleton, A. D. (2022). Dietary patterns of a versatile large carnivore, the puma (Puma concolor). *Ecology and Evolution, 12*(6), e9002. https://doi.org/10.1002/ece3.9002

Kattan, G., Hernández, O. L., Goldstein, I., Rojas, V., Murillo, O., Gómez, C., Restrepo, H., & Cuesta, F. (2004). Range fragmentation in the spectacled bear Tremarctos ornatus in the northern Andes. *Oryx, 38*(2), 155–163. https://doi.org/10.1017/s0030605304000298

Kelly, M. J., Noss, A. J., Di Bitetti, M., Maffei, L., Arispe, L. R., Paviolo, A., De Angelo, C. D., & Di Blanco, Y. E. (2008). Estimating puma densities from camera trapping across three study sites: Bolivia, Argentina, and Belize. *Journal of Mammalogy, 89*(2), 408–418.

Khorozyan, I., Ghoddousi, A., Soofi, M., & Waltert, M. (2015). Big cats kill more livestock when wild prey reaches a minimum threshold. *Biological Conservation, 192*, 268–275. https://doi.org/10.1016/j.biocon.2015.09.031

Kimbrough, L. (2022). Beef is still coming from protected areas in the Amazon, study shows. *Mongabay*. https://news.mongabay.com/2022/10/beef-is-still-coming-from-protected-areas-in-the-amazon-study-shows/

Kissling, W. D., Fernández, N., & Paruelo, J. M. (2009). Spatial risk assessment of livestock exposure to pumas in Patagonia, Argentina. *Ecography, 32*, 807–817. https://doi.org/10.1111/j.1600-0587.2009.05781.x

Laundré, J. W., & Hernández, L. (2010). What we know about pumas in Latin America. In M. Hornocker & S. Negri (Eds.), *Cougar ecology and conservation* (pp. 76–90). University of Chicago Press.

Lizcano, D., González, T. M., Sandoval-A, S., Peña, J., Galvis, C., Mosquera-Guerra, F., Mantilla-Meluk, H., Mosquera-Guerra, F., Trujillo, F., Lizcano, D., Sandoval, S., González, T. M., & Londoño, M. C. (2017). Características de la danta de montaña (Tapirus pinchaque).Corporación Autónoma Regional del Quindío (CRQ), Armenia, Quindío, Colombia. In *Plan de manejo para la conservación de la danta de montaña (Tapirus pinchaque) en el departamento del Quindío* (pp. 10–15).

Llanos, R. P., & Travaini, A. (2020). Diet of puma (Puma concolor) in sheep ranches of Central Patagonia (Argentina). *Journal of Arid Environments, 177*, 104145. https://doi.org/10.1016/j.jaridenv.2020.104145

López-González, C., & González-Romero, G. (1998a). A synthesis of current literature and knowledge about the ecology of the puma (Puma concolor). *Acta Zoologica Mexicana, 75*, 171–190.

López-Ordóñez, J. P., Gómez-Lora, I., & Ramírez-Chaves, H. E. (2020). A record of scavenging behavior of the Andean bear (Tremarctos ornatus) on mountain tapir (Tapirus pinchaque) in southwestern Colombia. *International Bear News (2020), 29*(2), 27–28.

Lucherini, M., & Merino, M. J. (2008). Perceptions of human–carnivore conflicts in the high Andes of Argentina. *Mountain Research and Development, 28*(1), 81–85. https://doi.org/10.1659/mrd.0903

Marchini, S., & Macdonald, D. W. (2012). Predicting ranchers' intention to kill jaguars: Case studies in Amazonia and Pantanal. *Biological Conservation, 147*(1), 213–221. https://doi.org/10.1016/j.biocon.2012.01.002

Mazzini, F., Relva, M. A., & Malizia, L. R. (2018). Impacts of domestic cattle on forest and woody ecosystems in southern South America. *Plant Ecology, 219*, 913–925. https://doi.org/10.1007/s11258-018-0846-y

Mazzolli, M., Graipel, M. E., & Dunstone, N. (2002). Mountain lion depredation in southern Brazil. *Biological Conservation, 105*(1), 43–51. https://doi.org/10.1016/S0006-3207(01)00178-1

Meza Mori, G., Barboza Castillo, E., Torres Guzmán, C., Cotrina Sánchez, D. A., Guzman Valqui, B. K., Oliva, M., Bandopadhyay, S., Salas López, R., & Rojas Briceño, N. B. (2020). Predictive modelling of current and future potential distribution of the spectacled bear (Tremarctos ornatus) in Amazonas, Northeast Peru. *Animals, 10*, 1816. https://doi.org/10.3390/ani10101816

Michalski, F., Boulhosa, R. L. P., Faria, A., & Peres, C. A. (2006). Human-wildlife conflicts in a fragmented Amazonian forest landscape: Determinants of large felid depredation on livestock. *Animal Conservation, 9*(2), 179–188. https://doi.org/10.1111/j.1469-1795.2006.00025.x

Miller, J. R. B. (2015). Mapping attack hotspots to mitigate human–carnivore conflict: Approaches and applications of spatial predation risk modeling. *Biodiversity and Conservation, 24*, 2887–2911. https://doi.org/10.1007/s10531-015-0993-6

Miranda, E. B. P., de Jácomo, A. T. A., Tôrres, N. M., Alves, G. B., & Silveira, L. (2018). What are jaguars eating in a half-empty forest? Insights from diet in an overhunted Caatinga reserve. *Journal of Mammalogy, 99*(3), 724–731. https://doi.org/10.1093/jmammal/gyy027

Muller, R., Muller, D., Schierhorn, F., Gerold, G., & Pacheco, P. (2012). Proximate causes of deforestation in the Bolivian lowlands: An analysis of spatial dynamics. *Regional Environmental Change, 12*, 445–459.

Nanni, A. S., Teel, T., & Lucherini, M. (2020). Predation on livestock and its influence on tolerance toward pumas in agroecosystems of the Argentine Dry Chaco. *Human Dimensions of Wildlife, 26*(5), 429–444. https://doi.org/10.1080/10871209.2020.1843742

National Wildlife Federation. (2023). A path towards zero deforestation. https://international.nwf.org/cattle/

Negrões, N., Sarmento, P., Cruz, J., Eira, C., Revilla, E., Fonseca, C., Sollmann, R., Tôrres, N. M., Furtado, M. M., Jácomo, A. T. A., & Silveira, L. (2010). Use of camera-trapping to estimate puma density and influencing factors in Central Brazil. *Journal of Wildlife Management, 74*(6), 1195–1203. https://doi.org/10.2193/2009-256

Nepstad, D., McGrath, D., Stickler, C., Alencar, A., Azevedo, A., Swette, B., Bezerra, T., DiGiano, M., Shimada, J., Seroa da Motta, R., Armijo, E., Castello, L., Brando, P., Hansen, M. C., McGrath-Horn, M., Carvalho, O., & Hess, L. (2014). Slowing Amazon deforestation through public policy and interventions in beef and soy supply chains. *Science (New York, N.Y.), 344*(6188), 1118–1123. https://doi.org/10.1126/science.1248525

Nepstad, D., Soares-Filho, B. S., Merry, F., Lima, A., Moutinho, P., Carter, J., et al. (2009). The end of deforestation in the Brazilian Amazon. *Science, 326*(5958), 1350–1351.

Nielsen, C., Thompson, D., Kelly, M., & Lopez-Gonzalez, C. A. (2015). Puma concolor (errata version published in 2016). The IUCN Red List of Threatened Species 2015: e.T18868A97216466. https://doi.org/10.2305/IUCN.UK.2015-4.RLTS.T18868A50663436.en

Nunez, R., Miller, B., & Lindzey, F. (2000). Food habits of jaguars and pumas in Jalisco, Mexico. *Journal of Zoology, 252*(3), 373–379. https://doi.org/10.1111/j.1469-7998.2000.tb00632.x

Ometto, J. P., Gorgens, E. B., de Souza Pereira, F. R., Sato, L., de Assis, M. L. R., Cantinho, R., Longo, M., Jacon, A. D., & Keller, M. (2023). A biomass map of the Brazilian Amazon from multisource remote sensing. *Scientific Data, 10*(1), 668. https://doi.org/10.1038/s41597-023-02575-4

Paisley, S. (2001). *Andean bears and people in Apolobamba, Bolivia: culture, conflict and conservation*. Dissertation, Durrel Institute of Conservation and Ecology, University of Kent, Canterbury, UK.

Palmeira, F. B. L., Crawshaw, P. G., Haddad, C. M., Ferraz, K. M. P. M. B., & Verdade, L. M. (2008a). Cattle depredation by puma (Puma concolor) and jaguar (Panthera onca) in Central-Western Brazil. *Biological Conservation, 141*(1), 118–125. https://doi.org/10.1016/j.biocon.2007.09.015

Palmeira, F. B., Trinca, C. T., & Haddad, C. M. (2015). Livestock predation by puma (Puma concolor) in the highlands of a southeastern Brazilian Atlantic Forest. *Environmental Management, 56*(4), 903–915. https://doi.org/10.1007/s00267-015-0562-5

Palomares, F., de Paula, T. A. R., & Srbek-Araujo, A. C. (2023). First capture of a Jaguar using a minimally invasive capture system for GPS tracking in an isolated patch of Atlantic Forest in Southern Brazil. *Animals, 13*(21), 3314. https://doi.org/10.3390/ani13213314

Perovic, P. G. (2002) *Ecología de la comunidad de félidos en las selvas nubladas del noroeste argentino*. Tesis Doctoral. Universidad Nacional de Córdoba.

Peyton, B. (1980). Ecology, distribution, and food habits of spectacled bears, Tremarctos ornatus, in Peru. *Journal of Mammalogy, 61*(4), 639–652. https://doi.org/10.2307/1380309

Peyton, B. (1981). Spectacled bears in Peru. *Oryx, 16*(1), 48–56. https://doi.org/10.1017/S0030605300016756

Peyton, B. (1999). Spectacled bear conservation action plan. In C. Servheen, S. Herrero, & B. Peyton (Eds.), *Bears. Status survey and conservation action plan* (pp. 157–164). IUCN/SSC Bear and Polar Bear Specialist Groups.

Pisso-Florez, G. A., Gómez-Lora, I., Vela-Vargas, I. M., Pizo, H., Bedoya Dorado, I., & Ramírez-Chaves, H. E. (2021). What's on the menu? A presumed attack of Andean bear on a Mountain tapir at the Puracé National Natural Park, Colombia. *Neotropical Biology and Conservation, 16*(1), 19–25. https://doi.org/10.3897/neotropical.16.e57140

Pitman, M. R. P. L., Oliveira, T. D., Paula, R. D., & Indrusiak, C. (2002). *Manual de identificação, prevenção e controle de predação por carnívoros* (p. 83). Edições IBAMA.

Polisar, J., Maxit, I., Scognamillo, D., Farrell, L., Sunquist, M. E., & Eisenberg, J. F. (2003a). Jaguars, pumas, their prey base, and cattle ranching: Ecological interpretations of a management problem. *Biological Conservation, 109*(2), 297–310. https://doi.org/10.1016/s0006-3207(02)00157-x

Quigley, H., Foster, R., Petracca, L., Payan, E., Salom, R., & Harmsen, B. (2017). Panthera onca (errata version published in 2018). The IUCN Red List of Threatened Species 2017: e.T15953A123791436. https://doi.org/10.2305/IUCN.UK.2017-3.RLTS.T15953A50658693.en

Rajão, R. G., Moutinho, P., & Soares, L. (2017). The rights and wrongs of Brazil's forest monitoring systems. *Conservation Letters, 10*(4), 495–496. https://doi.org/10.1111/conl.12350

Rodriguez, A., Gomez, R., Moreno, A., Cuellar, C., & Lizcano, D. J. (2014). Record of a mountain tapir attacked by an Andean bear on a camera trap. *Tapir Conservation, 23*, 25–26.

Rojas-VeraPinto, R., Butrón, R., & Martel, C. (2020). Reports of feeding incidents of cattle by andean bear (Tremarctos ornatus) in Central Peru. *Revista Mexicana de Mastozoología, nueva época, 10*(2), 25–32.

Rojas-VeraPinto, R., Zegarra, R. E., Gutiérrez, R., & Beraún, Y. (2019). Living with the Andean Bear in Peru: Diagnosis and Guidelines for the Amnejo of Human-Bear Conflicts. FZS Peru: Cusco, Peru.

Ruiz Garcia, M. (2003). Molecular population genetic analysis of the spectacled bear (Tremarctos ornatus) in the northern Andean area. *Hereditas, 138*, 81–93.

Salas, E. B. (2024). Forest land area in Latin America & the Caribbean 2010–2021. https://www.statista.com/statistics/1196711/forest-land-area-latin-america-caribbean-region/

Schulz, F., Printes, R. C., & Oliveira, L. R. (2014). Depredation of domestic herds by pumas based on farmer's information in Southern Brazil. *Journal of Ethnobiology and Ethnomedicine, 10*, 73. https://doi.org/10.1186/1746-4269-10-73

Scognamillo, D., Maxit, I. E., Sunquist, M., & Polisar, J. (2003a). Coexistence of jaguar (Panthera onca) and puma (Puma concolor) in a mosaic landscape in the Venezuelan llanos. *Journal of Zoology, 259*(3), 269–279. https://doi.org/10.1017/s0952836902003230

Silva Junior, C. H. L., Pessôa, A. C. M., Carvalho, N. S., & Reis, J. B. C. (2021). Anderson, L.O. & Aragão, L.E.O.C. 2021. The Brazilian Amazon deforestation rate in 2020 is the greatest of the decade. *Nature Ecology & Evolution, 5*, 144–145. https://doi.org/10.1038/s41559-020-01368-x

Silveira, L., Boulhosa, R. L., Astete, S., & Almeida, A. (2008). Management of domestic livestock predation by Jaguars in Brazil. Cat News Spec. Issue. 4.

Slough, T., Kopas, J., & Urpelainen, J. (2021). Satellite-based deforestation alerts with training and incentives for patrolling facilitate community monitoring in the Peruvian Amazon. *Proceedings of the National Academy of Sciences of the United States of America, 118*(29), e2015171118. https://doi.org/10.1073/pnas.2015171118

Sollmann, R., Furtado, M. M., Hofer, H., Jácomo, A. T. A., Tôrres, N. M., & Silveira, L. (2012a). Using occupancy models to investigate space partitioning between two sympatric large predators, the jaguar and puma in Central Brazil. *Mammalian Biology, 77*(1), 41–46. https://doi.org/10.1016/j.mambio.2011.06.011

Sunquist, M., & Sunquist, F. (2002a). *Wild cats of the world*. University of Chicago Press.

Sutherland-Smith, M. (2015). Suidae and Tayassuidae (wild pigs, peccaries). *Fowler's Zoo and Wild Animal Medicine, 8*, 568–584. https://doi.org/10.1016/B978-1-4557-7397-8.00058-X

Taber, A. B., Novaro, A. J., Neris, N., & Colman, F. (1997). The food habits of sympatric Jaguar and puma in the Paraguayan Chaco. *Biotropica, 29*(2), 204–213. https://doi.org/10.1111/j.1744-7429.1997.tb00025.x

Torres, D., Lobo, A., Ascanio, R., & Lobo, G. (1995). Monitoring the spectacled bear (Tremarctos ornatus) populations in the watershed of the Capaz River, Mérida, Venezuela. *Sociedad de Ciencias NaturalesLa Salle, 143*, 25–40.

Velez-Liendo, X., & García-Rangel, S. (2017). Tremarctos ornatus (errata version published in 2018). The IUCN Red List of Threatened Species 2017: e.T22066A123792952. https://doi.org/10.2305/IUCN.UK.2017-3.RLTS.T22066A45034047.en

Velez-Liendo, X. (2010). *Conservation of Andean bears (Tremarctos ornatus) in a fragmented landscape—Habitat models, potential distribution and patch connectivity*. University of Antwerp.

Vieira Ventura, R., Silva, F. E. F., Manuel Yáñez, J., & Brito, L. F. (2020). Opportunities and challenges of phenomics applied to livestock and aquaculture breeding in South America. *Animal Frontiers: the Review Magazine of Animal Agriculture, 10*(2), 45–52. https://doi.org/10.1093/af/vfaa008

Viteri, M. P. (2007). Conservation genetics of Andean bears (Tremarctos ornatus) in Northeastern Ecuador: Molecular tools, genetic diversity and population size. University of Idaho, Moscow, Idaho, USA.

Voiland, A. (2020). Tracking Amazon Deforestation from Above. https://earthobservatory.nasa.gov/images/145988/tracking-amazon-deforestation-from-above

Weckel, M., Giuliano, W. M., & Silver, S. C. (2006). Jaguar (Panthera onca) feeding ecology: Distribution of predator and prey through time and space. *Journal of Zoology, 270*, 25–30. https://doi.org/10.1111/j.1469-7998.2006.00106.x

West, A. P. T., Rausch, L., Munger, J., & Gibbs, H. K. (2022). Protected areas still used to produceBrazil'scattle. *Conservation Letters, 15*. https://doi.org/10.1111/conl.12916

Wirsing, A. J., Johnston, A. N., & Kiszka, J. J. (2022). Foreword to the special issue on 'The rapidly expanding role of drones as a tool for wildlife research'. *Wildlife Research, 49*, i-v. https://doi.org/10.1071/WR22006

Wordley, C. (2021). The secret bears of Bolivia's lost dry forests. Mongabay. https://news.mongabay.com/2021/06/the-secret-bears-of-bolivias-lost-dry-forests/

World Economic Forum. (2024). How tracing cattle in Brazil can prevent deforestation in the Amazon. https://www.weforum.org/agenda/2024/01/tracing-cattle-brazil-prevent-deforestation-amazon/

World Wildlife Fund. (2023). Living with Jaguars. Stories of Co-Existence. https://wwf.panda.org/discover/wildlife/living_with_jaguars_stories_of_coexistence/

World Wildlife Fund. (2024). Unsustainable cattle ranching. https://wwf.panda.org/discover/knowledge_hub/where_we_work/amazon/amazon_threats/unsustainable_cattle_ranching/

Zia, M., Hansen, J., Hjort, K., & Valdes, C. (2019). Brazil once again becomes the world's largest beef exporter. https://www.ers.usda.gov/amber-waves/2019/july/brazil-once-again-becomes-the-world-s-largest-beef-exporter/

Zimmermann, A., Walpole, M. J., & Leader-Williams, N. (2005a). Cattle ranchers' attitudes to conflicts with jaguar Panthera onca in the Pantanal of Brazil. *Oryx, 39*(4), 406–412. https://doi.org/10.1017/S0030605305000992

Zukowski, B., & Ormsby, A. (2016). Andean bear livestock depredation and community perceptions in northern Ecuador. *Human Dimensions of Wildlife, 21*, 111–126.

GPS, Animal Tracking and Conservation Biology

Arispe, R., Rumiz, D., & Venegas, C. (2007). *Censo de jaguares (Panthera onca) y otros mamíferos con trampas cámara en la Concesión Forestal El Encanto*. Santa Cruz.

Augliere B. (2015). Beyond GPS: *The next high-tech frontier in wild animal tracking*. Retrieved from https://blogs.scientificamerican.com/guest-blog/beyond-gps-the-nexthigh-tech-frontier-in-wild-animal-tracking/

Avery, M. (2004). *Managing depredation and nuisance problems caused by vultures*. United States Department of Agriculture.

Azevedo, F. C. (2008). Food habits and livestock depredation of sympatric jaguars and pumas in the Iguacu National Park area, South Brazil. *Biotropica, 40*(4), 494–500.

Bailey, T. N. (1993). *The African leopard: A study of the ecology and behavior of a solitary felid*. Columbia University Press.

Beier, P. (1993). Determining minimum habitat areas and habitat corridors for cougars. *Conservation Biology, 7*(1), 94–108.

Beier, P., Riley, S. P. D., & Sauvajot, R. M. (2010). Mountain lions (*Puma concolor*). In S. D. Gehrt, S. P. D. Riley, & B. L. Cypher (Eds.), *Urban carnivores: Ecology, conflict, and conservation* (pp. 140–155). John Hopkins University Press.

Belant, J. L. (1997). Gulls in the urban environment: Landscape level management to reduce conflict. *Landscape and Urban Planning, 38*, 245–258.

Benson, E. P. (1998). The lord, the ruler: Jaguar symbolism in the Americas. In N. J. Saunders (Ed.), *Icons of power: Feline symbolism in the Americas* (pp. 53–76). Routledge.

Bird, D. M., Garland, D. E., & Negro, J. (1996). *Raptors in human landscapes adaptations to built and cultivated environments*. Academic Press.

Boron, V., Tzanopoulos, J., Gallo, J., Barragan, J., Jaimes-Rodriguez, L., Schaller, G., et al. (2016). Jaguar densities across human-dominated landscapes in Colombia: The contribution of unprotected areas to long term conservation. *PLoS One, 11*(5), e0153973. https://doi.org/10.1371/journal.pone.0153973

Brodie, J. F. (2009). Is research effort allocated efficiently for conservation? Felidae as a global 343 case study. *Biodiversity and Conservation, 18*, 2927–2939.

Brown, D. E., & López-González, C. A. (2001). *Borderland Jaguars: Tigres de la Frontera*. University of Utah Press.

Burdett, C. L., Crooks, K. R., Theobald, D. M., Wilson, K. R., Boydston, E. E., Lyren, L. M., Fisher, R. N., Winston Vickers, I., Morrisson, A., & Boyce, W. M. (2010). Interfacing models of wildlife habitat and human development to predict the future distribution of puma habitat. *Ecosphere, 1*, 4.

Campbell, M., & Lancaster, B. (2010). Public attitudes towards black bears (*Ursus americanus*) and cougars (*Puma concolor*) on Vancouver Island. *Journal of Society and Animals, 10*(1), 47–51.

Campbell, M., & Torres-Alvarado, M. (2011). Public perceptions of jaguars (*Panthera onca*), pumas (*Puma concolor*) and coyotes (*Canis latrans*) in El Salvador. *Area, 43*(3), 250–256.

Campbell, M. (2009). The impact of habitat characteristics on bird presence and the implications for wildlife management in the environs of Ottawa, Canada. *Urban Forestry and Urban Greening, 8*, 87–95.

Campbell, M (2012a). The relevance of age and gender for public attitudes to brown bears (*Ursus arctos*), black bears (*Ursus americanus*), and cougars (*Puma concolor*) in Kamloops, British Columbia. *Journal of Society and Animals, 21*(4), 341–359.

Campbell, M. (2012b). Urban planning for cougar presence in North America: Practices, challenges and benefits. In M. P. Amado (Ed.), *Urban planning: Practices, challenges and benefits* (pp. 19–32). Nova Science Publishers.

Campbell, M. (2017b). *Biological conservation in the 21st century: A conservation biology of large wildlife*. Nova Science Publishers.

Campbell, M. (2018). The ecology and conservation of the West African lion. In *Advances in animal science and zoology*. Nova Science Publishers.

Carter, N., Jasny, M., Gurung, B., & Liu, J. (2015). Impacts of people and tigers on leopard spatiotemporal activity patterns in a global biodiversity hotspot. *Global Ecology and Conservation, 3*, 149–162. https://doi.org/10.1016/j.gecco.2014.11.013

Caso, A., Lopez-Gonzalez, C., Payan, E., Eizirik, E., de Oliveira, T., Leite-Pitman, R., Kelly, M., & Valderrama, C. (2008). *Panthera onca. IUCN Red list of threatened species. Version 2011.1*. International Union for Conservation of Nature.

Cavalcanti, S. M. C. (2008). *Predator-Prey Relationships and Spatial Ecology of Jaguars in the Southern Pantanal, Brazil: Implications for Conservation and Management*. https://doi.org/10.26076/dc41-1ba6

Chiari, C. G., & Seeland, K. (2004). Are urban green spaces optimally distributed to act as places for social integration? Results of a geographical information system GIS approach for urban forestry research. *Forest Policy and Economics, 6*, 3–13.

Conforti, V. L., & Azevedo, F. C. C. (2003b). Local perceptions of jaguars Panthera once and pumas (*Puma concolor*) in the Iguacu National Park area South Brazil. *Biological Conservation, 111*, 215–221.

Cornelis, J., & Hermy, M. (2004). Biodiversity relationships in urban and suburban parks in Flanders. *Landscape and Urban Planning, 69*, 385–401.

Davis, H., Wellwood, D., & Ciarniello, L. M. (2001). *"Bear Smart" community program: Background report*. British Columbia Ministry of Environment, Lands and Parks.

De Angelo, C. (2009). El paisaje del Bosque Atlántico del Alto Paraná y sus efectos sobre la distribución y estructura poblacional del jaguar (Panthera onca) y el puma (Puma concolor). PhD thesis, Universidad de Buenos Aires, Buenos Aires, Argentina.

Decker, D. J., Brown, T. L., Hustin, D. L., Clarke, S. H., & O'Pezio, J. (1981). Public attitudes toward black bears in the Catskills. *New York Fish and Game Journal, 28*(1), 1–20.

Dickson, B. G., & Beier, P. (2002). Home-range and habitat selection by adult cougars in southern California. *Journal of Wildlife Management, 66*, 1235–1245.

Dickson, B. G., Jenness, J. S., & Beier, P. (2005). Influence of vegetation, topography, and roads on cougar movement in southern California. *Journal of Wildlife Management, 69*, 264–276.

Donadio, E., & Buskirk, S. W. (2006). Diet, morphology, and interspecific killing in Carnivora. *The American Naturalist, 167*(4), 524–536. https://doi.org/10.1086/501033

Emmons, L. H. (1987). Comparative feeding ecology of felids in a neotropical rainforest. *Behavioral Ecology and Sociobiology, 20*, 271–283.

Emmons, L. H. (1991). Body size and feeding tactics. In J. Seidensticker & S. Lumpkin (Eds.), *Great cats, 62*. Rodale Press.

Erickson, D. L. (2004). The relationship of historic city form and contemporary greenway implementation: A comparison of Milwaukie Wisconsin USA and Ottawa Ontario Canada. *Landscape and Urban Planning, 68*, 199–221.

Farrell, L. E., Roman, J., & Sunquist, M. E. (2000). Dietary separation of sympatric carnivores identified by molecular analysis of scats. *Molecular Ecology, 9*, 1583–1590.

Fernandez-Juricic, E. (2000). Avifaunal use of wooded streets in an urban landscape. *Conservation Biology, 14*(2), 513–521.

Fonturbel, F. E., & Simonetti, J. A. (2011). Translocations and human-carnivore conflicts: Problem solving or problem creating. *Wildlife Biology, 17*, 217–224.

Freedman, A. H., Portier, K. M., & Sunquist, M. E. (2003). Life history analysis for black bears (*Ursus americanus*) in a changing demographic landscape. *Ecological Modeling, 167*(1), 47–64.

Guggisberg, C. A. W. (1975). *Wild cats of the world*. Taplinger Publishing.

Gutiérrez-González, C. E., & López-González, C. A. (2017b). Jaguar interactions with pumas and prey at the northern edge of jaguars' range. *Peer J, 5*, e2886. https://doi.org/10.7717/peerj.2886

Gutiérrez-González, C. E., Gómez-Ramírez, M. Á., & López-González, C. A. (2012). Estimation of the density of the Near Threatened jaguar Panthera onca in Sonora, Mexico, using camera trapping and an open population model. *Oryx, 46*(3), 431–437. https://doi.org/10.1017/s003060531100041x

Harmsen, B. J, Foster, R. J., Silver, S. C., Ostro, L. E. T., & Doncaster, P. (2009b). Spatial and temporal interactions of sympatric jaguars (*Panthera onca*) and pumas (*Puma concolor*) in a Neotropical forest. *Journal of Mammalogy, 90*(3), 612–620. https://doi.org/10.1644/08-MAMM-A-140R.1.

Harness, R. E., & Wilson, K. R. (2001). Utility structures associated with raptor electrocutions in rural areas. *Wildlife Society Bulletin, 29*, 612–623.

Hayward, M. W., Kamler, J. F., Montgomery, R. A., Newlove, A., Rostro-García, S., Sales, L. P., & Van Valkenburgh, B. (2016b). Prey preferences of the jaguar *Panthera onca* reflect the post-Pleistocene demise of large prey. *Frontiers in Ecology and Evolution, 3*, 1–19. https://doi.org/10.3389/fevo.2015.00148

Herrero, S., & Higgins, A. (1999). Human injuries inflicted by bears in British Columbia: 1960 - 1997. *Ursus, 11*, 209–218.

Höss, M., Kohn, M., Pääbo, S., Knauer, F., & Schröder, W. (1992). Excrement analysis by PCR. *Nature, 359*, 199.

Iriarte, J. A., Franklin, W. L., Johnson, W. E., & Redford, K. H. (1990b). Biogeographic variation of food habits and body size of the America puma. *Oecologia, 85*, 185–190.

Jaksic, F. M., Greene, H. W., & Yáñez, J. L. (1981). The guild structure of a community of predatory vertebrates in Central Chile. *Oecologia, 49*, 21–28.

Jokimaki, J. (1999). Occurrence of breeding bird species in urban parks: Effects of park structure and broad-scale variables. *Urban Ecosystems, 3*, 21–34.

Jokimaki, J. (2003). Spatial similarity of urban bird communities: A multiscale approach. *Journal of Biogeography, 30*, 1183–1193.

Karanth, K. U., & Chellam, R. (2009). Carnivore conservation at the crossroads. *Oryx, 43*, 1–2.

Kellert, S. R. (1994). Public attitudes towards bears and their conservation. *International Conference on Bear Research and Management, 9*(1), 43–50.

Kellert, S. R., Black, M., Rush, C. R., & Bath, A. J. (1996). Human culture and large carnivore conservation in North America. *Conservation Biology, 10*(4), 977–990.

Kertson, B. N., Spencer, R. D., & Grue, C. E. (2011a). Cougar prey use in a wildlandurban environment in western Washington. *Northwestern Naturalist, 92*, 175–185.

Kertson, B. N., Spencer, R. D., & Grue, C. E. (2013). Demographic influences on cougar residential use and interactions with people in western Washington. *Journal of Mammalogy, 94*(2), 269–281.

Kertson, B. N., Spencer, R. D., Marzluff, J. M., Hepinstall-Cymerman, J., & Grue, C. E. (2011b). Cougar space use and movements in the wildland–urban landscape of western Washington. *Ecological Applications, 21*, 2866–2881.

Kuhn, M. (2003). Greenbelt and greenheart: Separating and integrating landscapes in European city regions. *Landscape and Urban Planning, 64*, 19–27.

Linnell, J. D. C., Swenson, J. E., & Anderson, R. (2001). Predators and people: Conservation of large carnivores is possible at high human densities if management policy is favorable. *Animal Conservation, 4*, 345–349.

Logan, K. A., & Sweanor, L. L. (2001). *Desert puma: Evolutionary ecology and conservation of an enduring carnivore*. Island Press.

López-González, C., & González-Romero, A. (1998b). A synthesis of current literature and knowledge about the ecology of the puma (*Puma concolor*). *Acta Zoologica Mexicana, 75*, 171–190.

Lynch, E., Northrup, J. M., McKenna, M. F., Anderson, C. R., Jr., Angeloni, L., & Wittemyer, G. (2014). Landscape and anthropogenic features influence the use of auditory vigilance by mule deer. *Behavioral Ecology, 26*, 75–82.

Maehr, D. S., Shindle, D. B., & Hoctor, T. S. (2002). Florida panther dispersal and conservation. *Biological Conservation, 106*, 187–197.

Maffei, L., Polisar, J., Garcia, R., Moreira, J., & Noss, A. J. (2011). *Mesoamericana, 15*, 49–49.

McCain, E. B., & Childs, J. L. (2008). Evidence of resident jaguars (*Panthera onca*) in the southwestern United States and the implications for conservation. *Journal of Mammalogy, 89*, 1–10.

McDougal, C. (1987). The man-eating tiger in geographical and historical perspective. In R. I. Tilson & U. S. Seal (Eds.), *Tigers of the world* (pp. 435–448). Noyes Publications.

Merkle, J. A., Robinson, H. S., Krausman, P. R., & Alaback, P. (2013). Food availability and foraging near human developments by black bears. *Journal of Mammalogy, 94*, 378–385.

Mitchell, M. S., & Hebblewhite, M. (2012). Carnivore habitat ecology: Integrating theory and application. In L. Boitani & R. A. Powell (Eds.), *Carnivore ecology and conservation* (pp. 218–255). Oxford University Press.

Morrison, J. C., Sechrest, W., Dinerstein, E., Wilcove, D. S., & Lamoreux, J. F. (2007). Persistence of large mammal faunas as indicators of global human impacts. *Journal of Mammalogy, 88*, 1363–1380.

Morzillo, A. T., Mertig, A. G., Garner, N., & Liu, J. (2007). Resident attitudes toward black bears and population recovery in East Texas. *Human Dimensions of Wildlife, 12*(6), 417–428.

Novack, A. J., Main, M. B., Sunquist, M. E., & Labisky, R. F. (2005). Foraging ecology of jaguar (*Panthera onca*) and puma (*Puma concolor*) in hunted and non-hunted sites within the Maya Biosphere Reserve, Guatemala. *Journal of Zoology, 267*, 167–178. https://doi.org/10.1017/S0952836905007338

Nowell, K., & Jackson, P. (Eds.). (1996). *Wild cats: Status survey and conservation action plan*. Gland: IUCN/SSC Cat Specialist Group.

Odden, M., Wegge, P., & Fredriksen, T. (2010). Do tigers displace leopards? If so, why? *Ecological Research, 25*, 875–881. https://doi.org/10.1007/s11284-010-0723-1

de Oliveira, T. (2002). Evaluación del estado de conservación del jaguar em el este de La Amazonia y noreste de Brasil. In R. A. Medellín, C. Equihua, C. L. B. Chetkiewitcz, P. G. Crawshaw Jr., A. Rabinowitz, K. H. Redford, J. G. Robinson, E. W. Sanderson, & A. B. Taber (Eds.), *El Jaguar em El Nuevo Milenio* (pp. 419–436). Fondo de Cultura Económica, México, Universidad Nacional Autónoma de México, Wildlife Conservation Society. [Evaluation of the state of conservation of the jaguar in the east of the Amazon and northeast of Brazil. In *The Jaguar in The New Millennium*.

de Oliveira, T. G. (1994). *Neotropical cats: Ecology and conservation*. EDUFMA.

Palmeira, F. B. L., Crawshaw, P. G., Haddad, C. M., Ferrazd, K. M. P. M. B., & Verdadee, L. M. (2008b). Cattle depredation by puma Puma concolor and jaguar Panthera onca in central western Brazil. *Biological Conservation, 141*, 118–125.

Paviolo, A., De Angelo, C. D., Di Blanco, Y. E., & Di Bitetti, M. S. (2008). Jaguar *Panthera onca* population decline in the Upper Paraná Atlantic Forest of Argentina and Brazil. *Oryx, 42*, 554–561.

Paxinos, E., McIntosh, C., Ralls, K., & Fleischer, R. (1997). A non-invasive method for distinguishing among canid species: Amplification and enzyme restriction of DNA from dung. *Molecular Ecology, 6*, 483–486.

Pelton, M., Scott, C., & Burghardt, G. M. (1976). Attitudes and opinions of persons experiencing property damage and/or injury by black bears in the Great Smoky Mountains National Park. *International Conference of Bear Research and Management, 3*, 157–167.

Perry, T. (1994). Living in cougar country. *Los Angeles Times*, December 19, A3, A38. A39.

Pirnat, J. (2000). Conservation and management of forest patches and corridors in suburban landscapes. *Landscape and Urban Planning, 52*, 135–143.

Polisar, J., Maxit, I., Scognamillo, D., Farrell, L., Sunquist, M., & Eisenberg, J. F. (2003b). Jaguars, pumas, their prey base, and cattle ranching: Ecological interpretations of a management problem. *Biological Conservation, 109*, 297–310. https://doi.org/10.1016/S0006-207(02)00157-X

Quiroga, V. A., Boaglio, G. I., Noss, A. J., & Di Bitetti, M. S. (2013). Critical population status of the jaguar *Panthera onca* in the Argentine Chaco: Camera-trap surveys suggest recent collapse and imminent regional extinction. *Oryx, 48*, 1–8. https://doi.org/10.1017/S0030605312000944

Rabinowitz, A., & Nottingham, B. (1986). Ecology and behavior of the jaguar in Belize, Central America. *Journal of Zoology, 210*, 149–159.

Rabinowitz, A., & Zeller, K. A. (2010). A range-wide model of landscape connectivity and conservation for the jaguar, Panthera onca. *Biological Conservation, 143*, 939945.

Ramesh, T., Kalle, R., Sankar, K., & Qureshi, Q. (2012). Spatio-temporal partitioning among large carnivores in relation to major prey species in Western Ghats. *Journal of Zoology, 287*(4), 269–275. https://doi.org/10.1111/j.1469-7998.2012.00908.x

Ríos-Uzeda, B., Gómez, H., & Wallace, R. B. (2007). A preliminary density estimate for Andean bear using camera-trapping methods. *Ursus, 18*(1), 124–128. https://doi.org/10.2192/1537-6176(2007)18[124:apdefa]2.0.co;2

Romero-Muñoz, A., Maffei, L., Cuéllar, E., & Noss, A. J. (2010). Temporal separation between jaguar and puma in the dry forests of southern Bolivia. *Journal of Tropical Ecology, 26*(3), 303–311. https://doi.org/10.1017/S0266467410000052

Rosas-Rosas, O. C., & Valdez, R. (2010). The role of landowners in jaguar conservation in Sonora, Mexico. *Conservation Biology, 24*(2), 366–371. https://doi.org/10.1111/j.1523-1739.2009.01441.x

Rosas-Rosas, O. C., & Bender, L. C. (2012). Population status of jaguars (*Panthera onca*) and pumas (*Puma concolor*) in northeastern Sonora, Mexico. *Acta Zoologica Mexicana, 28*(1), 86–101.

Rosas-Rosas, O. C., Bender, L. C., & Valdez, R. (2008). Jaguar and puma predation on cattle calves in northeastern Sonora, Mexico. *Rangeland Ecology and Management, 61*, 554–560.

Ruth, T. K., & Murphy, K. (2010). Competition with other carnivores for prey. In M. Hornocker & S. Negri (Eds.), *Cougar Ecology & Conservation* (pp. 163–172). University of Chicago Press.
Ruth, T. K. (1991). *Mountain Lion Use of an Area of High Recreational Development in Big Bend National Park Texas*. MS. Thesis, Texas A&M University, College Station.
Saenz, J. C., & Carrillo, E. (2002). Jaguares depredadores de ganado en Costa Rica: Un problema sin solucion? *El Jaguar en el Nuevo Milenio*, 127–138. Mexico: A Taber Universidad Nacional Autonoma de Mexico and Wildlife Conservation Society. [Jaguars predators of cattle in Costa Rica: A problem without solution? *The Jaguar in the New Millennium*, 127–138. Mexico: A Taber National Autonomous University of Mexico and Wildlife Conservation Society.]
Salom-Pérez, R., Camillo, E., Sáenz, J. C., & Mora, J. M. (2007). Critical condition of the jaguar Panthera onca population in Corcovado National Park, Costa Rica. *Oryx, 41*, 51–56.
Sanderson, E. W., Redford, R. H., Chetkiewicz, C. B., Medellin, R., Rabinowitz, A., Robinson, J., & Taber, A. (2002). Planning to save a species: The jaguar as a model. *Conservation Biology, 16*, 58–72.
Santos, F. R., Jácomo, A. T. A., & Silveira, L. (2008). Humans and jaguars in five Brazilian biomes: Same country, different perceptions. *Cat News, 4*, 21–25.
Sarmiento, R. (2004). *Population estimation methods for jaguar (Panthera onca), Corcovado National Park, Costa Rica*. Master of Science Thesis. Universidad Nacional, Heredia, Costa Rica.
Saunders, N. J. (1998). *Icons of power: Feline symbolism in the Americas*. Routledge.
Savard, J. P. L., Clergeau, P., & Mennechez, G. (2000). Biodiversity concepts and urban ecosystems. *Landscape and Urban Planning, 48*, 131–142.
Scognamillo, D., Maxit, I. E., Sunquist, M., & Polisar, J. (2003b). Coexistence of jaguar (*Panthera onca*) and puma (*Puma concolor*) in a mosaic landscape in the Venezuelan Llanos. *Journal of Zoology (London), 259*, 269–279.
Shuey, M. L. (2008). *The influence of landscape on Cougar-human encounter Hazard in Boulder and Colorado Springs, Colorado*. ETD Collection for Texas State University. Retrieved from https://digital.library.txstate.edu/handle/10877/3004.
Silver, S. C., Ostro, L. E. T., Marsh, L. K., Maffei, L., Noss, A. J., Kelly, M. J., Wallace, R. B., Gómez, H., & Ayala, G. (2004). The use of camera traps for estimating jaguar *Panthera onca* abundance and density using capture-recapture analysis. *Oryx, 38*(2), 148–154.
Soisalo, M. K., & Cavalcanti, S. M. C. (2006). Estimating the density of a jaguar population in the Brazilian Pantanal using camera-traps and capture–recapture sampling in combination with GPS radio-telemetry. *Biological Conservation, 129*, 487–496.
Sollmann, R., Gardner, B., & Belant, J. L. (2012b). How does spatial study design influence density estimates from spatial capture-recapture models? *PLoS One, 7*, e34575.
Sollmann, R., Tôrres, N. M., & Silveira, L. (2008). Jaguar conservation in Brazil: The role of protected areas. *Cat News, 4*, 15–20.
Spencer, R. D., Pierce, D. J., Schirato, G. A., Dixon, K. R., & Richards, C. B. (2001). *Mountain lion home range, dispersal, mortality and survival in the Western Cascade Mountains of Washington*. Washington Department of Fish and Wildlife.
Sunquist, M., & Sunquist, F. (2002b). *Wild cats of the world*. The University of Chicago Press.
Thornton, C. T., & Quinn, M. S. (2009). Coexisting with cougars: Public perceptions, attitudes and awareness of cougars on the urban-rural fringe of Calgary, Alberta, Canada. *Human–Wildlife Conflicts, 3*(2), 282–295.
Tobler, M. W., Carrillo-Percastegui, S. E., Zúñiga Hartley, A., & Powell, G. V. N. (2013). High jaguar densities and large population sizes in the core habitat of the southwestern Amazon. *Biological Conservation, 159*, 375–381. https://doi.org/10.1016/j.biocon.2012.12.012
Tôrres, N. M., De Marco Jr, P., Diniz-Filho, J. A. F., & Silveira, L. (2008). Jaguar distribution in Brazil: Past, present and future. *Cat News Special Issue, 4*, 4–8.
Tôrres, S. G., Mansfield, T. M., Foley, J. E., Lupo, T., & Brinkhaus, A. (1996). Mountain lion and human activity in California: Testing speculations. *Wildlife Society Bulletin, 24*, 451–460.

Tortato, F. R., Layme, V. M. G., Crawshaw, P. G., Jr., & Izzo, T. J. (2015). The impact of herd composition and foraging area on livestock predation by big cats in the Pantanal of Brazil. *Animal Conservation, 18*(6), 539–547. https://doi.org/10.1111/acv.12207

Van Dyke, F. G., Brooke, R. H., Shaw, H. G., Ackerman, B. B., Hemker, T. P., & Lindzey, F. G. (1986). Reactions of mountain lion to logging and human activity. *Journal of Wildlife Management, 50*, 95–102.

Vanak, A. T., Fortin, D., Thaker, M., Ogden, M., Owen, C., Greatwood, S., & Slotow, R. (2013). Moving to stay in place: Behavioral mechanisms for coexistence of African large carnivores. *Ecology, 94*(11), 2619–2631. https://doi.org/10.1890/13-0217.1

Wallace, R. B., Gomez, H., Ayala, G., & Espinoza, F. (2003). Camera trapping capture frequencies for jaguar (*Panthera onca*) in the Tuichi Valley, Bolivia. *Mastozoologia Neotropical, 10*(1), 133–139.

Wasser, S. K., Houston, C. S., Koehler, G. M., Cadd, G. G., & Fain, S. R. (1997). Techniques for application of faecal DNA methods to field studies of Ursids. *Molecular Ecology, 6*, 1091–1097.

Weaver, J. L., Paquet, P. C., & Ruggiero, L. F. (1996). Resilience and conservation of large carnivores in the Rocky Mountains. *Conservation Biology, 10*, 964–976.

White, K. R., Koehler, G. M., Maletze, B. T., & Wielgus, R. B. (2011). Differential prey use by male and female cougars in Washington. *Journal of Wildlife Management, 75*, 1115–1120.

Wilmers, C. C., Nickel, B., Bryce, C. M., Smith, J. A., Wheat, R. E., & Yovovich, V. (2015). The golden age of bio-logging: How animal-borne sensors area advancing the frontiers of ecology. *Ecology, 96*, 1741–1753.

Wolch, J., Gullo, A., & Lassiter, U. (1997). Changing attitudes toward California's Cougars. *Society and Animals, 5*(2), 95–116.

Yáñez, J. L., Cárdenas, J. C., Gezelle, P., & Jaksic, F. M. (1986). Food habits of the southernmost mountain lions (*Felis concolor*) in South America: Natural versus livestocked ranges. *Journal of Mammalogy, 67*, 604–606.

Zeller, K. (2007). *Jaguars in the new millennium data base update: The state of the Jaguar in 2006*. Wildlife Conservation Society-Jaguar Conservation Program.

Zimmermann, A., Walpole, M. J., & Leader-Williams, N. (2005b). Cattle ranchers' attitudes to conflicts with jaguar (*Panthera onca*) in the Pantanal of Brazil. *Oryx, 39*, 406–412.

Conclusions

The relationship between cattle herding and agricultural development, and large, mostly charismatic carnivore conservation, and research methods assessing and analyzing these developments, are critical issues in the modern world. Currently, cattle breeding, management, and production are among the most important issues for global food production. Cattle are by far the most important livestock in modern industry, challenged only by alternative food sources. Currently, most cattle are reared in organized, well-managed farms, but substantial numbers are also managed as free-ranging herds in semi-rural and even natural landcover, on land near to large carnivore habitats. In North America and Europe, increasingly sophisticated agricultural systems have been developed, which are linked to the industrial world's linkages and markets. In Asia, South America, and Africa, modernizing trends are evident, especially in the urbanized areas, but there are also larger areas where small farmers rear cattle and other small livestock occupy natural, sometimes degraded landcover.

Cattle production is determined by human population growth rates, which may increase demand, or by the promotion of alternatives, such as poultry or vegetarian products, which may reduce demand. Regardless of trend, the cattle industry is currently dominant, and occupies vast areas of land, in competition with protected areas and urban development on all continents. The focus of this book, and of vast areas of cattle herding, is on agroecology. This concerns how cattle herding may interface with natural or semi-natural ecosystems, especially those that are important for charismatic or important wildlife. With the emergence of international interest in sustainable development, agroecology has increasingly contributed to discourses on agricultural management. As recorded in this book, there have been major impacts on cattle management, especially in western Europe and northern North America, but there are still serious political and economic problems that may affect the environmental implications of such management.

© The Author(s), under exclusive license to Springer Nature Switzerland AG 2025

M. O. Campbell, *Cattle, Their Predators and Geomatics Research*,
https://doi.org/10.1007/978-3-031-97363-5_10

Wildlife conservation is also an emerging trend in modern geopolitics, including the preservation of large charismatic carnivores such as big cats, wild canids, bears and large birds of prey, and wild cattle, and their primal habitats are increasingly important for global ecology and human development. Such large animals require large spaces for sustenance, therefore extensive conservation projects may conflict with preexisting or expanding agricultural landcover. These predators are the principal killers of cattle, and thus contribute to conflicts between the developers of cattle ranges and supporters of larger protected areas. Increased protected areas are promoted internationally, hence such conflicts exist across the five continents. In North America and Europe at least, these policies have been successful, and the increased populations of large predators have contributed to increased livestock predation. Ameliorative developments include payments to livestock owners for losses to predation, and even killing of such carnivores.

Computer-based technologies, including geomatics-based methods, based on tracking, image-based assessment, multiple-parameter modeling, including advanced GPS, GIS, LiDAR, Satellite Images, and drones, have also emerged as dominant methods of ecological and human resource use and land management. The power of these technologies has altered human-animal relationship assessments, and in some cases altered global perceptions of environmental, including agricultural and conservation, changes and possible trends. These technologies are related to the systems termed Big Data, which is based on vast computer-based information databases used for analyses, and Big Tech, which includes communication systems in social media, and are important tools for communicating with large numbers of people, and thus can influence vast numbers of people globally on agricultural and environmental issues.

Biological technologies have also emerged as important for the study of the evolution of cattle domestication, including the use of DNA methods, such as Y-DNA and Mt-DNA. Such studies examine and increase the ancestry of domestic cattle, which is important for knowledge of the success of domestication and cattle breeding. Cattle diseases are important, especially in Africa, and in North America and Europe, breeding commonly attempts to increase the physiology of cattle for increased production. Cattle populations in Asia have also been interbred with wild cattle species, and such hybrids may be the dominant breeds. Recent studies have found the genetic background of cattle breeds vital for cattle resistance to diseases and to extreme temperatures. Some of these resistances are due to ancient adaptation and modern breeding. Breeding also reduced the size of cattle breeds compared to ancestral breeds such as the aurochs, and wild species such as water buffalos, yaks, and gaur, which can affect the domestic cattle's ability to defend against predators.

Therefore, the future of cattle-predator relations depends on developments in the biological sciences (genetic and ecological research), habitat and agricultural land development (geomatics and related ecological assessment), agricultural and financial economics (the relative costs and benefits of agricultural and conservation management), consumer demand (the future demand for beef products), and the integration of these issues into effective management, possibly through the emerging methods of Big Data and Big Tech. The current trajectory of change in these dynamics may be considered at the interface of changing human values, organization of methods and management, scientific and more subjective knowledge, technological development, and the environment, the last including changing ecologies under natural and human-constructed parameters. These associations may be expressed by the model shown in Fig. 1, and the descriptions in Tables 1, 2, 3, 4, 5, and 6, which present the model of social, economic, and environmental coevolution. The concept of coevolution adequately describes the current developments concerning the political, social, economic, ecological, and biological developments relevant to cattle management and large carnivore conservation.

Further research for the future development of an ecologically sound cattle industry and effective large carnivore conservation may focus on the extent to which the factors for these two issues may coevolve. Based on current trends, large carnivore conservation may be strengthened with national and international support, including expanded protected areas and supportive legal instruments that may prohibit the killing of large wildlife and protect their habitats. Cattle herding, although threatened by alternatives such as vegetarianism and alternative meat sources, may prevail in the foreseeable future. Predictive models, supported by analyses of wider issues, may assist in the analysis of such changes.

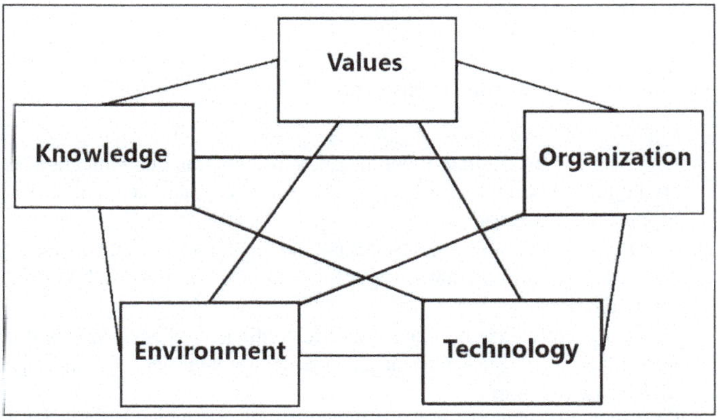

Fig. 1 Norgard's model of socioenvironmental coevolution. Here, each of the components coevolves with all the others, in terms of affecting and being affected by all the other components, back and forth through time, as shown in Table 1, in terms of the topic of large carnivore conservation and cattle development. (Source: Adapted from Norgard (1994))

Table 1 Coevolving matrix

Component	Environment	Values	Organization	Knowledge	Technology
Environment	Competing environmental uses. Conservation, cattle herding	Public attitudes to conservation, agriculture, impacts on environment	Landuse management, environmental consequences	Increased research affecting land management, nature	Technological advancements, creating environmental management
Values	Environmental change, affecting public values	Competing values. Conservation, agriculture	Political, economic organizations affecting society	Increased research affecting society	Technological advancements, affecting social values, contacts
Organization	Environmental change, influencing politics, business	Values impact on political and private organizations	Competing political, socioeconomic organizations	Increased research affecting politics, business	Technological advancements, creating effective organizations
Knowledge	Environmental change, influencing research	Value-driven knowledge research	Political, economic organizations affecting science	Competing scientific paradigms	Technological advancements, creating more accurate data
Technology	Environmental change, influencing technology	Value-driven technological development	Political, economic organizations affecting tech	Increased research affecting, tech, advancements	Technology for ecology, vs agriculture

Source: Adapted from Norgard (1994) and Campbell (2017)
The component at the top of the column affects the component in the row. Details are in the following tables

Table 2 Environment impact on other components

Component	Impact of environment
Values	Large carnivores decline or extinction, deforestation, desertification, pollution, predation of livestock affect economic, political, social, and cultural values, and support for conservation
Organization	Political organizations at international, national, and local levels may change in response to environmental change, forming organizations such as protected areas and plantations
Knowledge	Environmental change, such as wildlife conflicts, and environmental degradation may influence new research on such changing dynamics, with new specialisms such as agroecology
Technology	Environmental change, such as wildlife conflicts, and environmental degradation may require new assessment technology such as geomatics-based methods, Big Data, Big Tech

Table 3 Values impact on other components

Component	Impact of values
Environment	Changing human values, political, social, and economic, may affect the management of the environment, creating conservation movements that affect the form of environment change, protection
Organization	Changing human values may increase public political, economic, and social support for some organizations, such as conservation agencies, and reduce support for others, such as industry
Knowledge	Changing human values may influence changes in public demand and support for some types of research, such as polluting technologies, which may be replaced by ecological research
Technology	Changing human values may increase or decrease the use and support for some technologies, such as social media, engineering, and ecological applications

Table 4 Organizational impact on other components

Component	Impact of organization
Environment	Organizational design may affect the focus on ecological, economic, and political aspects of land management organizations, and the management of the consequences of these changes
Values	Organizational design may have impacts that affect the values of the various stakeholders in the agricultural development and/or carnivore conservation systems.
Knowledge	Organizational design may require changes in the types and sources of knowledge for management. Organizations may affect relevant environments and species, requiring new research
Technology	Organizational design may have various impacts, that may require new technologies, such as mechanization, surveying equipment, cattle breeding, or fencing.

Table 5 Knowledge impact on other components

Component	Impact of knowledge
Environment	Knowledge sources on environmental variables, including wild and domestic animals, may determine or contribute to effective or flawed environmental assessments, and affect conflicts
Values	Knowledge may produce information that changes stakeholder and public values. Gaps in knowledge on the impact predators such as wolves affect human values concerning conservation
Organization	Knowledge facilitates effective organizations, and ecologically sensitive knowledge sources may positively affect the organization of the ameliorative actions reducing carnivore conflicts.
Technology	Knowledge sources support the design, application, and monitoring of technology used to reduce conflicts, such as geomatics-based survey equipment and computer-based assessments

Table 6 Technology impact on other components

Component	Impact of technology
Environment	New technology may ameliorate the problems created by older technology. For example, cattle herding that is more aware of predator presence may reduce conflicts
Values	Technology that improves the environmental situation may increase human support for its application, while technology that increases conflicts, such as poor herding practices may negatively affect values
Organization	Efficient and effective technology that may reduce conflicts may support effective organizations of agricultural systems. Poor technology, e.g., deforestation, increases organizational problems
Knowledge	Technological improvements require more effective knowledge, for example of the ecology of the carnivores and possible temporal, spatial, and seasonal variations in conflicts, requiring knowledge

References

Campbell, M. (2017). *Conservation biology in the 21st century: A conservation biology of large wildlife*. Nova Science Publishers, New York.

Norgard, R. B. (1994). *Development betrayed: The end of progress and a co-evolutionary revisioning of the future*. Routledge.

Index

© The Editor(s) (if applicable) and The Author(s), under exclusive license to Springer Nature Switzerland AG 2025
M. O. Campbell, *Cattle, Their Predators and Geomatics Research*,
https://doi.org/10.1007/978-3-031-97363-5

The manufacturer's authorised representative in the EU is Springer Nature Customer Service Centre GmbH, Europaplatz 3, 69115 Heidelberg, Germany. If you have any concerns regarding our products, please contact ProductSafety@springernature.com

Printed and bound by CPI Group (UK) Ltd, Croydon, CR0 4YY
30/09/2026
02225564-0016